Errata

Page 17, problem 1.9: in the figure the lines running down right
  from A and up from C should meet at B.
Page 56, Fig. 3.20: in the tetragonal column the label on the fifth
  diagram should read 4mm and not $\bar{4}$mm; the label on the sixth
  diagram should read $\bar{4}$2m and not 2m.

McKie/McKie: *Essentials of Crystallography* 1986

Essentials of Crystallography

CRYSTALLINE SOLIDS

VOLUME I
Essentials of Crystallography

VOLUME II
Crystal Chemistry and Crystal Physics

Essentials of Crystallography

DUNCAN McKIE / CHRISTINE McKIE
UNIVERSITY OF CAMBRIDGE

BLACKWELL SCIENTIFIC PUBLICATIONS

OXFORD LONDON EDINBURGH

BOSTON PALO ALTO MELBOURNE

© 1986 by Blackwell Scientific Publications
Editorial offices:
Osney Mead, Oxford, OX2 0EL
8 John Street, London, WC1N 2ES
23 Ainslie Place, Edinburgh, EH3 6AJ
52 Beacon Street, Boston
    Massachusetts 02108, USA
667 Lytton Avenue, Palo Alto
    California 94301, USA
107 Barry Street, Carlton
    Victoria 3053, Australia

First published 1986

Set by Eta Services Ltd, Beccles, Suffolk and printed
and bound in Great Britain by Butler & Tanner
Ltd, Frome and London

DISTRIBUTORS

USA and Canada
    Blackwell Scientific Publications Inc
    PO Box 50009, Palo Alto
    California 94303

Australia
    Blackwell Scientific Publications
    (Australia) Pty Ltd
    107 Barry Street,
    Carlton, Victoria 3053

British Library
Cataloguing in Publication Data

Library of Congress
Cataloging in Publication Data

Contents

Preface

Some depth of understanding of crystallography and some knowledge of the methods employed in the study of crystalline solids are central to the contemporary study of many of the active areas in chemistry, the earth sciences, materials science, and physics. In this volume we explain the basic concepts of crystallography and discuss the principal modes of study of crystalline solids, the diffraction of X-rays, electrons, and neutrons. In our experience students require a thorough, rather than a superficial, understanding of the basic concepts if they are to make effective progress in the study of crystalline solids and that is what we have set out to provide in this volume. In a subsequent volume we shall discuss the next stage in the study of crystalline solids, crystal chemistry and crystal physics.

This volume stems from the first nine chapters of our *Crystalline Solids* (Nelson, London, and Wiley, New York, 1974). Significant changes in treatment and extensive widening of the scope will be immediately discernible to those who used that book. We have adopted a vector treatment of lattice and diffraction geometry: the elegant methods of spherical trigonometry have been almost totally discarded, albeit with regret, in favour of vector methods which are eminently suitable for computer usage. The reciprocal lattice and the reflecting sphere are now utilized throughout our treatment of diffraction by crystals.

A new chapter on crystal structure determination outlines the contemporary approach to this permanently important aspect of the subject: the chapter is concerned with principles and, by design, no attempt is made to present a detailed account of modern methods of structure determination.

The volume concludes with a new chapter on electron diffraction and microscopy in which we have concentrated on principles and kept technological detail to the minimum. We have deliberately restricted the treatment to the kinematical theory of diffraction and avoided excursions into dynamical theory in order not to overload the chapter with theory. It is our experience that once the student has grasped the essentials of the kinematical theory it is not difficult to make the next step to an understanding of the dynamical theory.

These two new chapters take the reader to direct methods of structure determination and to high-resolution electron images of unit-cells. The unifying threads throughout the ten chapters of this volume are the lattice concept and symmetry; they are explored in abstract in the earlier chapters and their significance for the diffraction of X-rays, electrons, and neutrons is the basis of the later chapters.

We find crystallography immensely exciting and delight in its essential elegance; we hope to infect some of our readers with our own enthusiasm. In the writing of this volume our enthusiasm has been guided by our experience of teaching all the topics contained in it to students at Cambridge over the past twenty-five years. In our choice of the problems at the end of each chapter we have attempted to provide some interest for the chemist, the geologist, the materials scientist, and the physicist. It is our hope that these problems will provide students with the opportunity to test their understanding of the subject as their reading of the book proceeds.

We owe a debt of gratitude to colleagues from the Department of Metallurgy and Materials Science and from our own Department of Earth Sciences who have been involved in teaching crystallography in Cambridge where the courses we teach are for chemists and physicists as well as for materials scientists and geologists. We are greatly indebted to those who reviewed our preliminary proposals for the revision of the earlier book:

Dr P. Day (Inorganic Chemistry, University of Oxford)
Professor N. L. Paddock (Chemistry, University of British Columbia)
Professor B. Ralph (Metallurgy and Materials Science, University College, Cardiff)
Professor D. W. A. Sharp (Chemistry, University of Glasgow)
Professor J. P. Simons (Physical Chemistry, University of Nottingham)
Professor D. G. W. Smith (Geology, University of Alberta, Edmonton)

Our thanks are also due to our editor, Navin Sullivan, for his continuing interest in our work from the inception of the first edition of *Crystalline Solids*. To Dr Trevor Page (Metallurgy and Materials Science, Cambridge) we are indebted for Fig 10.11 and to Dr Ross Angel (Earth Sciences, Cambridge, and Earth and Space Sciences, Stony Brook, New York) for Figs 10.5 and 10.10. Our debt to Sheila Tuffnell, who converted our manuscript into impeccable typescript, is very special. We are indebted too to Judith Ginifer, who helped in many ways.

Cambridge,   Duncan McKie
December 1985   Christine McKie

1
Crystal Lattices

A crystalline solid is essentially a solid whose atoms are disposed in regular three-dimensional array. The atoms in a solid are not static: each atom possesses thermal energy and vibrates about its mean position. It is the mean positions of the constituent atoms that are regularly arranged in space in a crystalline solid. Such regularity of mean atomic positions corresponds to a state of minimum free energy and is the fundamental characteristic of the crystalline state.

In its early development crystallography was confined to the study of *single crystals*, that is solid bodies bounded by natural plane faces within which the mean positions of all the constituent atoms are related to a single regular three-dimensional array of points. But there are in addition many other solid crystalline substances which can never, or only with difficulty, be obtained in single crystal form; such are the common metals, brass and steel, which are aggregates of interlocking randomly oriented crystals of varying shape and size. Such *polycrystalline* substances belong just as surely to the crystalline state as do the single crystals which exclusively formed the subject of the science of crystallography in its early days. Not all solids are crystalline however; glasses and other amorphous solids have, like liquids, only severely localized volumes of atomic order involving merely hundreds or thousands of atoms. Examples of solids with two-dimensional or one-dimensional atomic periodicity are known and are regarded as special cases within the crystalline state.

In this first chapter we develop the principles of geometrical crystallography by consideration of *perfect single crystals*. For a perfect single crystal, the regular arrangement of atoms in the crystal can be completely described by definition of a fundamental *repeat unit* coupled with a statement of the translations necessary to build the crystal from the repeat unit. For geometrical simplicity we exemplify this basic crystallographic concept first by consideration of a two-dimensional case.

The arrangement of atoms in a layer of graphite (the crystalline form of carbon stable at room temperature and atmospheric pressure) is shown in Fig 1.1. The carbon atoms, represented as small solid circles in the figure, are in a honeycomb pattern. The distance between the centres of adjacent hexagons of the 'honeycomb' is 2.46 Å so that a layer of area about 1 mm^2 will contain about $(4.10^6)^2 = 1.6 . 10^{13}$ hexagons; the array of atoms in a layer of this size is thus effectively infinite. The repeat unit of the two-dimensional structure, containing two carbon atoms, is shown in the top left-hand corner of the figure enclosed in a parallelogram whose corners lie at the

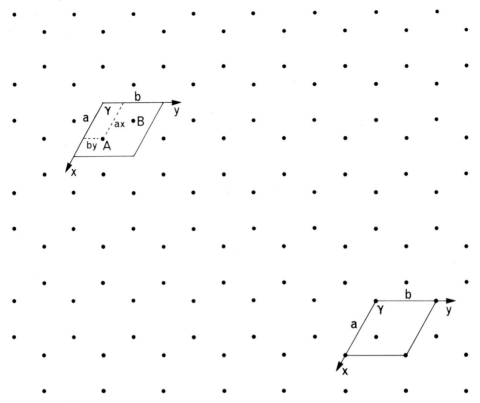

Fig 1.1   The arrangement of carbon atoms in one layer of the graphite structure. Each carbon atom is represented by a small solid circle. Two reasonable unit-meshes are outlined and labelled with the axial vectors **a**, **b**, and the inter-axial angle γ.

centres of four adjacent hexagons. The atomic pattern of the layer can be reconstructed by repeating this parallelogram in a regular manner so as to fill the plane of the atomic layer completely. The parallelogram, known as the *unit-mesh* of the layer, is completely specified by designating two of its sides as the reference axes x and y, stating the interaxial angle, and specifying the lengths of its edges. It is conventional to denote the lengths of the edges of the unit-mesh parallel to the x and y axes as a and b respectively and to denote the angle between the x and y axes as γ. In graphite the unit-mesh has $a = b = 2\cdot46\,\text{Å}$, $\gamma = 120°$. A variety of parallelograms, all of the same area, could have been chosen as the unit-mesh of a graphite layer; but it is in general conventional and convenient to select a unit-mesh with a and b as short as possible and the angle $\geqslant 90°$. It is immaterial where the corners of the unit-mesh are placed in relation to the atoms of the graphite layer; the shape of the conventional unit-mesh is controlled by the atomic pattern to be constructed from it, a change of origin merely affecting the coordinates of the atoms within the unit-mesh. For the purpose of defining the positions of the atoms of the repeat unit within the unit-mesh we employ a coordinate system which has the edges of the unit-mesh as axes, the unit of length along each axis being taken as the length of the corresponding edge; atomic coordinates are thus given as fractions of the lengths of the edges of the unit-mesh referred conventionally to an origin at the top left-hand corner of the unit-mesh. The origin of each of the unit-meshes in Fig 1.1 is differently disposed with respect to the atomic array, but the reference axes are parallel and the area is the same in each case. The unit-mesh on the left of the figure contains an atom A with coordinates $\frac{2}{3}, \frac{1}{3}$ and an atom B with coordinates $\frac{1}{3}, \frac{2}{3}$. The periodic nature of the atomic arrangement naturally implies that an atom situated at a point with coordinates x, y, that is at a vector distance $x\mathbf{a} + y\mathbf{b}$ from the origin, will be repeated at vector distances $(m+x)\mathbf{a} + (n+y)\mathbf{b}$ from the origin, where m and n are integers; in this case the atom A at $\frac{2}{3}, \frac{1}{3}$ is repeated at $(m+\frac{2}{3})\mathbf{a}, (n+\frac{1}{3})\mathbf{b}$ and the atom B at $(m+\frac{1}{3})\mathbf{a}$, $(n+\frac{2}{3})\mathbf{b}$. The presence of an atom at the origin of the unit-mesh on the right of the figure implies the presence of other atoms of the same element at points with coordinates 1, 0; 0, 1; 1, 1; 2, 1; and so on: a statement of any one such pair of coordinates is sufficient for reconstruction of the structure. In the unit-mesh on the left of the figure the atom B has coordinates $\frac{1}{3}, \frac{2}{3}$ and there will be necessarily an equivalent atom with coordinates $-1+\frac{1}{3}, -1+\frac{2}{3}$, i.e. $-\frac{2}{3}, -\frac{1}{3}$, corresponding to the coordinates of the atom A in this unit-mesh with change of sign. The positions of the two carbon atoms in unit-mesh I can thus be neatly specified as $\pm(\frac{2}{3}, \frac{1}{3})$. In terms of this unit-mesh the structure of a layer of graphite can be completely specified by stating the dimensions of the unit-mesh, $a = b = 2\cdot46$ Å, $\gamma = 120°$, and the coordinates of the carbon atoms within it, $\pm(\frac{2}{3}, \frac{1}{3})$; the atomic layer can then be reconstructed by repetition of the unit-mesh in two non-parallel directions.

We now pass on to the next stage of complexity and consider in general terms a three-dimensional structure. Here the repeat unit can always be enclosed within a parallelepiped, known as the *unit-cell*, and the effectively infinite structure can be built up by repetition of the unit-cell in three non-coplanar directions which are conventionally taken as the reference axes x, y, and z. The lengths of the unit-cell edges parallel to the x, y, and z axes are respectively denoted a, b, and c. It is conventional also to take the positive directions of the reference axes so that the axial system is right-handed and the interaxial angles $\alpha = y \wedge z$, $\beta = z \wedge x$, $\gamma = x \wedge y$ are all three $\geqslant 90°$ as exemplified in Fig 1.2.[1] As in the two-dimensional example considered earlier the coordinates of atomic positions are conventionally stated as fractions of the unit-cell edges.

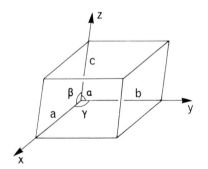

Fig 1.2   Unit-cell nomenclature. The reference axes x, y, z are right-handed, the length of the unit-cell edge parallel to each reference axis is respectively a, b, c, and the interaxial angles are denoted α, β, γ.

It is difficult to make easily intelligible perspective drawings of three-dimensional structures unless they are very simple and the task is virtually impossible for really complicated structures. It has consequently become common practice to use structural plans where the three-dimensional structure is projected down one of the reference

---

[1] Only very occasionally is it convenient to modify this simple convention.

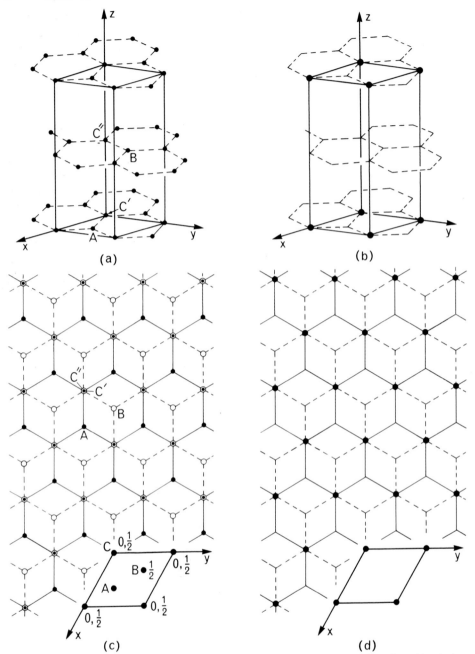

Fig 1.3 The structure of graphite. (a) and (b) are perspective drawings to show how identical two-dimensional layers are stacked to make the three-dimensional structure; in (a) the positions of carbon atoms are shown as small solid circles and C_6 rings are outlined; in (b) the C_6 rings are again outlined and lattice points are shown as large solid circles; in both (a) and (b) the unit-cell is outlined. The coordinates of the carbon atoms in the graphite unit-cell, are $0, 0, 0$; $0, 0, \frac{1}{2}$; $\frac{2}{3}, \frac{1}{3}, 0$; $\frac{1}{3}, \frac{2}{3}, \frac{1}{2}$. (c) and (d) are projections down the z-axis on to the xy plane; in (c) the carbon atoms with $z = 0$ are shown as small solid circles and the C_6 rings are outlined with solid lines while the carbon atoms of the superimposed layer at $z = \frac{1}{2}$ are shown as open circles and their linkage into C_6 rings is indicated by broken lines; in (d) the C_6 rings of the $z = 0$ and $z = \frac{1}{2}$ layers are similarly represented and lattice points are shown as large solid circles; in the lower right-hand corners of both (c) and (d) the unit-cell is shown in projection.

axes on to the plane containing the other two axes, which may or may not be perpendicular to the axis of projection. Atomic coordinates in the direction of the axis of projection are marked on the plan beside the symbol representing the atomic position. In Fig 1.3(a) and (c) a perspective drawing and a plan of the three-dimensional graphite structure are shown. The atom labelled A lies in the x, y plane and at distances $m\mathbf{c}$, where m is a positive or negative integer, above or below the plane. When the coordinate parallel to the axis of projection of an atom, such as A, is zero it is customary to omit the coordinate from the plan of the structure; an atom with no coordinate written beside it is to be taken as lying in the plane of projection. The atom labelled B lies at $\frac{1}{2}\mathbf{c}$ above the plane of projection and this is indicated by writing $\frac{1}{2}$ next to the symbol for the atom on the plan. At C, and related positions, two atoms, C' and C'', are superimposed in projection, one with $z = 0$ and the other with $z = \frac{1}{2}$; in such a case it is customary to write both coordinates beside the symbol for the atom as $0, \frac{1}{2}$.

Lattices

Some crystal properties of interest and importance are dependent only on the shape of the unit-cell, that is to say they depend only on the way in which repeat units are related to one another. It is consequently useful to have a simple way of describing the periodicity of a crystal structure and for this purpose the concept of the *lattice* is introduced. The way in which the crystal structure is built up by repetition of the repeat unit can be completely, and very simply, described by replacing each repeat unit by a *lattice point* placed at an exactly equivalent point in each and every repeat unit. All such lattice points have the same environment in the same orientation and are indistinguishable from one another. We return to two dimensions to exemplify this matter in the first instance and again take as our example a layer of the graphite structure (Fig 1.4). Figure 1.4a shows a layer of the graphite structure with carbon atoms labelled A, B, C, ..., a, b, c, ... and a conventional unit-mesh outlined. The lattice of this structure can be constructed by placing a lattice point at the carbon atom A and at all equivalent points, that is at B, C, D, E, F, G, H, I, etc. The resultant two-dimensional lattice is shown in Fig 1.4(b). If a lattice point is placed at A, then it is not permissible to place a lattice point at a because, although A and a both represent carbon atoms they are not identically situated; both lie at the centroid of a triangle formed by their three nearest neighbours, but the triangles about A and a are disposed at 60° to each other so that although both atoms have identical environments, their environments are not similarly oriented. Either the carbon atoms at A, B, C, ... or the atoms at a, b, c, ..., but not both sets of atoms, may be taken as lattice points.

Figure 1.4(c) represents a layer of the structure of boron nitride, BN, boron atoms being represented by solid circles and nitrogen atoms by open circles. The two-dimensional lattices of graphite and BN are evidently identical except for the small difference in their unit-mesh dimensions: for graphite $a = b = 2\cdot46$ Å, while for BN $a = b = 2\cdot51$ Å. The repeat unit in graphite however consists of two carbon atoms, while in boron nitride it consists of one boron and one nitrogen atom.

In a lattice every repeat unit of the structure is represented by a lattice point. A graphite layer, for instance, can be built up by placing the repeat unit of two carbon atoms in the same orientation at each lattice point in such a manner that the corresponding point of every repeat unit is placed at a lattice point. It is of no consequence which point of the repeat unit is sited at the lattice point so long as it is the same point for every repeat unit. The lattice thus has, in two dimensions, the same unit-mesh as the structure to which it refers and is completely specified by a statement of the repeat lengths *a* and *b* parallel to its *x* and *y* axes and its interaxial angle *γ*.

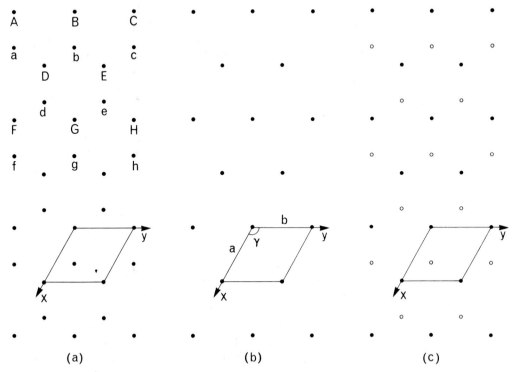

(a)   (b)   (c)

Fig 1.4 The two-dimensional lattice of a graphite layer. (a) shows the arrangement of carbon atoms (solid circles) in one layer of the graphite structure (Figs 1.1, 1.3); identically situated carbon atoms are labelled A, B, C, D, . . . ; the carbon atoms labelled a, b, c, d, . . . have a differently oriented environment but are each identically situated; the unit-mesh is outlined and it is apparent that the repeat unit consists of two carbon atoms, such as A and a. (b) shows the corresponding two-dimensional *lattice* with the dimensions $a = b$, $γ$ of the unit-mesh indicated. (c) shows the structure of a layer of boron nitride, BN, which has the same lattice with $γ = 120°$ and $a = b$, but *a* is slightly different from *a* for graphite, the difference being too small to show on the diagram; boron and nitrogen atoms are represented respectively as solid and open circles.

A three-dimensional lattice can be derived in an exactly analogous manner. For instance, the three-dimensional structure of graphite has a repeat unit containing four carbon atoms (Fig 1.3(a) and (c)). The lattice of this structure may be simply obtained by placing lattice points at the site of the carbon atom C' (Figs 1.3(a) and (c)) and at all equivalent points. Inspection of the figure shows that the atom B cannot be related to the atom C' by a lattice translation; both atoms have identical environments in their own layer, but their environments in adjacent layers are different. The unit-cell of the graphite lattice has dimensions $a = b = 2.46\,\text{Å}$, $c = 6.80\,\text{Å}$, $α = β = 90°$, $γ = 120°$. The lattice of the graphite structure is shown in perspective and in plan in Figs 1.3(b) and (d) respectively.

Having exemplified a crystal lattice, we are now ready to make a formal definition of a lattice as *an array of points in space such that each lattice point has exactly the same environment in the same orientation.* It follows immediately that any lattice point is related to any other by a simple lattice translation.

A plane passing through three non-colinear lattice points is known as a *lattice plane.* Since all lattice points are equivalent there will be equivalent parallel planes passing through all the other points of the lattice. Such a set of planes is known as a *set of lattice planes*; several sets are illustrated in Fig 1.5. A set of lattice planes divides each edge of the unit-cell into an integral number of equal parts; this property forms the basis of the very useful system of indexing of lattice planes developed by W. H. Miller, Professor of Mineralogy in the University of Cambridge from 1832 to 1880. If the lattice repeats along the x, y, z axes are respectively a, b, c and if the first plane out from the origin (at a lattice point) of a set of lattice planes makes intercepts a/h, b/k, c/l, where h, k, l are integers, on the x, y, z axes respectively, then the *Miller indices* of this set of lattice planes are (hkl), the three factors h, k, l being conventionally enclosed in round brackets. A set of lattice planes (hkl) thus divides a into $|h|$ parts, b into $|k|$ parts, and c into $|l|$ parts. The set of lattice planes labelled I in Fig 1.5 has Miller indices (122).

The equations to a set of lattice planes can be written in intercept form as $(hx/a)+(ky/b)+(lz/c) = n$, where n is an integer. If n is zero the lattice plane passes through the origin; if $n = 1$ the plane of the set makes intercepts a/h, b/k, c/l on the x, y, z axes respectively; if $n = 2$ the intercepts are $2a/h$, $2b/k$, $2c/l$; and if $n = -1$ the intercepts are $-a/h$, $-b/k$, $-c/l$. Thus the set of lattice planes (hkl) includes the plane with indices $(\bar{h}\bar{k}\bar{l})$, which makes intercepts $-a/h$, $-b/k$, $-c/l$ on the reference axes and is commonly spoken of as the 'bar h, bar k, bar l' plane.

Of course some sets of lattice planes will make intercepts that are not all positive or all negative: for instance the first plane out from the origin (taken as the front lower right-hand corner) of the set labelled II in Fig 1.5 makes intercepts $-a/2$, $-b$, $c/2$ on the x, y, z axes respectively so that the indices of this set are $(\bar{2}1\bar{2})$. If a plane is parallel to one of the reference axes, its intercept on that axis is at infinity and the corresponding Miller index is zero; thus set III in Fig 1.5 being parallel to the z-axis has c/l infinite so that l must be zero and the Miller indices of the set are $(\bar{2}10)$. The set of planes labelled IV in Fig 1.5 is parallel to the x and z axes so that $h = l = 0$; since the intercept of the first plane out from the origin on the y-axis is b, the indices of the set are (010). In terms of Miller indices the unit-cell can be described as the parallelepiped bounded by adjacent lattice planes of the sets (100), (010), (001).

The line of intersection of any two non-parallel lattice planes is the row of lattice points common to both planes. The intersections of two sets of lattice planes will thus be a set of parallel rows of lattice points; for instance sets III and IV in Fig 1.5 intersect in lines parallel to the z-axis. It is convenient to index such rows by reference to the parallel row through the origin, which is itself the intersection of the lattice planes through the origin belonging to each of the two sets. The coordinates of the lattice points in such a row are 0, 0, 0 for the lattice point at the origin; Ua, Vb, Wc, where U, V, W are integers with no common factor other than unity, for the next lattice point out from the origin; and nUa, nVb, nWc, where n is an integer, for the other lattice points of the row. Such a row of lattice points is completely specified by the three integers U, V, W, which are conventionally enclosed in square brackets as $[UVW]$ in order to distinguish them from Miller indices for lattice planes, conventionally enclosed in round brackets as (hkl). The symbol $[UVW]$ represents not only the lattice point row passing through the origin and through the lattice point with coordinates Ua, Vb, Wc but all **parallel lattice point rows**; the

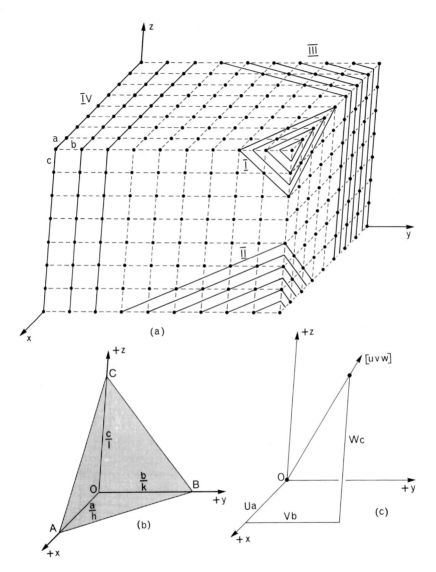

Fig 1.5 Lattice planes and zone axes. The array of lattice points exposed on the three visible faces of a parallelepiped whose edges are parallel to those of the unit-cell is displayed in (a) with solid circles to represent lattice points and thin broken lines parallel to axial directions. Four sets of lattice planes are indicated by thick solid lines representing the intersection of lattice planes with the visible faces of the parallelpiped; their Miller indices are I (122), II ($\bar{2}\bar{1}2$), III ($\bar{2}10$), IV (010). In (b) the definition of Miller indices is illustrated: the shaded plane (hkl) makes intercepts a/h, b/k, c/l on the x, y, z axes, a, b, c being the lattice repeat along each axis and h, k, l being integers. In (c) the definition of the zone axis symbol is illustrated: $[UVW]$ is the direction parallel to the line through the origin and the point Ua, Vb, Wc.

lattice direction $[UVW]$ is thus parallel to the *lattice vector* $U\mathbf{a}+V\mathbf{b}+W\mathbf{c}$, where U, V, W are integers and $\mathbf{a}$, $\mathbf{b}$, $\mathbf{c}$ are the base vectors of the lattice.

It is necessary for us to explore further the geometry and notation of lattice planes and rows because a thorough understanding of these matters is essential not only for the description of the external shape of single crystals but also, and more importantly, for the interpretation of the diffraction of X-radiation by crystals (chapter 6). In both these fields it is only the angular disposition of lattice planes and rows with respect to the reference axes of the unit-cell that is significant; the actual position in space of a given plane or row is of no consequence. For this reason the Miller indices (hkl) may be taken to represent a set of parallel lattice planes and the *lattice vector* $[UVW]$ to represent a set of parallel lattice point rows (the alternative description of $[UVW]$ as the *zone axis symbol* must await the definition of the term 'zone' on p. 13). The symbol (hkl) thus denotes any plane of the set of lattice planes which satisfy the equation $(hx/a)+(ky/b)+(lz/c) = n$, where n is integral, with one qualification which enables a distinction to be made between planes of the set which lie on opposite sides of the origin. If it is desired to make this distinction, as is often the case, the symbol (hkl) is reserved for those planes of the set which make intercepts on the same side of the origin as the plane whose intercepts on the x, y, z axes are respectively a/h, b/k, c/l and the symbol $(\bar{h}\bar{k}\bar{l})$ is used to denote those planes of the set which make intercepts on the same side of the origin as the plane whose intercepts on the x, y, z axes are respectively $-a/h$, $-b/k$, $-c/l$. The plane $(\bar{h}\bar{k}\bar{l})$ is said to be the *opposite* of (hkl), the superscript 'bar' representing, as is usual in crystallography, a minus sign. In a precisely analogous way the zone axis symbol $[UVW]$ represents all directions parallel to the vector from the origin to the lattice point with coordinates Ua, Vb, Wc, with the proviso that opposite directions may be distinguished as $[UVW]$ and $[\bar{U}\bar{V}\bar{W}]$; $[UVW]$ is taken to be in the same sense as the vector from the origin to Ua, Vb, Wc and $[\bar{U}\bar{V}\bar{W}]$ in the same sense as the vector from the origin to the lattice point at $-Ua$, $-Vb$, $-Wc$. In general, of course, the indices in the symbols (hkl) or $[UVW]$ need not all be of the same sign.

The condition for the lattice vector $[UVW]$ to be parallel to a plane (hkl) can be derived by reference to Fig. 1.5(b). Any vector $\mathbf{r}$ parallel to the set of planes (hkl) will be given by

$$\mathbf{r} = \lambda\mathbf{AB}+\mu\mathbf{AC},$$

where λ and μ are scalars

Since
$$\mathbf{AB} = -\frac{1}{h}\mathbf{a}+\frac{1}{k}\mathbf{b}$$

and
$$\mathbf{AC} = -\frac{1}{h}\mathbf{a}+\frac{1}{l}\mathbf{c},$$

$$\mathbf{r} = -\frac{1}{h}(\lambda+\mu)\mathbf{a}+\frac{\lambda}{k}\mathbf{b}+\frac{\mu}{l}\mathbf{c}.$$

So, for the lattice vector $[UVW]$ to be parallel to the set of planes (hkl)

$$U\mathbf{a}+V\mathbf{b}+W\mathbf{c} = -\frac{1}{h}(\lambda+\mu)\mathbf{a}+\frac{\lambda}{k}\mathbf{b}+\frac{\mu}{l}\mathbf{c},$$

i.e.
$$hU = -(\lambda+\mu), \quad kV = \lambda, \quad \text{and } lW = \mu,$$

whence
$$hU+kV+lW = 0.$$

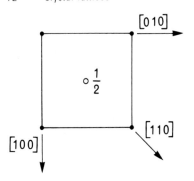

Fig 1.6 The structure of CsCl shown in plan
on (001). The directions [100] and [010] are
equivalent to each other but not to the direction
[110]. Solid circle = Cs; open circle = Cl.

This equation $hU + kV + lW = 0$ is known as the *zone equation* for reasons which
will be discussed in the next section; it is simply a statement of the condition that
the planes (hkl) are parallel to the lattice vector $[UVW]$.

The external shape of crystals

The regular nature of the spatial arrangement of the atoms within a crystal, whether
simple or complicated, leads directly to the consequence that different directions in
the crystal may not be equivalent. We take a very simple example, the structure of
caesium chloride illustrated in Fig 1.6. The unit-cell of caesium chloride is a cube and
thus has $a = b = c$, $\alpha = \beta = \gamma = 90°$; if a caesium atom (solid circle) is situated at
0, 0, 0, then a chlorine atom (open circle) lies at $\frac{1}{2}, \frac{1}{2}, \frac{1}{2}$. The direction parallel to the
x-axis, [100], is evidently equivalent to the direction parallel to the y-axis [010], but
neither of these in any way corresponds to the direction [110]. There is thus no
apparent reason why any directional property of the crystal should have equal

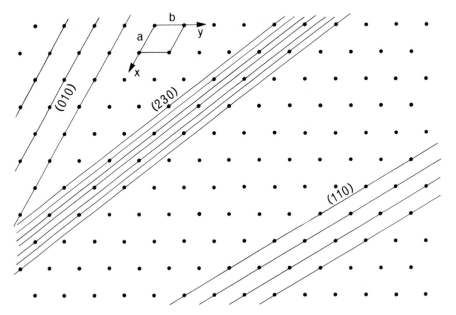

Fig 1.7 Projection of the lattice of graphite down the z-axis on to the xy plane to illustrate
the decrease in the density of lattice points per unit area of lattice planes ($hk0$) as their indices
h and k increase. The density of lattice points per unit length in projection decreases from
(010) to (110) to (230).

magnitude in the directions [100] and [110]. The reasons why some directional properties do and others do not have equal magnitudes in such crystallographically distinct directions will be developed in volume 2: it suffices here to restrict the argument to the growth of crystals. There is no obvious physical reason why atoms should attach themselves to the growing crystal as readily in one direction as in a crystallographically distinct direction, and indeed one would intuitively expect that not to be so; if it were so, crystals would tend to grow towards a spherical shape and that is not found experimentally to happen. Crystals tend to grow with plane faces which are parallel to lattice planes, especially to lattice planes with a high density of lattice points per unit area. A high density of lattice points per unit area of a lattice plane implies a large interplanar spacing (Fig 1.7) and consequently large intercepts a/h, b/k, c/l on the reference axes; the indices of commonly well-developed faces on crystals thus tend to have small values of h, k, and l. Not only are crystal faces parallel to lattice planes, but in practice the Miller indices (hkl) of the faces on a natural crystal rarely involve an integer greater than six. Moreover it is customary to index crystal faces with reference to an origin within the crystal so that the faces (hkl) and $(\bar{h}\bar{k}\bar{l})$ are parallel faces on opposite sides of the crystal. The recognition of the comparative geometrical simplicity of the angular relations of the faces of a crystal goes back to the writings of the Abbé René Just Haüy (1743–1822). Although Haüy's manner of explaining the simplicity of natural crystal forms has not withstood the test of time, it is apparent that his ideas were a helpful influence on those who came later with more powerful experimental tools to develop the modern science of structural crystallography.[2]

Two crystal faces intersect in an edge which, since the faces are parallel to lattice planes, must be parallel to a lattice point row $[UVW]$. Commonly a crystal displays several faces whose mutual edges of intersection are all parallel; such faces, which must all be parallel to a common lattice point row, are said to lie in a *zone* and the common direction of their edges of mutual intersection is known as a *zone axis*. Since the faces in a zone are all parallel to the zone axis their normals from any point, must be coplanar, a geometrical consequence that will be developed in chapter 2. The concept of zones has obvious significance for the study of the external shapes of crystals, that is *morphological crystallography*, and it is important too in a study of the diffraction of X-radiation by crystals (chapters 6–9). A zone is geometrically characterized by the symbol of its axis $[UVW]$, which may be used to indicate the group of faces whose edges of mutual intersection are parallel to the direction $[UVW]$ as well as the direction of the lattice point row $[UVW]$; this dual interpretation of the zone axis symbol $[UVW]$ will be developed in succeeding chapters.

We consider first a crystal of the utmost simplicity of form, a parallelepiped whose six faces are parallel to the faces of the chosen unit-cell of the lattice. The faces (Fig 1.8) of such a crystal have indices:

(100) parallel to the y and z axes and intersecting the positive x-axis;
($\bar{1}$00) parallel to the y and z axes and intersecting the negative x-axis;
(010) parallel to the x and z axes and intersecting the positive y-axis;
(0$\bar{1}$0) parallel to the x and z axes and intersecting the negative y-axis;

[2] A useful account of the relationship of Haüy's work to modern crystallography is to be found in Phillips (1971). For a critique of Haüy's work in its historical setting the reader is referred to Gillispie (1972).

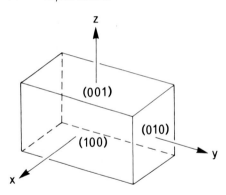

(001) parallel to the x and y axes and intersecting the positive z-axis;
(00$\bar{1}$) parallel to the x and y axes and intersecting the negative z-axis.

The faces (100), (010), ($\bar{1}$00), (0$\bar{1}$0) are all parallel to the z-axis and thus lie in the zone which has the z-axis as its zone axis, that is the zone [001]. Likewise the faces (010), (001), (0$\bar{1}$0), (00$\bar{1}$) are all parallel to the x-axis so that they belong to the [100] zone, while the faces (100), (001), ($\bar{1}$00), (00$\bar{1}$), being parallel to the y-axis, lie in the [010] zone. The reference axes x, y, z are in this particular case and in general the [100], [010], [001] zone axes.

In general the condition for the crystal face (hkl) to lie in the zone characterized by the zone axis symbol [UVW] is the condition for lattice planes of the set (hkl) to be parallel to the lattice vector $U\mathbf{a} + V\mathbf{b} + W\mathbf{c}$. We have already shown that this condition is the *zone equation*

$$Uh + Vk + Wl = 0.$$

In particular any face in the [001] zone must satisfy the zone equation

$$0.h + 0.k + 1.l = 0$$

and so must be of the type ($hk0$). For instance the faces (110), (2$\bar{1}$0), (1$\bar{2}$0), (320) belong to the [001] zone.

The zone axis symbol [UVW] for the zone containing the two generalized faces ($h_1 k_1 l_1$) and ($h_2 k_2 l_2$) is obtained by solving the simultaneous equations

$$h_1 U + k_1 V + l_1 W = 0$$

$$h_2 U + k_2 V + l_2 W = 0$$

for U, V, W. The solution is conveniently expressed in determinant form as

$$\frac{U}{\begin{vmatrix} k_1 & l_1 \\ k_2 & l_2 \end{vmatrix}} = \frac{V}{\begin{vmatrix} l_1 & h_1 \\ l_2 & h_2 \end{vmatrix}} = \frac{W}{\begin{vmatrix} h_1 & k_1 \\ h_2 & k_2 \end{vmatrix}}$$

i.e. $[UVW] = [k_1 l_2 - k_2 l_1, \ l_1 h_2 - l_2 h_1, \ h_1 k_2 - h_2 k_1].$

U, V, W are then chosen so as to have no common factor other than unity. A simple way of evaluating such two-by-two determinants is by using the *cross-multiplication* format: the indices of the first face are written twice in the upper line, the indices of the second face are written directly below them twice in the second line, the first and last columns are ignored; the first determinant is evaluated by cross-multiplying the second and third columns, the second determinant by cross-multiplying the third and fourth columns and the third determinant by cross-multiplying the fourth and fifth columns, i.e.

$$
\begin{array}{ccccccc}
\begin{vmatrix} h_1 & k_1 \\ h_2 & k_2 \end{vmatrix} & \times & \begin{array}{c} l_1 \\ l_2 \end{array} & \times & \begin{array}{c} h_1 \\ h_2 \end{array} & \times & \begin{vmatrix} k_1 & l_1 \\ k_2 & l_2 \end{vmatrix} \\
\hline
& & k_1 l_2 - k_2 l_1, & l_1 h_2 - l_2 h_1, & h_1 k_2 - h_2 k_1 & &
\end{array}
$$

It is convenient to isolate the first and last columns by drawing strong vertical lines to separate them off. By way of example we take (Fig 1.9) the zone containing the faces (210) and (011). Cross-multiplication

$$
\begin{array}{ccccccc}
\begin{array}{c|c} 2 & 1 \\ 0 & 1 \end{array} & \times & \begin{array}{c} 0 \\ 1 \end{array} & \times & \begin{array}{c} 2 \\ 0 \end{array} & \times & \begin{array}{c|c} 1 & 0 \\ 1 & 1 \end{array} \\
\hline
& 1 & & \bar{2} & & 2 &
\end{array}
$$

yields $[1\bar{2}2]$ for the zone axis symbol of the zone containing the faces (210) and (011).

It is occasionally convenient to represent a zone by a statement of the Miller indices of two non-parallel faces lying in the zone; this is done by enclosing the face symbols within square brackets, thus $[(h_1 k_1 l_1), (h_2 k_2 l_2)]$. For example the zone $[1\bar{2}2]$ can be referred to alternatively as $[(210), (011)]$.

Analogously the indices of the face (hkl) and its opposite $(\bar{h}\bar{k}\bar{l})$ which lie in the zones $[U_1 V_1 W_1]$ and $[U_2 V_2 W_2]$ are given by the solution of the simultaneous equations

$$hU_1 + kV_1 + lW_1 = 0$$
$$hU_2 + kV_2 + lW_2 = 0$$

i.e.

$$
\frac{h}{\begin{vmatrix} V_1 & W_1 \\ V_2 & W_2 \end{vmatrix}} = \frac{k}{\begin{vmatrix} W_1 & U_1 \\ W_2 & U_2 \end{vmatrix}} = \frac{l}{\begin{vmatrix} U_1 & V_1 \\ U_2 & V_2 \end{vmatrix}}
$$

i.e. $(hkl) \equiv (V_1 W_2 - V_2 W_1, \ W_1 U_2 - W_2 U_1, \ U_1 V_2 - U_2 V_1)$.

Here too h, k, l must have no common factor other than unity and the solution of the two-by-two determinants is most conveniently achieved by cross-multiplication

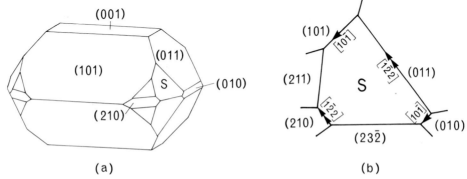

(a)   (b)

Fig 1.9   The *zone equation* and the *addition rule*. (a) is a perspective drawing of a crystal of the mineral *anglesite*, PbSO$_4$, to some faces of which Miller indices have been assigned and (b) is a plan of the face S parallel to its own plane with the neighbouring faces indexed. It is apparent from the plan that the edges of S against (210) and (011) are parallel and that the edges against (101) and (010) are also parallel so that S lies at the intersection of the zones $[10\bar{1}]$ and $[1\bar{2}2]$; the Miller indices of S are thus (232).

ignoring the first and last columns:

$$
\begin{array}{cccccc}
\left.\begin{array}{c|c} U_1 & V_1 \\ U_2 & V_2 \end{array}\right. & \times & \left.\begin{array}{c} W_1 \\ W_2 \end{array}\right. & \times & \left.\begin{array}{c} U_1 \\ U_2 \end{array}\right. & \times & \left.\begin{array}{c|c} V_1 & W_1 \\ V_2 & W_2 \end{array}\right.
\end{array}
$$

$$
\underline{ V_1 W_2 - V_2 W_1, \quad W_1 U_2 - W_2 U_1, \quad U_1 V_2 - U_2 V_1 }
$$

$$
\begin{array}{ccc}
\| & \| & \| \\
n.h & n.k & n.l
\end{array}
$$

By way of example (Fig 1.9) we take the intersection of the zones $[10\bar{1}]$ and $[1\bar{2}2]$:

$$
\begin{array}{ccccccc}
\left.\begin{array}{c|c} 1 & 0 \\ 1 & \bar{2} \end{array}\right. & \times & \left.\begin{array}{c} \bar{1} \\ 2 \end{array}\right. & \times & \left.\begin{array}{c} 1 \\ 1 \end{array}\right. & \times & \left.\begin{array}{c|c} 0 & \bar{1} \\ \bar{2} & 2 \end{array}\right.
\end{array}
$$

$$
\begin{array}{ccc}
\bar{2} & \bar{3} & \bar{2}
\end{array}
$$

The faces common to the zones $[10\bar{1}]$ and $[1\bar{2}2]$ are $(\bar{2}3\bar{2})$ and its opposite (232); the face labelled S in the figure is thus indexed as (232).

The Addition Rule

Very commonly in all branches of crystallography, including diffraction crystallography, one needs to be able to determine the indices of the faces that lie at the intersection of two zones, each of which is defined by the known indices of two faces belonging to it. If two faces $(h_1 k_1 l_1)$ and $(h_2 k_2 l_2)$ lie in the zone $[UVW]$,

$$h_1 U + k_1 V + l_1 W = 0$$

and $\qquad h_2 U + k_2 V + l_2 W = 0$

hence $\qquad (ph_1 + qh_2)U + (pk_1 + qk_2)V + (pl_1 + ql_2)W = 0$

so that the indices $(h_3 k_3 l_3)$ of any other face lying in the zone $[UVW]$ can be expressed as $(ph_1 + qh_2, pk_1 + qk_2, pl_1 + ql_2)$, where p and q are positive or negative integers. This is known as the *addition rule*. For instance the zone $[(101), (213)]$ includes the faces (314) for which $p = q = 1$, (112) for which $p = -1$, $q = 1$, (011) for which $p = -2$, $q = 1$, $(\bar{1}10)$ for which $p = -3$, $q = 1$, and so on. Thus it is quite a simple matter to determine the indices of a plane common to two zones, each defined by the Miller indices of two of its faces.

An example of the use of the addition rule is illustrated in Fig 1.9, where the face S lies at the intersection of the zones $[(210), (011)]$ and $[(010), (101)]$. Multiplication of the indices of the two faces which define the first zone by p_1 and q_1 respectively yields, on addition, indices for S $(2p_1, p_1 + q_1, q_1)$. Multiplication of the indices of the two faces which define the second zone by p_2 and q_2 respectively, followed by application of the addition rule yields indices for S $(q_2 p_2 q_2)$. Equating the alternative indices for S gives:

$$2p_1 = q_2, \quad p_1 + q_1 = p_2, \quad \text{and} \quad q_1 = q_2;$$

whence $p_2 = \frac{3}{2}q_2$ so that the indices (hkl) of S are such that $h:k:l = q_2:p_2:q_2 = 1:\frac{3}{2}:1$. Thus (hkl) is identified as (232) or $(\bar{2}3\bar{2})$ and the face S specifically as (232).

Problems

1.1 The coordinates of the carbon atoms in graphite may be stated as $\pm(0\,0\,\tfrac{1}{4})$, $\pm(\tfrac{1}{3}\tfrac{2}{3}\tfrac{1}{4})$, the dimensions of the unit-cell being $a = b = 2.46$ Å, $c = 6.80$ Å, $\alpha = \beta$

$= 90°$, $\gamma = 120°$. Draw a 2×2 array of four unit-cells of the structure projected on (001). Compare your projection with that shown in Fig. 1.3(c).

1.2 The unit-cell of $SrTiO_3$ is a cube with $a = b = c = 3\cdot904$ Å, $\alpha = \beta = \gamma = 90°$. The coordinates of the atoms in the unit-cell may be stated as Sr: 000, Ti: $\frac{1}{2}\frac{1}{2}\frac{1}{2}$, O: $0\frac{1}{2}\frac{1}{2}$, $\frac{1}{2}0\frac{1}{2}$, $\frac{1}{2}\frac{1}{2}0$. Draw a 2×2 array of four unit-cells of the structure projected on (001). Outline on your projection the (001) face of a unit-cell which has its origin at a Ti atom and write down the coordinates of all the atoms in this alternative unit-cell.

1.3 By inspection of Fig 1.5(a) find the indices of the zone axes of the zones containing (i) (100) and (010), (ii) (100) and ($2\bar{1}2$), (iii) (100) and (122), (iv) (001) and (010), (v) (001) and (122), (vi) (001) and ($\bar{2}10$). In each case use the zone equation to check your answer.

1.4 Index the zone axis of the zone containing (232) and (211). Fig 1.9 is relevant.

1.5 Make a sketch showing the (100), (010), and (001) faces of a unit-cell (as in Fig 1.8). Add to your sketch the traces of the intersections of ($\bar{2}13$) planes with the (100), (010), and (001) faces of the unit-cell. Find the zone axis symbols $[UVW]$ of the lines of intersection of ($\bar{2}13$) with (i) (100), (ii) (010), and (iii) (001) by application of the zone equation and confirm your indexing by reference to your drawing.

1.6 Find the zone axis of the zone containing the planes (121) and (231). Show that the plane (011) lies in this zone. Find the factors p and q which allow the indices (011) to be written as $(ph_1 + qh_2, pk_1 + qk_2, pl_1 + ql_2)$ where $(h_1k_1l_1)$ is (121) and $(h_2k_2l_2)$ is (231).

1.7 Find the zone axes of the zones containing the pairs of planes (i) (111) and ($\bar{1}\bar{1}1$), (ii) ($1\bar{1}0$) and (123). Use the zone equation to index the faces which are common to the zones $[(111), (\bar{1}\bar{1}1)]$ and $[(1\bar{1}0), (123)]$.

1.8 Draw a 2×2 array of four unit-meshes of a plane lattice with $a = 7\cdot5$ Å, $b = 13\cdot0$ Å, $\gamma = 90°$ on a scale of 10 mm to 1 Å. Add to your drawing the [12] direction and the trace of the plane (12). Calculate the angle between the direction [12] and the trace of the plane (12). What would that angle be if the unit-mesh had $a = b$, $\gamma = 90°$?

1.9 Find the Miller indices (hk) of the lines AB, BC, CD, DE, EF, and FA which bound the fragment of a close-packed plane of spherical atoms in the figure. Crystal faces are usually parallel to lattice planes with a high density of lattice points, that is to lattice planes with low values of h, k, and l. Compare the arrangement of spheres on the lines bounding this two-dimensional crystal and suggest an explanation of that observation.

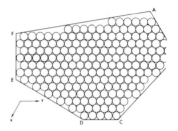

2
Representation in two dimensions: The stereographic projection

The interrelationships in space between lattice planes and lattice rows, crystal faces and zone axes are central to the study of crystal geometry. The point has already been made that angular relationships are of much greater importance than actual position in space in the limited range of topics discussed in chapter 1 and that is so generally. For graphical representation we have so far made use of perspective drawings and plans, but it is obvious that both these methods of representation will fail to give a clear picture of angular relationships in all but the simplest cases. Angular relationships are critical too in the study of crystal symmetry which we shall come to in chapter 3. Clearly then some means of representing three-dimensional angular relationships in two dimensions is needed. This need is met in crystallography by the *stereographic projection*.

We shall develop the stereographic projection in the context of crystal shape simply because a realistic perspective drawing of a crystal, showing faces and zone axes, can easily be made and compared with the angular morphological relations displayed in stereographic projection. But it must be emphasized that the stereographic projection is just as applicable and indeed more useful for the representation of angular relationships within a lattice; but for these a clearly intelligible perspective drawing cannot be made except in very simple cases. We shall therefore in this chapter, concerned with the technique of stereographic projection, confine ourselves to morphological examples; in later chapters we shall apply the stereographic projection in other and more significant fields.

The actual shape of a crystal of a given substance will depend on the conditions in which it grew. It is only in the rare circumstances of ideal conditions that a crystal will grow as a regular polyhedron or with equivalent faces equally developed: thus two crystals of the same substance may have quite different external appearances (Fig 2.1(a)). But crystal faces must be parallel to lattice planes, therefore the angle between any particular pair of faces will be the same in every crystal of the substance that exhibits those two faces. This is the *law of constancy of angle* discovered by Nicolaus Steno in 1669 and stated formally as: *In all crystals of the same substance the angles between corresponding faces have a constant value.* It is adequate therefore, if we are only concerned with angular relationships, to represent a crystal face by its normal drawn from an origin within the crystal. The normal to a given face may or may not intersect the face (Fig 2.1(b)), but this distinction is of no consequence. The whole crystal can thus be represented by a bundle of normals drawn from the origin, one to each face; angular relationships between the faces are preserved in the bundle of normals and the variation of aspect between different crystals of the same substance resulting from uneven development of faces is eradicated. It becomes convenient to quote the angle between the corresponding normals when the angle between two crystal faces is required; this is of course the supplement of the angle within the crystal which in single crystals is commonly a salient angle. Thus in Fig 2.1(b) the angle between (111) and (1$\bar{1}$1), usually written as (111):(1$\bar{1}$1), is quoted as 70° 32′; the angle between the face normals is 70° 32′ and the angle between the crystal faces is 180° − 70° 32′.[1]

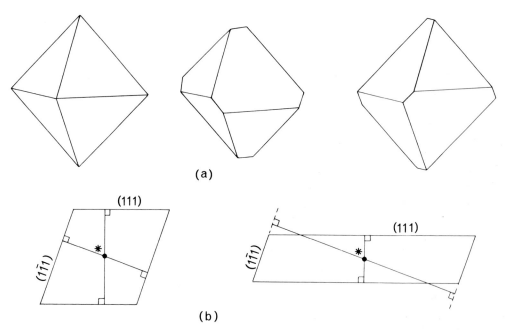

Fig 2.1   The law of constancy of angle. (a) shows a regular octahedron and two distorted octahedra: in the central drawing one pair of parallel faces is relatively overdeveloped and in the right-hand drawing the same pair is underdeveloped. (b) shows sections normal to (111) and (1$\bar{1}$1) through a regular octahedron and through an octahedron with one pair of parallel faces overdeveloped; in each case the angle marked * is 70°32′.

The stereographic projection cannot be drawn immediately from the bundle of face normals; an intermediate stage of *spherical projection* is necessary. The spherical projection of a crystal is made on a sphere circumscribing the crystal and having its centre at the origin from which the face normals are drawn. On the surface of the sphere the point of intersection of every face normal is marked. These points of intersection are the *face poles* (usually abbreviated to *poles*) of the spherical projection. Fig 2.2(b) shows the spherical projection of the crystal drawn in perspective in Fig 2.2(a). In spherical projection crystal faces are represented by

[1] The angle between two planes is measured as the angle between their lines of intersection with a plane parallel to their normals.

points on the surface of the sphere of projection. It is of course not easy to draw an accurate picture of this three-dimensional projection in two dimensions and so the stereographic projection is introduced to project the spherical projection on to a plane.

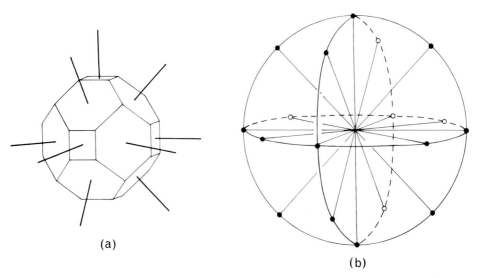

(a)

(b)

Fig 2.2   The spherical projection. (a) is a drawing of a crystal with face normals radiating from an origin within the crystal shown as bold lines. (b) shows the intersection of the resultant bundle of face normals with a sphere centred at the same origin: the array of intersections (shown as solid circles if on the front of the sphere, open circles if behind) is the spherical projection of the crystal drawn in (a).

The essentials of the stereographic projection are: a point of projection on the surface of the sphere and a plane of projection perpendicular to the diameter of the sphere passing through the point of projection. It is usual to select as the plane of projection the plane passing through the centre of the sphere and we shall always so place the plane of projection. By analogy with the earth the plane of projection is called the *equatorial plane* (Fig 2.3(a)), the projection point S is called the *south pole*, and the opposite end N of the diameter through S is called the *north pole*. The equatorial plane intersects the sphere in a circle known as the *primitive circle*. The line SP joining the pole P to the projection point S intersects the projection plane in p, which is the stereographic projection of the pole P. The stereographic projection of any pole lying above the equatorial plane, that is in the northern hemisphere, falls inside the primitive circle; the projection of a pole lying on the primitive circle is coincident with the pole itself (Fig 2.3(b)); the projection of a pole lying below the equatorial plane, that is in the southern hemisphere, falls outside the primitive circle (Fig 2.3(c)); and the projection of the south pole, the extreme case, is at infinity.

Poles lying in the southern hemisphere can more conveniently be projected within the primitive circle by taking the north pole as the projection point. Poles so projected from the north pole are distinguished on the stereogram from those projected from the south pole by representing the former as open circles and the latter as smaller solid circles.

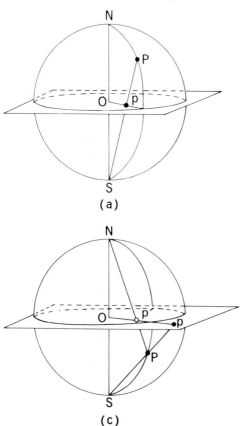

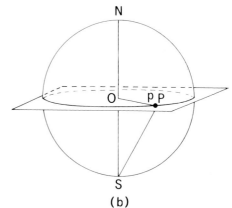

Fig 2.3 The stereographic projection. In each diagram the intersection of the sphere of the spherical projection with the equatorial plane is shown; this circle of intersection is known as the *primitive circle*; O is the centre of the sphere and the line SON is perpendicular to the equatorial plane; P is a pole in spherical projection and p is the corresponding pole in stereographic projection. (a) illustrates the stereographic projection p of a pole P in the northern hemisphere projected from the south pole. (b) illustrates the coincident projection p of a pole P on the primitive. (c) illustrates the projection of a pole P in the southern hemisphere projected from the south pole to p, lying outside the primitive, or projected from the north pole to p′ (represented by an open circle) within the primitive.

Figure 2.4(a) shows a central section through the sphere containing the projection point S and a pole P. The normal to the crystal face represented by the pole P is the radius OP. If this normal makes an angle ρ with the north–south diameter of the sphere, then $\widehat{NOP} = \rho$ and $\widehat{NSP} = \rho/2$. Therefore the distance of the projection p of the pole P from the centre of the primitive circle is $r \tan \rho/2$, where r is the radius of of the sphere and of the primitive circle. If the pole P′ lies in the southern hemisphere $180° > \rho > 90°$ and if the north pole is taken as the projection point, then $Op' = r \cot \rho/2$. If the crystal has two faces whose normals OP and OP′ are coplanar with and equally inclined to the north–south diameter NOS so that $\rho = 180° - \rho'$, then if both normals lie on the same side of NOS their poles will project within the primitive circle at the same point; such pairs of poles are represented by an open circle concentric with a smaller solid circle in Figs 2.4(c) and (d). The normals to the pair of parallel faces (hkl) and $(\bar{h}\bar{k}\bar{l})$ are represented in spherical projection by the poles P and P_0 which lie at opposite ends of a diameter of the sphere of projection, the pole P_0 being described as the *opposite* of the pole P. Inspection of the section of the sphere of projection containing P, S, and P_0 (Fig 2.4(e)) indicates that if P is projected as p using the south pole as the point of projection and if P_0 is projected as p_0 from the north pole, then $Op = Op_0$. The pole p and its opposite p_0 thus lie on a diameter pNp_0 of the primitive (Fig 2.4(f)), equidistant from the centre but on opposite sides; one being projected from the south pole is represented by a solid circle and the other projected from the north pole by an open circle.

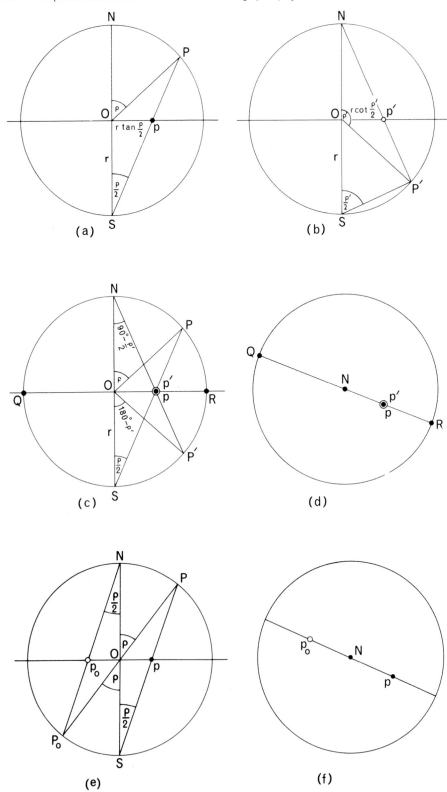

(a)

(b)

(c)

(d)

(e)

(f)

Fig 2.4 Construction of the stereographic projection p of a pole P. In (a) the pole P lies in the upper hemisphere and its projection from S is p; in (b) the pole P' lies in the lower hemisphere and its projection from N is p'. In (c) the projection (p and p') of the poles P and P', for which $\rho + \rho' = 180°$, respectively from S and N is shown: the resultant stereographic projection with p represented by a solid circle and p' by a concentric open circle is illustrated in (d). In (e) the projection (p and p_0) of the pole P and its opposite P_0 respectively from S and N is shown: the resultant stereographic projection with p represented by a solid circle and p_0 by an open circle with $p_0 N = Np$ is illustrated in (f).

In principle the faces of a crystal could be plotted stereographically by measuring for each face (i) the angle ρ between its normal and the diameter through the projection point, and (ii) the angle ϕ between a reference plane (defined by the north and south poles and the reference point R, i.e. NRS in Fig 2.5) and the plane containing the north–south diameter and the pole P of the face. It is however simpler and more practical to make use of a property of the stereographic projection, dealt with in the next few paragraphs, and of the relationships between crystal faces when plotting a stereogram of a crystal.

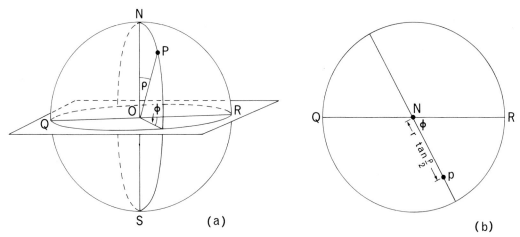

Fig 2.5 (a) serves to define the angle ρ between the normal OP to the plane (*hkl*) and the diameter NOS, and the angle ϕ between the plane NOP and a reference plane NRS. (b) represents the stereographic projection of P as p with coordinates $r \tan \rho/2$ and ϕ.

The property that is perhaps primarily responsible for making the stereographic projection attractive to crystallographers is that planes intersecting the sphere of projection project either as circles or as straight lines. Two types of planes are distinguished; a plane passing through the centre of the sphere of projection intersects the sphere in a *great circle*, while a plane that does not pass through the centre intersects the sphere in a *small circle* (Fig 2.6(a)). A great circle is a special case of a small circle. A small circle can be considered alternatively as the intersection with the sphere of a cone whose apex is at the centre of the sphere, the axis of the cone passing through the centre of the small circle and the semiangle of the cone being the angular radius of the small circle. The small circle is projected stereographically by drawing lines joining every point on the small circle to the projection point. These lines lie on the surface of an oblique cone (that is a cone whose base is not perpendicular to its axis) with a circular base, the small circle. The projection of the small circle (Fig 2.6(b)) is the intersection of this oblique cone with the equatorial

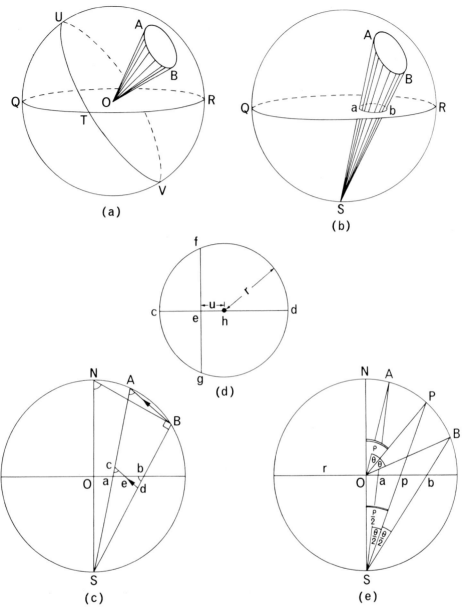

Fig 2.6   Stereographic projection of a small circle. In (a) QTR and UTV are examples of great circles and the cone AOB intersects the sphere in the small circle AB. In (b) the oblique cone ASB has a circular base, the small circle AB, and its apex is at the point of projection S; it intersects the equatorial plane QR in the figure ab. (c) is the section of the sphere of projection perpendicular to the equatorial plane and containing the diameter AB of the small circle; the stereographic projection of AB is ab. The circular section, shown in (d), of the oblique cone ASB is drawn parallel to the small circle AB and intersects the equatorial plane in feg. (e) is the same section of the sphere as (c) drawn to show the axis OP of the right cone AOB and the projection p of the centre of the small circle AB; in general ap ≠ pb.

plane. A section of the sphere of projection containing the north–south diameter and a diameter AB of the small circle is drawn in Fig 2.6(c): a and b are the projection of the points A and B so that if the small circle projects as a circle ab will be a diameter of the small circle in projection. We shall show that the projection of a small circle is itself circular by drawing a circular section cfdg parallel to the plane of the small circle to intersect the equatorial plane in feg (Fig 2.6(d)). It is evident from Fig 2.6(c) that $\triangle$SBN and $\triangle$SOb are similar, being right-angled with a common angle, and therefore

$$\widehat{SNB} = \widehat{SbO}.$$

But $\widehat{SNB} = \widehat{SAB}$, being the angle subtended at the circumference
by the arc SB,

and, since AB $\parallel$ cd,

$$\widehat{SAB} = \widehat{Scd}$$

Therefore $\widehat{Scd} = \widehat{SbO}$,

so $\triangle$ace and $\triangle$dbe are similar,

and $\dfrac{ce}{be} = \dfrac{ea}{ed}$;

i.e. $ce.ed = be.ea.$

Now let the radius of the circle cfdg be r (Fig. 2.6(d)) and let the distance of the point e from the centre h of the circle be u, then

$$ce.ed = (r-u)(r+u)$$
$$= r^2 - u^2$$
$$= ef^2.$$

This conclusion, that the square of a semi-chord is equal to the product of its intercepts on the diameter perpendicular to the chord, is a characteristic property of a circle.

But $ce.ed = be.ea$

therefore $be.ea = ef^2,$

and consequently afbg is a circle with centre h and diameter ba. The stereographic projection of a small circle is therefore a circle.

    The centre P of a small circle (Fig 2.6(e)) will not in general project at the geometrical centre of the projected small circle, for if the centre of the small circle is such that $\widehat{NOP} = \rho$ and if the angular radius of the small circle (that is, the semiangle of the right cone whose intersection with the sphere of projection is the small circle) is θ, then

$$ap = Op - Oa = r\tan\frac{\rho}{2} - r\tan\frac{\rho-\theta}{2}$$

and $pb = Ob - Op = r\tan\dfrac{\rho+\theta}{2} - r\tan\dfrac{\rho}{2}.$

In general therefore ap $\neq$ pb, but ap $=$ pb when $\rho = 0$. The projection of the centre of a small circle coincides with the geometrical centre of the projected small circle only when the centre of the small circle lies at the point of projection or its opposite, i.e. at S or N.

It is obvious from Fig 2.7(a) that a small circle which passes through the south pole projects as a straight line because the lines joining every point on the circle to the point of projection are then coplanar with the small circle.

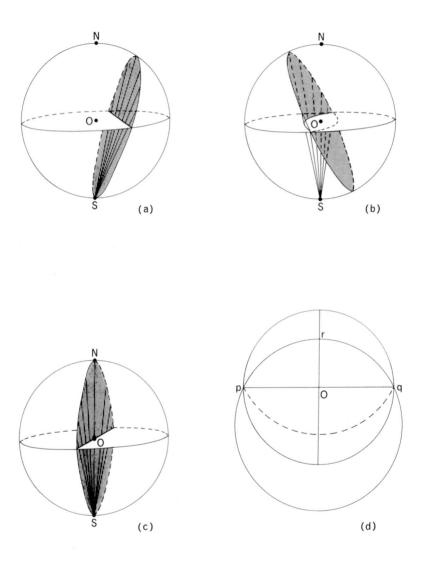

Fig 2.7 (a) shows that a small circle passing through the point of projection S projects as a straight line. (b) shows the projection of a great circle as an arc which intersects the primitive at the ends of a diameter of the primitive circle. (c) illustrates the special case of a great circle passing through the point of projection S; its projection is a diameter of the primitive. (d) is the stereographic projection of a great circle with S as the point of projection; it is customary to change the point of projection from S to N for that part of the great circle lying outside the primitive. Circles projected from S are shown conventionally as solid arcs and those projected from N as broken arcs.

Great circles, which are of course special cases of small circles, likewise project as circles; but since all great circles are coplanar with the centre of the sphere of projection they intersect the primitive circle at the opposite ends of a diameter of it (Fig 2.7(b)). If a great circle passes through the south pole (Fig 2.7(c)) it will project as a diameter of the primitive because the great circle is then coplanar with its projection. When a complete great circle is projected with the south pole as projection point, part of it must project outside the primitive circle. By changing the point of projection to the north pole this part can be brought within the primitive circle. It is customary to show great, or small, circles projected from the north pole as dashed arcs on the stereogram (Fig 2.7(d)).

It is appropriate at this point to mention that it is common practice to refer to the projections of poles and of great circles simply as poles and great circles. No confusion need arise if it is remembered that the stereographic projection is merely a device for representing the spherical projection in two dimensions.

The crystallographer rarely needs an accurately drawn stereogram; when he does so the geometrical constructions detailed in Appendix A can be employed. These constructions are dealt with at length in the appendix because, although they are little used in practice, study of their geometry can greatly help the reader towards a thorough understanding of the stereographic projection, which will be extensively used in subsequent chapters without additional explanation.

The stereographic net

The accuracy to which stereograms can be drawn by using the constructions of Appendix A will depend on the care with which the constructions are made in relation to the magnitude of the radius selected for the sphere of projection. For most purposes a stereogram of sufficient accuracy can be drawn with the aid of the *stereographic net*, sometimes called the Wulff net. A stereographic net is a stereographic projection of a set of great circles passing through a diameter of the primitive and inclined at 2° intervals to the equatorial plane and a set of small circles drawn with the same diameter of the primitive as their stereographic centre and with radii at 2° intervals. A stereographic net is illustrated in Fig 2.8. The planes of the small circles are normal to the equatorial plane and to the planes of the great circles. The intersections of two adjacent small circles with a great circle correspond to an angular distance of 2° measured in the plane of the great circle and the intersections of two adjacent great circles on a given small circle correspond to a rotation of 2° about the stereographic centre of the small circle. Nets are usually printed on opaque paper which may be stuck down on to a firm base. The stereogram is then drawn on a piece of tracing paper pinned through the centre of the net so that it can rotate freely relative to the fixed net.

A pole may be plotted with the net by placing the appropriate diameter of the stereogram along the 0°–0° or the 90°–90° diameter of the net and counting the necessary number of intersections with the traces of great or small circles. The angle between two poles may be measured by rotating the stereogram until both poles lie on the same great circle and then counting the number of intersections of small circles between them, the interval between intersections corresponding to 2°. The projection of a great circle which passes through two poles can simply be drawn by rotating the tracing paper until the two poles lie on the same great circle and then tracing the great circle from the underlying net.

Alternatively, but nowadays less commonly, nets printed on transparent paper are

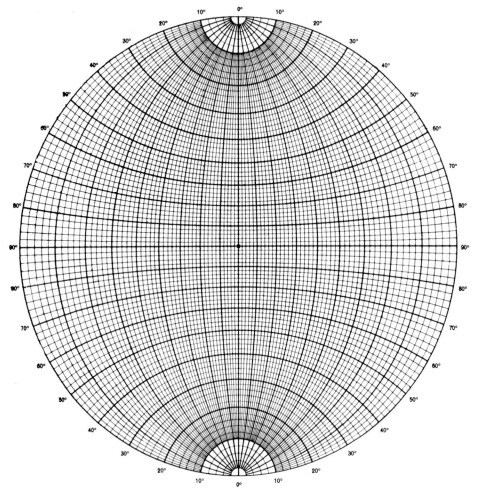

Fig 2.8   The stereographic net.

available. Such full-circle or half-circle nets may be superimposed on a stereogram drawn on opaque paper. The procedures are essentially the same as those outlined in the previous paragraph with poles or the traces of great circles pricked through the net on to the stereogram with a pin or a compass point. Such transparent nets have only a short useful life.

Use of the stereographic projection in crystallography

The stereographic projection is of value in the study of solids because it enables angular relationships between planes and directions to be represented. A plane is usually represented by its normal, although it is occasionally more convenient to represent it by the great circle to which it is parallel. It is necessary at this point to introduce the definition of the *pole of a great circle* as the direction normal to the plane of the great circle. The face pole representing the normal to a crystal face is then the pole of the great circle representing the face. (The two distinct uses of the term *pole* do not in practice lead to confusion.) A group of faces lying in a zone will all have one direction, the zone axis, in common and the normals to the faces will all lie in the plane normal to the zone axis. Their poles will therefore all lie on a great circle whose pole is the zone axis of the zone. Zonal relationships are therefore used extensively in plotting stereograms.

Figure 2.9(a) shows a crystal of copper, whose unit cell is a cube ($a = b = c = 3.61$ Å, $\alpha = \beta = \gamma = 90°$). The external shape of the crystal is a combination of a cube, an octahedron, and a rhombic dodecahedron (Fig 2.9(b), (c), (d)). The faces of the cube are parallel to the faces of the unit cell and have indices (100), (010), (001), and their opposites. The face (100) is parallel to the y and z axes. The normal to the face is therefore perpendicular to y and z and, because γ and β are right-angles, it is parallel to the x axis. Similarly the normal to (010) is parallel to y because $\alpha = \gamma = 90°$ and the normal to (001) is likewise parallel to z because $\alpha = \beta = 90°$. It is conventional to plot stereograms of crystals with $+z$ in the centre of the northern hemisphere, i.e. parallel to the north pole, and the face normal (010) at the right-hand end of the horizontal diameter of the primitive as drawn. A stereogram showing the faces of a cube is drawn in Fig 2.9(e). (001) and (00$\bar{1}$) plot in the centre of the primitive, (001) being represented by a small solid circle and (00$\bar{1}$) by a larger open circle. In such a case, when the projections of (hkl) and ($hk\bar{l}$) are coincident it is usual to write the indices (hkl) of the face in the northern hemisphere beside the pole, it being obvious that the indices of the face in the southern hemisphere are ($hk\bar{l}$). It is usual not to enclose face indices in brackets when they are written beside poles on the stereogram.

The faces of the rhombic dodecahedron are all parallel to one of the reference axes and are equally inclined to the other two. Therefore one index must be zero and, since $a = b = c$, the other two indices must be equal and are set equal to unity; the faces are thus (110), (101), (011), ($\bar{1}$10), and so on (Fig 2.9(d)). It is clear from Fig 2.9(f) that both the face (110) and its normal are equally inclined to x and y. The pole of (110) therefore projects on the primitive (since the face is parallel to z, its normal is perpendicular to z) at 45° to x and y. The other faces of the rhombic dodecahedron plot in similar positions (Fig 2.9(g)).

The faces of the octahedron are equally inclined to the x, y, and z axes and therefore have indices such as (111), ($\bar{1}$11), and so on. It is clear from Fig 2.9(a) that the face (111) lies in the zone containing (001) and (110), since the edge of intersection of (001) and (111) is visibly parallel to the edge of intersection of (111) and (110). Therefore the normal to the face (111) lies in the plane containing the normals to (001) and (110). The face (111) also lies in the zones [(100), (011)] and [(010), (101)]; its position can be plotted on the stereogram by drawing the great circles representing these zones and marking their point of intersection. Only two great circles are needed to plot (111) but in Fig 2.9(h) the three zones mentioned above have been drawn to emphasize zonal relationships and the application of the addition rule.

Figure 2.9(h) is a stereogram of the crystal of copper shown in Fig 2.9(a) in which the great circles necessary to show the zonal relationships between all the faces are drawn.

We turn now to another example to illustrate the use of the stereographic projection: the determination of the angle between (100) and (311) in a crystal of barium sulphate, the unit cell of which has dimensions $a = 8.85$ Å, $b = 5.44$ Å, $c = 7.13$ Å, and $\alpha = \beta = \gamma = 90°$. The pole of (100) can be plotted directly (Fig 2.10(a)); the (100) plane is parallel to y and z and therefore its normal is parallel to x. The pole (311) is most easily plotted by making use of zonal relationships. If the axes are orthogonal, as they are here, planes which have one of their indices zero lie in the plane containing the other two axes. The angles which such planes make with (100),

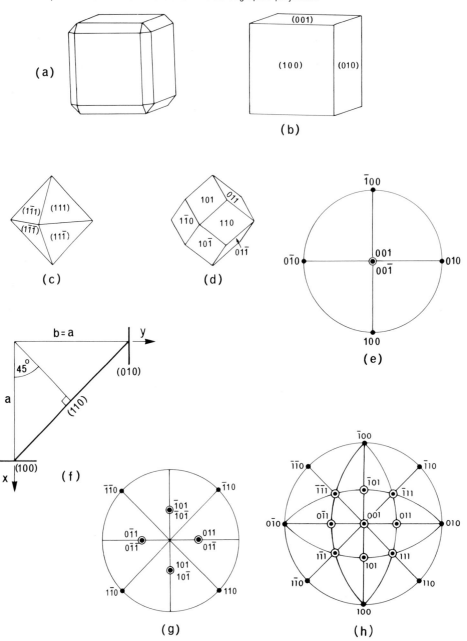

Fig 2.9 Stereographic projection of a cubic crystal. Copper forms crystals (a) which may be combinations of a cube (b), an octahedron (c), and a rhombic dodecahedron (d). A stereogram of the faces of a cube is shown in (e). The face (110) and its normal are equally inclined (f) to the x and y axes so that this and the other faces of the rhombic dodecahedron give rise to the stereographic projection (g). The projection of the octahedron faces is achieved by plotting intersecting zones with the aid of a stereographic net. The stereogram of the crystal drawn in (a) with all faces plotted but only those in the upper hemisphere indexed is shown as (h).

(010), and (001) can be calculated by elementary two-dimensional geometry. For example the plane $(hk0)$ is parallel to z and its normal therefore lies in the xy plane. The plane makes intercepts a/h on the x axis and b/k on the y axis. From Fig 2.10(b) it can be seen that since OP is normal to RPQ and x is normal to y, $\widehat{POQ} = 90°$ $- \widehat{OQR} = \widehat{ORQ}$. Since the normal to (100) is parallel to x,

$$(100):(hk0) = \widehat{POQ} = \widehat{ORQ} = \tan^{-1} \frac{a/h}{b/k}. \qquad (1)$$

Similarly,

$$(001):(h0l) = \tan^{-1} \frac{c/l}{a/h} \qquad (2)$$

and

$$(001):(0kl) = \tan^{-1} \frac{c/l}{b/k}. \qquad (3)$$

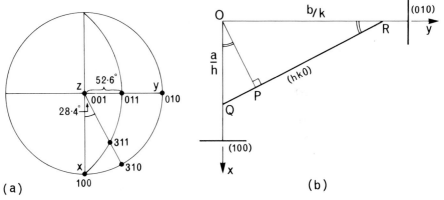

Fig 2.10 Stereographic projection of a face normal when the unit-cell has orthogonal but unequal axes. The pole (311) lies at the intersection of the zones [(100), (011)] and [(001), (310)] as shown in (a). In order to plot the stereographic projection of (311) it is necessary to determine the angles (100) : (310) and (001) : (011) by a simple geometrical calculation illustrated in (b).

By use of the addition rule it can be shown that in the zone containing (100) and (311) the plane with indices of the type $(0kl)$ is (011). Therefore from equation (3) $(001):(011) = \tan^{-1}(7·13)/(5·44) = 52·65°$. The zone $[(100), (011)]$ can therefore be plotted by first plotting (011) and then drawing the great circle containing the two poles (100) and (011). The position of (311) can be found by drawing a second zone on which (311) must lie. A zone which also contains (001) projects as a diameter of the primitive and is therefore easy to construct. By use of the addition rule the face of the type $(hk0)$ which lies in the zone $[(001), (311)]$ is found to be (310). From equation (1) the angle $(100):(310) = \tan^{-1}[8·85/(3 \times 5·44)] = 28·47°$. Therefore (310) can be plotted and the zone $[(001), (310)]$ drawn on the stereogram. (311) lies at the intersection of the zones $[(100), (011)]$ and $[(001), (310)]$. The angle (100):(311) can now be measured with the stereographic net or a more accurate value may be obtained by one of the methods of calculation which will be dealt with in chapter 5.

Measurements occasionally have to be made of the angles between the faces actually developed on a crystal. Before the discovery of the diffraction of X-rays by crystals this was the principal method of studying crystals but its importance has declined over the past sixty years as the techniques for studying the internal structure of crystals have become increasingly available. However measurement of the angles between faces remains useful for determining the orientation of a crystal of known unit-cell dimensions. Two simple devices for the measurement of interfacial angles are described in Appendix B.

The gnomonic projection

The gnomonic projection (Fig 2.11), in which the centre of the projection sphere is used as projection point and the tangent plane at the north pole is the plane of projection, has certain limited uses in crystallography. Its advantage is that all zones project as straight lines because every great circle passes through the projection

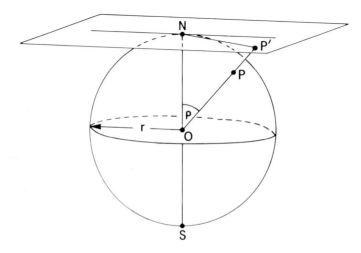

Fig 2.11   The gnomonic projection. The point of projection is the centre of the sphere and the plane of projection is tangential to the sphere.

point. Its main disadvantage is that a pole at an angle ρ from N projects at a distance $r \tan \rho$ from N and therefore only a small part of the surface of the projection sphere can be plotted within a manageable area. Poles lying on the equatorial plane project to infinity. Another disadvantage is that small circles do not project as circles.

Problems

2.1 Construct a stereogram to represent the earth in the following manner. Draw a circle of radius $2\frac{1}{2}$ inches as the primitive circle. Draw the horizontal diameter of the primitive as the equator and label its left- and right-hand ends as 90°W and 90°E, respectively. Draw the vertical diameter as the Greenwich meridian and label its upper and lower ends as 90°N and 90°S respectively. Lines of latitude are small circles about the North and South poles. Lines of longitude are great circles perpendicular to the equator.

Construct the poles which represent Entebbe (0°N, 32·5°E) and New Orleans (30°N, 90°W). What is the bearing (i.e. the angle between the Entebbe-North Pole and the Entebbe-New Orleans great circles) that an aircraft must set when leaving Entebbe to fly to New Orleans on the great circle route?

2.2 An aircraft leaves London ($51\frac{1}{2}°$N, $0°$W) with enough fuel to travel 5600 km. Using the same orientation as in problem 2.1 plot the position of London and construct a small circle corresponding to the aircraft's range from London. Has the aircraft enough fuel to reach Toronto ($44°$N, $79\frac{1}{2}°$W) or should it aim for New York ($41°$N, $74°$W)? The radius of the earth may be taken as 6371 km.

2.3 A basalt dyke outcropping on the coast gives rise to a line of rocks L_1, which runs precisely NW–SE on the surface of the sea. On shore it intersects a vertical cliff running $10°$E of N in a line L_2, which is inclined at $60°$ to the northerly horizontal. Draw an accurate stereogram (radius $2\frac{1}{2}$ inches) showing L_1 and L_2, construct the normal P to the plane of the basalt dyke, and measure the bearing and elevation of P. The land surface above the cliff is a gentle planar slope, whose maximum dip is $20°$ below the horizontal on a bearing of $165°$E of N. What is the angle between the outcrop of the dyke L_3 and the horizontal on this slope?

2.4 A crystal, having a unit-cell with $a = b = c$, $\alpha = \beta = \gamma = 90°$, displays the faces (100) and (012). Plot a stereogram (radius $2\frac{1}{2}$ inches) showing the normals to these faces and the great circles representing their planes. Locate the pole of the zone axis of the zone containing these two faces and, by application of the zone equation, determine its indices $[UVW]$. The intersection of the (111) plane may be apparent on these faces as slip lines (to be discussed in volume 2). Plot the pole of (111) and the great circle which represents the (111) plane. Mark the intersections of the (111) great circle with the (100) and (012) great circles. Hence determine the angles between the zone axis $[UVW]$ and the traces of the (111) plane on (i) the (100) face, and (ii) on the (012) face of the crystal.

2.5 Draw a sketch stereogram of a cubic crystal showing the poles of (100), (110), (111) and all related faces as in Fig 2.9(h). Which of the faces shown on the stereogram lie (i) in the zone containing (100) and (311), (ii) in the zone containing (110) and (311)? Plot the pole of (311) on the stereogram. Use the same procedure to locate ($\bar{1}$23) at the intersection of the zones $[(011), (\bar{1}23)]$ and $[(\bar{1}11), (\bar{1}23)]$.

2.6 On a stereogram of radius $2\frac{1}{2}$ inches plot the poles of (100), (010), (001) and their opposites for a crystal with orthogonal reference axes. Given that (001):(012) = $32°$ and (001):(201) = $50°$ plot the poles of (012) and (201). Draw the great circles representing the zones $[(100, (012)]$ and $[(010), (201)]$; index the plane (hkl) whose pole is common to both these zones.

2.7 A crystal has a unit-cell with $a = 9·85$, $b = 15·42$, $c = 7·71$ Å, $\alpha = \beta = \gamma = 90°$. On a stereogram, the radius of whose primitive is $2\frac{1}{2}$ inches, plot the poles of (100), (010), (001), (230) and (201). Locate the pole of (231). Use a stereographic net to determine the angles between (231) and (100), (010), and (001).

2.8 Index the faces *a* to *g* on the stereogram drawn below. Find the zone axis symbol of the zone containing these faces and plot the pole of the zone axis on the stereogram.

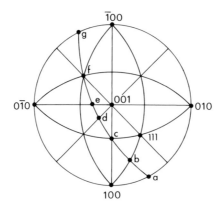

2.9 Draw an accurate stereogram of a cubic crystal showing the poles of (100), (110), and all related faces. Locate the pole of (111). Draw small circles of radii 35° and 55° such that their stereographic centres are coincident with the pole of (111). List the indices of those faces whose poles you have already constructed which lie on each of these small circles. Construct the great circle through (1$\bar{1}$0) and ($\bar{1}$01) and then locate its pole. Observe that for a cubic crystal [111] and (111) are coincident.

3
Crystal symmetry

We have already explored in chapter 1 the description of the regular arrangement in space of the atoms in a crystalline solid in terms of the lattice concept. The lattice is a regular array of imaginary points such that each lattice point has the same environment in the same orientation. The unit-cell of the lattice is defined in such a manner that every unit-cell has a lattice point at its origin[1] and the lattice is produced by stacking unit-cells in three dimensions. The whole structure is then completely described by stating the dimensions of the unit-cell and the nature and coordinates of every atom within the unit-cell. Any two atoms separated by a lattice translation must therefore be equivalent in every respect: not only must they be of the same element, but each must have the same atomic environment in the same orientation. Distinct from such *lattice repetition* is another kind of repetition known as *symmetry*, which is our prime concern in this chapter.

Axes of symmetry

When atoms of the same element in the unit-cell have identical atomic environment except for the orientation of the environment they are said to be related by symmetry. For example the hypothetical two-dimensional structure illustrated in Fig 3.1 contains atoms of two elements, one shown as solid circles and the other as open circles. The lattice translations are such that all the atoms labelled A are equivalent. The atoms of the same element labelled B are likewise all equivalent to one another and have the same environment as those labelled A; the only difference is that the environment of an atom on a B site has to be rotated through 180° to bring it into the same orientation as the environment of an atom on an A site. Moreover if the whole structure is rotated through 180° about any point equivalent to O_1, then the rotated structure will be coincident with the structure as shown. There thus exists a *rotation axis of symmetry* perpendicular to the plane of the structure through O_1 and similar axes through all points related to O_1 by lattice translation, i.e. O_2, O_3, O_4, etc. The rotation axes in this example yield coincidence by rotation through 180° or $2\pi/n$ radians where $n = 2$; such axes are described as twofold or *diad* axes of symmetry.

Before going on to define the various kinds of symmetry axis it is convenient to distinguish between two types of symmetry, that of a finite body and that of an

---

[1] This statement requires amplification in the case of non-primitive lattices, which will be dealt with in chapter 4.

infinite body. In the crystallographic field the symmetry displayed by a finite body is apparent in the external shapes of crystals and in the physical properties of crystals (volume 2) such as thermal and electrical conductivity and elastic properties. The shapes of crystals grown, either naturally or in the laboratory, under ideal conditions provide the clearest means of illustrating the symmetries possible in finite bodies; we shall therefore use crystal shape in the main to illustrate this chapter. It must always be borne in mind, however, that crystals commonly grow under non-ideal conditions even in nature so that symmetry related faces are unequally developed. In such cases the symmetry of the crystal will not be immediately apparent from its external shape (Fig 2.1); in this chapter we shall confine ourselves for purposes of illustration to perfectly developed crystals. In chapter 4 we shall be concerned with the symmetry of infinite bodies as exhibited by the arrangement of atoms in crystal structures.

An *n-fold rotation axis of symmetry* is defined as a line, rotation about which produces congruent positions (i.e. positions indistinguishable from the initial position) after rotation through $2\pi/n$. The crystal therefore comes into self-coincidence n times in a complete rotation through 2π. Rotation axes are described with reference to the value of n, which must of course be integral: onefold, twofold, threefold, fourfold, fivefold, sixfold, etc, rotation axes are known as monads, diads, triads, tetrads, pentads, hexads, etc. and are denoted by the symbols 1, 2, 3, 4, 5, 6, etc.

The monad, which brings the crystal into self-coincidence after rotation through 2π, is of course trivial. Crystals displaying diads, triads, tetrads, and hexads are shown in Fig 3.2, where the graphical symbol for each type of rotation axis is also

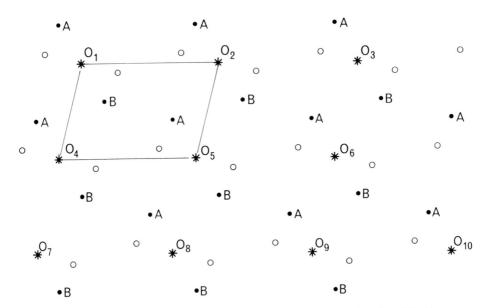

Fig 3.1 Hypothetical two-dimensional structure with identical atoms on sites A and B. Atoms of another element are represented by small open circles. The rotation diads, O_1, O_2, etc, are perpendicular to the plane of the diagram and related to one another by lattice translations. The unit-mesh is outlined.

Fig 3.2 Rotation axes of symmetry. The left-hand column shows the operation of a diad, a triad, a tetrad, a pentad and a hexad on a point; the central column shows crystals displaying each type of axis, other than the pentad; and the right-hand column shows stereograms illustrating the operation of the rotation axes on a general pole.

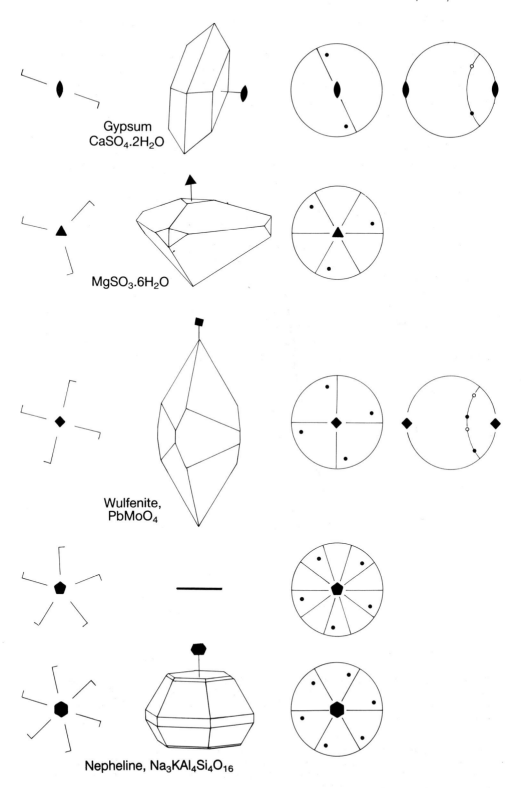

Gypsum
$CaSO_4.2H_2O$

$MgSO_3.6H_2O$

Wulfenite,
$PbMoO_4$

Nepheline, $Na_3KAl_4Si_4O_{16}$

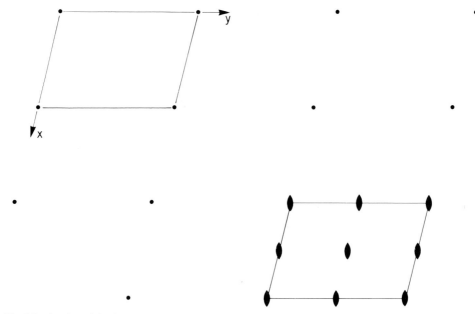

Fig 3.3 Lattice of the hypothetical two-dimensional structure shown in Fig 3.1. The unit-mesh, a general parallelogram is outlined upper left and the disposition of rotation diads on the unit-mesh is shown lower right.

displayed. The presence of an n-fold rotation axis of symmetry implies that a given plane or direction is repeated by being rotated through $2\pi/n$ radians about the axis, the operation being repeated until the initial position is reproduced. The operation of a rotation axis of symmetry on a single pole is conveniently displayed stereo-graphically and this is done too in Fig 3.2.

That the only values of n possible for rotation axes of symmetry operative on crystalline solids are 2, 3, 4, and 6 (the monad being trivial) is a direct consequence of the regularity of atomic arrangement in a crystal. For a crystal to have an n-fold rotation axis implies that the atomic arrangement in the substance must likewise have n-fold symmetry. Consequently the shape of the unit-cell of the lattice of the structure must be consistent with the presence of an n-fold rotation axis of symmetry. In short the only rotation axes of symmetry that can operate on a crystal structure are those that can operate on a lattice. For example the hypothetical two-dimensional structure illustrated in Fig 3.1 has diad symmetry; the lattice of this structure likewise has diad symmetry (Fig 3.3) with diads through the lattice points and midway between adjacent lattice points. Comparison of Figs 3.1 and 3.3 shows that structure and lattice have identical arrangements of diads, which are separated by halved lattice translations. The example demonstrates that a diad can operate on a lattice and moreover that the operation of a diad imposes no geometrical restrictions on the lattice plane perpendicular to itself; the unit-mesh of the lattice plane perpendicular to a diad is thus usually a general parallelogram.

If a crystal structure is to have a fivefold rotation axis of symmetry, then the lattice on which the structure is based must also display pentad symmetry. If a pentad passes through one lattice point, such as A in Fig 3.4, then there must be a pentad through every other lattice point, all lattice points being by definition equivalent; in particular there has to be a pentad through the lattice point B which is separated

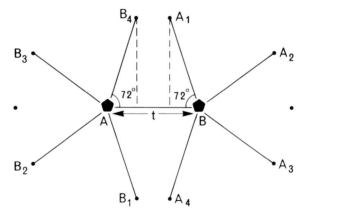

Fig 3.4 Generation of lattice points by the operation of a pentad through the lattice point A on the lattice point B and by a pentad through B operating on A. B_4A_1 is not an integral multiple of AB and therefore the array does not constitute a lattice.

from A by the lattice translation t. The operation of the pentad through A produces rows of lattice points passing through A identical to the row AB... and inclined at angles of $\frac{2}{5}\pi$, $\frac{4}{5}\pi$, $\frac{6}{5}\pi$, and $\frac{8}{5}\pi$ to the row AB. In Fig 3.4 only the lattice point equivalent to B in each of these rows, B_1, B_2, B_3, and B_4, is shown. Since there must likewise be a pentad through B, there must be rows of lattice points identical to the row BA passing through B and inclined to the row BA at angles of $\frac{2}{5}\pi$, $\frac{4}{5}\pi$, $\frac{6}{5}\pi$, and $\frac{8}{5}\pi$; again only the lattice points equivalent to A, that is A_1, A_2, A_3, and A_4, are shown in the figure. If the resultant array of points is to form a lattice, it must be a regular array; in particular the spacing of points on lines parallel to AB, such as B_4A_1, must be equal to t or some multiple thereof. It is evident from Fig 3.4 that

$$B_4A_1 = AB - AB_4 \cos 72° - BA_1 \cos 72°$$

and since

$$AB_4 = AB = BA_1$$

$$B_4A_1 = AB(1 - 2\cos 72°) = 0·38\,AB$$

Therefore the array of points generated by pentads through adjacent lattice points is not regular and consequently not itself a lattice. In short pentads cannot be repeated on a lattice and never occur in crystals.

The argument that we have used in the special case of the pentad can simply be generalized to determine what values of n are permissible for an n-fold rotation axis

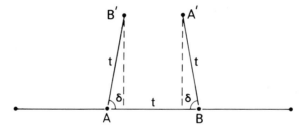

Fig 3.5 The point A' is one of the points generated by an n-fold rotation axis through B operating on the point A; and the point B' by a similar axis through A operating on B, the angle δ being equal to $2\pi/n$.

operating on a lattice. Consider a lattice with an n-fold rotation axis of symmetry. If the lattice point A is placed on an n-fold axis, then n-fold axes in the same orientation will pass through every lattice point and in particular through the adjacent lattice point B, selected so that the line AB is perpendicular to the axis of symmetry. The n-fold axis through A will repeat the lattice point B $n-1$ times in the plane perpendicular to the symmetry axis (Fig 3.5), the $(n-1)$th repetition being B' where $AB' = AB = t$ and $\widehat{BAB'} = 2\pi/n = \delta$. Likewise the n-fold axis through B will repeat the lattice point A $n-1$ times in the same plane, the first repetition being A' where $A'B = AB = t$ and $\widehat{ABA'} = \delta$. If the four points A, B, A', B' are to form a lattice, the spacing of points in the row containing A' and B' must be the same as that in the parallel row containing A and B; that is to say $B'A' = mt$ where m is an integer. Now

$$B'A' = AB - AB' \cos \delta - A'B \cos \delta$$

i.e. $$B'A' = t(1 - 2 \cos \delta)$$

Therefore the condition for formation of a lattice is

$$mt = t(1 - 2 \cos \delta)$$

i.e. $$\cos \delta = \frac{1 - m}{2}$$

But $-1 \leqslant \cos \delta \leqslant 1$, so that the limits of the integer $1 - m$ are $-2 \leqslant (1 - m) \leqslant 2$, i.e. possible values of $1 - m$ are -2, -1, 0, 1, 2. Possible values of n (Table 3.1) are therefore 2, 3, 4, 6, and the trivial monad with $n = 0$.

Table 3.1
Rotation axes of symmetry which can operate on a lattice

$1-m$	$\cos \delta$	δ	$n = 2\pi/\delta$	$B'A'$	Conventional unit mesh of lattice planes perpendicular to the axis
-2	-1	π	2	3AB	$a \neq b; \gamma \neq 90°$
-1	$-\frac{1}{2}$	$\pm\frac{2}{3}\pi$	3	2AB	$a = b; \gamma = 120°$
0	0	$\pm\frac{1}{2}\pi$	4	AB	$a = b; \gamma = 90°$
1	$\frac{1}{2}$	$\pm\frac{1}{3}\pi$	6	0	$a = b; \gamma = 120°$
2	1	0	—	—	

Note:
The symbol $\neq$ implies that equality is not required by symmetry.

    Thus the rotation axes of symmetry that can operate on a lattice are restricted to diads, triads, tetrads, and hexads; these are the only rotation axes displayed by crystals. The operation of three, four, or sixfold rotation axes on a lattice plane imposes restrictions on the arrangement of lattice points in the plane. For example the presence of a triad, for which $\delta = \frac{2}{3}\pi$, restricts the arrangement of lattice points in the plane normal to the axis to the pattern shown in Fig 3.6(a). It is conventional to choose a rhombus as the unit-mesh so that $a = b$ and the interaxial angle between the x and y axes, $\gamma = 120°$. The same arrangement of lattice points is necessary in the plane perpendicular to hexads operating on the lattice. When a tetrad, with $\delta = \frac{1}{2}\pi$, is repeated on a lattice, the arrangement of lattice points in the plane normal to the axis must be as shown in Fig 3.6(b); the conventional unit-mesh is here a square with $a = b$, $\gamma = 90°$.

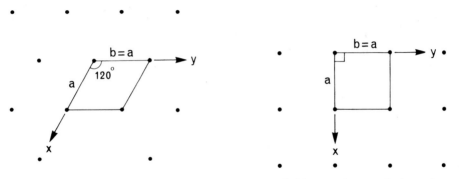

Fig 3.6 Two-dimensional lattices with (a) a triad or a hexad, (b) a tetrad perpendicular to the plane of the lattice.

Since the operation of a diad merely relates lattice points of the same row, the presence of diad symmetry places no restriction on the arrangement of lattice points in the plane perpendicular to itself; the unit-mesh is a general parallelogram with $a \neq b$, γ general. However, although symmetry places no restriction on the values of a, b, or γ, there may be fortuitous equality of a and b or γ may happen to have a special value. For example the hypothetical two-dimensional structure shown in Fig 3.7 has a square unit-mesh with $a = b$, $\gamma = 90°$, but there are no tetrads because the atomic arrangement shows only diad symmetry.

Combination of symmetry axes

A crystal may have more than one symmetry axis. When this is so the angular disposition of the axes relative to one another must be such that their operation is mutually consistent. We shall now proceed to establish all the combinations of rotation axes of symmetry that can operate on a lattice and can thus be displayed by a crystal.

It is a general principle that when two rotation axes are combined a third rotation axis is created. The principle is illustrated in Fig 3.8. The face I in Fig 3.8(a) is related to the face II and to four other faces that meet at a by a hexad. The face II is related to the face III by the diad which is perpendicular to the hexad and passes through the mid-point of the edge bc. The operation of the hexad on this diad will give rise to diads through the mid-points of the edges de, fg, hi, jk, and lm (hi and jk are at the back of the crystal and not shown in the figure) in the plane perpendicular to the hexad. Since the faces I and III are each equivalent to the face II they must be equivalent to each other; they are directly related by a diad perpendicular to the hexad and passing through the mid-point of the edge mb. This diad is inclined at 30° to the diad through the mid-point of the edge bc, both lying in the plane perpendicular to the hexad. In general then, the combination of a hexad and a diad perpendicular to it gives rise to five equivalent diads 60° apart in the plane perpendicular to the hexad and in addition a set of six diads in the same plane are inclined at 30° to those of the first set. Figure 3.8(b) shows a stereogram of the crystal drawn in Fig 3.8(a) and its rotation axes of symmetry.

The generation of additional rotation axes can be considered generally in terms of rotation axes A and B that respectively produce equivalence after rotation through $\delta_A = 2\pi/n_A$ and $\delta_B = 2\pi/n_B$ where n_A and n_B are integers. The operation of two such

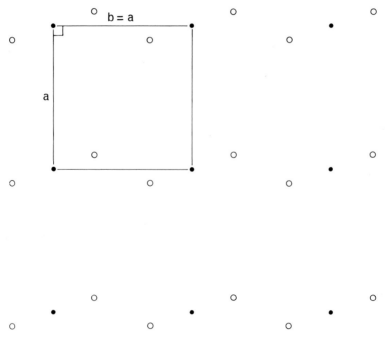

Fig 3.7 Hypothetical two-dimensional structure with two atomic types shown as solid and open circles respectively. The unit-mesh has $a = b$, $\gamma = 90°$ but the lattice has only diad symmetry.

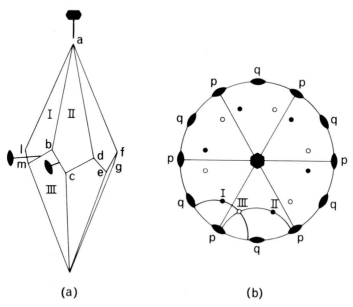

(a)   (b)

Fig 3.8 The combination of two rotation axes necessitates the presence of a third. (a) Faces I and II are related by the hexad; faces II and III are related by the diad through the edge bc; a diad through the edge mb therefore relates faces I and III. (b) A stereogram of the same crystal showing the disposition of the two sets (p and q) of six equivalent diads in the plane perpendicular to the hexad; the pole III is related to pole I by operation of a diad p and to pole II by operation of a diad q.

general axes on a pole P is illustrated in Fig 3.9, where for convenience the axis A is plotted at the centre of the stereogram. The pole P′ produced by the operation of A on P lies on the small circle that passes through P and has A as its stereographic centre; P′ also lies on the great circle AP′ which makes an angle δ_A with the great circle AP. The pole P″ produced by the operation of B on P′ lies on the small circle that passes through P′ and has its stereographic centre at B; P″ also lies on the great circle BP″ which makes an angle δ_B with the great circle BP′. (To avoid irrelevant complexity in the figure the other poles generated by the operation of A and B are not shown.) If the proposition that we discussed in specific terms in the preceding paragraph is generally true a third symmetry axis relates P to P″; but the position of this axis cannot be located from the stereogram of Fig 3.9 because the general poles P and P″ do not uniquely define a small circle.

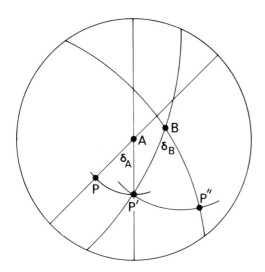

Fig 3.9 Successive operation on a pole P of a rotation axis A, rotating through δ_A, to yield the pole P′ and a rotation axis B, rotating through δ_B, to yield the pole P″. The position of the rotation axis relating P to P″ cannot be located from this stereogram.

To locate the position of the third, or derivative, rotation axis it is necessary to apply *Euler's construction*, which is based on the proposition that if three great circles intersect in the poles A, B, and C in such a manner (Fig 3.10(a)) that the angle between the great circles AB and AC is α, the angle between the great circles BA and BC is β, and the angle between the great circles CA and CB is γ, then rotation through the angle 2α in a clockwise sense about A followed by rotation through the angle 2β in a clockwise sense about B is equivalent to rotation through the angle 2γ in an *anti*clockwise sense about C. We shall now proceed to demonstrate the validity of this proposition in general and go on to establish all the possible combinations of rotational axes of symmetry that can operate on crystal lattices.

Let A and B (Fig 3.10(a)) be the poles of two rotation axes of symmetry, rotating respectively through the angles δ_A and δ_B. The consecutive operation of A and B will thus be equivalent to the single operation of a rotation axis of symmetry C whose pole lies at the intersection of the great circles through A and B which make angles of $\alpha = \frac{1}{2}\delta_A$ and $\beta = \frac{1}{2}\delta_B$ respectively with the great circle AB and lie on the same side of that great circle. If the rotation axis of symmetry whose pole is C produces equivalence by rotation through the angle δ_C, then the great circles AC and BC intersect at an angle $\gamma = \frac{1}{2}\delta_C$.

That this is so can be verified by operating A and B consecutively on the pole C

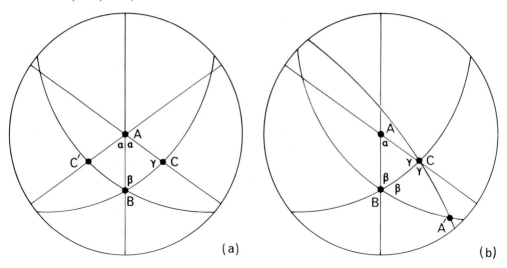

Fig 3.10   Euler's construction. The great circles AB, BC, CA are inclined to one another at angles $\alpha = \frac{1}{2}\delta_A$, $\beta = \frac{1}{2}\delta_B$, $\gamma = \frac{1}{2}\delta_C$. The poles A and B are rotation axes rotating respectively through 2α and 2β. (a) illustrates the successive clockwise operation of the axes A and B on the pole C. (b) illustrates the successive clockwise operation of the axes A and B on the pole A. C is shown to be a rotation axis rotating through 2γ.

(Fig 3.10(a)) and on the pole A (Fig 3.10(b)). When A operates on C in a clockwise sense the first equivalent pole produced will be C' such that $\widehat{AC} = \widehat{AC'}$ and $\widehat{BAC} = \widehat{BAC'} = \frac{1}{2}\delta_A$. Since the side $\widehat{AB}$ is common to both, the spherical triangles[2] ABC, ABC' are congruent and therefore $\widehat{ABC} = \widehat{ABC'} = \frac{1}{2}\delta_B$ and $\widehat{BC} = \widehat{BC'}$. Consequently operation of B in a clockwise sense will take the pole C' back to C. Consecutive clockwise rotation of C through δ_A about A and through δ_B about B thus leaves C unmoved; C is therefore itself the pole of a rotation axis of symmetry whose operation is equivalent to the consecutive operations of A and B.

Now consider the consecutive operation of A and B on the pole A (Fig 3.10(b)). The operation of the rotation axis whose pole is A will leave A unmoved. The first equivalent pole produced by the operation of B on the pole A in a clockwise sense will be A' such that $\widehat{BA} = \widehat{BA'}$ and $\widehat{ABC} = \widehat{A'BC} = \frac{1}{2}\delta_B$. Since the side $\widehat{BC}$ is common to both, the spherical triangles ABC, A'BC are congruent so that $\widehat{CA} = \widehat{CA'}$ and $\widehat{BCA} = \widehat{BCA'} = \frac{1}{2}\delta_C$. Therefore A' is the first equivalent pole produced by anti-clockwise operation of the symmetry axis whose pole is C and whose angle of rotation is δ_C. Consecutive clockwise rotation of A about itself and through δ_B about B is thus equivalent to anticlockwise rotation of A through δ_C about C. In general then one can say that if these symmetry axes have their poles at the vertices of a spherical triangle, each of whose angles is half the angle of rotation of the corresponding axis, then the consecutive clockwise operation of two axes is equivalent to the anticlockwise operation of the third axis.

A spherical triangle is uniquely determined by the three angles at its vertices; therefore if α, β, and γ are known, the angles between the three rotation axes can be evaluated from relations of the type (Appendix D)

$$\cos \widehat{BC} = \frac{\cos \alpha + \cos \beta \cos \gamma}{\sin \beta \sin \gamma}.$$

[2] A short account of spherical trigonometry is given in Appendix D.

It has already been shown that the only rotation axes of symmetry that can operate on a lattice are diads, triads, tetrads, and hexads. Therefore δ_A, δ_B, and δ_C are restricted to the values $\frac{2}{2}\pi$, $\frac{2}{3}\pi$, $\frac{2}{4}\pi$, and $\frac{2}{6}\pi$ and consequently the only possible values of α, β, and γ are $\frac{1}{2}\pi$, $\frac{1}{3}\pi$, $\frac{1}{4}\pi$, and $\frac{1}{6}\pi$. If A, B, and C are taken to be each type of rotation axis in turn, the twenty combinations shown in Table 3.2 result. Substitution of the magnitude of α, β, and γ for each combination in the expressions for the cosines of $\widehat{BC}$, $\widehat{CA}$, and $\widehat{AB}$ yields for the majority of combinations cosines outside the range $+1$ to -1, for a few combinations trivial solutions representing coincidence of axes, and for six combinations real solutions (rows 1 to 6 of Table 3.2 and Fig 3.11(a)–(f)). We do not propose to work laboriously through the derivation of Table 3.2, but merely to discuss the set of combinations which have a diad and a tetrad combined in turn with a diad, a triad, a tetrad, and a hexad. If we fix α and β at 90° and 45° respectively the expressions for the sides of the general spherical triangle (Appendix D) become

$$\cos \widehat{BC} = \frac{\cos 90° + \cos 45° \cos \gamma}{\sin 45° \sin \gamma} = \cot \gamma$$

$$\cos \widehat{CA} = \frac{\cos 45° + \cos \gamma \cos 90°}{\sin \gamma \sin 90°} = \frac{1}{\sqrt{2} \sin \gamma}$$

$$\cos \widehat{AB} = \frac{\cos \gamma + \cos 90° \cos 45°}{\sin 90° \sin 45°} = \sqrt{2} \cos \gamma$$

Table 3.2

A	B	C	$\widehat{BC}$	$\widehat{CA}$	$\widehat{AB}$	
2	2	2	90°	90°	90°	
2	2	3	90°	90°	60°	
2	2	4	90°	90°	45°	
2	2	6	90°	90°	30°	
2	3	3	70° 32′	54° 44′	54° 44′	
2	3	4	54° 44′	45°	35° 16′	
2	3	6	0	0	0	trivial
2	4	4	0	0	0	trivial
2	4	6	*	*	*	impossible
2	6	6	*	*	*	impossible
3	3	3	0	0	0	trivial
3	3	4	*	*	*	impossible
3	3	6	*	*	*	impossible
3	4	4	*	*	*	impossible
3	4	6	*	*	*	impossible
3	6	6	*	*	*	impossible
4	4	4	*	*	*	impossible
4	4	6	*	*	*	impossible
4	6	6	*	*	*	impossible
6	6	6	*	*	*	impossible

*indicates cosine > 1

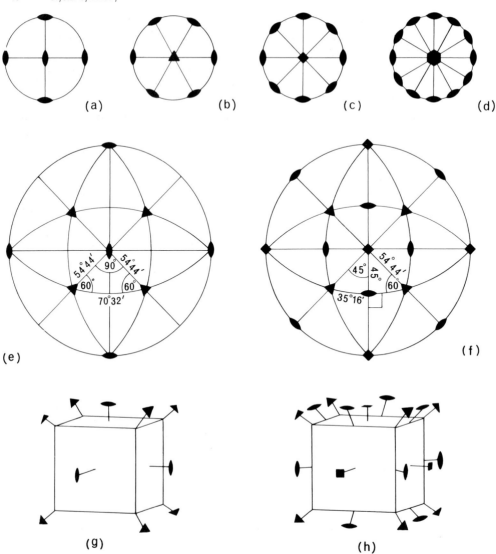

Fig 3.11 Combinations of rotation axes. (a) 222, three mutually perpendicular diads, (b) 223, a triad perpendicular to two diads inclined at 60° to each other, (c) 224, a tetrad perpendicular to two diads inclined at 45° to each other, (d) 226, a hexad perpendicular to two diads inclined at 30° to each other, (e, g) 233, a diad inclined at 54°44′ to two triads inclined at 70°32′ to each other, (f, h) 234, a diad inclined to a triad at 35°16′ and to a tetrad at 45°, the triad and tetrad being mutually inclined at 54°44′.

which on substitution of $\gamma = 90°$, 60°, 45°, 30° in turn yield values of the interaxial angles in the combinations 242 (224), 243 (234), 244, 246 (the symbols shown in brackets are those shown in Table 3.2). The first of these combinations, a tetrad combined with two sets of diads, has its diads perpendicular to the tetrad and inclined at 45° to one another as illustrated in Fig 3.11(c). The combination of a diad, a tetrad, and a triad, illustrated stereographically in Fig 3.11(f), has a disposition of symmetry axes that is simply related to the geometry of the cube (Fig 3.11(h)): three mutually perpendicular tetrads are normal to the cube faces, four triads lie along the body diagonals of the cube, and six diads join the mid-points of opposite edges of the cube. The angle θ between a triad and a tetrad and the angle ω between a triad and an adjacent diad can simply be evaluated by drawing a central section of a cube containing a face diagonal (Fig 3.12): if the cube edge is of length a, $\theta = \tan^{-1}(\sqrt{2}a)/a = 54°44'$ and $\omega = 90° - \theta = 35°16'$. The combination of a diad with two sets of triads has similar geometry (Fig 3.11(e) and (g)) with its diads perpendicular to cube faces and triads along the body diagonals of the cube. The combination of a diad with two independent tetrads is trivial in that all symmetry axes must be coincident. The combination of a diad, a tetrad, and a hexad is impossible because solution of the expressions for the interaxial angles yields $\cos^{-1}\sqrt{3}$, $\cos^{-1}\sqrt{2}$, $\cos^{-1}\sqrt{\frac{3}{2}}$, all of which are greater than unity.

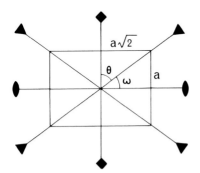

Fig 3.12   Section of a cube parallel to (110) showing coplanar symmetry elements in the combination 243. The length of the section is $a\sqrt{2}$ and its height is a.

   In the combination 234 (Fig 3.11(f) and (h)) there are three tetrads, but they are not independent, each being related to the others by the other symmetry axes present; the only independent axes in this combination are one diad, one triad, and one tetrad. In contrast the combination 244 has two independent tetrads, but this combination is trivial because all three symmetry axes have to be coincident. In general Euler's proposition is concerned only with combinations of independent rotation axes of symmetry.
   Inspection of Table 3.2 indicates that while a hexad can only be combined with two sets of diads, a tetrad and a triad can be combined either together with a diad or separately with two sets of diads. When a hexad, a tetrad, a triad, or a diad is combined with two independent diads, the diads lie in the plane normal to the axis of higher symmetry and are inclined to one another at an angle equal to half the rotation angle of that axis (Fig 3.11(a)–(d)).[3] When a diad and a triad are combined with either a triad or a tetrad the resultant disposition of symmetry axes is, as we have already indicated, related to the geometry of the cube (Fig 3.11(e)–(h)). These are the only six possible combinations of rotation axes of symmetry that can operate on crystal lattices.

Inversion axes of symmetry

A distinct type of symmetry axis is that which combines rotation about a line through $2\pi/n$ with inversion through a point. Such axes are known as *inversion axes of symmetry* and are designated as *inverse n-fold axes*, where $n = 1, 2, 3, 4, 6$. The operation of inversion can be considered at two levels. On inversion through the

---

[3] In the particular case of the combination of a triad (the only axis of odd order with which we are here concerned) with two diads (Fig 3.11(b)) the diads are not independent.

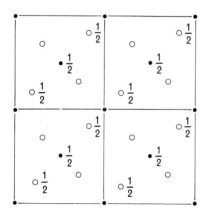
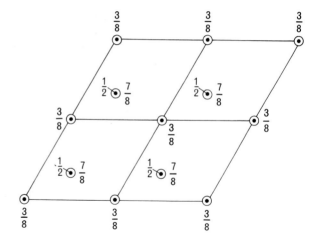

Fig 3.13 Centrosymmetric and non-centrosymmetric structures. (a) Rutile, TiO$_2$, with Ti atoms shown as solid circles and oxygens as open circles. (b) Wurtzite, ZnS, with Zn atoms shown as solid circles and sulphur atoms as open circles. Both structures are shown projected on the *xy* plane.

origin of coordinates every point with coordinates x, y, z becomes a point with coordinates $\bar{x}$, $\bar{y}$, $\bar{z}$. If in a crystal structure every atom with coordinates x, y, z, is duplicated by an atom of the same element with coordinates $\bar{x}$, $\bar{y}$, $\bar{z}$ the structure is said to possess a *centre of symmetry* at the origin.[4] The structure of rutile, TiO$_2$, which has a centre of symmetry at its origin is compared with that of wurtzite, which is non-centrosymmetric, in Fig 3.13. On the macroscopic scale a centre of symmetry causes the crystal faces (hkl) and ($\bar{h}\bar{k}\bar{l}$) to be equivalent so that in a perfectly developed crystal they will be equally developed (Fig 3.15(a)). The operation of a centre of symmetry on a general pole is illustrated in Fig 3.14, from which it is apparent that the operation of a centre of symmetry amounts to trivial rotation of the pole through $2\pi/1$ about any line through the centre followed by inversion through the centre. In conformity with the notation that we shall use for higher inversion axes of symmetry the centre of symmetry can be described as an inverse monad and assigned the symbol $\bar{1}$. In space group diagrams and occasionally in plans of crystal structures it is convenient to represent the positions of centres of symmetry; this is done by a small open circle.

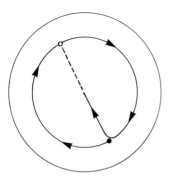

Fig 3.14 Stereogram to illustrate the operation of a centre of symmetry. The pole shown as a solid circle (in the upper hemisphere) is rotated through 360° and inverted through the centre to yield the pole shown as an open circle (in the lower hemisphere).

---

[4] It is worth noting that if the point x, y, z is equivalent to the point $\bar{x}$, $\bar{y}$, $\bar{z}$ then it must also be equivalent to the point $1-x$, $\bar{y}$, $\bar{z}$. Therefore if there is a centre of symmetry at 0, 0, 0 there must also be a centre of symmetry midway between x, y, z and $1-x$, $\bar{y}$, $\bar{z}$, i.e., at $\frac{1}{2}$, 0, 0, and likewise at 0, $\frac{1}{2}$, 0; 0, 0, $\frac{1}{2}$; and $\frac{1}{2}$, $\frac{1}{2}$, $\frac{1}{2}$; etc. Thus the spacing of centres of symmetry along any direction within the structure is half that of the lattice spacing in that direction.

The perfectly developed crystal of gypsum, $CaSO_4.2H_2O$, shown in Fig 3.15(a) has a centre of symmetry relating faces such as I and II. In addition the crystal, as drawn, displays a mirror image relationship between its left- and right-hand sides, that is to say the part of the crystal to the left of the plane abcd is a reflexion in that plane of the part of the crystal to the right of the plane. Such a crystal is said to possess a *plane of symmetry* or *mirror plane*. The mirror plane illustrated is an (010) plane and thus relates a face (*hkl*) to a face (*h̄kl*), these faces being mirror images of one another. Stereograms showing the operation of mirror planes perpendicular and parallel to the plane of the diagram are shown in Fig 3.15(b) and (c). In both cases it is apparent that the operation of the mirror plane amounts to rotation through $2\pi/2 = 180°$ about the pole D followed by inversion through the centre of the stereogram. In conformity with the notation that we shall use for higher inversion axes of symmetry the mirror plane can be described as an inverse diad and designated $\bar{2}$, the inverse diad D being normal to the mirror plane; it is however common practice to describe this symmetry operator as a mirror plane or plane of symmetry and to assign to it the symbol *m*. Mirror planes are conventionally represented on stereograms as boldly drawn great circles.

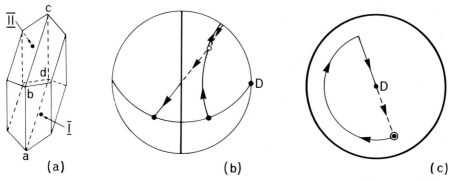

Fig 3.15   The operation of an inverse diad or mirror plane. (a) shows a crystal of gypsum; the faces I and II are related by a centre of symmetry; that part of the crystal to the left of the plane abcd is a reflexion in that plane of the part of the crystal to the right of the plane. (b) shows stereographically the operation of an inverse diad D in the plane of the diagram; the inverse diad rotates the pole, shown in the upper hemisphere on the right-hand side of the diagram, on the small circle about D through 180° to the position shown by the open circle and this is followed by inversion through the centre. (c) shows stereographically the operation of an inverse diad D perpendicular to the plane of the diagram. In (b) and (c) the corresponding mirror planes are respectively a vertical great circle and the primitive.

The centre of symmetry and the mirror plane are commoner and rather more important in crystallography than the higher inversion axes, the inverse triad, tetrad, and hexad. Stereograms displaying the operation of each of these higher inversion axes on a general pole are shown in Fig 3.16(a)–(c) and crystals showing these axes are illustrated in Fig 3.16(d)–(f). The inverse triad, tetrad, and hexad are conventionally represented as $\bar{3}$, $\bar{4}$, and $\bar{6}$ respectively. Their conventional graphical symbols are shown in Fig 3.16; these are difficult to draw on a small scale, especially the $\bar{6}$ symbol, and are often shown as open triangles, squares, and hexagons. The operation of the inverse triad is equivalent to the operation of a triad combined with a centre of symmetry (i.e. $\bar{3} = 3 + \bar{1}$); the operation of the inverse hexad is equivalent to that of a triad combined with a perpendicular mirror plane (i.e. $\bar{6} = 3 + m$); but the operation of the inverse tetrad is not equivalent to any combination of other rotation or inversion axes although it includes a rotation diad.

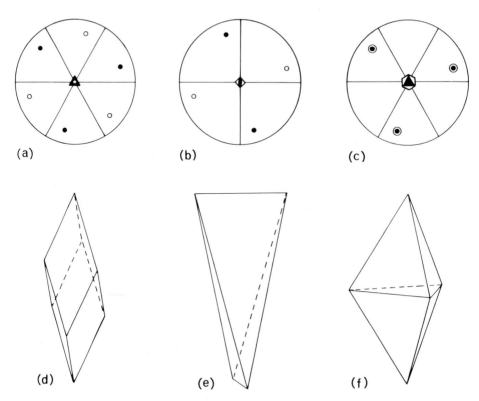

Fig 3.16 Inversion axes of symmetry. The stereograms show the operation of a $\bar{3}$, a $\bar{4}$, and a $\bar{6}$ axis on a general pole. The perspective drawings show crystals displaying inverse triad, tetrad, and hexad symmetry.

Since a lattice is a regular array of points in space every lattice point is associated with lattice points at vector distances $+t$ and $-t$ from it. That is to say all lattices are necessarily centrosymmetric with a centre of symmetry at every lattice point. A lattice consistent with fourfold symmetry thus has a centre of symmetry at every lattice point as well as a tetrad; the symmetry of such a lattice (Fig 3.17) can be described variously as a tetrad or an inverse tetrad combined with a perpendicular mirror plane or as a tetrad or an inverse tetrad combined with a centre of symmetry.

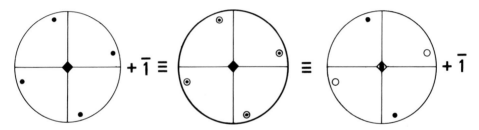

Fig 3.17 Stereograms to illustrate that the combination of a rotation tetrad with a centre of symmetry is equivalent to the combination of an inverse tetrad with a centre of symmetry and to the combination of a rotation tetrad with a perpendicular mirror plane.

The lattice is thus consistent with the presence not only of a rotation tetrad but also of an inverse tetrad.

In general a lattice consistent with the presence of an n-fold rotation axis is also consistent with the presence of an n-fold inversion axis.

Crystallographic point groups

Having established the crystallographic symmetry operators as 1, 2, 3, 4, 6, $\bar{1}$, $\bar{2}$ (m), $\bar{3}$, $\bar{4}$, and $\bar{6}$ we now propose to extend our earlier discussion of the combination of rotation axes to the combination of rotation and inversion axes and so to establish the combinations of symmetry operators that can operate on a lattice. We begin with two definitions. The symmetry elements (or operators) of a finite body must pass through a point, which is taken as the centre of the body; such a group (or combination) of symmetry elements is known as a *point group*. A point group operating on a crystalline solid must be such that every symmetry element of the group can operate on a lattice; such a point group is known as a *crystallographic point group*. A crystallographic point group is completely defined as a group of symmetry elements that can operate on an infinite three-dimensional lattice so as to leave one point unmoved.

It is convenient to group the crystallographic point groups into *crystal systems*, such that the point groups of one system all have some symmetry in common. Thus the *tetragonal* system is the group of all the crystallographic point groups that contain one tetrad, which may be either a rotation or an inverse tetrad; in certain of the tetragonal point groups the tetrad is combined with diads or mirror planes or both. We have already seen that the operation of a tetrad on a lattice requires the lattice points on planes normal to the tetrad to be arranged on a square unit-mesh. It is convenient then to take the reference axes of the tetragonal system as z parallel to the tetrad, x and y parallel to the sides of the square unit-mesh of a lattice plane perpendicular to the tetrad; the conventional unit-cell of the tetragonal system thus has $a = b \neq c$ and $\alpha = \beta = \gamma = 90°$.

In general it is convenient to select a unit-cell whose axes are, if possible, parallel or normal to symmetry axes for the good reason that this simplifies the description of symmetry relationships. The shape of the conventional unit-cell is then characteristic of the system to which it refers. In Table 3.3 the nomenclature of the crystal systems and the restrictions on the shape of the conventional unit-cell for each system are set out.

Table 3.3
The crystal systems

Name of system	Characteristic symmetry	Conventional unit-cell
Triclinic	Onefold symmetry only	$a \neq b \neq c$; $\alpha \neq \beta \neq \gamma$
Monoclinic	One diad ($\parallel y$)	$a \neq b \neq c$; $\alpha = \gamma = 90°$, $\beta > 90°$
Orthorhombic	Three mutually perpendicular diads ($\parallel x$, y and z)	$a \neq b \neq c$; $\alpha = \beta = \gamma = 90°$
Trigonal*	One triad ($\parallel [111]$)	$a = b = c$; $\alpha = \beta = \gamma < 120°$, $\neq 90°$
Tetragonal	One tetrad ($\parallel z$)	$a = b \neq c$; $\alpha = \beta = \gamma = 90°$
Hexagonal	One hexad ($\parallel z$)	$a = b \neq c$; $\alpha = \beta = 90°$, $\gamma = 120°$
Cubic	Four triads ($\parallel \langle 111 \rangle$)	$a = b = c$; $\alpha = \beta = \gamma = 90°$

The symbol $\neq$ implies that equality is not required by symmetry.
*The unit-cell of the hexagonal system is however commonly used for the trigonal system.

We turn now to consider what combinations of rotation and inversion axes are possible and so to derive all the distinct crystallographic point groups. We have already shown that a rotation axis of order 1, 2, 3, 4, or 6 can operate alone on a lattice and thus is itself a crystallographic point group. We have further shown that there are only six combinations of rotation axes capable of operating on a lattice and that each combination has a definite geometrical arrangement (Table 3.2); each of the combinations 222, 223, 224, 226, 233, and 234 is thus a distinct crystallographic point group.[5] We have also shown that the lattice consistent with the operation of an n-fold rotation axis is consistent with the operation of an inversion axis of the same order; each of the inversion axes, $\bar{1}$, $\bar{2}$ ($= m$), $\bar{3}$, $\bar{4}$, and $\bar{6}$, operating on its own thus constitutes a distinct crystallographic point group.

Since all lattices are necessarily centrosymmetric, a crystal may itself have a centre of symmetry. Further crystallographic point groups can thus be derived simply by adding a centre of symmetry to the point groups that we have already identified. Before doing this it may be helpful to the reader to be reminded of two points that have been made earlier in this chapter. Firstly, the operation of a rotation axis of odd order combined with a centre of symmetry is equivalent to an inversion axis of the same order (Figs 3.14 and 3.16); thus, trivially, $1 + \bar{1} = \bar{1}$ and, more significantly, $3 + \bar{1} = \bar{3}$. Secondly, the operation of a rotation axis of even order combined with a centre of symmetry is equivalent to the operation of an inversion axis of the same order combined with a centre of symmetry; we have shown this (Fig 3.17) for the tetrad and leave the reader to satisfy himself that it holds also for the diad and the hexad. Thus the combination of a centre of symmetry either with one of the rotation axes 2, 4, 6 or with one of the inversion axes, $\bar{2}$, $\bar{4}$, $\bar{6}$, yields a new crystallographic point group. Each of these point groups can alternatively be described as the combination of a 2, 4, or 6 rotation axis with a perpendicular mirror plane (Fig 3.17 illustrates one case); this alternative description is the basis of the conventional nomenclature for these three point groups $2/m$, $4/m$, $6/m$.

Before discussing the point groups obtained by substitution of inversion for rotation axes or by addition of a centre of symmetry in the point groups derived by means of Euler's proposition (Table 3.2) it is convenient to consider the question of *hand*. The operation of a rotation axis, being a simple rotation, is incapable of changing the hand of the object on which it operates. In contrast the operation of an inversion axis, because it involves inversion through the centre, must convert a right-handed object into a left-handed object. For example Fig 3.18(a) is a stereogram showing the operation of an inverse tetrad on a group of three poles. If the original group of poles is taken to be that in the upper right-hand quadrant a single operation of the inverse tetrad gives rise to the group in the lower left-hand quadrant; it is apparent from the figure that these two groups of poles cannot be superimposed, being mirror images of one another, the one right-handed and the other left-handed. Now in general terms an inversion axis A will produce a left-handed group P_L from a right-handed group P_R (Fig 3.18(b)) and a rotation axis B will produce a left-handed group P'_L from the left-handed group P_L. Therefore the axis consistent with A and B must be an axis C relating P_R to P'_L and, since a change of hand is produced, this must be an inversion axis. Thus in a combination of three axes either all three are rotation axes *or* one is a rotation axis and two are inversion axes; no other combination is possible.

[5] We shall postpone discussion of the conventional nomenclature for point groups and for the time being use a nomenclature related to that of Table 3.2.

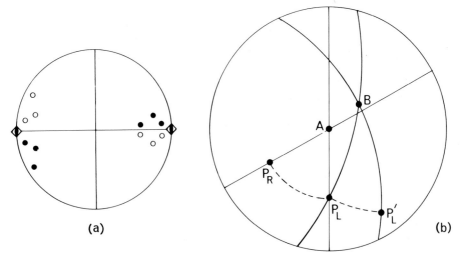

Fig 3.18   The significance of *hand* in the combination of axes. The stereogram (a) illustrates the change of hand resulting from the operation of an inverse tetrad. The stereogram (b) illustrates the general proposition that the successive operation of an inversion axis A, which produces the left-handed group P_L from the right-handed group P_R, and a rotation axis B, which produces the left-handed group P'_L from the left-handed group P_L, can only be consistent with the operation of an inversion axis producing the left-handed group P'_L from the right-handed group P_R.

We now proceed to apply the conclusion of the last paragraph to each of the point groups derived by means of Euler's proposition. From the point group 222 only one new crystallographic point group can be derived, $\overline{2}2\overline{2}$, which can alternatively be described as *mm*2.[6] This point group has two perpendicular mirror planes with a rotation diad along their line of intersection. From the point group 223 (conventionally described as 32) we derive $\overline{2}2\overline{3}$ (conventionally 3*m*) and $\overline{2}2\overline{3}$ (conventionally $\overline{3}m$), which are new point groups. From the point group 224 (conventionally 422) we derive $\overline{2}2\overline{4}$ (conventionally 4*mm*) and $\overline{2}2\overline{4}$ (conventionally $\overline{4}2m$), both of which are new.

Table 3.4
Derivation of the crystallographic point groups

Symmetry axes parallel to one direction only:

Rotation X	1	2	3	4	6
Inversion $\overline{X}$	$\overline{1}$	*m*	$\overline{3}$	$\overline{4}$	$\overline{6}$
$X + \overline{1}$	$(\overline{1})$	2/*m*	$(\overline{3})$	4/*m*	6/*m*

Symmetry axes in more than one direction:

XYZ	222	223 (32)	224 (422)	226 (622)	233 (23)	234 (432)
$\overline{X}\,\overline{Y}Z$	*mm*2	*mm*3 (3*m*)	*mm*4 (4*mm*)	*mm*6 (6*mm*)	$m\overline{3}3$ (*m*3)	$m\overline{3}4$ (*m*3*m*)
$\overline{X}Y\overline{Z}$	(*m*2*m*)	$m2\overline{3}$ ($\overline{3}m$)	$m2\overline{4}$ ($\overline{4}2m$)	$m2\overline{6}$ ($\overline{6}m2$)	(*m*3)	$m3\overline{4}$ ($\overline{4}3m$)
$X\,\overline{Y}\,\overline{Z}$	(2*mm*)	($\overline{3}m$)	($\overline{4}2m$)	($\overline{6}m2$)	(*m*3)	(*m*3*m*)
$XYZ + \overline{1}$	*mmm*	($\overline{3}m$)	4/*mmm*	6/*mmm*	(*m*3)	(*m*3*m*)

Note:
(1) Symbols in brackets standing alone refer to point groups higher up in the table.
(2) Symbols in brackets following symbols not in brackets are conventional point group symbols.

[6] To follow this and succeeding paragraphs the reader will find it helpful to refer to Table 3.4 and to look forward to Fig 3.20.

From the point group 226 (known conventionally as 622) we derive $\overline{2}26$ (conventionally 6mm) and $\overline{2}2\overline{6}$ (conventionally $\overline{6}m2$). From the point group 233 (conventionally known as 23) we derive $\overline{2}33$ and $23\overline{3}$, which turn out to be identical and are conventionally described as $m3$. From the point group 234 (conventionally 432) we derive $\overline{2}34$ and $2\overline{3}\overline{4}$, which are identical and known as $m3m$, and $\overline{2}3\overline{4}$ which is conventionally known as $\overline{4}3m$.

To complete our list of crystallographic point groups we have to add a centre of symmetry to each of the point groups derived from Euler's proposition and to each of those obtained in the preceding paragraph. This task can be simplified by bearing in mind that the combination of a centre of symmetry with a rotation triad is equivalent to an inverse triad and that the combination of a centre of symmetry with either a rotation or an inversion axis of even order is identical. Thus 222 and $mm2$ on combination with a centre of symmetry both yield the same new point group $\frac{2}{m}\frac{2}{m}\frac{2}{m}$, conventionally known as mmm. The addition of a centre of symmetry to 223 (32) and $\overline{2}23$ (3m) yields the same point group, which is identical with $\overline{2}2\overline{3}$ ($\overline{3}m$) and already accounted for. The addition of a centre of symmetry to 224 (422), or $\overline{2}24$ (4mm), or $\overline{2}2\overline{4}$ ($\overline{4}2m$) yields the same new point group $\frac{2}{m}\frac{2}{m}\frac{4}{m}$, conventionally known as 4/mmm. Likewise the addition of a centre of symmetry to 226 (622) or $\overline{2}26$ (6mm), or $\overline{2}2\overline{6}$ ($\overline{6}m2$) yields the same new point group $\frac{2}{m}\frac{2}{m}\frac{6}{m}$, conventionally known as 6/mmm. Addition of a centre of symmetry to 233 (23) yields the point group $\frac{2}{m}\overline{3}\,\overline{3}$ which has already been accounted for and designated $m3$. Finally addition of a centre of symmetry to 234 (432) or $\overline{2}3\overline{4}$ ($\overline{4}3m$) yields the point group $\frac{2}{m}\overline{3}\frac{4}{m}$ which has already been accounted for and designated $m3m$.

We have derived in all thirty-two crystallographic point groups, five consisting of a single rotation axis, another five consisting of a single inversion axis, three by combination of a centre of symmetry with a rotation axis, six directly derivative from Euler's proposition and a further ten by mixing inversion and rotation axes in these geometrical combinations, and three more by combining a centre of symmetry with these combinations.

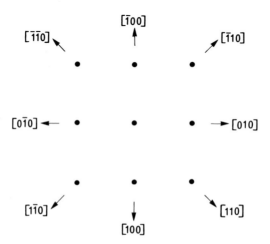

Fig 3.19   Plan of the (001) plane of the tetragonal lattice.

In the conventional nomenclature of point groups, which we have already used without explanation, emphasis is placed on the presence of mirror planes of symmetry. In point group symbols a mirror plane is associated with the direction of its normal whereas, naturally, a rotation or inversion axis is associated with its own direction in the crystal. Where all the symmetry elements of the point group are associated with a single direction the conventional point group symbol takes one of the forms X (an X-fold rotation axis only), $\bar{X}$ (an X-fold inversion axis only), or $\frac{X}{m}$ (an X-fold rotation axis combined with a perpendicular mirror plane). This last type of symbol is often printed for convenience of typesetting as X/m. Where the symmetry elements of the point group are associated with more than one direction those associated with each direction are stated in the conventional point group symbol in a standard order. By way of illustration we take the tetragonal system, where z is taken as the direction of the rotation or inversion tetrad and the x and y axes are chosen so as to be parallel to the shortest lattice repeats in the plane perpendicular to the tetrad; the unit-cell is thus a square prism with $a = b \neq c$. The geometry of the combinations of symmetry axes with which we are concerned in the tetragonal system is that shown for 224 in Table 3.2: a tetrad combined with two independent diads inclined at 45° to one another and both perpendicular to the tetrad. Inspection of the (001) plane of the tetragonal lattice (Fig 3.19) shows that the only possible diad directions[7] are $\langle 100 \rangle$ and $\langle 110 \rangle$. Where mirror planes occur in the point group in place of diads they are necessarily perpendicular to $\langle 100 \rangle$ or $\langle 110 \rangle$, that is parallel to $\{100\}$ or $\{110\}$. The conventional symbol for those tetragonal point groups which have symmetry elements associated with more than one direction consists of three terms: first, a statement of the symmetry elements associated with the $[001]$ direction; second, a statement of the symmetry elements associated with the $\langle 100 \rangle$ directions; and third, a statement of the symmetry elements associated with the $\langle 110 \rangle$ directions. Thus the symbol 422 refers to a point group (Figs 3.20 and 3.21) with a rotation tetrad parallel to $[001]$ and diads parallel to $\langle 100 \rangle$ and $\langle 110 \rangle$. The symbol $4mm$ refers to the point group with a rotation tetrad parallel to $[001]$ and mirror planes perpendicular to $\langle 100 \rangle$ and $\langle 110 \rangle$. In the symbol $\bar{4}2m$ however the $\langle 100 \rangle$ and $\langle 110 \rangle$ directions are distinguished: the inverse tetrad is parallel to $[001]$, diads are parallel to $\langle 100 \rangle$ and mirror planes are perpendicular to $\langle 110 \rangle$ in this point group. The symbol $\bar{4}m2$, which is illustrated in Fig 3.22, refers to the point group $\bar{4}2m$ in a different orientation; as far as point group symmetry is concerned the two are indistinguishable, being simply related by rotation through 45° about $[001]$. Only when it is known whether a diad or a mirror plane is associated with the shortest lattice repeat in the (001) plane is it realistic to distinguish between them; this takes us into the field of space groups, which will be discussed in chapter 4. Finally, the centrosymmetric point group $\frac{4}{m}\frac{2}{m}\frac{2}{m}$ has a rotation tetrad parallel and a mirror plane perpendicular to $[001]$ combined with diads parallel and mirror planes perpendicular to both the sets of directions $\langle 100 \rangle$ and $\langle 110 \rangle$; this point group is commonly represented by its *short symbol* 4/mmm.

Further discussion of point group nomenclature is postponed until the crystal

---

[7] $\langle UVW \rangle$ and $\{hkl\}$ represent respectively all the zone axes and all the faces derived from $[UVW]$ and (hkl) by the operation of the point group.

	Triclinic	Monoclinic	Tetragonal
X	1	2	4
X̄	1̄	m (=2̄)	4̄
X+1̄	–	2/m	4/m
		Orthorhombic	
X2	–	222	422
Xm	–	mm2	4̄mm
X̄m	–	–	2m
X2+1̄		mmm	4/mmm

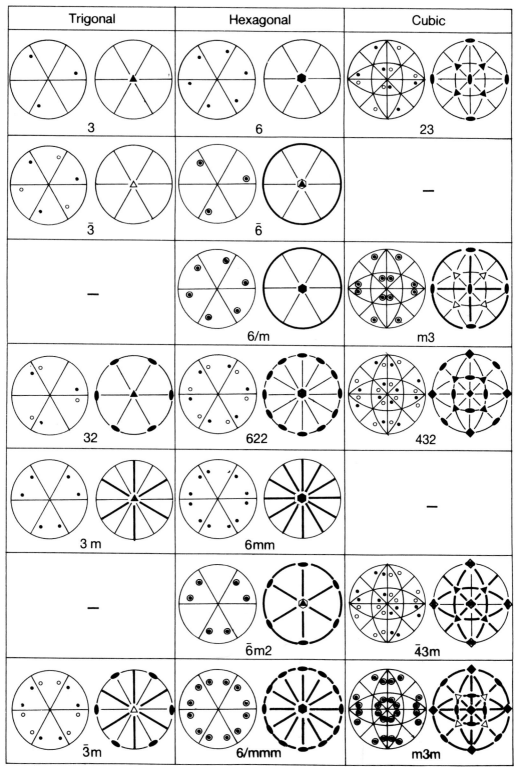

Trigonal	Hexagonal	Cubic
3	6	23
$\bar{3}$	$\bar{6}$	—
—	6/m	m3
32	622	432
3 m	6mm	—
—	$\bar{6}$m2	$\bar{4}$3m
$\bar{3}$m	6/mmm	m3m

Fig 3.20 The thirty-two crystallographic point groups. Each pair of stereograms shows, on the left, the poles of a general form and, on the right, the symmetry elements of the point group. Planes of symmetry are indicated by bold lines.

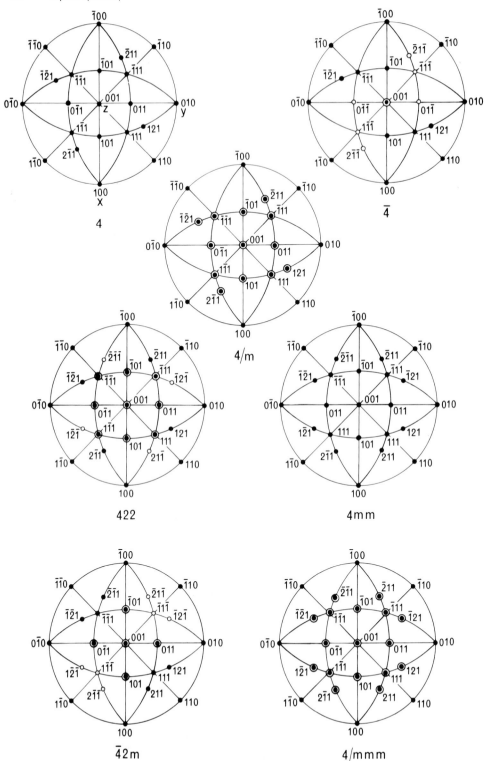

Fig 3.21 The tetragonal point groups. The stereograms show the poles of {100}, {001}, {110}, {101}, {111}, and {121} in each of the point groups. Where poles are superimposed only the pole in the upper hemisphere is indexed.

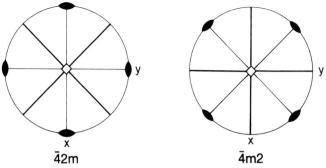

Fig 3.22   The point groups $\bar{4}2m$ and $\bar{4}m2$ which are related by rotation through 45° about [001].

systems have been explored and is then dealt with system by system. The basis of the nomenclature is summarized in Table 3.5.

Table 3.5
The conventions used for the symbols of the point groups

System	Directions associated with symmetry elements		
	First position	Second position	Third position
Triclinic		Centre only	
Monoclinic	[010]		
Orthorhombic	[100]	[010]	[001]
Trigonal	[0001]	$\langle 10\bar{1}0\rangle$	$\langle 21\bar{1}0\rangle$
Tetragonal	[001]	$\langle 100\rangle$	$\langle 110\rangle$
Hexagonal	[0001]	$\langle 10\bar{1}0\rangle$	$\langle 21\bar{1}0\rangle$
Cubic	$\langle 100\rangle$	$\langle 111\rangle$	$\langle 110\rangle$

Crystal classes

A crystalline solid that exhibits the symmetry of a particular point group is said to belong to the corresponding *crystal class,* which is denoted by the same symbol as the point group. For instance gypsum, $CaSO_4.2H_2O$, has point group symmetry $2/m$; it is said to belong to the crystal class $2/m$. A crystal class is defined as the group of substances that display the point group symmetry characteristic of the class. The term 'crystal class' is often incorrectly used as a synonym for 'point group'; the distinction, which may at first sight seem pedantic, is in practice useful.

A crystal exhibiting the largest possible number of symmetry elements for its system may be described as exhibiting the highest point group symmetry possible for the system and is said to belong to the *holosymmetric* class of the system. The point group symmetry of each holosymmetric class is that of the corresponding lattice (the trigonal system is, as we shall show, an exception). The symmetry of every other point group of the system is lower because the repeat unit of the crystal has lower symmetry than the lattice. Figure 3.23, in which three two-dimensional patterns based on a square lattice are shown, illustrates this point: the lattice has a tetrad perpendicular to the plane of the mesh and lines of symmetry every 45°, so does pattern (b), but pattern (c) is entirely lacking in lines of symmetry, and pattern (d), for which the tetrad descends to a diad, has lines of symmetry only parallel to the sides of the unit-mesh. The pattern shown in Fig 3.23(d) has mutually perpendicular lattice repeats of equal length but unrelated by the symmetry elements of the pattern; formally the lattice on which this pattern is based is rectangular with *a* only *accidentally* equal to *b*.

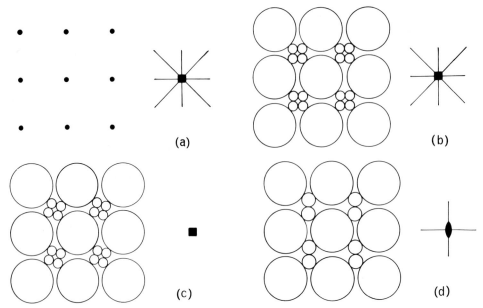

Fig 3.23 Two-dimensional patterns based on the square lattice (a). Pattern (b) has the full symmetry of the lattice; pattern (c) lacks lines of symmetry; and in pattern (d) the tetrad has become a diad and there are lines of symmetry parallel to the sides of the unit-mesh only.

Crystal systems

We have defined a crystal system as a group of point groups that have some symmetry in common and we have made use of the point groups of the tetragonal system to explain point group nomenclature. We now proceed to define each of the seven crystal systems and to specify and comment on all the point groups of each system.

Before examining each system in detail it may be helpful to name and make some general comments on the seven crystal systems. Three systems can be described as *orthogonal*, that is they can be referred to mutually perpendicular reference axes: these are the *cubic system*, for which the unit-cell is a regular cube, the *tetragonal system*, whose unit-cell is a square prism, and the *orthorhombic system*, whose unit-cell is a rectangular parallelepiped. Two systems can be referred to unit-cells that are 120° prisms: these are the *hexagonal* and *trigonal systems*. The remaining two systems are the least symmetrical, the *monoclinic system*, which has a unit-cell with $a \neq b \neq c$, $\alpha = \gamma = 90° \neq \beta$ conventionally, and the *triclinic system*, which has a unit-cell with $a \neq b \neq c, \alpha \neq \beta \neq \gamma$ in general.

In the course of exploring each system we shall deal with such special cases of nomenclature and representation as may arise. But there are two general matters of nomenclature that are most conveniently dealt with now. The group of faces produced by the operation of all the symmetry elements of a point group acting on one face (*hkl*) is known as a *form* and represented as {*hkl*}. Similarly all the zone axes produced by the operation of all the symmetry elements of a point group on one zone axis [*UVW*] is known as a *form of zone axes* and represented as ⟨*UVW*⟩. Thus in crystallography we distinguish between indices enclosed in four kinds of brackets: ( ) known simply as *brackets*, { } as *braces*, [ ] as *square brackets*, and ⟨ ⟩ as *carets*.

Triclinic system

The characteristic symmetry of the triclinic[8] system is a onefold axis. The only symmetry that a triclinic crystal can display is a centre of symmetry and it may not have even that. Since a centre of symmetry is inherent in any lattice, its presence places no restriction on the shape of the triclinic unit-cell, which is thus a general parallelepiped with $a \neq b \neq c$, $\alpha \neq \beta \neq \gamma$. There are just two point groups in the triclinic system: point group 1, in which neither any plane nor any direction is symmetrically repeated, and point group $\bar{1}$ (the holosymmetric point group of the system), in which the planes (hkl) and $(\bar{h}\bar{k}\bar{l})$ are equivalent, as are the directions $[UVW]$ and $[\bar{U}\bar{V}\bar{W}]$. The simplicity engendered by the absence of symmetry elements of order higher than one is offset in practice by the absence of right-angles in the geometry of the unit-cell.

Stereograms of triclinic crystals can conveniently be drawn by plotting the z-axis at the centre of the stereogram and the pole of (010) at the right-hand extremity of the horizontal diameter of the primitive (Fig 3.24). The x-axis then projects on the vertical diameter of the primitive at an angle β from the z-axis. The y-axis is located

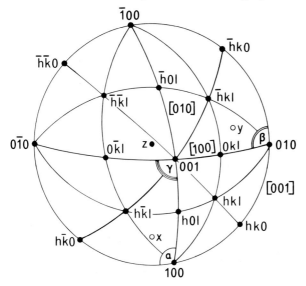

Fig 3.24 The triclinic system. The stereogram shows the relationship of various poles to the reference axes in a hypothetical crystal of class 1. In a crystal of class $\bar{1}$ symmetry would require the opposite of any pole shown to be shown also.

at the intersection of small circles of radius α and γ about the z and x axes respectively. The poles of (100) and (001) are respectively the poles of the great circles containing the y and z axes and the x and y axes.

Monoclinic system

The characteristic symmetry of the monoclinic system is a single axis of twofold symmetry, which is conventionally taken as the y-axis of the unit-cell. The axis of diad symmetry places no restriction on the shape of the unit-mesh of the lattice in the plane perpendicular to itself. It is reasonable therefore to take as the x and z axes directions in this plane so that a and c are as short as possible. The positive directions of the x and z axes are conventionally chosen so that the angle between them, β, is obtuse. The geometry of the monoclinic unit-cell is thus $a \neq b \neq c$, $\alpha = \gamma = 90°$, $\beta \geqslant 90°$ (Fig 3.25(a)).

[8] This system used to be known as the *anorthic system*. After many years of disuse the old name has recently been revived by some authors. Both triclinic and anorthic are currently in common use.

The plane (010) is by definition parallel to x and z; therefore, since $\alpha = \gamma = 90°$, the normal to the (010) plane is parallel to the y-axis. The plane (100) is necessarily parallel to the y and z axes, but, since $\beta \neq 90°$, its normal must be inclined to the x-axis (Fig 3.25(b)). Likewise the normal to the (001) plane must be inclined to the z-axis and perpendicular to the x and y axes. It is apparent from Fig 3.25(b) that $\widehat{ARC} = \widehat{AOC} = \beta$ and that $\widehat{ARC} + \widehat{POQ} = 180°$. Therefore $\widehat{POQ} = 180° - \beta$, that is to say the angle $(100):(001) = 180° - \beta$.

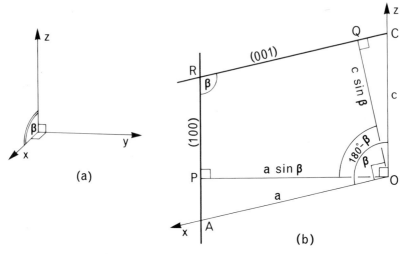

Fig 3.25 Monoclinic geometry: (a) shows the disposition of the reference axes and (b) is an (010) section through the origin. The angle between the normals to the planes (100) and (001) is $180° - \beta$, where β is the interaxial angle $x:z$.

The monoclinic system comprises three point groups. The point group 2 has a single symmetry element, a rotation diad parallel to the y-axis; the planes (hkl) and $(\bar{h}k\bar{l})$ are equivalent, as are the directions $[UVW]$ and $[\bar{U}V\bar{W}]$. The point group m has a mirror plane perpendicular to the y-axis, that is parallel to (010); the planes (hkl) and $(h\bar{k}l)$ are equivalent, as are the directions $[UVW]$ and $[U\bar{V}W]$. The point group $2/m$ is the holosymmetric point group of the system; its symmetry elements are a rotation diad parallel to the y-axis and a mirror plane parallel to (010), a combination that introduces a centre of symmetry. In point group $2/m$ the four planes (hkl), $(\bar{h}k\bar{l})$, $(h\bar{k}l)$, and $(\bar{h}\bar{k}\bar{l})$ are equivalent, as are the four directions $[UVW]$, $[\bar{U}V\bar{W}]$, $[U\bar{V}W]$, and $[\bar{U}\bar{V}\bar{W}]$. Stereograms showing selected forms in each monoclinic point group are presented in Fig 3.26.

It is immediately apparent from Fig 3.26 that certain forms in each point group comprise fewer faces than does the *general form* $\{hkl\}$. Thus in point group 2 the face-normal (010) is coincident with the diad so that the form $\{010\}$ comprises only the face (010), whereas the general form $\{hkl\}$ comprises the two faces (hkl) and $(\bar{h}k\bar{l})$. The form $\{0\bar{1}0\}$ likewise consists of a single face $(0\bar{1}0)$. The general form in point group m likewise comprises two faces (hkl) and $(h\bar{k}l)$; but a face of the type[9] $(h0l)$, whose pole lies in the mirror plane, is not repeated by the mirror plane so that the form $\{h0l\}$ consists of a single face. Forms such as $\{010\}$ and $\{0\bar{1}0\}$ in point group 2

---

[9] Now and subsequently we use the symbol $\{h0l\}$ to represent any form with $k = 0$, the indices h and l having any integral values including 0. The symbol $\{h0l\}$ thus includes $\{100\}$ and $\{001\}$; but in point groups where these have fewer faces than $\{h0l\}$ they receive special mention.

and $\{h0l\}$ in point group m are said to be *special forms*. In point group $2/m$ the general form $\{hkl\}$ comprises four faces (hkl), $(\bar{h}kl)$, $(h\bar{k}l)$, and $(\bar{h}\bar{k}l)$; special forms are of two kinds, that not affected by the diad, i.e. $\{010\}$ and those not affected by the mirror plane, i.e. $\{h0l\}$. In this case both kinds of special form have the same number of faces, but we shall see in other systems that different kinds of special form may have different numbers of faces.

We take as our definition of a special form the statement: *a special form is any form comprising fewer faces than the general form in the same point group*. It follows that the normal to any face of a special form must either be parallel to an axis of symmetry or lie in a mirror plane.[10]

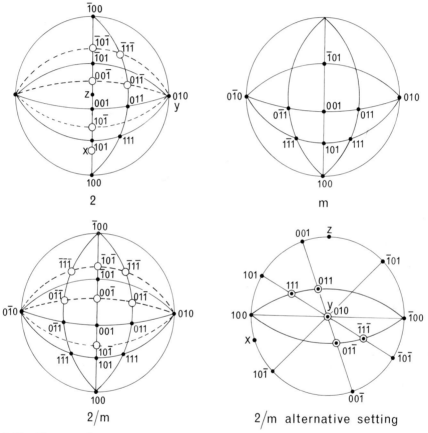

Fig 3.26 The monoclinic point groups. The stereograms show poles of the forms $\{100\}$, $\{010\}$, $\{001\}$, $\{101\}$, $\{\bar{1}01\}$, $\{011\}$, and $\{111\}$ in each of the point groups. The two lower stereograms illustrate alternative settings for $2/m$.

Stereograms of monoclinic crystals are usually plotted, as in Fig 3.26, with the z-axis at the centre of the stereogram and the y-axis at the right-hand end of the horizontal diameter of the primitive. The pole of the positive direction of the x-axis then projects on the vertical diameter of the primitive and, as β is obtuse, in the

[10] Other definitions of a special form have occasionally been employed by authors whose main concern was crystal morphology. The definition adopted here is to be preferred because it is analogous to the definition of a special equivalent position in a space group (chapter 4) and leads directly to the determination of multiplicity in powder photographs (chapter 7).

southern hemisphere. The pole of (010) is coincident with the pole of the *y*-axis. Since (100) is parallel to *y* and *z* its pole lies on the primitive at the lower extremity of the vertical diameter. Since (001) is parallel to *x* and *y* its pole lies at the intersection of the vertical diameter of the primitive with the great circle whose pole is *x*; the pole of (001) thus makes an angle $\beta - 90°$ with *z*.

An alternative setting for the monoclinic stereogram is occasionally employed. The *y*-axis is plotted in the centre of the stereogram so that the primitive represents the (010) plane, the mirror plane in point groups *m* and 2/*m*. The advantage of this setting in these two point groups is that the planes (*hkl*) and (*hk̄l*), which are related by the mirror plane, project so as to be superimposed. Some simplification is thus achieved; in the first mentioned setting symmetry related poles cannot be superimposed. The stereogram of a crystal of point group 2/*m* plotted in the alternative setting is shown in the lower right-hand diagram of Fig 3.26.

Orthorhombic system

The characteristic symmetry of the orthorhombic system is the presence of three mutually perpendicular axes of twofold symmetry. It is obviously convenient to place the *x*, *y*, and *z* axes parallel to symmetry axes so that the unit-cell is obliged to be a rectangular parallelepiped. None of the axes is related to either of the other two so the unit-cell has, in general, edges unequal in length. The shape of the unit-cell is thus given by $a \neq b \neq c, \alpha = \beta = \gamma = 90°$.

The plane (100) is by definition parallel to the *y* and *z* axes and, since *x*, *y*, and *z* are mutually perpendicular, its pole is coincident with the *x*-axis. Similarly the normals to the planes (010) and (001) are respectively coincident with the *y* and *z* axes.

The orthorhombic system comprises three point groups. Each is represented by a

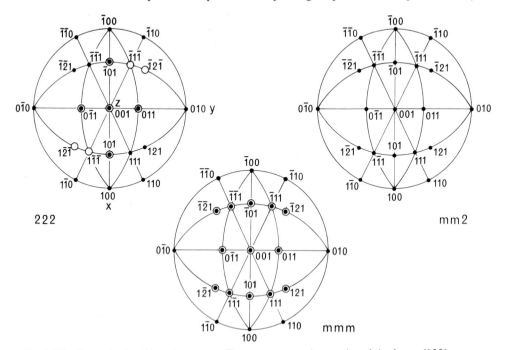

Fig 3.27 The orthorhombic point groups. The stereograms show poles of the forms {100}, {010}, {001}, {011}, {101}, {110}, {111}, and {121} in each of the point groups. Where poles are superimposed only the pole in the upper hemisphere is indexed.

symbol having three positions, which state in turn the symmetry elements associated with the x, y, and z axes.

The point group 222 has three mutually perpendicular rotation diads and is one of the point groups derived directly from Euler's proposition. The general form $\{hkl\}$ comprises four faces (hkl), $(h\bar{k}\bar{l})$, $(\bar{h}k\bar{l})$, and $(\bar{h}\bar{k}l)$; Fig 3.27 shows five such forms $\{011\}$, $\{101\}$, $\{110\}$, $\{111\}$ and $\{121\}$. The special forms in this point group are $\{100\}$, $\{010\}$, and $\{001\}$, each of which has its two faces perpendicular to a diad.

The point group $mm2$ has mirror planes perpendicular to two of the reference axes and a rotation diad parallel to the third. Conventionally the rotation diad is selected as the z-axis. The general form $\{hkl\}$ in this point group comprises four faces (hkl) and $(\bar{h}kl)$ related by the (100) mirror plane and $(h\bar{k}l)$ and $(\bar{h}\bar{k}l)$ related to the first two by the (010) mirror plane (Fig 3.27). The special forms $\{0kl\}$ and $\{h0l\}$ have poles lying in the (100) and (010) mirror planes respectively and each consists of two faces $(0kl)$ and $(0\bar{k}l)$ or $(h0l)$ and $(\bar{h}0l)$. The special forms, $\{001\}$ and $\{00\bar{1}\}$ have only a single face (001) and $(00\bar{1})$, the pole of which lies on the rotation diad at the intersection of the two mirror planes. Analogous to the special form $\{001\}$ is the special form of zone axes $\langle 001 \rangle$: the direction $[001]$ is not related by symmetry to the direction $[00\bar{1}]$ so that the positive and negative directions of the z-axis are unrelated by symmetry. In this point group therefore the z-axis is a *polar axis*, unlike the x and y axes whose positive and negative directions are related by mirror planes. A consequence of z being a polar axis is that poles in the northern and southern hemispheres of the stereogram are not symmetry related.

The holosymmetric point group of the orthorhombic system $\frac{2}{m}\frac{2}{m}\frac{2}{m}$ is derived by adding a centre of symmetry to 222 or to $mm2$, and has a rotation diad and a mirror plane associated with each of the reference axes, x, y, and z; the lattice of an orthorhombic crystal necessarily has the symmetry of this point group. This point group is commonly represented by its short symbol mmm, which sufficiently specifies the point group symmetry. The general form $\{hkl\}$ in point group mmm comprises the eight faces (hkl), $(\bar{h}kl)$, $(h\bar{k}l)$, $(\bar{h}\bar{k}l)$, $(hk\bar{l})$, $(\bar{h}k\bar{l})$, $(h\bar{k}\bar{l})$, $(\bar{h}\bar{k}\bar{l})$; Fig 3.27 shows two such forms $\{111\}$ and $\{121\}$. The special forms whose poles lie on mirror planes are $\{0kl\}$, $\{h0l\}$, and $\{hk0\}$; each comprises four faces. The special forms whose poles lie on diads at the intersection of mirror planes are $\{100\}$, $\{010\}$, and $\{001\}$; each comprises two faces, e.g. (100) and $(\bar{1}00)$, which are related by the (100) mirror plane.

Stereograms of orthorhombic crystals are conventionally drawn with the z-axis in the centre of the stereogram, the positive direction of the y-axis at the right-hand end of the horizontal diameter of the primitive, and the positive direction of the x-axis at the lower end of the vertical diameter of the primitive as in Fig 3.27.

Tetragonal system

Many of the comments that we would make for other systems at this point have already been made for the tetragonal system and illustrated in Figs 3.20 and 3.21. It suffices to add here some comments on general and special forms and on the conventional setting for stereograms of tetragonal crystals.

In point group 4 the general form $\{hkl\}$ consists of four faces (hkl), $(k\bar{h}l)$, $(\bar{h}\bar{k}l)$, and $(\bar{k}hl)$. The only special forms are $\{001\}$ and $\{00\bar{1}\}$, whose normal is parallel to the tetrad; each form consists of the single face. Point group $\bar{4}$ likewise has a general form $\{hkl\}$ consisting of four faces, but here they are (hkl), $(k\bar{h}\bar{l})$, $(\bar{h}\bar{k}l)$, and $(\bar{k}h\bar{l})$; the only special form is $\{001\}$, which here consists of the two parallel faces (001) and

$(00\bar{1})$. The rotation tetrad is thus a polar axis while the inverse tetrad is non-polar. In point group $4/m$ the general form $\{hkl\}$ consists of eight faces, the indices of which are simply derived from those of the general form in either 4 or $\bar{4}$ by adding a centre of symmetry, that is by including faces with indices opposite in sign to those already listed to give (hkl), $(\bar{h}\bar{k}\bar{l})$, $(k\bar{h}l)$, $(\bar{k}h\bar{l})$, $(\bar{h}\bar{k}l)$, $(hk\bar{l})$, $(\bar{k}hl)$, $(k\bar{h}\bar{l})$. There are two kinds of special form: $\{hk0\}$ consists of four faces whose poles lie in the mirror plane, $(hk0)$, $(k\bar{h}0)$, $(\bar{h}\bar{k}0)$, and $(\bar{k}h0)$, while $\{001\}$ consists of the two faces (001) and $(00\bar{1})$ so that in this point group the tetrad is not polar.

In point group 422 the general form $\{hkl\}$ consists of eight faces: operation of the $[100]$ diad on (hkl) yields $(h\bar{k}\bar{l})$ and operation of the tetrad on these two faces yields $(\bar{k}hl)$ and $(\bar{k}\bar{h}\bar{l})$, $(\bar{h}\bar{k}l)$ and $(\bar{h}k\bar{l})$, and $(k\bar{h}l)$ and $(kh\bar{l})$. There are two kinds of special form, $\{100\}$ and $\{110\}$, each consisting of four faces whose poles are parallel to diads, and a third special form $\{001\}$ consisting of two opposite faces whose poles are parallel to the tetrad, which is thus non-polar.

The general form of point group $4mm$ consists again of eight faces, all of which lie in the same hemisphere (Fig 3.21). Special forms, the poles of whose faces lie in mirror planes, are $\{h0l\}$ and $\{hhl\}$; each comprises four faces. The special forms $\{001\}$ and $\{00\bar{1}\}$ consist each of a single face, the tetrad being polar.

Point group $\bar{4}2m$ likewise has a general form consisting of eight faces (Fig 3.21). The special form $\{100\}$ consists of four faces whose poles are parallel to the $\langle 100 \rangle$ diads. The special form, which has $h = k$ and l unrestricted, that is $\{hhl\}$ likewise consists of four faces, the poles of its faces lying in mirror planes. The special form $\{001\}$ consists of two faces whose poles are parallel to the inverse tetrad.

The holosymmetric point group of the tetragonal system $4/mmm$ has a general form with more faces than that for any other point group of the system. Faces are superimposed on the northern and southern hemispheres of the stereogram by the (001) mirror plane so that the general form has the same faces as the general form in any of the point groups 422, $4mm$, or $\bar{4}2m$ duplicated by the (001) mirror plane to give sixteen faces (Fig 3.21). Special forms with eight faces are those whose poles lie in mirror planes, $\{h0l\}$, $\{hhl\}$, $\{hk0\}$; special forms with four faces are those whose poles are parallel to diads, $\{100\}$ and $\{110\}$; and $\{001\}$ is a special form consisting of two opposite faces whose poles are parallel to the tetrad.

This is a convenient point at which to make a brief digression into morphological crystallography. A well-developed crystal can be assigned to the correct system usually by inspection and always (with a few exceptions) by precise goniometric measurement. Whether it can be assigned unambiguously to a crystal class however depends on the sort of faces displayed. Reference to Fig 3.21 shows that the disposition of the faces of a general form such as $\{121\}$ enables the point group to be determined with certainty. But if the only faces present are those of the forms $\{001\}$ and $\{h0l\}$, unambiguous determination of class will not be possible unless the crystal belongs to class $\bar{4}$; the point groups 4 and $4mm$ will be indistinguishable as will the point groups $4/m$, 422, $\bar{4}2m$, and $4/mmm$. Now $\{001\}$ is a special form in all the tetragonal point groups but $\{h0l\}$ is a special form only in the point groups $4mm$ and $4/mmm$. Crystal morphologists have found it convenient to extend the definition of a special form to include forms whose faces are parallel to symmetry axes (i.e. poles of faces normal to axes). It then becomes possible to say that a crystal can only be assigned unambiguously to a class if it displays one or more general forms. On the extended definition $\{h0l\}$ is a special form in 422 and $\bar{4}2m$ so that the disposition of the faces of this form enables the three point groups, 4, $\bar{4}$, and $4/m$ in which $\{h0l\}$ is a general form to be distinguished. Since we are not primarily concerned with crystal morphology we shall make no further use of the extended definition of a special form.

Stereograms of tetragonal crystals are conventionally drawn, as in Figs 3.20 and 3.21, with the tetrad in the centre of the stereogram, the positive direction of the y-axis at the right-hand end of the horizontal diameter of the primitive, and the positive direction of the x-axis at the lower end of the vertical diameter of the primitive.

Cubic system

The cubic system is characterized by triads equally inclined to orthogonal reference axes so that x, y, and z are equivalent. The unit-cell must therefore be a cube; in consequence the angle between any pair of planes (hkl) and $(h'k'l')$ is independent of the magnitude of the unit-cell dimension a and the same for all cubic crystals, a point that we shall explore in detail later in this chapter.

The point groups of the cubic system (shown in Figs 3.20 and 3.28) are derived from the last two effective combinations of rotation axes listed in Table 3.2 and illustrated in Fig 3.11(e)–(h); triads are disposed parallel to the body diagonals of a cube in both, with diads parallel to cube edges in 233 and parallel to face diagonals of the cube in 234, which has tetrads parallel to cube edges. The cubic system, like the orthorhombic, lacks the sort of point group that merely has a single rotation or inversion axis with or without a centre of symmetry. The conventional symbol for a cubic point group consists of three terms: first a statement of the symmetry elements associated with the cube edges $\langle 100 \rangle$, second a statement of the symmetry elements associated with the body diagonals $\langle 111 \rangle$, and third a statement of the symmetry elements, if any, associated with face diagonals $\langle 110 \rangle$.

The cubic point group 23 is simply the Euler combination 233. Its symmetry elements are four triads parallel to $\langle 111 \rangle$ and three diads parallel to $\langle 100 \rangle$. The general form $\{hkl\}$ comprises twelve faces (Fig 3.28), whose indices are (hkl), (klh), (lhk), $(h\bar{k}\bar{l})$, $(k\bar{l}\bar{h})$, $(l\bar{h}\bar{k})$, $(\bar{h}\bar{k}l)$, $(\bar{k}\bar{l}h)$, $(\bar{l}\bar{h}k)$, $(\bar{h}k\bar{l})$, $(\bar{k}l\bar{h})$, $(\bar{l}h\bar{k})$. The special form $\{111\}$ consists of four faces whose poles are parallel to triads; it is a tetrahedron. A similarly shaped but distinct special form is $\{11\bar{1}\}$. The other special form $\{100\}$ consists of faces whose poles are parallel to diads and is a cube.

The point group[11] $m3$, which may be derived by addition of a centre of symmetry to 23, has four inverse triads parallel to $\langle 111 \rangle$ with three diads parallel to $\langle 100 \rangle$ and three mirror planes $\{100\}$. The general form (Fig 3.28) consists of twenty-four faces. The special form $\{111\}$ in this point group consists of eight faces and is an octahedron. The special form $\{100\}$ is again, as in all cubic point groups, the cube. The special form $\{hk0\}$ consists of twelve faces.

The remaining three cubic point groups are derived from the Euler combination 234, itself the point group 432. This point group has four triads parallel to $\langle 111 \rangle$, three tetrads parallel to $\langle 100 \rangle$, and six diads parallel to $\langle 110 \rangle$. The general form (Fig 3.28) consists of twenty-four faces. The special forms $\{111\}$ and $\{100\}$ are respectively the octahedron and the cube. A new kind of special form arises in this point group: $\{110\}$ which consists of twelve faces and is known as the rhombic dodecahedron (dodecahedron = a twelve-faced body; rhombic because in regular development each face is a rhombus). The rhombic dodecahedron can also occur in 23 and $m3$; in 23 it is the general form with $h = k$, $l = 0$ and in $m3$ it is the special form $\{hk0\}$ with $h = k$.

[11] Strictly the symbol for this point group is $m\bar{3}$, but $m3$ is always preferred. The same comment applies to the holosymmetric point group $m3m$.

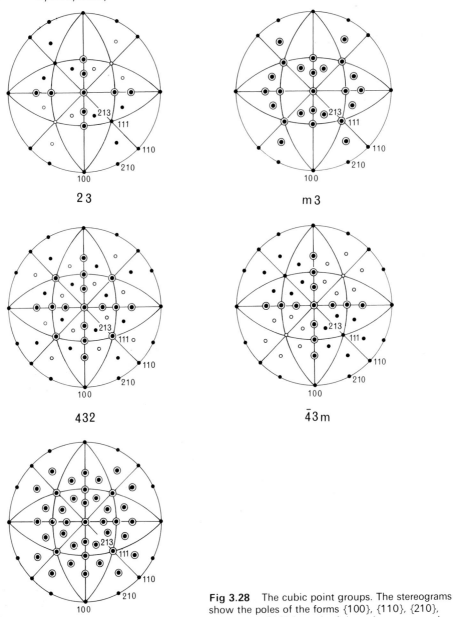

Fig 3.28   The cubic point groups. The stereograms show the poles of the forms {100}, {110}, {210}, {111}, and {213} in each of the point groups, only one face of each form being indexed.

The point group $\bar{4}3m$, obtained by allowing the tetrad and the diad in the Euler combination 234 to be inverse, has triads parallel to $\langle 111 \rangle$, inverse tetrads parallel to $\langle 100 \rangle$, and six mirror planes parallel to {110}. These {110} mirror planes, which are known as diagonal mirror planes, are inclined at 45° to two of the reference axes and parallel to the third reference axis; each plane of the form is parallel to two of the $\langle 111 \rangle$ triads as shown in Fig 3.29 so that each triad is a line of intersection of three diagonal mirror planes which are mutually inclined at 60°. The general form (Fig 3.28) consists of twenty-four faces. The special forms are {111} and {11$\bar{1}$}, tetrahedra as in 23; {100}, a cube; and {hhl}, which consists of twelve faces.

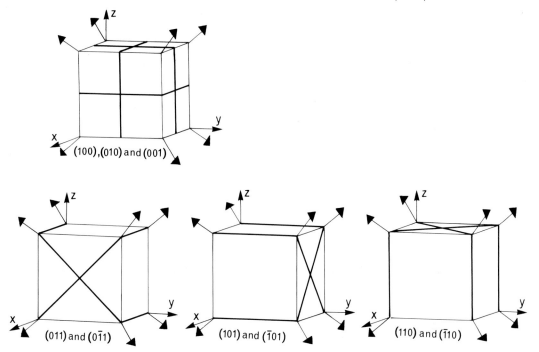

Fig 3.29 Perspective drawings of the cubic unit-cell to show {100} and {110} mirror planes (represented by bold lines). Each of the three lower drawings shows a pair of {110} mirror planes and their relationship to the ⟨111⟩ triads. Only the point groups $\bar{4}3m$ and $m3m$ have {110} mirror planes.

The holosymmetric point group of the cubic system, $m3m$, is derived by combining a centre of symmetry with either of the point groups 432 or $\bar{4}3m$. The essential symmetry elements of this point group, as specified in its short symbol, are ⟨111⟩ triads, {100} and {110} mirror planes; tetrads parallel to ⟨100⟩ and diads parallel to ⟨110⟩ may be regarded as consequential symmetry elements, as may the upgrading of the triads to inverse triads (Fig 3.20). The general form consists of forty-eight faces (Fig 3.28). The special forms {111}, {100}, and {110} are respectively the octahedron, the cube, and the rhombic dodecahedron. There are two other special forms whose poles lie in mirror planes, {h0l} and {hhl}, each consisting of twenty-four faces.

Stereograms of cubic crystals are conventionally drawn with the z-axis in the centre of the stereogram, the positive direction of the y-axis at the right-hand end of the horizontal diameter of the primitive, and the positive direction of the x-axis at the lower end of the vertical diameter of the primitive. Occasionally it is convenient to draw a cubic stereogram with a triad axis perpendicular to the plane of the diagram, but this orientation does not provide a particularly clear statement of point group symmetry.

Hexagonal system

The hexagonal system is characterized by the presence of a rotation or inverse hexad, which is taken as the z-axis of the unit-cell. The unit-mesh of the lattice planes perpendicular to the hexad can conveniently be chosen as a rhombus with $a = b$, $\gamma = 120°$; x and y for the unit-cell are taken to be parallel to the sides of this rhombus.

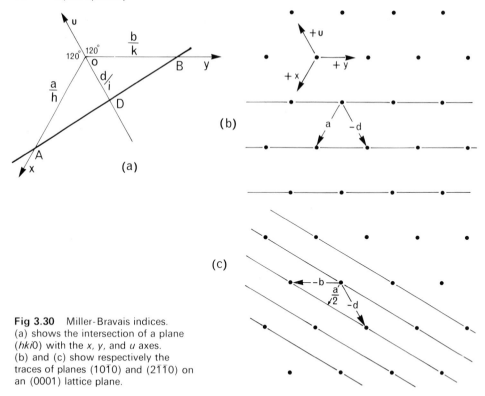

Fig 3.30 Miller-Bravais indices.
(a) shows the intersection of a plane
($hki0$) with the x, y, and u axes.
(b) and (c) show respectively the
traces of planes ($10\bar{1}0$) and ($2\bar{1}\bar{1}0$) on
an (0001) lattice plane.

The hexagonal unit-cell can thus be described as having $a = b \neq c$, $\alpha = \beta = 90°$, $\gamma = 120°$.

The hexagonal and trigonal systems differ from all other systems in that the operation of their principal symmetry axis generates a third axis equivalent to x and y, whose positive direction is inclined at $120°$ to $+x$ and to $+y$. This extra axis (Fig 3.30) is designated u, the lattice repeat along it being $d = a = b$. To take into account the generation of this extra axis a fourth index i is introduced into the symbol for a plane so that ($hkil$) represents a plane making intercepts a/h, b/k, d/i, c/l on the x, y, u, and z axes respectively. But three-dimensional geometry cannot be described in terms of four independent parameters so h, k, and i must be interrelated. Inspection of Fig 3.30(a) indicates that h, k, and i cannot all have the same sign; in the case illustrated h and k are positive while i is negative. The area of the triangle OAB is the sum of the areas of the triangles OAD and ODB, therefore

$$\frac{1}{2} \cdot \frac{a}{h} \cdot \frac{b}{k} \sin 120° = -\frac{1}{2} \cdot \frac{a}{h} \cdot \frac{d}{i} \sin 60° - \frac{1}{2} \cdot \frac{b}{k} \cdot \frac{d}{i} \sin 60°$$

since i is negative. But $a = b = d$,

therefore $\dfrac{1}{hk} = -\dfrac{1}{hi} - \dfrac{1}{ki}$

and $h + k + i = 0$.

Thus lattice planes parallel to y and z make equal intercepts on the $+x$ and $-u$ axes (Fig 3.30(b)); this set of planes is thus indexed as ($10\bar{1}0$). The set of planes whose normal is parallel to the x-axis (Fig 3.30(c)) makes intercepts $-b$ and $-d$ on the y

and u axes respectively, an intercept $a/2$ on the x-axis, and zero intercept on the z-axis; their indices are therefore ($2\bar{1}\bar{1}0$). Such four-digit indices, which are a modification of Miller indices to take the extra symmetry-related axis into consideration, are known as *Miller–Bravais indices*.

The reason for introducing Miller–Bravais indices can simply be demonstrated by considering the forms $\{10\bar{1}0\}$ and $\{11\bar{2}0\}$ in point group 6 (Fig 3.31). The stereogram is plotted in the conventional setting for a hexagonal crystal with z in the centre of the stereogram, $+y$ at the right-hand end of the horizontal diameter of the primitive, and $+x$, $+y$, $+u$ in anticlockwise sequence 120° apart. The poles of ($2\bar{1}\bar{1}0$), ($\bar{1}2\bar{1}0$), ($\bar{1}\bar{1}20$) are coincident with the poles of $+x$, $+y$, $+u$ respectively and their opposites are coincident with the poles of $-x$, $-y$, $-u$. The faces ($0\bar{1}10$), ($10\bar{1}0$), ($\bar{1}100$) and their opposites are all parallel to z and respectively parallel to x, y, and u. The symmetry relationship between the faces of each of these forms is immediately obvious in Miller–Bravais indices: the indices of the faces of each form contain the same quartet of numbers regularly interchanged in position and sign. In contrast Miller indices wholly fail to make the symmetry relationship between the faces of a form immediately apparent as is evident from Table 3.6. In the sequel we shall invariably use Miller–Bravais in preference to Miller indices when discussing the hexagonal system.

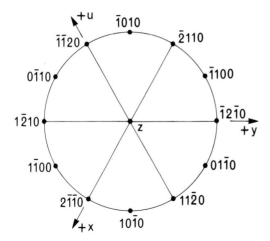

Fig 3.31   Stereogram of the forms $\{10\bar{1}0\}$ and $\{11\bar{2}0\}$ in point group 6 to illustrate Miller–Bravais indexing.

Table 3.6
Comparison of Miller–Bravais indices and Miller indices of the forms $\{10\bar{1}0\}$ and $\{11\bar{2}0\}$

Miller–Bravais	Miller	Miller–Bravais	Miller
($10\bar{1}0$)	(100)	($2\bar{1}\bar{1}0$)	($2\bar{1}0$)
($01\bar{1}0$)	(010)	($11\bar{2}0$)	(110)
($\bar{1}100$)	($\bar{1}10$)	($\bar{1}2\bar{1}0$)	($\bar{1}20$)
($\bar{1}010$)	($\bar{1}00$)	($\bar{2}110$)	($\bar{2}10$)
($0\bar{1}10$)	($0\bar{1}0$)	($\bar{1}\bar{1}20$)	($\bar{1}\bar{1}0$)
($1\bar{1}00$)	($1\bar{1}0$)	($1\bar{2}10$)	($1\bar{2}0$)

When using Miller–Bravais indices the addition rule still holds for tautozonal faces: thus ($11\bar{2}0$) lies in the same zone and between ($10\bar{1}0$) and ($01\bar{1}0$). But when using cross-multiplication to determine the indices of a zone axis it is necessary to get rid of the superfluous index i. For this purpose the Miller–Bravais index is written as $(hk.l)$ and the resultant zone axis symbol as $[UV\dagger W]$.

Analogous to Miller–Bravais face indices there is a system of four-digit zone axis symbols, known as *Weber symbols*; but these cannot be converted to Millerian three-digit zone axis symbols simply by omitting the superfluous index. The symbol $[UVW]$ specifies a line through the origin passing through a point with coordinates Ua, Vb, Wc; in three dimensions three indices are necessarily adequate. In contrast the four-digit symbol $[uvtw]$ contains an unnecessary index so that unless some condition is imposed to link u, v, and t a direction in three-dimensional space will not be uniquely represented by a four-axis symbol. For example if no such condition were imposed the x-axis could variously be described as $[1000], [01\bar{1}0], [21\bar{1}0], [2\bar{1}\bar{1}0]$ as illustrated in Fig 3.32(a). The condition applied in the Weber nomenclature is $u+v+t=0$, analogous to $h+k+i=0$ for Miller–Bravais face indices; the positive direction of the x-axis is then represented by $[2\bar{1}\bar{1}0]$. The symbol $[UVW]$ represents the vector $U\mathbf{a}+V\mathbf{b}+W\mathbf{c}$ and the symbol $[uvtw]$ represents the vector $u\mathbf{a}+v\mathbf{b}+t\mathbf{d}+w\mathbf{c}$.

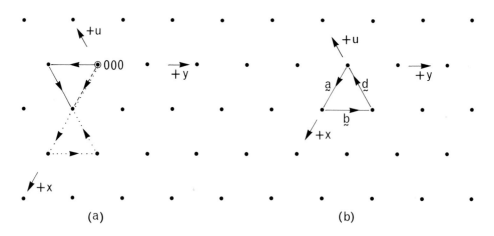

(a)   (b)

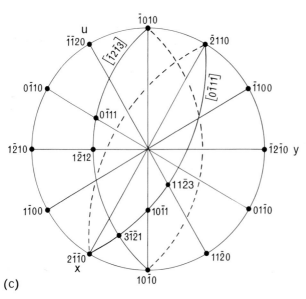

(c)

Fig 3.32 Weber symbols. (a) and (b) show an (0001) net of lattice points: (a) illustrates the various ways in which the $+x$ direction can be indexed in four-digit nomenclature and (b) the condition $\mathbf{a}+\mathbf{b}+\mathbf{d}=0$. The stereogram (c) illustrates the condition for the plane $(hkil)$ to lie in the zone $[uvtw]$, $hu+kv+it+lw=0$.

For these two vectors to be identical

$$U\mathbf{a} + V\mathbf{b} + W\mathbf{c} = u\mathbf{a} + v\mathbf{b} + t\mathbf{d} + w\mathbf{c}$$

Since the x, y, and u axes are inclined at $120°$ to one another and $a = b = d$,

$$\mathbf{a} + \mathbf{b} + \mathbf{d} = 0 \qquad \text{(Fig 3.32(b))}$$

Hence $U\mathbf{a} + V\mathbf{b} + W\mathbf{c} = (u - t)\mathbf{a} + (v - t)\mathbf{b} + w\mathbf{c}$

and $U = u - t$

$$V = v - t$$
$$W = w$$

These equations are adequate to convert Weber symbols to Millerian zone axis symbols, but to convert in the opposite direction it is necessary to apply the condition

$$u + v + t = 0$$

whence $U = 2u + v$

$$V = u + 2v$$
$$W = w$$

so that $u = \dfrac{2U - V}{3}$

$$v = \dfrac{2V - U}{3}$$

$$t = -\dfrac{U + V}{3}$$

$$w = W.$$

The condition for the plane (hkl) to lie in the zone $[UVW]$ is $hU + kV + lW = 0$, which becomes when the equivalent Weber symbol $[uvtw]$ is used

$$h(u - t) + k(v - t) + lw = 0$$

i.e. $hu + kv - (h + k)t + lw = 0.$

And if the Miller–Bravais index i is introduced, where $h + k + i = 0$, this equation becomes

$$hu + kv + it + lw = 0,$$

the condition for the plane $(hkil)$ to lie in the zone $[uvtw]$. For example the zone $[\bar{1}2\bar{1}3]$ contains the planes $(10\bar{1}0)$, $(0\bar{1}11)$, $(1\bar{2}12)$ and the zone $[0\bar{1}11]$ contains the planes $(2\bar{1}\bar{1}0)$, $(10\bar{1}1)$, $(11\bar{2}3)$. The addition rule yields the indices of the planes at the intersection of these two zones as $(3\bar{1}21)$ and its opposite $(\bar{3}12\bar{1})$ as shown in Fig 3.32(c).

To determine the indices of a plane at the intersection of two zones whose axes are represented by Weber symbols it is necessary to convert to three-digit zone axis symbols before cross-multiplying. Thus $[\bar{1}2\bar{1}3]$ and $[0\bar{1}11]$ become respectively $[011]$ and $[\bar{1}2\bar{1}]$. Cross-multiplication

$$\begin{array}{ccccccc} \dfrac{0}{\bar{1}} & \dfrac{1}{2} & \times & \dfrac{1}{1} & \dfrac{0}{\bar{1}} & \times & \dfrac{0}{\bar{1}} & \dfrac{1}{2} & \times & \dfrac{1}{2} & \dfrac{1}{1} \\ & 3 & & & \bar{1} & & & 1 & & & \end{array}$$

yields $(3\bar{1}.1)$ i.e. $(3\bar{1}2\bar{1})$ and its opposite $(\bar{3}12\bar{1})$ for the faces common to both zones. Similarly the Weber symbol for the zone containing the faces $(2\bar{1}\bar{1}1)$ and $(01\bar{1}1)$ is determined by first cross-multiplying with the i index omitted

$$
\begin{array}{cccc}
\dfrac{2}{0}\Big|\dfrac{\bar{1}}{1} & \times & \dfrac{1}{1}\Big|\dfrac{2}{0} & \times & \dfrac{2}{0}\Big|\dfrac{\bar{1}}{1} & \times & \dfrac{\bar{1}}{1}\Big|\dfrac{1}{1} \\
\hline
\bar{2} & & \bar{2} & & 2
\end{array}
$$

to give the three-digit symbol $[\bar{2}\bar{2}2]$ which reduces to $[\bar{1}\bar{1}1]$, and then converting to the Weber symbol $[\bar{\frac{1}{3}}\bar{\frac{1}{3}}\bar{\frac{2}{3}}1]$ which reduces to $[\bar{1}\bar{1}23]$.

The reason for introducing the Weber symbol is that it provides a clear expression of symmetry relationships between zone axes in the hexagonal system just as Miller–Bravais indices do for faces. But, as we have seen, conversion to normal three-digit symbols is a prerequisite to calculation. We shall therefore mostly use three-digit symbols $[UVW]$ modified in one respect. To avoid ambiguity about which two of the three symmetry-related axes x, y, and u are in use, it is convenient to indicate the position of the digit referable to the omitted axis by a dagger $[UV{\dagger}W]$. For example the symbol $[11{\dagger}0]$ represents the vector $\mathbf{a}+\mathbf{b}$, whereas the symbol $[{\dagger}110]$ represents the vector $\mathbf{b}+\mathbf{d}$. Symmetry relationships can satisfactorily be displayed, although this is not often done, by varying the omitted axis. For instance an inverse hexad makes the zone axes $[21{\dagger}0]$, $[\bar{1}1{\dagger}0]$, $[\bar{1}\bar{2}{\dagger}0]$ equivalent (Fig 3.33); the equivalence becomes apparent in the symbols on rewriting as $[1{\dagger}\bar{1}0]$, $[\bar{1}1{\dagger}0]$, $[{\dagger}\bar{1}10]$. One final comment has to be made on hexagonal zone axis symbols: it is conventional, convenient, and in no way misleading to represent the z-axis as $[0001]$ whatever symbolic notation is being used for other zone axes.

The point groups of the hexagonal system correspond exactly in nomenclature to those of the tetragonal system, which we have already explored. Those having symmetry elements associated only with the z-axis are denoted 6, $\bar{6}$, $6/m$. Those derived from the Euler combination 226 are denoted by symbols having three positions: first a statement of whether a rotation or inverse hexad is parallel to z, second a statement of whether a diad or a mirror plane is associated with the x, y, and u axes, and third a statement of whether a diad or a mirror plane is associated with $\langle 1\bar{1}{\dagger}0 \rangle$ directions. The directions referred to in the third position are equally

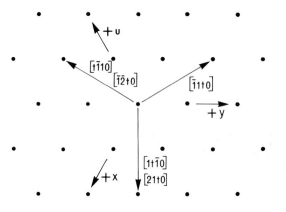

Fig 3.33 Symmetry relationships using three-digit zone axis symbols in the hexagonal system displayed by varying the axis omitted.

inclined in the (0001) plane to adjacent reference axes, thus $[1\bar{1}\bar{\imath}0]$ makes angles of 30° with $+x$ and $-y$.

In point group 6 the general form consists of six faces and the only special forms are $\{0001\}$ and $\{000\bar{1}\}$ which each consists of a single face; the hexad is thus polar. In point group $\bar{6}$ the general form again consists of six faces, but the special form $\{0001\}$ here has two faces; and, because the $\bar{6}$ axis includes a mirror plane (0001), $\{hk.0\}$ is a special form of three faces. Point group $6/m$ has a general form consisting of twelve faces and special forms $\{hk.0\}$, consisting of six faces, and $\{0001\}$ consisting of two faces. Stereograms showing selected forms in these three point groups are shown in Fig 3.34.

The Euler combination 226 becomes on rearrangement of its symbol into the conventional form the point group 622. This point group has a rotation hexad [0001] and two sets of rotation diads, one set parallel to x, y, and u, the other parallel to $\langle 1\bar{1}\bar{\imath}0 \rangle$. The general form (Fig 3.34) consists of twelve faces; the special forms $\{2\bar{1}\bar{1}0\}$ and $\{10\bar{1}0\}$ each have six faces; and the special form $\{0001\}$ consists of the parallel faces (0001) and (000$\bar{1}$).

Point group $6mm$ has a rotation hexad [0001] and two sets of mirror planes $\{2\bar{1}\bar{1}0\}$ and $\{10\bar{1}0\}$. The general form once again comprises twelve faces, but here they all lie in the same hemisphere (Fig 3.34). The two special forms whose faces have poles lying in mirror planes are $\{2h\bar{h}\bar{h}l\}$ and $\{h0\bar{h}l\}$; each comprises six faces. The special forms $\{0001\}$ and $\{000\bar{1}\}$ in this point group each has a single face; the hexad is therefore polar.

Point group $\bar{6}m2$ has an inverse hexad [0001], mirror planes $\{2\bar{1}\bar{1}0\}$ and diads $\langle 1\bar{1}\bar{\imath}0 \rangle$. The general form (Fig 3.34) comprises twelve faces, which appear on the stereogram in superimposed pairs because the $\bar{6}$ axis includes a mirror plane (0001) as is evident also from the stereogram for point group $\bar{6}$. Special forms related to mirror planes are thus $\{h0\bar{h}l\}$ and $\{hk.0\}$, each comprising six faces. There are two special forms associated with diads, $\{10\bar{1}0\}$ and its opposite $\{\bar{1}010\}$ each of which has three faces. The only other special form, of two faces, is $\{0001\}$. In this point group, as in $\bar{4}2m$, there are two orientations $\bar{6}m2$ and $\bar{6}2m$ depending on whether the x-axis is taken perpendicular to a mirror plane or parallel to a diad. The distinction here too can only be made on the basis of which gives x parallel to the shortest lattice repeat in the (0001) plane.

The holosymmetric point group of the hexagonal system has a rotation hexad and perpendicular mirror plane associated with [0001] combined with a diad and a mirror plane associated both with the x, y, u axes and with the directions $\langle 1\bar{1}\bar{\imath}0 \rangle$. The full symbol $\dfrac{6}{m}\dfrac{2}{m}\dfrac{2}{m}$ is commonly abbreviated by omission of the diads and written $6/mmm$. The general form (Fig 3.34) here consists of twenty-four faces. Special forms associated with mirror planes are $\{2h\bar{h}\bar{h}l\}$, $\{h0\bar{h}l\}$, and $\{hk.0\}$, each comprising twelve faces. Special forms associated with diads, $\{2\bar{1}\bar{1}0\}$ and $\{10\bar{1}0\}$, have the number of their faces again halved to six. The special form $\{0001\}$ comprises two parallel faces.

It is conventional to draw stereograms of hexagonal crystals, as in Fig 3.34, with z in the centre of the stereogram and the positive direction of the y-axis at the right-hand end of the horizontal diameter of the primitive. The lower end of the vertical diameter of the primitive is then $[21\bar{\imath}0]$.

Trigonal system

The characteristic symmetry of the trigonal system is the presence of a rotation or

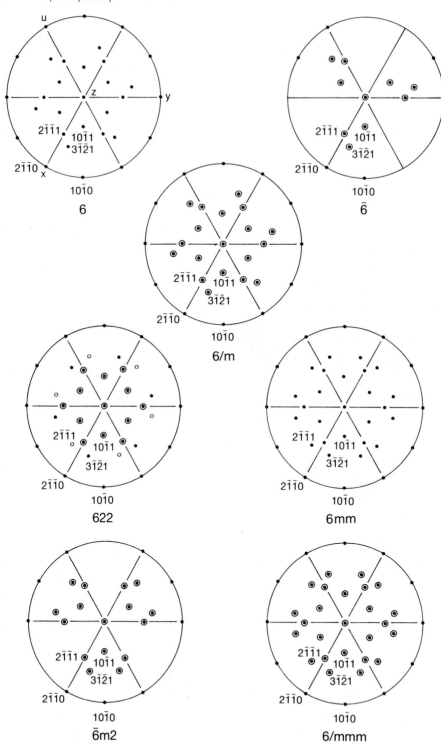

Fig 3.34 The hexagonal point groups. The stereograms show the poles of the forms {0001}, {10$\bar{1}$0}, {2$\bar{1}\bar{1}$0}, {10$\bar{1}$1} {2$\bar{1}\bar{1}$1}, and {3$\bar{1}\bar{2}$1}. Only one face of each form is indexed; (0001) is not indexed.

inverse triad. It is common practice to describe trigonal crystals in terms of the same unit-cell as we have used for the hexagonal system. The rotation or inverse triad is parallel to the z-axis; the x and y axes are disposed at $120°$ to one another in the plane normal to the z-axis. As in the hexagonal system it is customary to introduce an additional axis u such that $+u$ is inclined at $120°$ to $+x$ and $+y$ in the plane normal to the triad. Miller–Bravais indices, Weber symbols, and zone axis symbols of the form $[UV\dag W]$ are used as in the hexagonal system.

The point groups of the trigonal system (Fig 3.20) follow a sequence different from that of the hexagonal and tetragonal systems because we are here concerned with an axis of odd order. The combination of a rotation triad with a centre of symmetry is equivalent to an inverse triad. The combination of a triad and a mirror plane associated with the same direction, that is $3/m$ or $\bar{3}/m$, has the symmetry of point group $\bar{6}$, higher symmetry than is admissible in this system. There are only two point groups with symmetry elements associated with a single direction, 3 and $\bar{3}$ (Fig 3.35). In point group 3 the general form comprises three faces and the only special forms are $\{0001\}$ and $\{000\bar{1}\}$, each of which has a single face; the triad is therefore polar. In point group $\bar{3}$ the general form comprises six faces and the special form $\{0001\}$ two faces.

The remaining point groups of the trigonal system are derived from the Euler combination 223, which is itself the point group 32. For each of these point groups a symbol having two positions specifies the combination of symmetry elements completely: the first position states whether $[0001]$ is a rotation or inverse triad, the second position states whether a diad, a mirror plane or both is associated with the x, y, and u axes. Point group 32 has a rotation triad parallel to $[0001]$ and diads parallel to x, y, and u. The general form (Fig 3.35) in point group 32 comprises six faces. There are special forms $\{2\bar{1}\bar{1}0\}$ and $\{\bar{2}110\}$, each of three faces, and $\{0001\}$ of two faces.

Point group $3m$ has a rotation triad parallel to $[0001]$ and mirror planes $\{2\bar{1}\bar{1}0\}$ perpendicular to the x, y, and u axes. The general form (Fig 3.35) contains six faces. There are two kinds of special form: $\{h0\bar{h}l\}$ with three faces and $\{000l\}$, where $l = \pm1$, with a single face. The triad is therefore polar.

The holosymmetric point group of the trigonal system is derived by adding a centre of symmetry to either 32 or $3m$. Its full symbol is $\bar{3}\dfrac{2}{m}$ which states the presence of an inverse triad parallel to the z-axis, diads parallel to x, y, u and mirror planes perpendicular to x, y, u. It is usually known by its short symbol $\bar{3}m$. The general form (Fig 3.35) consists of twelve faces. There are three kinds of special form: $\{h0\bar{h}l\}$ of six faces, $\{2\bar{1}\bar{1}0\}$ also of six faces, and $\{0001\}$ of two faces.

We have tacitly assumed that the rotation or inverse diads in the (0001) plane are parallel to the shortest lattice repeats in that plane. Structurally, there is no reason why that should be so; the lattice has diad symmetry about $\langle 1\bar{1}\dag 0 \rangle$ as well as about $\langle 10\dag 0 \rangle$. If the relationship of the diad axes to the shortest lattice repeat is known a third position is introduced into the point group symbol to indicate the symmetry associated with $\langle 1\bar{1}\dag 0 \rangle$ directions as in the hexagonal point groups. Thus 321 or $3m1$ or $\bar{3}m1$ indicates that it is known that the rotation or inverse diads are parallel to x, y, and u, while 312 or $31m$ or $\bar{3}1m$ indicates that the axes of diad symmetry are parallel to $\langle 1\bar{1}\dag 0 \rangle$. The bottom row of diagrams in Fig 3.35 shows $\bar{3}m1$ and $\bar{3}1m$.

Our discussion of the trigonal system so far has been in terms of the unit-cell of the hexagonal system, but the trigonal system has a unit-cell peculiar to itself. This is the

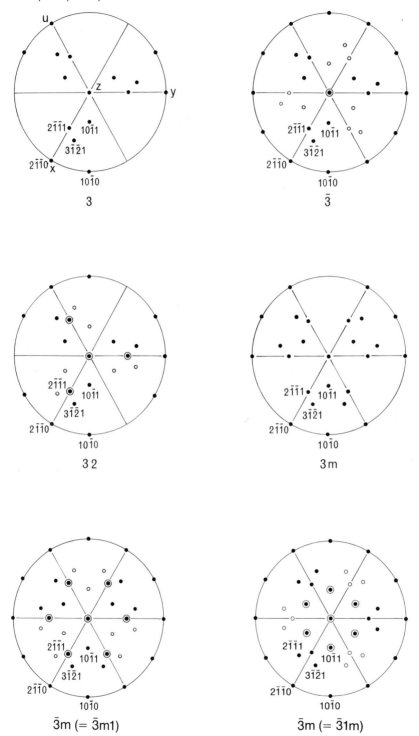

Fig 3.35 The trigonal point groups. The stereograms show the poles of the forms {0001}, {10$\bar{1}$0}, {2$\bar{1}$$\bar{1}$0}, {10$\bar{1}$1}, {2$\bar{1}$$\bar{1}$1}, and {3$\bar{1}$$\bar{2}$1}. Only one face of each form is indexed; (0001) is not indexed.

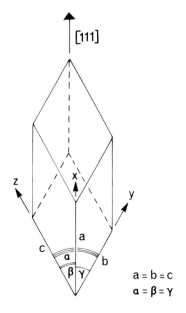

Fig 3.36   The rhombohedral unit-cell: $a = b = c$, $\alpha = \beta = \gamma \neq 90° < 120°$.

$a = b = c$
$\alpha = \beta = \gamma$

unit-cell illustrated in Fig 3.36; it has its x, y, and z axes equally inclined to the triad with $a = b = c$, $\alpha = \beta = \gamma < 120°$. A solid body of such a shape is known as a *rhombohedron*. This is the only conventional unit-cell in crystallography which has all its symmetry axes necessarily non-parallel to its reference axes. For this reason and because of its inconvenient geometry the rhombohedral unit-cell is little used for the description of trigonal crystals; the hexagonal unit-cell is generally preferred.

For a trigonal crystal indexed on the rhombohedral lattice angular relationships between faces are dependent only on the interaxial angle $\dot\alpha$ and independent of the unit-cell edge a. Stereograms are usually plotted with the triad $[111]$ in the centre of the stereogram and the x-axis on the lower half of the vertical diameter of the primitive at the appropriate inclination to the triad (Fig 3.37(a)). The zones $[100]$,

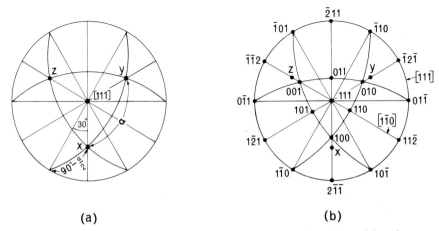

(a)   (b)

Fig 3.37   Rhombohedral indexing. The stereogram (a) shows the disposition of the reference axes about the triad, which is plotted centrally. The stereogram (b) shows how faces may be indexed on rhombohedral axes by use of intersecting zones.

[010], [001] can then be plotted by drawing the great circles whose poles are respectively x, y, and z (Fig 3.37(b)). It is evident from the symmetry of the stereogram that the normal to (111) is parallel to the triad [111] and faces (hkl) with $h+k+l=0$ lie on the primitive. The position of any pole can thereafter be found by intersection of zones: for instance (110) lies at the intersection of the zones [001] and [1$\bar{1}$0].

The plotting of trigonal crystals in terms of the hexagonal lattice needs no comment; it is exactly the same as for hexagonal crystals.

Interplanar and interzonal angles

To conclude this chapter we begin to consider in general terms how the angle between a pair of planes (hkl) and ($h'k'l'$) or between a pair of zones [UVW] and [$U'V'W'$] is related to the dimensions of the unit-cell in each of the crystal systems. At this stage it is possible only to set down certain basic equations and to indicate explicitly how certain simple calculations may be performed; it is only after the powerful vector methods have been introduced in chapter 5 that the general problem can be solved in any system.

We begin with the least symmetrical system, the triclinic system, and progress to more symmetrical systems, obtaining more immediately useful results as the symmetry increases. Consider the triclinic stereogram shown in Fig 3.38(a) on which the poles of the general plane (hkl) and of the planes (100), (010), (001) and the zones containing these planes in pairs are plotted. The symbols of the axes of the zones shown in the figure are [0$\bar{l}$k], [l0$\bar{h}$], [$\bar{k}$h0], [100], [010], and [001]. The zone axes [0$\bar{l}$k], [010], and [001] are all parallel to the plane (100). The disposition of these three zone axes in the (100) plane is shown in Fig 3.38(b): the interaxial angle between the y and z axes is α (shown on the stereogram as the angle between the great circles whose poles are

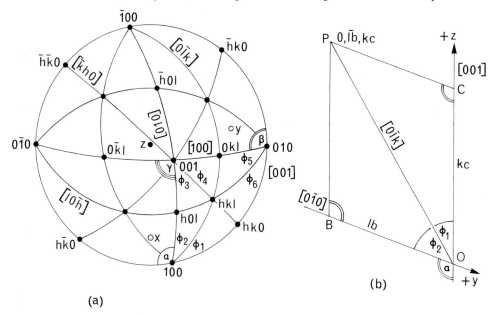

(a)   (b)

Fig 3.38 Triclinic geometry. The stereogram (a) shows the principal zones and the zones which serve to define the angles $\phi_1 \ldots \phi_6$. The disposition of zone axes in the (100) plane is shown in the lattice section (b), from which the relationship between ϕ_1, ϕ_2 and the axial ratio b/c can be demonstrated.

these axes) and $[0\bar{l}k]$ is drawn from the origin through a point with coordinates $0, -lb, kc$. Let the interzonal angles $[001]:[0\bar{l}k]$ and $[0\bar{1}0]:[0\bar{l}k]$ be respectively ϕ_1 and ϕ_2. Then $\widehat{OPC} = \widehat{BOP} = \phi_2$ so that in the triangle OCP

$$\frac{\sin \phi_1}{\sin \phi_2} = \frac{lb}{kc} = \frac{b/k}{c/l}.$$

Now, returning to Fig 3.38(a), the angle between the great circle through the poles of (100) and (010) and the great circle through the poles of (100) and (hkl) is the angle ϕ_1; and likewise the angle between the great circle through the poles of (100) and (hkl) and the great circle through the poles of (100) and (001) is the angle ϕ_2. The analogous relationships

$$\frac{\sin \phi_3}{\sin \phi_4} = \frac{a/h}{b/k}$$

and
$$\frac{\sin \phi_5}{\sin \phi_6} = \frac{c/l}{a/h}$$

can be derived by identical arguments. All six angles ϕ_1 to ϕ_6 are marked on Fig 3.38(a). The reader will observe that the form of these three relationships is such that they are particularly easy to remember.

This is as far as relationships between interzonal angles and unit-cell dimensions can conveniently be taken in the triclinic system without recourse to the methods of calculation to be dealt with in chapter 5.

In the *monoclinic system* some simplification is achieved because the interaxial angles α and γ are right-angles. Consequently (Fig 3.39) $\phi_1 + \phi_2 = \phi_3 + \phi_4 = 90°$ so that the first and second relationships for the triclinic system become

$$\tan \phi_1 = \frac{b/k}{c/l},$$

$$\tan \phi_3 = \frac{a/h}{b/k}.$$

And since the y-axis is normal to (010), the great circle on which the poles of planes in the [010] zone lie is perpendicular to the pole of (010). Therefore

$$\phi_5 = (001):(h0l)$$
$$\phi_6 = (h0l):(100).$$

Therefore the third triclinic relationship becomes

$$\frac{\sin (001):(h0l)}{\sin (h0l):(100)} = \frac{c/l}{a/h}$$

for the monoclinic system. This expression enables the angle between any pair of faces $(h0l)$ and $(h'0l')$ in the [010] zone to be evaluated from a knowledge of a and c if it is borne in mind that $(001):(100) = 180° - \beta$.

Once the angles ϕ_1, ϕ_3 and $(001):(h0l)$ have been calculated from known unit-cell dimensions and the appropriate great circles plotted on the stereogram the pole of any other face $(h'k'l')$ can be located either by intersecting zones or by calculation and the angle $(hkl):(h'k'l')$ determined by measurement with the stereographic net.

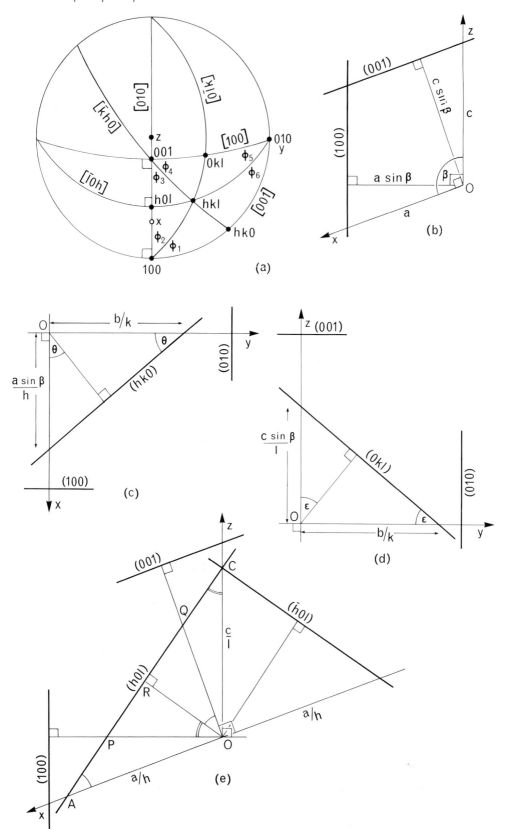

Fig 3.39 Monoclinic geometry. The stereogram (a) shows the principal zones and zones through the general pole (hkl); the angles $\phi_1+\phi_2 = \phi_3+\phi_4 = 90°$ and $\phi_5+\phi_6 = 180°-\beta$. The lattice sections (b)–(e) are referred to in detail in the text.

For more precise evaluation of the angle $(hkl):(h'k'l')$ the vector methods detailed in chapter 5 are required.

Further simplification is achieved in the *orthorhombic system* (Fig 3.40) where $\alpha = \beta = \gamma = 90°$. The three general expressions here become

$$\tan\phi_1 = \frac{b/k}{c/l}, \quad \tan\phi_3 = \frac{a/h}{b/k}, \quad \tan\phi_5 = \frac{c/l}{a/h}$$

and, since the great circle representing the [100] zone is perpendicular to the pole of the (100) face, $\phi_1 = (0kl):(010)$

therefore $\tan(0kl):(010) = \dfrac{b/k}{c/l}$

and similarly

$$\tan(hk0):(100) = \frac{a/h}{b/k}$$

and $\tan(h0l):(001) = \dfrac{c/l}{a/h}.$

These expressions enable the angle between any pair of faces in the [100] or [010] or [001] zone to be evaluated from known unit-cell dimensions.

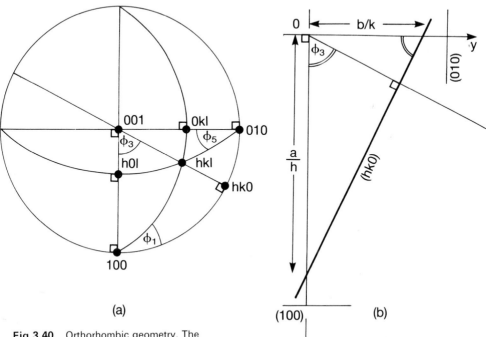

(a)

(b)

Fig 3.40 Orthorhombic geometry. The stereogram (a) shows the principal zones and zones through the general pole (hkl). The lattice section (b) in the (001) plane illustrates the relationship $\tan(hk0):(100) = (a/h)/(b/k)$.

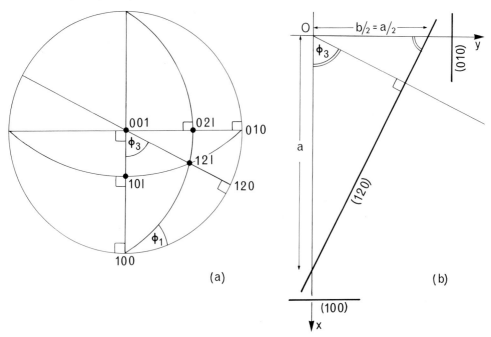

Fig 3.41 Tetragonal geometry. The stereogram (a) shows the principal zones and zones through the general pole (*hkl*) taken as (121). The lattice section (b) in the (001) plane illustrates the relationship tan (*hk*0):(100) = *k/h* for the plane (120).

In the *tetragonal system* (Fig 3.41) the further simplification $a = b$ is introduced so that the three orthorhombic equations become

$$\tan(0kl){:}(010) = \frac{a/k}{c/l}$$

$$\tan(hk0){:}(100) = \frac{k}{h}$$

$$\tan(h0l){:}(001) = \frac{c/l}{a/h}$$

It is immediately apparent from the second of these that interplanar angles in the [001] zone are independent of the unit-cell dimensions, that is to say the angle (*hk*0):(100) is the same for all tetragonal crystals and in particular (110):(100) = 45°. Angles in the [100] and [101] zones are symmetry related so that (0*kl*):(001) = (*h*0*l*):(001) when $h = k$.

In the *cubic system* (Fig 3.42) $a = b = c$ so that

$$\tan(0kl){:}(010) = \frac{l}{k}$$

$$\tan(hk0){:}(100) = \frac{k}{h}$$

$$\tan(h0l){:}(001) = \frac{h}{l}.$$

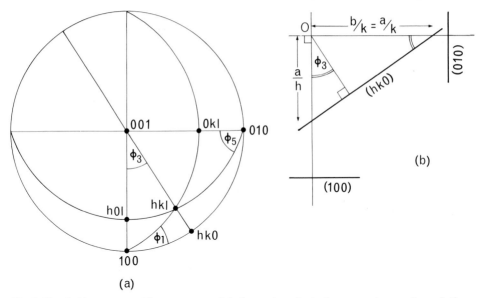

Fig 3.42 Cubic geometry. The stereogram (a) shows the principal zones and zones through the general pole (*hkl*), taken as (321). The lattice section (b) in the (001) plane illustrates the relationship tan (*hk*0) : (100) = *k/h* for the plane (320).

Interplanar angles in each of the zones [100], [010], [001] are thus independent of the magnitude of the unit-cell edge *a*; that the same is true for any zone [*UVW*] in a cubic crystal should be self-evident, but will be shown explicitly in chapter 5.

We shall make no mention of the trigonal and hexagonal systems at this point except to say that this type of approach is not especially fruitful in these systems.

In the three systems that have orthogonal axes, certain interplanar angles can be calculated quite efficiently and straightforwardly by plane geometry. In the *orthorhombic system* the poles of planes in the [001] zone, that is planes with indices (*hk*0), lie in the (001) plane, which of course contains the *x* and *y* axes (Fig 3.40(a)). Therefore, since the (*hk*0) plane makes intercepts *a/h* and *b/k* on the *x* and *y* axes and since the normals to (100) and (010) are respectively the *x* and *y* axes (Fig 3.40(b)),

$$\tan (hk0){:}(100) = \frac{a/h}{b/k}.$$

The expressions for (0*kl*):(010) and (*h0l*):(001) can be obtained by analogous arguments.

The geometrical argument is precisely the same in the *tetragonal* and *cubic systems*, where it yields the simplified results previously noted.

In the *monoclinic system* (Fig 3.39(b)) the perpendicular distance of the (100) plane from the origin is $a \cos(\beta-90°) = a \sin\beta$. Therefore any plane (*hk*0) parallel to the *z*-axis will make an intercept $a/h . \sin\beta$ on the normal to (100). Now the planes (100), (*hk*0), (010) lie in the zone [001] so their normals are coplanar (Fig 3.39(c)). Moreover the normal to (010) is the *y*-axis, which is perpendicular to the normal to (100). Therefore

$$\tan (100){:}(hk0) = \frac{a}{h} \sin\beta \left/ \frac{b}{k} \right. .$$

Similarly by consideration of the normals to the faces (001), (0kl), (010), which lie in the plane normal to [100] it can be shown (Fig 3.39(d)) that

$$\tan(001){:}(0kl) = \frac{c}{l}\sin\beta\left|\frac{b}{k}\right.$$

These two expressions are sometimes of greater practical utility than the corresponding expressions $\tan\phi_1 = (b/k)/(c/l)$ and $\tan\phi_3 = (a/h)/(b/k)$ derived earlier for the monoclinic system.

Angular relationships in the [010] zone in the monoclinic system are less conveniently established by plane geometry, but nevertheless useful results can be obtained. The x and z axes and the normals to the planes (100), (h0l), (001) lie in the (010) plane (Fig 3.39(e)). The (h0l) plane makes intercepts a/h and c/l on the x and z axes respectively. If OR is the normal to (h0l) and OQ the normal to (001),

$$\widehat{RQO} + \widehat{QOR} = \widehat{AQO} + \widehat{OAQ} = 90°$$

and therefore $\widehat{QOR} = \widehat{OAQ} = (001){:}(h0l)$

Similarly $\widehat{OPR} + \widehat{ROP} = \widehat{OPC} + \widehat{PCO} = 90°$

hence $\widehat{ROP} = \widehat{PCO} = (h0l){:}(100)$

In the triangle OAC

$$\frac{\sin\widehat{OAC}}{\sin\widehat{ACO}} = \frac{c/l}{a/h}$$

therefore $\dfrac{\sin(001){:}(h0l)}{\sin(h0l){:}(100)} = \dfrac{c/l}{a/h}.$

If the plane $(\bar{h}0l)$, also shown in Fig 3.39(e), is similarly considered the analogous relationship

$$\frac{\sin(001){:}(\bar{h}0l)}{\sin(\bar{h}0l){:}(\bar{1}00)} = \frac{c/l}{a/h}$$

is obtained; of course $(001){:}(\bar{h}0l) \neq (001){:}(h0l)$. These relationships are the same as those obtained earlier by consideration of interzonal angles.

We discuss finally *hexagonal* crystals and *trigonal* crystals indexed on the hexagonal unit-cell where the simple geometrical approach yields useful results. The normals to the faces $(2\bar{1}\bar{1}0)$, $(2h, \bar{h}, \bar{h}, l)$, (0001) are coplanar (Fig 3.43(a)) with the x and z axes. The $(2h, \bar{h}, \bar{h}, l)$ plane makes intercepts $a/2h$ and c/l on the x and z axes respectively (Fig 3.43(b)). Therefore

$$\tan(2h, \bar{h}, \bar{h}, l){:}(0001) = \frac{c/l}{a/2h}$$

A face $(h0\bar{h}l)$ makes intercepts a/h on the $+x$ and $-u$ axes (Fig 3.43(c)). It must therefore make an intercept $a/h.\cos 30°$ on the normal to $(10\bar{1}0)$. Since the z-axis is coplanar with the normals to $(h0\bar{h}l)$ and $(10\bar{1}0)$, as illustrated in Fig 3.43(d).

$$\tan(h0\bar{h}l){:}(0001) = \frac{c}{l}\left|\frac{a}{h}\right.\cos 30°.$$

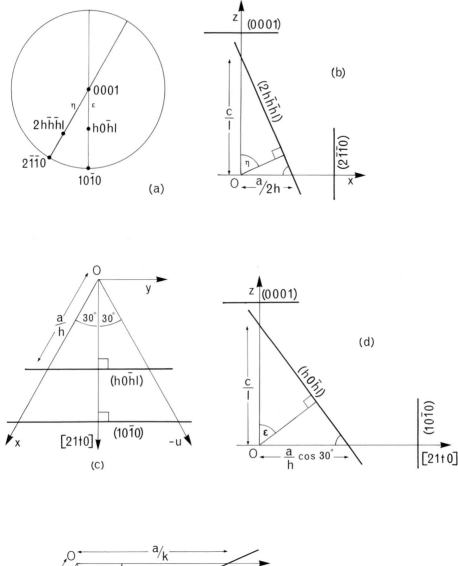

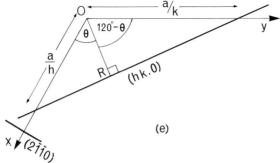

Fig 3.43 Hexagonal geometry. The stereogram (a) serves to define the angles η and ε used in the lattice sections (b)–(e); (b) is in the plane of the x and z axes, (c) in the (0001) plane, (d) in the plane of the z and $[21\bar{1}0]$ axes, and (e) in the (0001) plane.

A face $(hk.0)$ makes intercepts a/h and a/k on the $+x$ and $+y$ axes (Fig 3.43(e)) and its normal is coplanar with these axes. If OR is the normal to $(hk.0)$ and θ the angle $(2\bar{1}10){:}(hk.0)$, then

$$OR = \frac{a}{h}\cos\theta = \frac{a}{k}\cos(120° - \theta)$$

hence

$$\frac{\cos\theta}{h} = -\frac{\cos\theta}{2k} + \frac{\sqrt{3}\sin\theta}{2k}$$

and

$$\tan(2\bar{1}10){:}(hk.0) = \frac{h+2k}{h\sqrt{3}}.$$

Thus the angular relationships of faces in the [0001] zone are independent of the unit-cell dimensions of the hexagonal or trigonal lattice and are the same for all crystals referable to these systems.

In the next chapter we explore the symmetry of the internal structure of crystals and in chapter 5 resume consideration of the evaluation of interplanar and interzonal angles.

Two-dimensional point groups

By way of preparation for chapter 4 we now deal briefly with the two-dimensional point groups, that is the combinations of symmetry elements that can operate on a two-dimensional lattice. The symmetry operators for a two-dimensional lattice are the rotation axes 1, 2, 3, 4, and 6 and the *line of symmetry* (analogous to the plane of symmetry in three-dimensions and represented by the same symbol *m*). Rotation axes are necessarily perpendicular to the plane of the two-dimensional lattice and therefore cannot be combined with one another. Each type of rotation axis can however be combined with a line of symmetry in the plane of the lattice. There are thus ten combinations of symmetry elements that can operate on a two-dimensional lattice, the ten two-dimensional point groups; these comprise each of the five rotation axes on its own and combined with a line of symmetry. The ten two-dimensional point groups are listed in Table 3.7. When a rotation axis of even order (2, 4, or 6) is combined with a mirror line a set of mirror lines is produced equally inclined in the

Table 3.7
The two-dimensional point groups

Combination of symmetry elements	Conventional symbol	Angle between mirror lines of primary and secondary sets	Conventional unit-mesh of lattice	
1	1	—	$a \neq b, \gamma \neq 90°$	Oblique
$1+m$	$1m$	—	$a \neq b, \gamma = 90°$	Rectangular
2	2	—	$a \neq b, \gamma \neq 90°$	Oblique
$2+m$	$2mm$	$90°$	$a \neq b, \gamma = 90°$	Rectangular
3	3	—	$a = b, \gamma = 120°$	Hexagonal
$3+m$	$3m$	—	$a = b, \gamma = 120°$	Hexagonal
4	4	—	$a = b, \gamma = 90°$	Square
$4+m$	$4mm$	$45°$	$a = b, \gamma = 90°$	Square
6	6	—	$a = b, \gamma = 120°$	Hexagonal
$6+m$	$6mm$	$30°$	$a = b, \gamma = 120°$	Hexagonal

Note:
The symbol $\neq$ implies that equality is not required by symmetry

plane of the lattice to those of the primary set; these are noted in the third place of the conventional symbol for the point group, i.e. $4mm$ rather than just $4m$. In the final column of Table 3.7 the shape of the conventional unit-mesh of the two-dimensional lattice is noted, oblique, rectangular, square, or hexagonal; this is a point that will be taken up in the next chapter.

Twinning

Some substances commonly crystallize as composite crystals of a sort known as *twinned crystals* or, colloquially, as *twins*. Well-known examples are copper, diamond, fluorite (CaF_2), and calcite ($CaCO_3$). A twinned crystal consists of two or more individual single crystals joined together in some definite mutual orientation; the lattice of one individual is related to that of the other individual or individuals in the composite crystal by some simple symmetry operation.

Twinned crystals may be produced in various ways. As a crystal grows from its initial nucleus some accident of growth may cause it to twin, such accidents being for a variety of reasons very much more probable in some structures than in others. Twinning may alternatively provide a means of relieving the strain induced by some applied stress. Twinning may also be produced as the result of polymorphic transformations when a structure of higher symmetry is converted to a structure of lower symmetry on cooling. These are the three principal types of twins and they are known respectively as *growth twins*, *deformation* (or glide) *twins*, and *transformation* (or inversion) *twins*. The anti-phase domains produced when a disordered alloy orders on cooling (volume 2) are a special sort of transformation twinning. Here we shall concern ourselves with the geometry of twinning rather than with its physical origin and most of our examples will be growth twins.

The mutual relationship between the two components of a twinned crystal is described by a statement of the symmetry operation necessary to bring the lattice of one component into coincidence with that of the other component. The necessary operation is very commonly either a rotation through 180° about a direction known as the *twin axis* or reflexion in a plane known as the *twin plane*. A twin axis is always a zone axis or the normal to a lattice plane and a crystal twinned about such an axis is known as a *rotation twin*; a twin plane is always a lattice plane and a crystal twinned on such a plane is known as a *reflexion twin*. In a rotation twin where the twin axis is a zone axis rotation may be through 60°, 90°, 120°, or 180°, the first three cases being of very much less common occurrence than the last.

It is obvious that a twin axis cannot be parallel to a symmetry axis of even order in the point group of the crystal, nor can a twin plane be parallel to a mirror plane of the point group. Thus a diad, tetrad, or hexad cannot be a twin axis (at least not the common sort of twin axis rotating through 180°), but a twin axis may be parallel to a triad. In the case of a twin axis parallel to a triad, the twinning operation can variously be described as a 60°, a 180°, or a 240° rotation about the twin axis; it is conventional and convenient however always to consider such a twinning operation as a rotation of 180°.

Where the two components of a twinned crystal are joined in a plane, the crystal is called a *contact twin* and the plane of mutual contact, which is a lattice plane, is known as the *composition plane*. In general it is true to say that if the twin axis is a zone axis, then the composition plane is parallel to the twin axis, but if the twin axis is the normal to a lattice plane then the composition plane is normal to the twin axis; in reflexion twins the composition plane is parallel to the twin plane. Figure 3.44

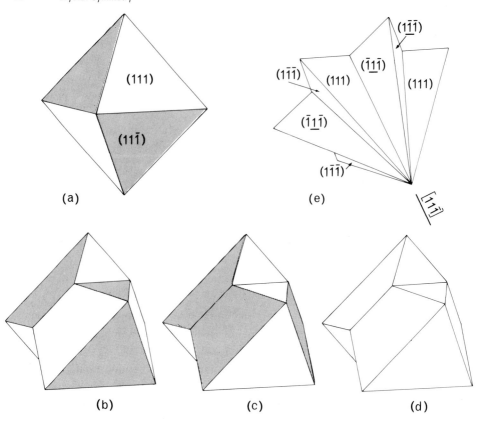

Fig 3.44   A crystal of a cubic substance of point group $\bar{4}3m$ exhibiting the forms {111} and {11$\bar{1}$} is shown in (a). Twinning of a similar crystal by rotation about the normal to (11$\bar{1}$) is shown in (b) and by reflexion in (11$\bar{1}$) in (c). In (a)–(c) faces of the form {111} are shown clear and of the form {11$\bar{1}$} shaded. A twin of a crystal of the centrosymmetric class $m3m$ which exhibits the form {111} and is twinned on (11$\bar{1}$) is shown in (d). An interpenetrant rotation twin, twin axis [11$\bar{1}$], of a crystal of class $\bar{4}3m$ exhibiting only the form {111} is shown in (e): the face ($\bar{1}$11), which lies at the back of the crystal as shown, is normal to the twin axis and is in consequence coplanar with ($\bar{1}\bar{1}$1). Single crystals cannot display re-entrant angles; but twinned crystals, such as those shown here, frequently do so.

illustrates contact twinning in a cubic crystal of point group $\bar{4}3m$ which exhibits the two complementary forms {111} and {11$\bar{1}$}. The combination of these two tetrahedral forms in a truly single crystal is shown in Fig 3.44(a). A contact twin in which twinning is by rotation about the normal to (11$\bar{1}$), which is of course parallel to the zone axis [11$\bar{1}$] in the cubic system, is shown in Fig 3.44(b). Another contact twin in which twinning is by reflexion in (11$\bar{1}$) is shown in Fig 3.44(c). Inspection of figures (b) and (c) shows that although the two twinned crystals are alike in shape, they differ in the disposition of symmetry-related faces; the difference is clearly displayed when the twinning operations are represented in stereographic projection. In Fig 3.45(a) the pole $\underline{N}$ is related to the pole N by the twin operation of rotation through 180° about the twin axis P. In Fig 3.45(b) the pole $\underline{N}'$ is related to the pole N by the twin operation of reflexion in the twin plane whose pole is P. Since N and $\underline{N}$ are related by 180° rotation about P, it follows that N, P, and $\underline{N}$ are coplanar, that is they lie on the same great circle, and N:P = P:$\underline{N}$. Since $\underline{N}'$ is the reflexion of N in the twin plane, whose normal is P, it follows that N, P, and $\underline{N}'$ lie in a plane normal to the

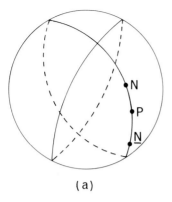

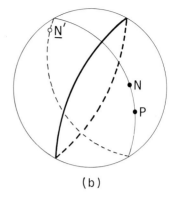

(a)   (b)

Fig 3.45 Stereograms to illustrate the twinning of a face whose normal is N by rotation (a) about a twin axis P to yield the face N̲ or by reflexion (b) in the plane (shown bold) whose normal is P to yield the face N̲'. Since N̲' is the opposite of N̲ twinning by reflexion and by rotation are indistinguishable in a centrosymmetric crystal.

twin plane and the angles which N and N̲' make with the twin plane are both equal to $90° - N{:}P$ so that $\underline{N}'{:}N = 180° - 2(N{:}P)$. Therefore $\underline{N}'{:}\underline{N} = (\underline{N}'{:}N) + (N{:}\underline{N}) = (\underline{N}'{:}N) + 2(N{:}P) = 180°$, so that N̲' is the opposite of N̲ and they are in general only equivalent if the crystal is centrosymmetric. The reader will recall that we have earlier shown that the operation of a mirror plane followed by the operation of a centre of symmetry generates a rotation diad normal to the mirror plane. Thus in a centrosymmetric crystal twinning by rotation about a twin axis and twinning by reflexion in a plane normal to the twin axis are indistinguishable operations; in these circumstances it is sufficient to state that the crystal is 'twinned on (hkl)' and unnecessary to specify whether by reflexion in the plane (hkl) or by diad rotation about the normal to (hkl). The twin of a cubic substance of class $m3m$ illustrated in Fig 3.44(d) may thus be simply described as 'twinned on $(11\bar{1})$'.

In contrast to contact twins, where the two components of the twinned crystal are joined only on an interface parallel to a lattice plane, in *interpenetrant twins* the interface between the twin components is irregular and the twin components in such twinned crystals are often intimately intergrown as illustrated in Fig 3.44(e).

The angular relationships between the faces of twinned crystals can, just as for truly single crystals, conveniently be displayed on a stereogram. One twin component is plotted in the standard orientation for its crystal system and the other in the orientation determined by the twinning operation (Fig 3.46). In plotting the poles of the faces of a twinned centrosymmetric crystal it is usually convenient to regard the second component as derived by rotation twinning from the first component because rotation about an inclined axis is very much more easily performed than reflexion in an inclined plane in the stereographic projection. The faces of each component of the twinned crystal are indexed separately in terms of the conventional orientation of the crystallographic reference axes within that component, the indices of the faces of one component being distinguished by underlining so that (hkl) refers to one component and $(\underline{hkl})$ to the other.

So far we have restricted our discussion to twinned crystals containing only two components; but *multiple twins* consisting of three or more components also occur. In some multiple twins the twinning operations which relate adjacent components are all identical; then the components tend to take the form of lamellae parallel to the

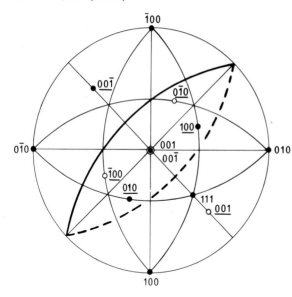

Fig 3.46 The stereogram shows the poles of the faces of a crystal exhibiting the form {100} in the point group m3m and the poles of the faces {100} of its twin on (111). The faces of the twin have been indexed on the assumption that it is a rotation twin about the normal to (111); on the alternative assumption that the twinning is by reflexion in (111) the poles of the form {100} would be identically placed but the signs of their indices would be reversed. The plane (111) is shown as a bold great circle.

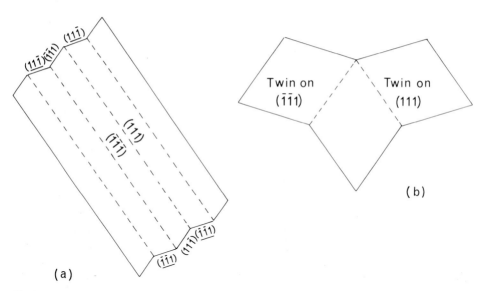

Fig 3.47   Polysynthetic and multiplet twinning. (a) is a section parallel to ($1\bar{1}0$) through an octahedral crystal of point group m3m twinned polysynthetically on (111). (b) is a section parallel to ($1\bar{1}0$) through an octahedral crystal of point group m3m twinned on the symmetry related planes (111) and ($\bar{1}\bar{1}1$) so as to become a triplet. In each case the traces of composition planes are shown as broken lines.

composition plane so that such twins are known as *lamellar* or *polysynthetic twins* (Fig 3.47(a)). Polysynthetic twins may be on a macroscopic, microscopic, or sub-microscopic scale.

Another sort of multiple twin is the *multiplet*, where several components are produced by the operation of symmetry-related twin planes or twin axes. For instance if a cubic crystal of point group m3m can twin on (111), it can also twin on other planes of the form {111}, such as ($\bar{1}11$), ($1\bar{1}1$), ($\bar{1}\bar{1}1$), and may actually do so. In such circumstances some twinned crystals will consist of three or more components, each pair of components being related by a different twinning operation. Such multiplets may be distinguished as *triplets* when they contain three components, *quartets* when they contain four, and so on. A triplet produced by twinning on (111) and ($\bar{1}\bar{1}$1) in a cubic substance of point group $m3m$ is illustrated in Fig 3.47(b).

Occasionally the geometry of the twinning operation may be such that the twinned crystal appears to have higher point group symmetry than a single crystal of the same substance. Such *mimetic twinning* is very well displayed by the orthorhombic form of $CaCO_3$, the mineral *aragonite*. The point group of aragonite is *mmm* and its unit-cell dimensions are $a = 4·95$ Å, $b = 7·95$ Å, $c = 5·73$ Å. Twinning occurs on {110} and, since the interfacial angle (110):($1\bar{1}0$) = 63°48′ is sufficiently close to 60°, multiplets will appear to have a hexad parallel to [001]. Contact and interpenetrant multiplets are quite common in aragonite; an interpenetrant multiplet is shown in Fig 3.48.

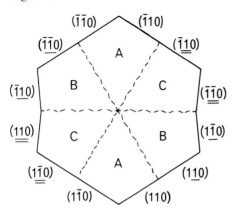

Fig 3.48   Mimetic twinning. Aragonite is orthorhombic, point group *mmm*, and twins on {110} to produce interpenetrant triplets of pseudo-hexagonal shape. A is the parent individual, B (indices underlined) is derived by rotation twinning about the normal to (110), and C (indices doubly underlined) is derived by rotation twinning about the normal to ($1\bar{1}0$). The sinuous broken lines indicate possible traces of the composition planes in this interpenetrant twin.

The sort of twinning which we have just described in aragonite is primary in that it has occurred in the course of the growth of the crystal. The other common form of $CaCO_3$, the mineral *calcite*, also displays multiple twinning; but the polysynthetic twinning quite commonly observed in calcite is secondary in origin, being due to the deformation of single crystals after their growth was complete. In both these examples and in general, no matter what caused twinning to take place, the twin elements (twin axis or plane and composition plane) are determined by the crystal structure of the substance. Discussion of mechanisms of twinning, about which much is known, lies outside our scope here; the interested reader is referred to the survey by Bloss (1971).

Problems

3.1 Theoretical chemists describe symmetry axes in terms of rotation axes and rotation-reflection axes rather than rotation axes and inversion axes. The operation of a rotation–reflection axis S_n consists of a rotation through $2\pi/n$ followed by reflection in a plane normal to the rotation–reflection axis. Draw sketch stereograms showing the operation of 1-, 2-, 3-, 4- and 6-fold rotation–reflection axes on a single pole in a general position. Deduce the inversion axis to which each is equivalent.

3.2 In which classes of the cubic system is the form {111} (i) a tetrahedron, (ii) an octahedron?

3.3 In which classes of the hexagonal system is it possible for a crystal to show faces of one form $\{hkil\}$ only?

3.4 In a certain orthorhombic crystal the planes (hkl), $(hk\bar{l})$, $(\bar{h}kl)$, and $(\bar{h}k\bar{l})$ are equivalent. Draw a sketch stereogram showing the poles of these planes, identify the symmetry elements present, and state the conventional point group symbol.

3.5 Draw sketch stereograms showing the arrangement of symmetry elements in the crystal classes $\bar{3}m$ and $6/mmm$. Add to each stereogram the poles of all the planes of the forms $\{2\bar{1}\bar{1}0\}$ and $\{2\bar{1}\bar{1}1\}$. Show that in a crystal displaying only these two forms the two clases are morphologically indistinguishable.

3.6 Synthetic crystals of diamond (cubic) are often twinned by rotation through 180° about [111]. Draw a sketch stereogram showing all the poles of the forms $\{100\}$ and $\{\underline{100}\}$.

3.7 Use Euler's proposition to derive all the possible combinations of rotation axes of symmetry where the axis of highest order is a pentad and one of the other axes is a diad.

3.8 Barium titanate $BaTiO_3$ is cubic, class m3m, at temperatures above 120°C and tetragonal, class 4mm, in the temperature range 120°C to 5°C. How many different ways are there of selecting the direction [001] of 4mm from the point group m3m?

4
Internal structure of crystalline matter

In chapter 3 the shape of unit-cell appropriate to each crystal system was established and the crystallographic point groups were derived. We now proceed to relate these concepts in the two-dimensional or *plane lattices* as a preliminary to the development of the types of crystallographic three-dimensional or *Bravais lattices*. It is convenient that many of the significant features of the Bravais lattices are simply exemplified in the plane lattices. The reader is reminded of the restrictions imposed by symmetry on unit-cell shape, listed in Table 3.3.

Plane lattices
We now investigate systematically the operation of the two-dimensional point groups (listed in Table 3.7) on a lattice to establish the types of plane lattice. The most generalized unit-mesh ($a \neq b$; γ general) has, as has already been said, a diad perpendicular to the plane of the mesh through every lattice point and midway between adjacent lattice points; whether the hypothetical two-dimensional crystal has point group symmetry 1 or 2 depends on the atomic arrangement and not on the nature of the lattice. This lattice type is known as *oblique* and since the unit-mesh contains only one lattice point it is said to be a *primitive* lattice; the so-called *oblique* p-*lattice* is shown in Fig 4.1(a).

Next we examine the restrictions placed on lattice geometry by the presence of one set of parallel lines of symmetry. It is evident from Fig 4.2 that in a one-dimensional lattice, that is a lattice point row, lines of symmetry must either be of type I, passing through lattice points (Fig 4.2(a)), or of type II, passing midway between lattice points (Fig 4.2(b)). A two-dimensional lattice plane will be generated either by an array of rows all of type I, or all of type II, or an alternation of the two types (Fig 4.2(c), (d), and (e)); no other arrangement will preserve the essential lattice criterion that every lattice point must have the same environment in the same orientation. If all the rows are of the same type, whether I or II, identical lattices are generated, the unit-mesh being a rectangle with $a \neq b$, $\gamma = 90°$; this is the *rectangular* p-*lattice* (Figs 4.1(b) and 4.2(f)). The manner in which we were obliged to arrange the point rows has introduced additional symmetry elements so that the rectangular p-lattice has mutually perpendicular sets of lines of symmetry, one set perpendicular to x and the other perpendicular to y, each set being separated by half the appropriate

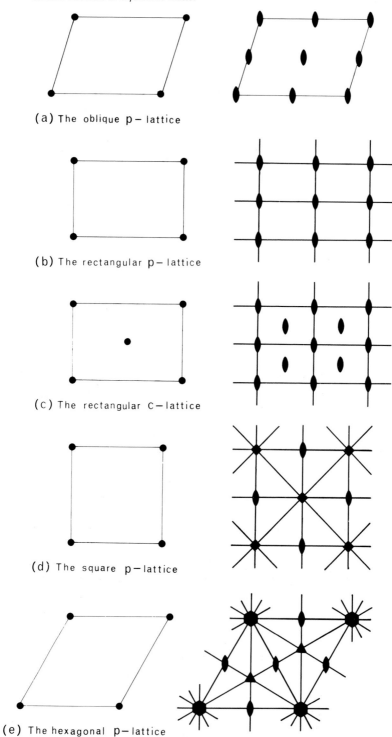

(a) The oblique p—lattice

(b) The rectangular p—lattice

(c) The rectangular c—lattice

(d) The square p—lattice

(e) The hexagonal p—lattice

Fig 4.1   Unit-cells of the five plane lattice types.

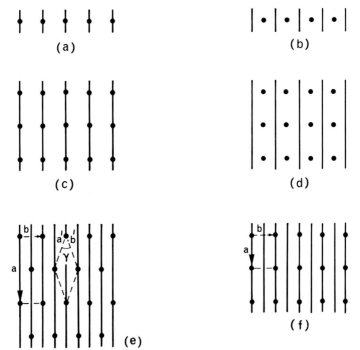

Fig 4.2   Restrictions on plane lattice geometry imposed by one set of parallel lines of symmetry. Lines of symmetry of two types operating on lattice point rows are shown in (a) and (b). The remaining diagrams show plane lattices generated by an array of rows all of type I (c), all of type II (d), or by alternation of types I and II (e). The plane lattice shown in (e) is rectangular-c and that shown in (f) is rectangular-p.

lattice spacing, $a/2$ and $b/2$ respectively; also of course a diad occurs perpendicular to the lattice plane at every intersection of lines of symmetry. The point group symmetry of this lattice is $2mm$; only the atomic arrangement in the hypothetical structure determines whether the crystal has point group symmetry $1m$ or $2mm$.

We now have to return to the lattice produced by alternation (Fig 4.2(e)). The smallest unit-mesh is a rhombus with $a' = b' = \sqrt{\{(a/2)^2 + (b/2)^2\}}$ and $\gamma = 2\tan^{-1} b/a$, which is not in general a special angle. But it must be noted that the symmetry of this lattice is closely related to that of the rectangular p-lattice: it has mutually perpendicular sets of lines of symmetry, one set $\perp x$ and $a/2$ apart, the other $\perp y$ and $b/2$ apart, a diad at every intersection of lines of symmetry, and in addition a diad at the mid-point of each side of the smallest unit-mesh (Fig 4.1(c)). But the relationship between lattice points and symmetry elements is different: whereas in the rectangular p-lattice alternate lines of symmetry of either set are devoid of lattice points, in this *rectangular c-lattice* every line of symmetry passes through lattice points. The lattice symmetry is again $2mm$ and hypothetical two-dimensional crystals of point group symmetry $1m$ or $2mm$ may have this lattice type. It is conventional, and convenient, in this lattice type to take as the unit-mesh not the smallest unit-mesh but a rectangular unit-mesh with edges perpendicular to each set of lines of symmetry and edges equal to the separation of lattice points in these directions. Such a unit-mesh has lattice points not only at its corners but also at its centre; it is described as *non-primitive* because the lattice it generates has more than one lattice point per unit-mesh. The larger area of a non-primitive unit-mesh is in some respects disadvantageous, but for the purposes of indexing lattice planes and specifying symmetry-related coordinates the advantages of a unit-mesh with edges more closely related to symmetry elements outweigh the disadvantages: for instance a certain set of symmetry-related lines would be indexed as (hk), $(\bar{h}k)$, $(\bar{h}\bar{k})$, $(h\bar{k})$ in terms of the conventional non-primitive unit-mesh and, less obviously, as (hk), (kh), $(\bar{h}\bar{k})$, $(\bar{k}\bar{h})$ in terms of the smallest (primitive) unit-mesh. The nomenclature of this lattice type requires a word of explanation: we have designated the axes in the lattice plane as x and y, the corresponding edges of the conventional unit-mesh being a and b, so that the axis perpendicular to the lattice plane is z and the two-dimensional unit-mesh is the (001) face of a three-dimensional lattice of infinite repeat in the z direction; the unit-mesh is a rectangle and it is centred by a lattice point; therefore the lattice type is described as *rectangular*-c.

We now return to the oblique lattice and suppose a tetrad to pass through each lattice point. The effect of the tetrad is to make the edges of the unit-mesh equal and perpendicular to one another; the unit-mesh is thus a square with $a = b$, $\gamma = 90°$. Introduced symmetry elements are tetrads in the centre of each square unit-mesh, lines of symmetry with separation $a/2$ perpendicular to x and perpendicular to y, lines of symmetry diagonal to the square with separation $\frac{1}{2}a\sqrt{2}$ (Fig 4.1(d)). This is known as the *square* p-*lattice*. The lattice symmetry is $4mm$ and it is applicable to hypothetical two-dimensional crystals whose atomic arrangement is consistent with either of the point groups $4mm$ or 4.

If either a hexad or a triad passes through each lattice point of the oblique p-lattice, the unit-mesh becomes a 60° rhombus with $a = b$, $\gamma = 120°$. Introduced symmetry elements are sets of lines of symmetry at 30° to one another such that one member of every set passes through each lattice point and a triad at the centroid of each of the two equilateral triangles that make up the unit-mesh (Fig 4.1(e)). The point group symmetry of the lattice, the *hexagonal* p-*lattice*, is $6mm$, but atomic arrangements of inadequate symmetry may produce degeneration of the symmetry of the structure to the point groups 6, $3m$, or 3 with retention of this lattice.

The relationship of the 5 plane lattices to the 10 two-dimensional point groups is summarized in Table 4.1.

Bravais lattices

There are fourteen three-dimensional or Bravais lattice types differentiated one from another by the symmetry of the arrangement of their lattice points, the actual dimensions of the lattice being of course unimportant. Before proceeding to the development of the Bravais lattices by considering the variety of ways in which plane lattices can be stacked so as to be consistent with three-dimensional point group

Table 4.1
The 5 plane lattices

Lattice type	Point group of lattice	Possible crystal point groups	Shape of conventional unit-mesh
Oblique p	2	1, 2	$a \neq b$; γ general
Rectangular p ⎫ Rectangular c ⎭	$2mm$	$1m$, $2mm$	$a \neq b$; $\gamma = 90°$
Square p	$4mm$	4, $4mm$	$a = b$; $\gamma = 90°$
Hexagonal p	$6mm$	6, $6mm$ / 3, $3m$	$a = b$; $\gamma = 120°$

symmetry it may be helpful to the reader if we first recapitulate some statements made in earlier chapters, in some cases developing their implications for this topic.

In chapter 3 we developed seven lattice types, each primitive, each consistent with the symmetry requirements of one crystal system, and each having unit-cell edges parallel to the crystallographic reference axes of the system. We now have to investigate whether other arrangements of lattice points may be consistent with crystal symmetry in any system. We shall, for example, show that there are two distinct lattice types consistent with tetragonal symmetry (Fig 4.7(b) and (c)). Taking orthogonal axes with z parallel to the tetrad, in conformity with the usual convention for the tetragonal system, the lattice shown in Fig 4.7(b) has a unit-cell with lattice points only at its corners, that is to say it is a P-lattice;[1] but the lattice shown in Fig 4.7(c) has a unit-cell with lattice points not only at its corners but in addition a lattice point at its centre. This latter type is known as the tetragonal I-lattice, I being the initial letter of the German word for body-centred, *innenzentrierte*. Of course the tetragonal I-lattice could alternatively be described in terms of a primitive unit-cell, but that would not have the characteristic tetragonal shape, a square prism, and consequently would not embody the characteristic symmetry of the lattice. The practical advantages to be gained by using such non-primitive unit-cells in general far outweigh the disadvantages that stem from their larger volume (the tetragonal I-cell contains two lattice points and so has twice the volume of the corresponding primitive unit-cell).

The task of discovering how many different arrangements of lattice points are possible is simplified by taking into account from the start a fundamental property of lattices: that they are centrosymmetric and that every lattice point lies at a centre of symmetry. Since a lattice is a regular array of points in space, it follows that if a lattice point A lies at a vector distance t from a lattice point B, then a lattice point C must lie at a vector distance $-t$ from B; therefore the lattice point B is a centre of symmetry of the lattice and so is every other lattice point. Thus our discussion of possible lattice types can be limited to lattices consistent with the eleven centro-symmetric point groups, $\bar{1}$, $2/m$, mmm, $\bar{3}$, $\bar{3}m$, $4/m$, $4/mmm$, $6/m$, $6/mmm$, $m3$, and $m3m$.

Our task is further simplified by making use of the observation that a lattice plane normal to a threefold, a fourfold, or a sixfold symmetry axis must contain lines of symmetry, which become planes of symmetry parallel to the axis in the three-dimensional lattice. Thus a lattice consistent with point group symmetry $4/m$ must display the higher symmetry of point group $4/mmm$. The point groups $\bar{3}$, $4/m$, $6/m$, and $m3$ can thus be struck off our list and we are left with the seven point groups $\bar{1}$, $2/m$, mmm, $\bar{3}m$, $4/mmm$, $6/mmm$, and $m3m$, which are the holosymmetric point groups of the seven crystal systems. It follows that in deriving the Bravais lattices for the tetragonal system, for example, all we have to do is to discover the types of lattice consistent with the presence of a single tetrad; all such lattices will have point group symmetry $4/mmm$ and be characteristic of the tetragonal system.

We now consider the variety of ways in which plane lattices can be superimposed, or stacked, to produce three-dimensional lattices consistent with the characteristic symmetry of each crystal system. We start with the triclinic and proceed in sequence of increasing symmetry, leaving the trigonal and the hexagonal systems, which pose special problems, to the end. We shall consistently use t to represent the *stacking vector*

---

[1] We follow here the practice of *International Tables for X-ray Crystallography*, vol I (1969) in using lower case letters for plane lattice types, p and c, and capital letters for Bravais lattice types, P, C, I, F, etc.

between adjacent lattice planes and define **t** as the vector from a lattice point in one plane to a lattice point in the immediately superimposed plane.

Triclinic system

Suppose an oblique p-lattice plane is superimposed on another that is identical in such a manner that the diads perpendicular to the two planes do not coincide and that this mode of superposition is repeated indefinitely. The result is a primitive triclinic lattice; if the dimensions of the unit-mesh of the original lattice plane are $a \neq b$, γ general and the stacking vector has magnitude c and makes angles α and β with the directions of y and x respectively, then the dimensions of the general parallelpiped that is the unit-cell of this *triclinic P-lattice* will be $a \neq b \neq c$, $\alpha \neq \beta \neq \gamma$. Since the holosymmetric class of the triclinic system has only a centre of symmetry, which is the symmetry element common to all lattice types, there is no advantage to be gained by selecting non-primitive unit-cells in this system.

Monoclinic system

Again our starting point is the primitive oblique plane lattice, the axes of which we shall relabel as x and z, interaxial angle β, in order to produce the monoclinic Bravais lattices in conventional orientation. We consider the various ways in which two such lattice planes of identical unit mesh can be superimposed so that their diads are coincident. It is apparent from Fig 4.4(a) that there are four types of diad normal to the plane of any oblique p-lattice; these are labelled I–IV and are such that diads of type I pass through lattice points, diads of type II lie midway between lattice points that are a distance a apart, diads of type III lie midway between lattice points that are a distance c apart, and diads of type IV pass through the centre of each unit-mesh. There are four ways in which such plane lattices can be stacked so as to achieve coincidence of diads and these are illustrated in Fig 4.4: (i) the stacking vector **t** has no component in the xz plane so that each diad is superimposed on one of its own kind, i.e. I on I, II on II, III on III, IV on IV, (ii) the stacking vector **t** has a component $\frac{1}{2}\mathbf{a}$ in the xz plane so that diads are superimposed according to the scheme I on II, II on I, III on IV, IV on III, (iii) the stacking vector **t** has a component $\frac{1}{2}\mathbf{c}$ in the xz plane so that diads are superimposed I on III, II on IV, III on I, IV on II, and (iv) the stacking vector **t** has a component $\frac{1}{2}(\mathbf{a}+\mathbf{c})$ in the xz plane so that the stacking scheme is I on IV, II on III, III on II, IV on I.

The stacking sequence (i) with $\mathbf{t} = \mathbf{b}$ (Fig 4.3(a)) generates a primitive monoclinic lattice, the unit-cell of which has lattice points only at its corners and dimensions $a \neq b \neq c$, $\alpha = \gamma = 90°$, β obtuse. As in the oblique plane lattice (Fig 4.4(a)) there are four types of diads with coordinates (Fig 4.4(b)) $0, y, 0$; $\frac{1}{2}, y, 0$; $0, y, \frac{1}{2}$; $\frac{1}{2}, y, \frac{1}{2}$; where y is a variable. Mirror planes lie parallel to (010) and are of two types, the members of each type being separated from one another by **b**: one type comprises the (010) planes passing through points with coordinates $0, y, 0$, where y is integral and the planes of this set contain all the lattice points; the other type comprises the (010) planes passing through points $0, y, 0$, where $y = (2n+1)/2$. The point group symmetry of the lattice is $2/m$, but the point group symmetry of the crystal referred to it may be 2 or m or $2/m$. This lattice type is known as the *monoclinic P-lattice*.

The stacking sequence (ii) has a stacking vector of the form $\mathbf{t} = \frac{1}{2}\mathbf{a} + q\mathbf{b}$ where q is a simple fraction (Fig 4.3(b)). The stacking vector from the zeroth layer to the second layer will then be $2\mathbf{t} = \mathbf{a} + 2q\mathbf{b}$; this has an integral coefficient of **a** and alternate layers will then be directly superimposed when viewed down the y-axis. The resultant

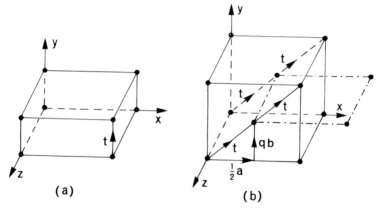

Fig 4.3   Stacking sequences for monoclinic P and C lattices.

non-primitive monoclinic lattice (Fig 4.4(c)) has a stacking vector $\mathbf{t} = \frac{1}{2}\mathbf{a} + \frac{1}{2}\mathbf{b}$ and two lattice points per unit-cell, one with coordinates 0, 0, 0 (i.e. $\frac{1}{8}$ of a lattice point at each of the 8 corners is associated with the chosen unit-cell) and the other with coordinates $\frac{1}{2}, \frac{1}{2}, 0$ (i.e. a lattice point lies at the centre of each of the two (001) faces and $\frac{1}{2}$ of each is associated with the chosen unit-cell). Since the (001) faces of the unit-cell are centred, this is known as the *monoclinic* C-*lattice*. Diads in this lattice type are of only two types: type I, with coordinates 0, y, 0 and $\frac{1}{2}$, y, 0, pass through lattice points and type II, with coordinates 0, y, $\frac{1}{2}$ and $\frac{1}{2}$, y, $\frac{1}{2}$, lie midway between lattice points that are $\mathbf{c}$ apart. Mirror planes are of one type only, parallel to (010) and $\frac{1}{2}\mathbf{b}$ apart. The point group symmetry of the monoclinic C-lattice is again $2/m$.

Stacking sequence (iii) with $\mathbf{t} = q\mathbf{b} + \frac{1}{2}\mathbf{c}$ likewise generates a lattice with alternate layers directly above and below each other when viewed along [010] because $2\mathbf{t} = 2q\mathbf{b} + \mathbf{c}$ and, in this case, the coefficient of $\mathbf{c}$ is integral so that $q = \frac{1}{2}$. The resultant lattice type (Fig 4.4(d)) has lattice points at the corners (0, 0, 0) and at the centres of the (100) faces (0, $\frac{1}{2}$, $\frac{1}{2}$) of a conventionally shaped monoclinic unit-cell. This is the monoclinic A-lattice, but it is not a distinct lattice type because the x and z axes, which are not restricted to particular directions by symmetry, can be interchanged to convert it into a monoclinic C-lattice.

Stacking sequence (iv) with $\mathbf{t} = \frac{1}{2}\mathbf{a} + q\mathbf{b} + \frac{1}{2}\mathbf{c}$ generates a monoclinic lattice with lattice points at the corners (0, 0, 0) and at the body centre ($\frac{1}{2}, \frac{1}{2}, \frac{1}{2}$) of each unit-cell, q being equal to $\frac{1}{2}$. That this is not a distinct lattice type is evident from Fig 4.4(e), where it is shown that diagonal axes may be selected to define either an A-cell, with (100) faces centred, or a conventional C-cell, with (001) faces centred.

We have now exhausted all the stacking possibilities of the oblique plane lattice and seen that generalized stacking gives rise to loss of diads and the production of a triclinic lattice, while stacking with the restriction of coincidence of diads gives rise variously to monoclinic P- and C-lattices.

Orthorhombic system
Here we are concerned with stacking the two rectangular plane lattice types (Figs 4.5(a), 4.6(a)) which have symmetry $2mm$. The basic criterion that has to be satisfied is that lattice planes normal to x, y, and z in the three-dimensional lattice should have symmetry $2mm$. This can be achieved by superimposing either type of rectangular plane lattice so that diads, parallel to z, at the intersection of lines of symmetry are

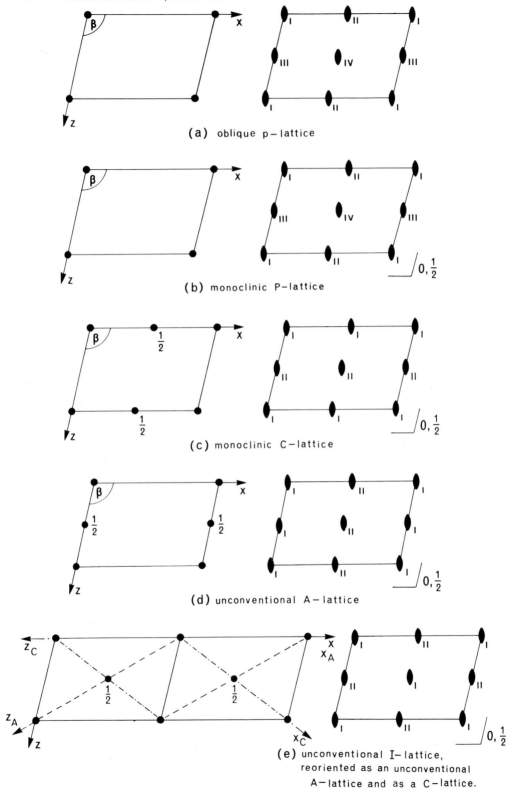

(a) oblique p-lattice

(b) monoclinic P-lattice

(c) monoclinic C-lattice

(d) unconventional A-lattice

(e) unconventional I-lattice,
reoriented as an unconventional
A-lattice and as a C-lattice.

coincident; the lines of symmetry parallel to x and y then generate (010) and (100) mirror planes and, because the lattice is necessarily centrosymmetric, [100] and [010] diads and (001) mirror planes are introduced; the resultant three-dimensional lattice types then have the required symmetry *mmm*.

The rectangular p-lattice has four types of diad, labelled I–IV on Fig 4.5(a), and four types of line of symmetry, labelled p, q, r, s. As for the oblique p-lattice there are four types (i)–(iv) of stacking sequence that maintain coincidence of diads, all of which lie at intersections of lines of symmetry. The stacking vector $\mathbf{t} = \mathbf{c}$ implies coincidence of diads of the same type in successive layers and gives rise to a three-dimensional lattice, the unit-cell of which is a parallelepiped $a \neq b \neq c$, $\alpha = \beta = \gamma = 90°$. This is a primitive unit-cell (Fig 4.5(b)) and has the shape characteristic of the orthorhombic system. The lattice, which is known as the *orthorhombic P-lattice*, has diads of four types parallel to [001] with coordinates $0, 0, z; \frac{1}{2}, 0, z; 0, \frac{1}{2}, z; \frac{1}{2}, \frac{1}{2}, z$ respectively. The presence of lines of symmetry parallel to [010] in each (001) plane of lattice points generates mirror planes parallel to (100); these are of two types, p which pass through lattice points and constitute the faces of the unit-cell, and q, which pass centrally through each unit-cell. In precisely the same way mirror planes of two types, r and s, are generated parallel to (010). The lattice thus has mirror planes perpendicular to x and perpendicular to y and diads parallel to z and, since it is necessarily centrosymmetric, its point group symmetry is that of the holosymmetric class of the orthorhombic system, *mmm*. Detailed discussion of the full symmetry of this lattice type is postponed until after the introduction of space groups.

The other three ways of stacking rectangular p-lattices give rise to centred orthorhombic lattices in a manner generally corresponding to the monoclinic cases discussed in detail earlier. The stacking sequence (ii), with $\mathbf{t} = \frac{1}{2}\mathbf{a} + r\mathbf{c}$, where r is a fraction, superimposes diads of type I on those of type II and type III on type IV. Since lattice points on alternate lattice planes lie directly above or below each other along the z-axis the value of r is determined as $\frac{1}{2}$. The resultant three-dimensional lattice has lattice points at $0, 0, 0$ and $\frac{1}{2}, 0, \frac{1}{2}$, that is to say the (010) or B-face of the unit-cell (Fig 4.5(c)) is centred; this is the *orthorhombic B-lattice*. This lattice type has two types of diad parallel to z: type I pass through lattice points and type III[2] lie midway between lattice points a distance $\mathbf{b}$ apart. The (100) mirror planes are all of type p and pass through lattice points, whereas the (010) mirror planes are again of two types, r and s.

The stacking sequence (iii), with $\mathbf{t} = \frac{1}{2}\mathbf{b} + r\mathbf{c}$ superimposes diads of type I on those of type III and type II on those of type IV. The resultant three-dimensional lattice (Fig 4.5(d)) has lattice points at $0, 0, 0$ and $0, \frac{1}{2}, \frac{1}{2}$, that is to say the (100) or A-face of the unit-cell is centred; this is the *orthorhombic A-lattice*, which has the same symmetry elements as the B-lattice reoriented by interchange of x and y.

Stacking sequence (iv) with $\mathbf{t} = \frac{1}{2}\mathbf{a} + \frac{1}{2}\mathbf{b} + r\mathbf{c}$ superimposes diads of type I on those of type IV and those of type II on type III. The resultant orthorhombic unit-cell (Fig 4.5(e)) has lattice points at $0, 0, 0$ and $\frac{1}{2}, \frac{1}{2}, \frac{1}{2}$; it is a body-centred unit-cell and the lattice type is described as the *orthorhombic I-lattice*. Since in the orthorhombic system the reference axes are required to be parallel to the orthogonal diads, the

Fig 4.4   Unit-cells of the monoclinic Bravais lattices generated by stacking plane oblique p-lattices. In (e) the axes x_A, z_A refer to the A-cell and the axes x_C, z_C to the C-cell.

[2] We retain here the labels of Fig 4.5(a).

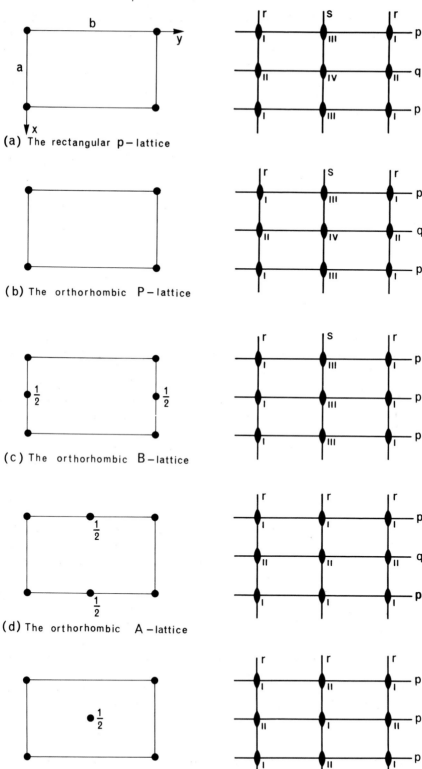

(a) The rectangular p-lattice

(b) The orthorhombic P-lattice

(c) The orthorhombic B-lattice

(d) The orthorhombic A-lattice

(e) The orthorhombic I-lattice

I-lattice is not equivalent either to the A- or the B-lattice; it will be recalled that, contrariwise, the monoclinic I-lattice was shown to be describable as an A- or a C-lattice. The body-centred orthorhombic lattice has two types of [001] diad: type I passes through lattice points while type II lies midway between lattice points on the x and y axes. All the (100) mirror planes pass through lattice points and are styled type p. Likewise all the (010) mirror planes pass through lattice points.

We now consider the variety of ways in which the rectangular c-lattice can be stacked with coincidence of diads. In this plane lattice type lines of symmetry are of two types, one perpendicular to x and designated type p, the other perpendicular to y and designated type r. This plane lattice type (Fig 4.6(a)) has two types of diad, I and II, at the intersection of lines of symmetry and two other types, III and IV, that do not lie on lines of symmetry. If diads of type I are superimposed on diads of type III, i.e. $\mathbf{t} = \frac{1}{4}\mathbf{a} + \frac{1}{4}\mathbf{b} + r\mathbf{c}$, planes of symmetry parallel to (100) and (010) will not be generated and the resultant three-dimensional lattice with $r = \frac{1}{2}$ would merely have diads parallel to z, that is to say it would be a monoclinic C-lattice in an unconventional orientation (Fig 4.6(b)). The dimensions of this monoclinic lattice would be such that two directions, x' and z' in the figure, would be of equal magnitude although unrelated by symmetry. A similar situation arises if diads of type I are superimposed on diads of type IV, i.e. $\mathbf{t} = \frac{3}{4}\mathbf{a} + \frac{1}{4}\mathbf{b} + r\mathbf{c}$.

Of the remaining stacking schemes possible for the rectangular c-lattice, one superimposes diads type for type, i.e. $\mathbf{t} = \mathbf{c}$, to yield an *orthorhombic C-lattice* (Fig 4.6(c)). This lattice type is identical with the orthorhombic A- and B-lattices; one can be converted to another merely by interchange of axial labels.

The fourth, and final, way in which rectangular c-lattices can be stacked is by superimposition of diads of type I on diads of type II; there are two possible stacking vectors $\mathbf{t} = \frac{1}{2}\mathbf{a} + r\mathbf{c}$ and $\mathbf{t} = \frac{1}{2}\mathbf{b} + r\mathbf{c}$, which, because the plane lattice is centred, are equivalent. Alternate layers are directly above and below one another in the z direction so that $r = \frac{1}{2}$. The resulting lattice has every face of its unit-cell centred, lattice points being sited at $0,0,0; 0,\frac{1}{2},\frac{1}{2}; \frac{1}{2},0,\frac{1}{2}; \frac{1}{2},\frac{1}{2},0$ (Fig 4.6(d)). This lattice type is all-face-centred and is known as the *orthorhombic F-lattice*. Here the [001] diads are restricted to two types (I and III); the (100) and (010) mirror planes are each of one type (p or r respectively).

By regular stacking of rectangular plane lattices we have derived six three-dimensional lattice types each of which has orthorhombic symmetry and can therefore be described in terms of the conventional orthorhombic unit-cell, $a \neq b \neq c$, $\alpha = \beta = \gamma = 90°$. The orthorhombic A-, B-, and C-lattices differ only in the labelling of their reference axes and together constitute a single Bravais lattice type; conventionally axes are chosen so that it is the (001) face that is centred and this lattice type is known as the orthorhombic C-lattice. The remaining three lattice types P, I, and F are clearly distinct. Of course any of the three non-primitive orthorhombic lattice types can be described in terms of primitive unit-cells which are dimensionally monoclinic or triclinic and do not reflect the lattice symmetry. Except for certain specialized computational purposes, there is no advantage to be gained by using a primitive unit-cell of lower symmetry and many disadvantages.

Fig 4.5   Unit-cells of the orthorhombic Bravais lattices generated by stacking plane rectangular p-lattices. In this and in the immediately following figures symmetry elements parallel to the plane of the figure are not shown.

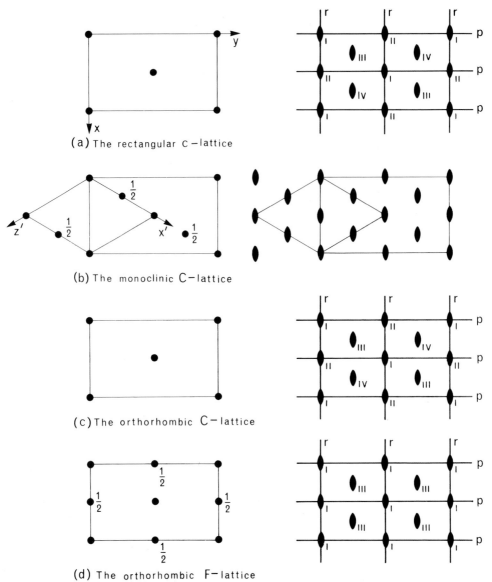

(a) The rectangular C–lattice

(b) The monoclinic C–lattice

(c) The orthorhombic C–lattice

(d) The orthorhombic F–lattice

Fig 4.6   Unit-cells of the orthorhombic and monoclinic Bravais lattices generated by stacking plane rectangular c-lattices.

Tetragonal system

The next plane lattice to which we turn our attention is the square p-lattice (Fig 4.7(a)). This plane lattice type has tetrads of two kinds, type I which pass through lattice points, and type II, which pass through the centre of the unit-mesh. This is the only plane lattice type that has tetrad symmetry and in consequence lattice planes normal to the tetrad in three-dimensional tetragonal lattices must be of this type. In deriving the tetragonal lattice types it is adequate to consider stacking sequences that involve coincidence of tetrads and all other symmetry elements can safely be ignored; the resultant lattices will necessarily display the point group symmetry of the holosymmetric class of the tetragonal system, $4/mmm$. Since there are two tetrad types

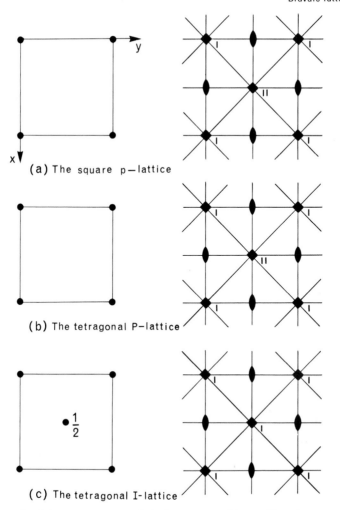

Fig 4.7 The square p—lattice

(a) The square p—lattice

(b) The tetragonal P—lattice

(c) The tetragonal I-lattice

Fig 4.7   Unit-cells of the tetragonal Bravais lattices generated by stacking plane square p-lattices.

in the square p-lattice two stacking sequences are possible: (i) tetrads are superimposed type for type, (ii) tetrads of type I are superimposed on tetrads of type II. The first of these has a stacking vector $\mathbf{t} = \mathbf{c}$ and gives rise to a primitive unit-cell (Fig 4.7(b)) with dimensions $a = b \neq c$, $\alpha = \beta = \gamma = 90°$ and lattice points at its corners. In this *tetragonal P-lattice* the tetrads are, as in the plane lattice, of two non-equivalent types. The other stacking sequence has $\mathbf{t} = \frac{1}{2}\mathbf{a} + \frac{1}{2}\mathbf{b} + r\mathbf{c}$ with $r = \frac{1}{2}$, as in the corresponding orthorhombic case, and gives rise to a body-centred lattice with lattice points at 0, 0, 0, and $\frac{1}{2}, \frac{1}{2}, \frac{1}{2}$. This *tetragonal I-lattice* (Fig 4.7(c)) has equivalent tetrads at 0, 0, z and $\frac{1}{2}, \frac{1}{2}, z$. These two lattices, P and I, are the only distinct tetragonal lattice types; both have {100} and {110} mirror planes and indeed all the symmetry of the holosymmetric class $4/mmm$.

Cubic system
Once again our starting point is the square p-lattice and our task is to discover all the stacking sequences consistent with cubic symmetry, the essential characteristic of

which is the presence of four triads parallel to the $\langle 111 \rangle$ directions in a cubic unit-cell with $a = b = c$, $\alpha = \beta = \gamma = 90°$. Obviously the stacking vectors that generate the two tetragonal lattice types, P and I, will generate a *cubic P-lattice* and a *cubic I-lattice* when the additional restriction $a = c$ is applied (Fig 4.8). In both these cases the characteristic cubic triads lie in the mirror planes generated by the diagonal $\{11\}$ lines of symmetry of the square plane lattice. But if the cubic triads lie in the mirror planes generated by the $\{10\}$ lines of symmetry of the plane lattice, then the x and y axes of the cubic unit-cell have to be rotated through $45°$ relative to the x and y axes of the square plane lattice; the P-lattice becomes a C-lattice and the I-lattice an F-lattice and their unit-cells will be cubes if $c = a\sqrt{2}$ (the conventional tetragonal orientation of axes is preserved here). The resultant C-lattice has a unit-cell of cubic shape, but is not a cubic lattice type because the presence of the $[111]$ triad requires that if the (001) face of the unit-cell is centred, then the (100) and (010) faces must likewise be centred and that is not so. The F-lattice is however consistent with the essential requirements of cubic symmetry and exists as a distinct lattice type, the *cubic F-lattice*.

All three cubic lattice types, P, I, and F, have holosymmetric cubic symmetry, *m3m*, and crystals of any class of the cubic system are referable to them.

Hexagonal system

Our concern now is with the last of the plane lattice types, the hexagonal p-lattice (Fig 4.1(e)). The unit mesh of this lattice contains only one hexad axis, which passes through the origin in a direction normal to the plane of the lattice. Thus only one stacking scheme, that with stacking vector $\mathbf{t} = \mathbf{c}$, can generate a three-dimensional lattice having hexagonal symmetry. The resultant unit-cell has dimensions $a = b \neq c$, $\alpha = \beta = 90°$, $\gamma = 120°$; the unit-cell is primitive with point group symmetry $6/mmm$ and the lattice type is known as the *hexagonal P-lattice*.

Trigonal system

We saw in the last chapter that the characteristic symmetry of this system is the presence of a single direction with threefold symmetry, that direction being conventionally taken as the z-axis to yield a unit-cell with $a = b \neq c$, $\alpha = \beta = 90°$, $\gamma = 120°$. Our starting point is again the hexagonal p-lattice, which is the only plane lattice type that can be consistent with trigonal symmetry. The hexagonal p-lattice (Fig 4.9(a)) has triads at $\frac{2}{3}, \frac{1}{3}$ and $\frac{1}{3}, \frac{2}{3}$, which are not lattice points, and hexads through the lattice points at the corners of the unit-mesh. Each hexad represents in essence the coincidence of the diad that necessarily passes through any lattice point of a plane lattice and a triad which is the basic symmetry element of this lattice type. There are thus two kinds of triad in the hexagonal p-lattice: type I passing through the lattice point at the origin and type II with coordinates $\frac{2}{3}, \frac{1}{3}$ and $\frac{1}{3}, \frac{2}{3}$, the two triads of type II being related by the diad through the centre of the unit-mesh and by the diad through the origin.

Two stacking sequences are possible; $\mathbf{t} = \mathbf{c}$ which superimposes triads type for type and $\mathbf{t} = \frac{2}{3}\mathbf{a} + \frac{1}{3}\mathbf{b} + r\mathbf{c}$ which superimposes triads of type I on the type II triad at $\frac{2}{3}, \frac{1}{3}$. We discuss the second of these first and illustrate the way in which a three-dimensional lattice is produced in Fig 4.9(b). Every third lattice plane is directly superimposed since $3\mathbf{t} = 2\mathbf{a} + \mathbf{b} + 3r\mathbf{c}$; therefore $r = \frac{1}{3}$. The point group symmetry of the resultant three-dimensional lattice is $\bar{3}m$ which is that of the holosymmetric class of the trigonal system. This lattice type is known variously as the *rhombohedral lattice*

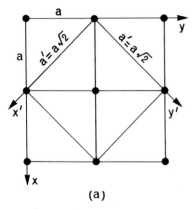

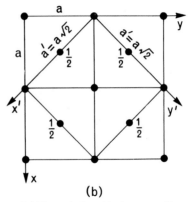

(a) The relationship between the tetragonal P-lattice (a=b≠c) and the tetragonal C-lattice (a'=b'≠c); the former becomes the cubic P-lattice when a=b=c.

(b) The relationship between the tetragonal I-lattice (a=b≠c) and the tetragonal F-lattice (a'=b'≠c); the former becomes the cubic I-lattice when a=b=c and the latter becomes the cubic F-lattice when a'=b'=c.

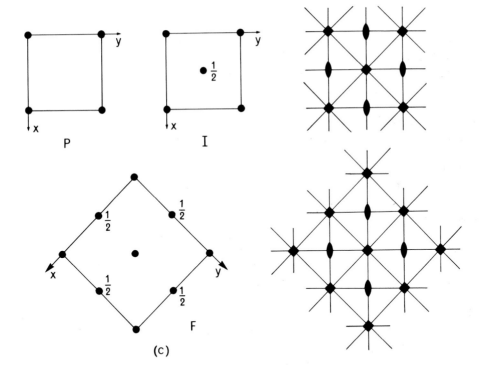

(c) The diagrams on the left show the arrangement of lattice points in the unit-cells of the cubic P, I, and F lattices; those on the right show symmetry elements parallel to [001] for the P and I lattices in the upper diagram, for the F-lattice in the lower diagram.

Fig 4.8   The cubic Bravais lattices generated by stacking plane square p-lattices in such ways as to satisfy the essential requirement for cubic symmetry, the presence of triads parallel to ⟨111⟩.

or the *trigonal* R-*lattice*; since its unit-cell is the same shape as that employed for the hexagonal system but contains (Fig 4.9) three lattice points (at $0, 0, 0$; $\frac{2}{3}, \frac{1}{3}, \frac{1}{3}$; $\frac{1}{3}, \frac{2}{3}, \frac{2}{3}$) instead of one (at $0, 0, 0$) it is known as the *triple hexagonal unit-cell* of the rhombohedral lattice. That the lattice generated by the symmetry related stacking sequence $t = \frac{1}{3}\mathbf{a} + \frac{2}{3}\mathbf{b} + r\mathbf{c}$ is not a distinct type is evident from Fig 4.9; the two arrays of lattice points can be brought into coincidence by rotation of the reference axes x and y through $60°$ or $180°$. These two orientations of the triple-hexagonal unit-cell are known as the *obverse* orientation, with stacking vector $t = \frac{2}{3}\mathbf{a} + \frac{1}{3}\mathbf{b} + \frac{1}{3}\mathbf{c}$ and lattice points at $0, 0, 0$; $\frac{2}{3}, \frac{1}{3}, \frac{1}{3}$; $\frac{1}{3}, \frac{2}{3}, \frac{2}{3}$, and the *reverse* orientation, with stacking vector $t = \frac{1}{3}\mathbf{a} + \frac{2}{3}\mathbf{b} + \frac{1}{3}\mathbf{c}$ and lattice points at $0, 0, 0$; $\frac{1}{3}, \frac{2}{3}, \frac{1}{3}$; $\frac{2}{3}, \frac{1}{3}, \frac{2}{3}$. In subsequent comment we shall generally prefer the obverse orientation.

Although the triple hexagonal unit-cell, in one orientation or the other, is commonly employed for the description of the rhombohedral lattice and of structures derivative therefrom, this lattice type does have a primitive unit-cell which embodies its characteristic symmetry and is in shape a rhombohedron. The reference axes for the *rhombohedral unit-cell* of the trigonal R-lattice are equally inclined to the triad axis and are parallel to the directions $[21\bar{1}1]$, $[\bar{1}1\bar{1}1]$, and $[\bar{1}\bar{2}\bar{1}1]$ of the triple hexagonal unit-cell in the obverse orientation (Fig 4.9); the dimensions of the rhombohedral unit-cell are $a = b = c$, $\alpha = \beta = \gamma < 120°$. That the triple hexagonal unit-cell is usually preferred, in spite of being non-primitive, is because it has the same shape as the hexagonal unit-cell and moreover has two of its interaxial angles, α and β, right-angles.

Up to this point in our treatment of the trigonal system we have confined ourselves to the rhombohedral lattice type with stacking sequence $t = \frac{2}{3}\mathbf{a} + \frac{1}{3}\mathbf{b} + \frac{1}{3}\mathbf{c}$ or $t = \frac{1}{3}\mathbf{a} + \frac{2}{3}\mathbf{b} + \frac{1}{3}\mathbf{c}$. It is now time to consider the other stacking sequence $t = \mathbf{c}$. The three-dimensional lattice so generated retains the hexagonal symmetry of the hexagonal p-lattice and is identical with the hexagonal P-lattice, whose point group symmetry is $6/mmm$; but it has status too as a trigonal lattice type because the arrangement of atoms about each lattice point may be consistent with trigonal and not with hexagonal symmetry. Such use of the same lattice type by two systems is unique and gives rise to ambiguities of practice in different schools of crystallography. Some choose to stress the applicability of hexagonal reference axes to the unit-cells, whether single or triple, of all hexagonal and trigonal substances by regarding the trigonal system as a mere subdivision of the hexagonal system. Others prefer to stress the restriction of the rhombohedral lattice type, with point group symmetry $\bar{3}m$, to crystals of the classes 3, $\bar{3}$, 32, $3m$, $\bar{3}m$ and to maintain the trigonal system, with two lattice types, trigonal-R and hexagonal-P, distinct from the hexagonal system, which has a single lattice type, hexagonal-P. Either approach has its inherent difficulties, but we regard the latter as practically the more convenient.

In conclusion it must be pointed out that when trigonal crystals with the hexagonal P-lattice are described in terms of a rhombohedral unit-cell, that unit-cell is non-primitive and has lattice points at $0, 0, 0$; $\frac{1}{3}, \frac{1}{3}, \frac{1}{3}$; $\frac{2}{3}, \frac{2}{3}, \frac{2}{3}$, i.e. the cell diagonal parallel to the triad is three times the lattice repeat in that direction. In such cases quite obviously the use of a primitive hexagonal unit-cell is preferable.[3] In the case of trigonal crystals with a rhombohedral lattice the use of the triple hexagonal unit-cell

---

[3] In chapter 3 (cf. Table 3.3) we took the rhombohedron as the conventional unit-cell for the trigonal system; that is correct in the morphological context of chapter 3 and moreover the reasons why the rhombohedral unit-cell has fallen into disuse could not be properly explained at that stage.

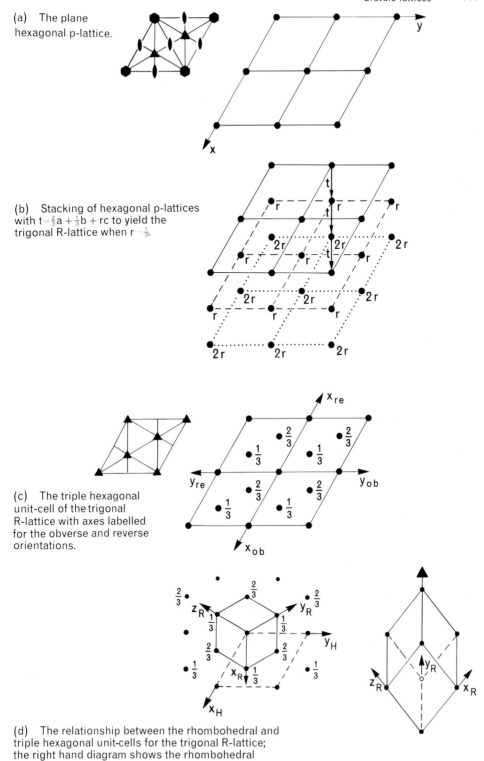

(a)   The plane hexagonal p-lattice.

(b)   Stacking of hexagonal p-lattices with $t=\frac{2}{3}a+\frac{1}{3}b+rc$ to yield the trigonal R-lattice when $r=\frac{1}{3}$.

(c)   The triple hexagonal unit-cell of the trigonal R-lattice with axes labelled for the obverse and reverse orientations.

(d)   The relationship between the rhombohedral and triple hexagonal unit-cells for the trigonal R-lattice; the right hand diagram shows the rhombohedral unit-cell in perspective.

Fig 4.9   The rhombohedral Bravais lattice.

is generally preferable and indeed the rhombohedral unit-cell, which was obviously useful in the days of morphological crystallography, is now of little more than historical interest.

We have now derived fourteen three-dimensional lattice types, each distinguished from the others by the symmetry of arrangement of lattice points. These fourteen, and there are no more, are commonly called the *Bravais lattices* after Auguste Bravais (1811–63) the French physicist who first listed them. Information about the fourteen Bravais lattices is summarized in Table 4.2 and Fig 4.10. That every Bravais lattice is necessarily centrosymmetric and has the point group symmetry of the holo-symmetric class of the system to which it belongs is emphasized in the last column of the table. The immediately preceding column shows the number of lattice points in the conventional unit-cell of each type. At the end of this chapter we complete the description of Bravais lattice symmetry after the introduction of non-translational symmetry elements.

Table 4.2
The 14 Bravais lattices

System	Lattice symbol	Conventional unit-cell	Number of lattice points	Point group symmetry of lattice
Triclinic	P	$a \neq b \neq c, \alpha \neq \beta \neq \gamma$	1	$\bar{1}$
Monoclinic	P C(A)	$a \neq b \neq c, \alpha = \gamma = 90° = \beta$ $(\beta > 90°)$	1 2	$2/m$
Orthorhombic	P C(A, B) I F	$a \neq b \neq c, \alpha = \beta = \gamma = 90°$	1 2 2 4	mmm
Tetragonal	P I	$a = b \neq c, \alpha = \beta = \gamma = 90°$	1 2	$4/mmm$
Cubic	P I F	$a = b = c, \alpha = \beta = \gamma = 90°$	1 2 4	$m3m$
Trigonal	R	$a = b \neq c, \alpha = \beta = 90°, \gamma = 120°$ or $a = b = c, \alpha = \beta = \gamma < 120°$	3 1	$\bar{3}m$
Hexagonal	P	$a = b \neq c, \alpha = \beta = 90°, \gamma = 120°$	1	$6/mmm$

Before leaving the subject of Bravais lattices it may be instructive to consider the structural implications of non-primitive lattices in general in terms of two simple cubic structures. Suppose the unit-cell has lattice points at $0, 0, 0$; X_1, Y_1, Z_1; X_2, Y_2, Z_2, etc. and that an atom of a certain element has coordinates x, y, z, then atoms of the same element must occur at points with coordinates $X_1 + x, Y_1 + y, Z_1 + z$; $X_2 + x, Y_2 + y, Z_2 + z$; and so on. Our first example is one of the forms of metallic iron which is known to have a cubic I-lattice with two atoms of iron per unit-cell: if the origin is taken at the site of one iron atom, then the other iron atom must lie at $\frac{1}{2}, \frac{1}{2}, \frac{1}{2}$. Our second example is diamond, which has a cubic F-lattice with eight atoms of carbon per unit-cell; since an F-cell contains four lattice points, the

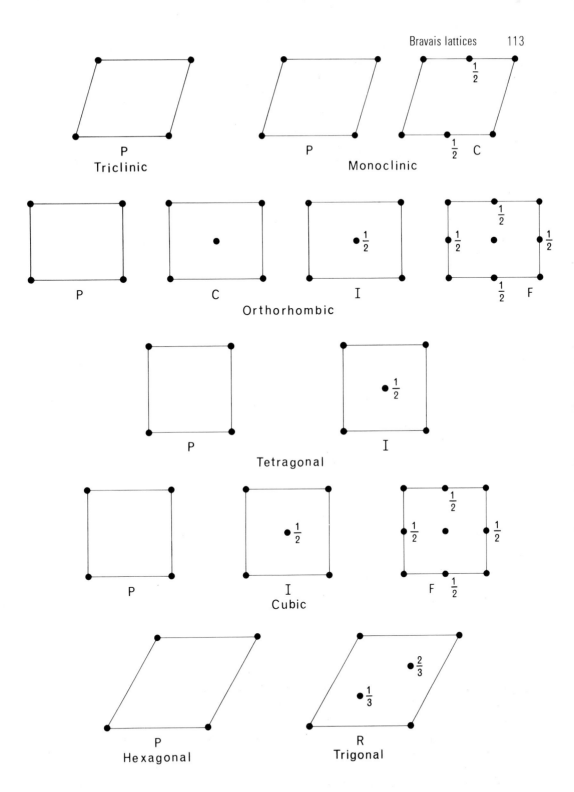

Fig 4.10   The unit-cells of the fourteen Bravais lattices.

repeat unit in this structure consists of two carbon atoms. It is known that in diamond the two carbon atoms of the repeat unit are separated by a vector of magnitude $\frac{\sqrt{3}}{4}a$ in the direction of a body diagonal of the unit-cell. Thus if the origin is taken at one carbon atom, the other atom of the repeat unit lies $\frac{\sqrt{3}}{4}a$ out from the origin along [111], that is at $\frac{1}{4}, \frac{1}{4}, \frac{1}{4}$. The coordinates of the remaining six atoms in the unit-cell are derived from 0, 0, 0 and $\frac{1}{4}, \frac{1}{4}, \frac{1}{4}$ by reference to the coordinates of the four lattice points of the F-cell, 0, 0, 0; 0, $\frac{1}{2}, \frac{1}{2}$; $\frac{1}{2}$, 0, $\frac{1}{2}$; $\frac{1}{2}, \frac{1}{2}$, 0; that is to say four carbon atoms lie at the lattice points and the remaining four are disposed at $\frac{1}{4}, \frac{1}{4}, \frac{1}{4}$ from each lattice point, i.e. $\frac{1}{4}, \frac{1}{4}, \frac{1}{4}$; $\frac{1}{4}, \frac{3}{4}, \frac{3}{4}$; $\frac{3}{4}, \frac{1}{4}, \frac{3}{4}$; $\frac{3}{4}, \frac{3}{4}, \frac{1}{4}$.

It is evident from the plan of the diamond structure, Fig 4.11, that every carbon atom is in fourfold coordination, its nearest neighbours lying at the apices of a regular tetrahedron. But there is a distinction between the four carbon atoms related by F-translations to that at 0, 0, 0 and the four related to the carbon atom at $\frac{1}{4}, \frac{1}{4}, \frac{1}{4}$ in the orientation of their coordination tetrahedra: the interatomic vectors radiating

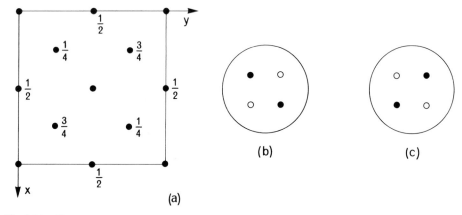

(b)   (c)

(a)

Fig 4.11   The crystal structure of diamond. (a) is a plan on (001) of the structure which has a cubic F-lattice with carbon atoms at 0, 0, 0; $\frac{1}{4}, \frac{1}{4}, \frac{1}{4}$ etc. (b) and (c) are stereograms of the interatomic vectors of nearest neighbours from the carbon atoms at 0, 0, 0 and $\frac{1}{4}, \frac{1}{4}, \frac{1}{4}$ respectively.

from 0, 0, 0 and like atoms are disposed parallel to [111], [$\bar{1}\bar{1}$1], [$\bar{1}$1$\bar{1}$], [1$\bar{1}\bar{1}$] whereas those radiating from $\frac{1}{4}, \frac{1}{4}, \frac{1}{4}$ and like atoms are disposed parallel to [11$\bar{1}$], [$\bar{1}$1$\bar{1}$], [$\bar{1}$11], [1$\bar{1}$1]. It is important to note that, although the environment of every carbon atom in the structure is the same, the orientation of this environment is of two kinds depending on whether the atom in question is related by lattice translations to the atom of the repeat unit at the origin or to that at $\frac{1}{4}, \frac{1}{4}, \frac{1}{4}$.

Symmetry elements involving translation: screw axes and glide planes
So far we have restricted our discussion of symmetry to those symmetry elements that occur in point groups, that is the rotation and inversion axes. Symmetry elements of these kinds are such that their continued operation inevitably leads to a return to the initial position: for example two operations of a diad or three operations of a triad on a crystal face yield a face coincident with the original face. Where the angular disposition of crystal faces or of lattice planes is concerned such non-translational symmetry elements are sufficient, but when we are concerned with the symmetry of arrangement of atoms within a unit-cell an additional sort of symmetry element has to be introduced. This is the translational symmetry element which by its continued operation on a point cannot yield a point coincident with the original

point but, after an appropriate number of operations, yields a point distant from the original point by an integral number of lattice translations. The simplest example of a translational symmetry element is the *screw diad*, whose operation is that of a diad (rotation through 180°) combined with translation through half the lattice repeat in the direction of the diad. Figure 4.12 contrasts the operation of a diad and a screw diad, each parallel to [001] through the origin; an atom at x, y, z is repeated by the diad at $\bar{x}$, $\bar{y}$, z and by the screw diad at $\bar{x}$, $\bar{y}$, $\frac{1}{2}+z$. That the translation component is on the scale of a lattice repeat implies that the screw nature of the [001] diad is not discernible from observation of the macroscopic symmetry of the crystal; indeed a crystal that on the macroscopic scale has a diad parallel to [001] may on the lattice scale have diads or screw diads or both in that direction.

We now proceed to develop the variety of types of translational symmetry element, their nomenclature and conventional representation.

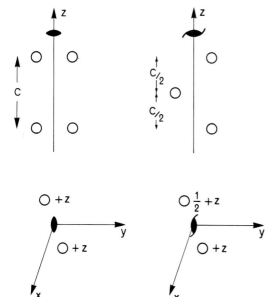

Fig 4.12   The contrasting effect of the operation of a diad (left-hand diagrams) and a screw diad (right-hand diagrams) parallel to [001] through the origin. In the upper diagrams the diad and screw diad (and the points on which they operate) lie parallel to the plane of the diagram; in the lower diagrams they are perpendicular to that plane.

Screw axes

An n-fold rotation axis is such that each operation of it is a rotation through $2\pi/n$, where $n = 1, 2, 3, 4, 6$. For an n-fold screw axis it is necessary to specify the value of n and in addition the magnitude of the pitch τ, which is restricted by the condition that the accumulated pitch after n operations, $n\tau$, must be an integral multiple m of the lattice repeat $\mathbf{t}$ in the direction of the axis, i.e. $n\tau = m\mathbf{t}$. When $m = 0$ there is no translation and the n-fold rotation axis appears as the special case of the n_m screw axis with $m = 0$. In general possible values of τ are given by $(m/n)\mathbf{t}$ where n may have the values given above and $0 < m < n$; the latter condition arises directly from the lattice concept which implies the presence of translations $\mathbf{t} + z\mathbf{t}$, $2\mathbf{t} + z\mathbf{t}$, etc. when the translation $z\mathbf{t}$ is specified. The symbol n_m can be used to give a complete description of a screw axis by adding to what has already been said the convention that the rotation through $2\pi/n$ is in the anticlockwise sense when the associated translation $(m/n)\mathbf{t}$ is in the positive direction along the screw axis (i.e. a right-handed screw).

All possible crystallographic screw axes are illustrated in Fig 4.13, from which we take as our examples for more detailed study the three screw tetrads, 4_1, 4_2, and 4_3. All the axes are taken to be parallel to z through the origin and symmetry related points (represented by open circles) are shown in projection on the xy plane. Beside each such *equivalent position* its z coordinate is written in terms of the code: $+ =$ a position at a height $+z$ above the xy plane, $1/p+ =$ a position at a height $(1/p)+z$ above the xy plane, where $1/p$ and z are, as in structure plans, fractions of the lattice repeat c in the direction of the z axis (Fig 4.14). For instance the screw tetrads shown in Fig 4.13 have equivalent positions indicated variously as $+$, $\frac{1}{4}+$,

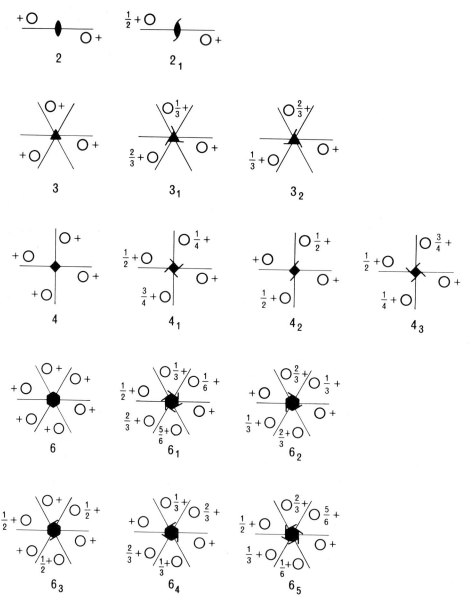

Fig 4.13 The operation of, and conventional symbols for, rotation and screw axes parallel to [001] shown in plans on (001).

$\frac{1}{2}+, \frac{3}{4}+$ to represent heights above the xy plane of z, $\frac{1}{4}+z$, $\frac{1}{2}+z$, $\frac{3}{4}+z$ respectively.

The rotation tetrad shown in Fig 4.13 operates on a position with coordinates x, y, z to yield equivalent positions with coordinates $\bar{y}$, x, z after one operation, $\bar{x}$, $\bar{y}$, z after the second operation, y, $\bar{x}$, z after the third operation, and with the *fourth* operation returns to x, y, z. We have taken the rotation in the anticlockwise sense, but where there is no translation the sense does not affect the total pattern of equivalent positions. The 4_1 axis (Fig 4.14) rotates anticlockwise through $2\pi/4 = \frac{1}{2}\pi$ and simultaneously translates through $c/4$ so that the position xyz yields equivalent positions $\bar{y}$, x, $\frac{1}{4}+z$ by its first operation, $\bar{x}$, $\bar{y}$, $\frac{1}{2}+z$ by its second, y, $\bar{x}$, $\frac{3}{4}+z$ by its third, and by its fourth operation x, y, $1+z$, which is necessarily equivalent to the initial position being one lattice repeat removed from it. The 4_2 axis likewise rotates anticlockwise through $\frac{1}{2}\pi$ and simultaneously translates through $2c/4 = \frac{1}{2}c$ so that the xyz position yields in succession equivalent positions at $\bar{y}$, x, $\frac{1}{2}+z$; $\bar{x}$, $\bar{y}$, $1+z$; y, $\bar{x}$, $1\frac{1}{2}+z$ and returns to superposition on the original position at x, y, $2+z$; this introduces a new situation in that it now becomes necessary to reduce the equivalent positions produced by direct operation of the 4_2 screw axis by subtraction of integral lattice repeats parallel to z to bring them within one positive lattice repeat of the origin, i.e. $\bar{x}$, $\bar{y}$, $1+z$ becomes $\bar{x}$, $\bar{y}$, z and y, $\bar{x}$, $1\frac{1}{2}+z$ becomes y, $\bar{x}$, $\frac{1}{2}+z$, these positions being themselves produced by direct operation of the 4_2 axis on the unit-cell immediately below the origin. The 4_3 axis rotates anticlockwise through $\frac{1}{2}\pi$ and simultaneously translates through $3c/4$ to yield from x, y, z successive equivalent positions $\bar{y}$, x, $\frac{3}{4}+z$; $\bar{x}$, $\bar{y}$, $1\frac{1}{2}+z$; y, $\bar{x}$, $2\frac{1}{4}+z$, which reduce to x, y, z; $\bar{y}$, x, $\frac{3}{4}+z$; $\bar{x}$, $\bar{y}$, $\frac{1}{2}+z$; y, $\bar{x}$, $\frac{1}{4}+z$ as shown in Fig 4.13.

Comparison of the diagrams in Fig 4.13 for the equivalent positions generated from x, y, z by 4_1 and 4_3 axes reveals that both can be regarded as screws of pitch $\frac{1}{4}c$, the former right-handed and the latter left-handed. It follows that a 4_1 and a 4_3 axis produce sets of equivalent positions that are mirror images of one another and cannot be superposed; such a pair of screw axes are said to be *enantiomorphous*. The 4_2 screw axis, like the rotation tetrad, is without hand, the same disposition of equivalent positions being obtained by an anticlockwise as by a clockwise rotation. In general one can say that the screw of smallest pitch that can be used to describe the disposition of equivalent positions related by a screw axis n_m is right-handed if $m < \frac{1}{2}n$, left-handed if $m > \frac{1}{2}n$, and without hand if $m = 0$ or $m = \frac{1}{2}n$.

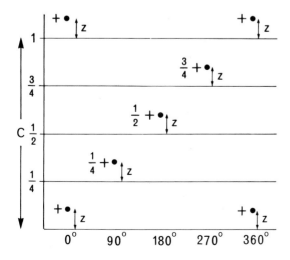

Fig 4.14 Equivalent positions generated by a 4_1 axis. The angle of rotation about a 4_1 axis parallel to [001] is represented as the horizontal coordinate and the translation in fractions of the lattice repeat c is plotted vertically.

It has already been stated that only screw axes involving 2, 3, 4, or 6-fold rotation elements can apply in crystals and it is appropriate at this point to justify that assertion. In general the operation of a screw axis on a group of atoms yields equivalent groups in identical environments, but differently orientated. For instance if a 4_1 axis is to be repeated on a lattice, a lattice point can be placed on that 4_1 axis and it follows that a 4_1 axis passes through every lattice point in the same direction. Consider two such lattice points P and Q, each of which represents a group of fo ir atoms related by the 4_1 axis through the relevant lattice point (Fig 4.15): the 4_1 ax s through P relates the atoms at A, B, C, and D and the 4_1 axis at Q relates the aton at α, β, γ, and δ. But the 4_1 axis through P relates the atom at α to that at T, the aton. at β to that at U, the atom at γ to that at V, and the atom at δ to that at W; in short a third lattice point arises at R related to that at Q by a *rotation tetrad* through P. This result may be generalized: when screw axes are repeated on a lattice, the lattice exhibits the symmetry of the rotation axis of the same order. This apparent anomaly arises from the fact that lattice points represent groups of atoms in the same environment in the same orientation—the basis of the lattice concept—and therefore if we are to be strictly rigorous a symmetry axis should be operated not on the lattice point itself but on the group of atoms that it represents; for rotation axes and mirror planes the distinction is immaterial, but for translational symmetry elements (screw axes and glide planes) the translational component is lost and the lattice exhibits only the symmetry of the corresponding rotation axis or mirror plane. It follows that only screw axes derived from diads, triads, tetrads, or hexads can operate on atomic groupings in crystal structures.

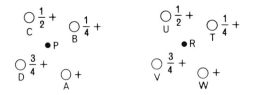

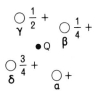

Fig 4.15   A lattice on which a screw tetrad 4_1 is repeated exhibits the symmetry of a rotation tetrad. The solid circles P, Q, R represent lattice points; the open circles, A, α, T, etc, represent atoms. Screw tetrads 4_1 perpendicular to the plane of the diagram through P and Q relate the atoms A, B, C, D and α, β, γ, δ respectively. The 4_1 axis through P relates the atomic groups $\alpha\beta\gamma\delta$ and TUVW so that the lattice points Q and R are related by a *rotation* tetrad through P perpendicular to the plane of the diagram.

Glide planes

A glide plane is a translational symmetry element representing simultaneous reflexion, as in a mirror plane, and translation through half a lattice repeat in a direction parallel to the plane. As in the case of screw axes, continued operation of a glide plane does not produce coincidence, but positions separated by a whole lattice repeat; for instance a glide plane parallel to (010) with a translation of $\frac{1}{2}c$ yields after two operations a position at a distance c in the direction of the z-axis from the original position. The magnitude of the translation associated with a glide plane is restricted, by the requirement of consistency with repetition on a lattice, to half a lattice spacing. The direction of the translation may be either parallel to a unit-cell edge (an *axial glide*) or parallel to a face-diagonal or body diagonal of the unit-cell (a *diagonal glide*). The translation of a diagonal glide plane is one half of the length of the relevant diagonal of the unit-cell except in the special case of the *diamond glide* where it is one quarter of the length of the diagonal.[4] The nomenclature of symmetry planes, mirror and glide, and the types of translation permitted in the latter are set out in Table 4.3.

Table 4.3
Symmetry planes

	Symbol	Translation
Mirror	m	none
Axial glide	a	$a/2$
	b	$b/2$
	c	$c/2$
Diagonal glide	n	$\dfrac{a+b}{2}, \dfrac{b+c}{2}, \dfrac{a+c}{2}$
		$\dfrac{a+b+c}{2}$[*]
Diamond glide	d	$\dfrac{a\pm b}{4}, \dfrac{b\pm c}{4}, \dfrac{a\pm c}{4}$
		$\dfrac{a\pm b\pm c}{4}$[*]

[*] cubic and tetragonal systems only

The graphical representation of glide planes on structural plans needs some explanation because a distinction has to be made between planes lying parallel and planes perpendicular to the plane of the diagram. A symmetry plane parallel to the plane of the diagram is conventionally represented (Fig 4.16) by the symbol ⌐ beyond the top right-hand corner of the outline of the unit-cell and an arrow is incorporated in the symbol to indicate the direction of glide; the appropriate fraction is written beside the symbol to indicate height above the reference plane when the glide plane is not coincident with the reference plane. A symmetry plane perpendicular to the plane of the diagram is represented conventionally by a bold line (Fig 4.16) which is unbroken for a mirror plane, dashed for a glide plane with translation in the plane of the diagram, dotted for a glide plane with translation perpendicular to the plane of the diagram and alternate dashes and dots for a diagonal glide.

Equivalent positions are again represented by open circles, but here we have to distinguish between the hand of the atom group occupying each equivalent position

---

[4] Diamond glide planes are restricted to certain orientations in certain Bravais lattices such that a lattice point lies at the mid-point of the diagonal concerned so that the translation remains one half of the lattice repeat in the direction of the diagonal. These conditions are satisfied only in I and F lattices with the exception of the orthorhombic I lattice which cannot have planes of symmetry parallel to the body diagonals of the unit-cell, $\langle 111 \rangle$. A full account of the operation of diamond glides may be found in Buerger (1956).

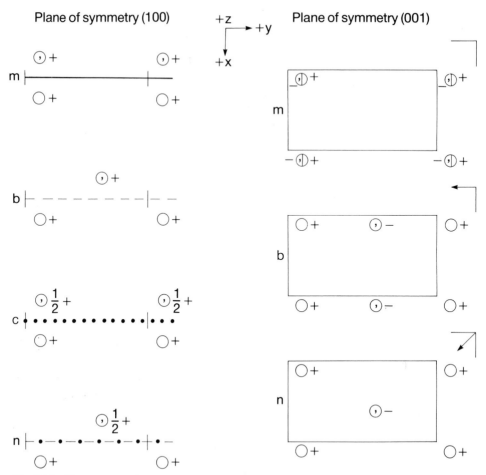

Fig 4.16 The representation and operation of mirror and glide planes. The orientation of axes is displayed centrally at the top of the figure. The left-hand and right-hand columns respectively show the operation of symmetry planes parallel to (100) and to (001), that is perpendicular and parallel to the plane of the diagram.

because the operation of a symmetry plane, unlike a rotation axis, produces a reversal of hand. The accepted convention is an open circle for the original position, at x, y, z, and an open circle enclosing a comma for positions related to it by an odd number of reflexions. Superimposition in projection of positions of different hand is represented by adjacent half circles. The heights of equivalent positions are denoted according to the convention explained in detail in the section on screw axes.

That the spacing of mirror and glide planes is necessarily half the lattice repeat in the direction perpendicular to the plane is illustrated in Fig 4.17, which shows two projections of a unit-cell with an (001) mirror plane passing through x, y, $\frac{1}{4}$. The position x, y, z on reflexion becomes x, y, $\frac{1}{2} - z$, which is itself related to the position x, y, $1 + z$ by a parallel mirror plane x, y, $\frac{3}{4}$.

Space groups

We have already seen that there are thirty-two groups of non-translational symmetry elements that can be repeated on a lattice, the so-called crystallographic *point groups*. When translational symmetry elements, screw axes and glide planes, are taken into

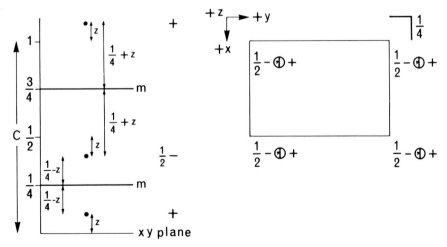

Fig 4.17   The generation of an (001) mirror plane at $z = \frac{3}{4}$ by an (001) mirror plane at $z = \frac{1}{4}$. The presence of an (001) mirror plane at $z = \frac{1}{4}$ requires an atom at a height z above the xy plane to be repeated at a height $\frac{1}{2} - z$ with the same x and y coordinates; the atom at $\frac{1}{2} - z$ is related to the lattice-repeated atom at $1 + z$ by an (001) mirror plane at $z = \frac{3}{4}$.

account the number of groups of symmetry elements that can be repeated on a lattice increases to two hundred and thirty, the so-called crystallographic *space groups*. The derivation and the complete listing of the 230 space groups are outside the scope of this book, and we attempt only an outline of nomenclature and graphical representation and, through examples, the use of space groups in the interpretation of crystal structures. For the reader who wishes to explore the fundamentals of this topic, excellent accounts of space group derivation are to be found in Buerger (1956) and Hilton (1963). An exhaustive tabulation of space groups with a most thorough explanatory introduction is provided by *International Tables for X-ray Crystallography*, vol I (1969). In what follows we adopt the conventions laid down in *International Tables* which are the current usage of most laboratories.

A space group symbol consists of two parts: first a letter to indicate lattice type and that is followed by a statement of the essential symmetry elements present. This second part of the space group symbol is of the same form as a point group symbol but may include reference to translational symmetry elements; if any translational elements referred to in the space group symbol are replaced by the corresponding non-translational elements, the point group, and thence the system, to which the space group belongs can immediately be read off. For instance the symbol C2/c represents a space group of the point group 2/m, which is one of the point groups of the monoclinic system; the lattice type is monoclinic C; there is a rotation diad parallel to y and a c-glide perpendicular to y. Taking another example, the space group Pmcn belongs to point group *mmm* (orthorhombic), the lattice type is orthorhombic P, a mirror plane lies perpendicular to x, a c-glide perpendicular to y, and a diagonal glide perpendicular to z. In both examples other symmetry elements are present in addition, but those stated in the symbol are adequate for the complete description of the symmetry of the space group.

For the graphical representation of a space group two diagrams are employed: one shows the distribution of symmetry elements in the unit-cell as a plan and the other shows, on a plan in the same orientation, all the positions generated by the operation of the symmetry elements on a position with general coordinates x, y, z. The first plan shows all the symmetry elements, consequent as well as essential, in the unit-cell. The second plan, showing the *general equivalent positions* for the space group, uses the mode of representation that we discussed in detail in the section on screw axes and glide planes. It is customary not to indicate axial directions when space group plans are drawn as projections on (001) with the origin in the top left-hand corner and the positive direction of the y-axis pointing to the right; when other orientations are employed it is advisable to specify axial directions. It is conventional, and usually convenient, to take the origin of a centrosymmetric space group at a centre of symmetry because the coordinates of the general equivalent positions can then be simply written in pairs as $\pm(x, y, z)$ etc.: there is no straightforward convention for the siting of the origin of non-centrosymmetrical space groups. We have already shown that symmetry elements are repeated at half lattice spacings: all symmetry elements perpendicular to (001) are of course shown on the conventional space group diagrams, but those parallel to (001) are only marked with the height above the reference plane of the lowest of the pair, for instance the symbol $\neg\frac{1}{4}$ implies the presence of (001) mirror planes at $z = \frac{3}{4}$ as well as at $z = \frac{1}{4}$.

Figure 4.18 shows the conventional diagrams for the space group Pmcn, which will be taken to illustrate the points made in the previous paragraph. The (100) mirror planes are shown on the right-hand diagram intersecting the x-axis at $\frac{1}{4}a$ and $\frac{3}{4}a$; the (010) c-glide planes are shown intersecting the y-axis at $\frac{1}{4}b$ and $\frac{3}{4}b$; but the (001) diagonal glide planes, which intersect the z-axis at $\frac{1}{4}c$ and $\frac{3}{4}c$, are indicated briefly as $\frac{1}{4}$. Consequent symmetry elements are three non-intersecting sets of screw diads, one parallel to each reference axis, and eight centres of symmetry represented in the diagram by small open circles. Since the centres of symmetry, at $0, 0, 0$; $\frac{1}{2}, 0, 0$; $0, \frac{1}{2}, 0$; $0, 0, \frac{1}{2}$; $0, \frac{1}{2}, \frac{1}{2}$; $\frac{1}{2}, 0, \frac{1}{2}$; $\frac{1}{2}, \frac{1}{2}, 0$; $\frac{1}{2}, \frac{1}{2}, \frac{1}{2}$; have $z = 0, \frac{1}{2}$ the height above the reference plane is not indicated. The graphical representation of diads and screw diads when parallel to the reference plane requires explanation: diads are shown as full arrows and screw diads as one-armed arrows outside the unit-cell outline with the height, if not 0 or $\frac{1}{2}$, written alongside. It is worth noting at this point that although the space group symbol Pmcn makes explicit reference only to planes of symmetry perpendicular to each reference axis the presence of diad axes, in this case all screw diads, parallel to each reference axis is clearly implied; the point group to which this space group belongs is *mmm* which has three mutually perpendicular diads as well as three mutually perpendicular mirror planes and therefore the space group Pmcn must have three mutually perpendicular sets of rotation or screw diads as well as three mutually perpendicular sets of symmetry planes.

The left-hand diagram shows the disposition of the general equivalent positions for the space group. The number of general equivalent positions in a space group with a P lattice is equal to the number of planes in the general form $\{hkl\}$ of the corresponding point group; in this space group Pmcn the general equivalent positions have eightfold *multiplicity* as does the general form $\{hkl\}$, e.g. $\{123\}$, in point group *mmm*. In the case of a C or I lattice the multiplicity of the general form of the point group has to be doubled and for an F lattice quadrupled, e.g. the general form in point group *m3m* has multiplicity 48 and the general equivalent positions in space group Fd3m have multiplicity 192.[5] The coordinates of the general equivalent positions in space group Pmcn are listed in Table 4.4. It is important to notice that the determination of the

---

[5] Correspondingly the number of sets of symmetry elements parallel to a given direction in a space group is equal to the number of lattice points in the unit-cell. This point is illustrated in the discussion of C2/c which follows.

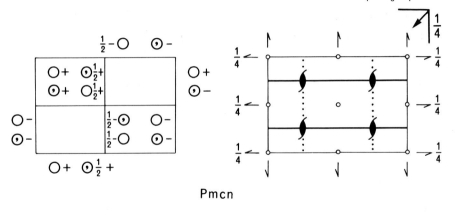

Pmcn

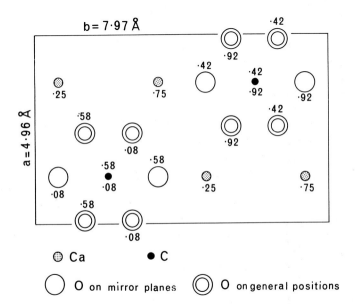

● Ca   ● C

O on mirror planes   ◎ O on general positions

Fig 4.18 The crystal structure of aragonite, $CaCO_3$. The two upper diagrams are the conventional diagrams for the space group P*mcn* oriented so that $+z$ is upwards perpendicular to the plane of the figure, $+y$ is directed to the right and $+x$ downwards in the plane of the figure. The upper left-hand diagram shows the disposition of general equivalent positions and the upper right-hand diagram the disposition of symmetry elements in the space group. The lower diagram shows the structure of aragonite in plan on (001) in the same orientation: all the calcium and carbon atoms and four of the oxygen atoms lie on special equivalent positions on mirror planes with $x = \pm\frac{1}{4}$ and the remaining eight oxygen atoms lie on one set of general equivalent positions.

three parameters x, y, and z is sufficient to fix the coordinates of eight atoms of the same element occupying one set of general equivalent positions and that the whole set of eight atoms is represented by a single lattice point.

When the coordinates of a position are such that it lies on a non-translational symmetry element, two or more equivalent positions coalesce and the multiplicity of the set of equivalent positions is reduced. Such a set of *special equivalent positions* arises in P*mcn* when $x = \frac{1}{4}$, all the positions in this set then lie on mirror planes. Substitution of $x = \frac{1}{4}$ in the list of coordinates of the general equivalent positions yields immediately $\pm(\frac{1}{4}, y, z)$, $\pm(\frac{1}{4}, \frac{1}{2} - y, \frac{1}{2} + z)$; this set of special equivalent positions

Table 4.4
Coordinates of equivalent positions in P*mcn*

	Multiplicity	Point symmetry	Coordinates
General	8	1	$\pm(x, y, z)$; $\pm(\frac{1}{2}-x, y, z)$; $\pm(x, \frac{1}{2}-y, \frac{1}{2}+z)$; $\pm(\frac{1}{2}-x, \frac{1}{2}-y, \frac{1}{2}+z)$
Special	4	m	$\pm(\frac{1}{4}, y, z)$; $\pm(\frac{1}{4}, \frac{1}{2}-y, \frac{1}{2}+z)$
		$\bar{1}$	$0, 0, 0$; $\frac{1}{2}, 0, 0$; $0, \frac{1}{2}, \frac{1}{2}$; $\frac{1}{2}, \frac{1}{2}, \frac{1}{2}$
		$\bar{1}$	$0, \frac{1}{2}, 0$; $\frac{1}{2}, \frac{1}{2}, 0$; $0, 0, \frac{1}{2}$; $\frac{1}{2}, 0, \frac{1}{2}$

has fourfold multiplicity, some positions lying on one mirror plane and some on the other. The only other non-translational symmetry elements in P*mcn* are the centres of symmetry. Substitution of the coordinates of the centre at the origin yields $0, 0, 0$; $\frac{1}{2}, 0, 0$; $0, \frac{1}{2}, \frac{1}{2}$; $\frac{1}{2}, \frac{1}{2}, \frac{1}{2}$; another set of special equivalent positions, again of multiplicity 4. The remaining four centres of symmetry likewise constitute a set of special equivalent positions, $0, \frac{1}{2}, 0$; $\frac{1}{2}, \frac{1}{2}, 0$; $0, 0, \frac{1}{2}$; $\frac{1}{2}, 0, \frac{1}{2}$. It is important to note that these two sets, each of four centres of symmetry, are not related to one another by symmetry; thus one set may be occupied by atoms of a certain element while the other set may either be occupied by atoms of a different element or be empty. The special equivalent positions associated with each type of non-translational symmetry element in space group P*mcn* have now been defined; all the remaining symmetry elements are translational in character and so cannot produce coalescence of equivalent positions.

We conclude our study of space group P*mcn* with some comments on a structure referable to this space group and take as our example the orthorhombic form of $CaCO_3$, the mineral *aragonite*. Measurements of the unit-cell dimensions and density of aragonite indicate that the unit-cell contains $4CaCO_3$. The calcium and carbon atoms must therefore each be situated on a set of special equivalent positions. The twelve oxygen atoms may be arranged either on three different sets of special positions or on one set of general positions and one set of special positions. Structure analysis of aragonite has shown that calcium, carbon, and four oxygens lie on mirror planes, y and z being determined for each set, and the remaining eight oxygens lie on general positions, for which x, y, and z have been determined. The resulting coordinates of all twenty atoms in the unit-cell are listed in Table 4.5 and the structure is shown in plan on Fig 4.18.

Table 4.5
Atomic coordinates in aragonite ($CaCO_3$)

Ca	4(m)	$\pm(\frac{1}{4}, 0{\cdot}42, 0{\cdot}75)$;	$\pm(\frac{1}{4}, 0{\cdot}08, 0{\cdot}25)$
C	4(m)	$\pm(\frac{1}{4}, 0{\cdot}75, \overline{0{\cdot}08})$;	$\pm(\frac{1}{4}, 0{\cdot}75, 0{\cdot}42)$
O(1)	4(m)	$\pm(\frac{1}{4}, \overline{0{\cdot}08}, \overline{0{\cdot}08})$;	$\pm(\frac{1}{4}, 0{\cdot}58, 0{\cdot}42)$
O(2)	8(1)	$\pm(0{\cdot}48, 0{\cdot}67, \overline{0{\cdot}08})$;	$\pm(0{\cdot}02, 0{\cdot}67, \overline{0{\cdot}08})$;
		$\pm(0{\cdot}48, 0{\cdot}83, 0{\cdot}42)$;	$\pm(0{\cdot}02, 0{\cdot}83, 0{\cdot}42)$

We take as our second example a non-primitive space group, $C2/c$, which belongs to the point group $2/m$ and has a monoclinic C lattice. Again the space group is centrosymmetrical and the origin is taken at a centre of symmetry (Fig 4.19). The (010) *c*-glide planes intersect the *y*-axis at the origin and at $y = \frac{1}{2}$. The [010] diads have $x = 0, \frac{1}{2}$ and $z = \frac{1}{4}, \frac{3}{4}$. The centring of the (001) face of the unit-cell introduces (010) diagonal glides at $y = \frac{1}{4}, \frac{3}{4}$ and [010] screw diads with $x = \frac{1}{4}, \frac{3}{4}$ and $z = \frac{1}{4}, \frac{3}{4}$;

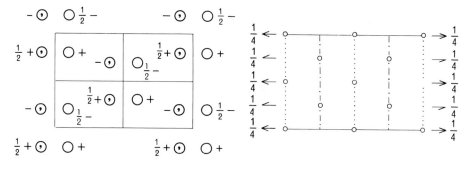

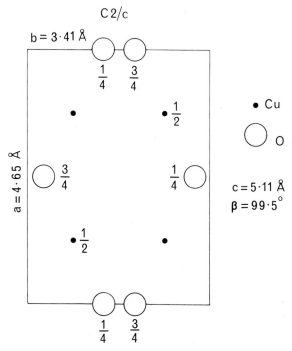

Fig 4.19 The crystal structure of tenorite, CuO. The two upper diagrams are the conventional diagrams for the monoclinic space group C2/c. The lower diagram is a plan of the structure of CuO on (001).

there are in all sixteen centres of symmetry. In a non-primitive space group a conveniently abbreviated way of listing the coordinates of equivalent positions is to state the coordinates of the several lattice points at the head of the list and then to give only the coordinates of the equivalent positions associated with one lattice point; this abbreviation is adopted in Table 4.6.

The only non-translational symmetry elements present are [010] diads and centres of symmetry. There is only one set of special equivalent positions on diads (with y as a variable parameter). The sixteen centres of symmetry are split into four groups, each of four symmetry related positions.

CuO, the mineral tenorite, has space group C2/c with only 4CuO in the unit-cell so that copper and oxygen must each occupy one set of special equivalent positions. The copper atoms are found to lie on centres of symmetry at $\frac{1}{4}$, $\frac{1}{4}$, 0 etc and the oxygen atoms lie on diads at 0, y, $\frac{1}{4}$ etc, where $y = 0.416$.

Table 4.6
Coordinates of equivalent positions in C2/c

	Multiplicity	Point symmetry	Coordinates
	$(0, 0, 0; \frac{1}{2}, \frac{1}{2}, 0)+$		
General	8	1	$\pm(x, y, z); \pm(\bar{x}, y, \frac{1}{2}-z)$
Special	4	2	$\pm(0, y, \frac{1}{4})$
	4	$\bar{1}$	$0, 0, 0; 0, 0, \frac{1}{2}$
	4	$\bar{1}$	$0, \frac{1}{2}, 0; 0, \frac{1}{2}, \frac{1}{2}$
	4	$\bar{1}$	$\frac{1}{4}, \frac{1}{4}, 0; \frac{3}{4}, \frac{1}{4}, \frac{1}{2}$
	4	$\bar{1}$	$\frac{1}{4}, \frac{1}{4}, \frac{1}{2}; \frac{3}{4}, \frac{1}{4}, 0$

In the tabulation of space groups various nomenclatorial and orientational preferences have to be expressed; this point is well illustrated by the two space groups we have just considered. It is evident from Fig 4.19 that the space group there represented could equally well be symbolized as C2/c, C2$_1$/c, C2$_1$/n, or C2/n without reorientation of axes. Since a choice has to be made, C2/c is arbitrarily preferred. In the orthorhombic system the same point may arise, but not in Pmcn, and moreover we are free to relabel the reference axes provided a right-handed axial system is maintained. For instance if the x, y, z axes of Pmcn are relabelled x′, y′, z′ according to the scheme $x \to y'$, $y \to z'$, $z \to x'$, the axes remain right-handed, the (100) mirror plane becomes (010), the (010) axial glide becomes (001) with a glide component $\frac{1}{2}a$, and the (001) diagonal glide becomes (100) so that the new space group symbol is Pnma. Alternative axial transformations lead to four more space group symbols Pbnm, Pnam, Pmnb, and Pcmn; from these six orientations of the symmetry elements of the space group Pnma is chosen arbitrarily as the *standard setting*. In compilations of data as in tabulation of space groups it is of course necessary to adhere rigidly to standard settings, but in discussion, especially comparative discussion, of actual structures it is often convenient to use non-standard settings; for instance in our brief discussion of the aragonite structure we chose the non-standard Pmcn in order to have the carbonate groups parallel to the plane of a plan drawn with the conventional axial orientation for space group diagrams.

A note of the rules used in selecting standard settings of space groups in *International Tables for X-ray Crystallography* and of the conventions used in *Crystal Data*, the principal compilation of information about the unit-cells of real substances, is given in Appendix C.

Symmetry of the Bravais lattices

Earlier in this chapter it was pointed out that each Bravais lattice was distinguished from all others by the symmetry of its arrangement of lattice points. It becomes possible at this stage to argue the point more closely. We do so by considering first the two monoclinic Bravais lattices, P and C, whose unit-cells and symmetry elements are displayed in conventional space group orientation in Fig 4.20. It is to be noted that the C-cell has all the symmetry elements of the P-cell and in addition a-glides interleave the mirror planes, screw diads alternate with the [010] rotation diads, and another eight centres of symmetry appear. It is quite general in centred lattices that all the symmetry elements of the corresponding primitive lattice are present and in addition appropriate glide planes and screw axes relate the centring lattice points, whether C or I or F, to the lattice point at the origin.

The symmetry elements of the four orthorhombic Bravais lattices are displayed

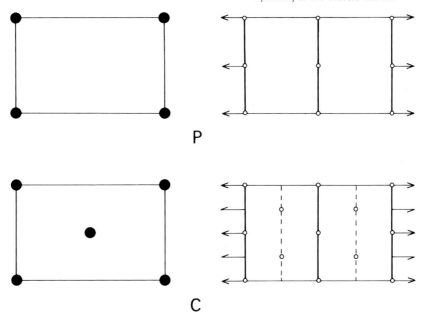

Fig 4.20   Unit-cells and symmetry elements of the two monoclinic Bravais lattices.

as Fig 4.21. The symmetry elements of the orthorhombic P-lattice are present in each of the non-primitive lattice types. The P-lattice has diads parallel to x, y, and z, mirror planes perpendicular to x, y, and z, and centres of symmetry at and midway between lattice points. In the centred lattices there are in addition translational symmetry elements and centres of symmetry that relate the lattice point at the origin to the centring lattice points. For instance in the I-lattice the lattice point at $\frac{1}{2}, \frac{1}{2}, \frac{1}{2}$ is related to the lattice point at the origin by a screw diad at $x, \frac{1}{4}, \frac{1}{4}$; this screw diad lies midway between the diads at $x, 0, 0$ and $x, \frac{1}{2}, \frac{1}{2}$, which are present also in the P-lattice; an n-glide interleaves the (100) mirror planes of the P-lattice and likewise relates the two lattice points; screw diads and n-glides correspondingly oriented with respect to the y and z axes similarly relate the two lattice points; finally, centres of symmetry at $\frac{1}{4}, \frac{1}{4}, \frac{1}{4}$, etc appear and likewise relate the lattice points at the origin and the body-centre. A significant distinction between the P and the I lattices is thus that, whereas in the former there is only one kind of diad parallel to and one kind of symmetry plane normal to each reference axis, there are in the latter two kinds of diad (2 and 2_1) parallel to and two kinds of symmetry plane (m and n) perpendicular to each reference axis; and moreover the number of centres of symmetry in the unit-cell increases from 8 to 16.

In the C-lattice an analogous situation exists. The centring of (001) faces gives rise to screw diads parallel to x and y, to additional rotation diads parallel to z, to b-glides and a-glides interleaving the (100) and (010) mirror planes respectively, and to eight additional centres of symmetry at $\frac{1}{4}, \frac{1}{4}, 0$, etc. The (001) n-glides that one would expect to be introduced are coincident with the mirror planes inherited from the P-lattice.

The F-lattice, like the P and the I lattices, has of course the same symmetry elements associated with each reference axis. Associated with the x-axis, for example, there are [100] rotation diads at $x, 0, 0$ and $x, \frac{1}{4}, \frac{1}{4}$, etc, and parallel screw diads at $x, 0, \frac{1}{4}$ and $x, \frac{1}{4}, 0$, etc, (100) mirror planes coincident with n-glides through $0, y, z$, etc, and b- and c-glides through $\frac{1}{4}, y, z$, etc; the additional translational symmetry elements relate the

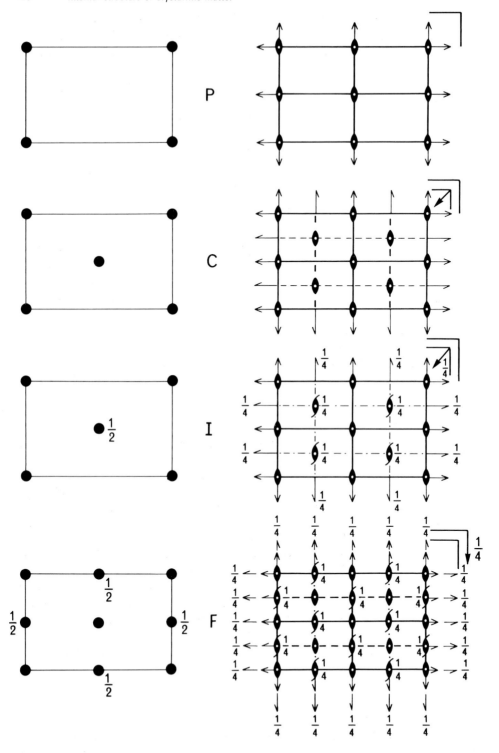

Fig 4.21   Unit-cells and symmetry elements of the four orthorhombic Bravais lattices.

face centring lattice points at $0, \frac{1}{2}, \frac{1}{2}; \frac{1}{2}, 0, \frac{1}{2}; \frac{1}{2}, \frac{1}{2}, 0$, to the lattice point at the origin and to one another. Centres of symmetry increase in number from 8 in the P-lattice to 32 in the F-lattice, where the additional centres have coordinates $0, \frac{1}{4}, \frac{1}{4}; \frac{1}{4}, 0, \frac{1}{4}; \frac{1}{4}, \frac{1}{4}, 0$; etc, midway between adjacent lattice points.[6]

Each orthorhombic Bravais lattice thus has a characteristic set of symmetry elements as it has a characteristic disposition of lattice points. The same holds for the Bravais lattices of the remaining systems and, without exploring each in turn in tedious detail, we arrive at the general conclusion that a Bravais lattice is characterized not only by the disposition of its lattice points in space, but also by the nature and disposition of its symmetry elements.

Problems

4.1 The crystal structure of NaCl has a cubic F lattice with one sodium ion and one chloride ion, $\frac{1}{2}\mathbf{a}$ apart, associated with each lattice point. Placing sodium ions at the lattice points and chloride ions $\frac{1}{2}\mathbf{a}$ away from the lattice points draw a plan of the structure projected on (001). In calcium carbide CaC_2 the carbon atoms in the C_2 group are 1·27 Å apart and the axis of each C_2 group is parallel to the z-axis of the unit-cell; the centre of a C_2 group corresponds to the site of a chloride ion and the calcium to a sodium ion in the NaCl structure, the cubic unit-cell being distorted to a unit-cell with $a = b = 5\cdot47$ Å, $c = 6\cdot37$ Å, $\alpha = \beta = \gamma = 90°$. Draw a 2×2 array of four unit-cells of the CaC_2 structure projected on (001). Identify the conventional unit-cell and state the crystal system and lattice-type. Calculate the dimensions of the conventional unit-cell and write down the coordinates of the atoms associated with one lattice point.

4.2 Many metals are cubic close-packed, a statement which implies a cubic F lattice with one atom associated with each lattice point. Draw a plan of one unit-cell on (001) and mark on the plan the centres of all the tetrahedral interstices in the unit-cell (that is the points equidistant from four metal atoms situated at the corners of a regular tetrahedron).

Some simple cubic ionic structures AX may be described as cubic close-packed arrays of one ion A with the oppositely charged ion X occupying some or all of the tetrahedral interstices within the array. The three common structures of this sort are (i) all tetrahedral interstices occupied, e.g. CaF_2, (ii) half of the tetrahedral interstices are occupied in such a manner that the environment of every A ion is identical, e.g. ZnS, (iii) one quarter of the interstices are occupied in such a manner that X ions are $\frac{1}{2}(\mathbf{a}+\mathbf{b}+\mathbf{c})$ apart, e.g. Cu_2O. For each case state the cubic lattice type and the coordinates of the atoms associated with one lattice point.

4.3 Platinum sulphide, PtS, is tetragonal. Its crystal structure has been described on a unit-cell with $a = 4\cdot90$ Å, $c = 6\cdot10$ Å and atomic coordinates

---

[6] Coincident glide planes, or coincident mirror and glide planes, perpendicular to the plane of the diagram cannot conveniently be shown on space group diagrams. Only that which comes higher in the conventional order of precedence, m, a, b, c, n is shown; thus in the F-lattice diagram of Fig 4.21 mirror planes are preferred to n-glides parallel to (100) at $x = 0, \frac{1}{2}$ and b-glides are preferred to c-glides parallel to (100) at $x = \pm\frac{1}{4}$. No such problem arises in the representation of symmetry planes parallel to the plane of the diagram.

$$Pt \quad 000, \tfrac{1}{2}\tfrac{1}{2}0, \tfrac{1}{2}0\tfrac{1}{2}, 0\tfrac{1}{2}\tfrac{1}{2},$$

$$S \quad \tfrac{131}{444}, \tfrac{311}{444}, \tfrac{133}{444}, \tfrac{313}{444}.$$

Draw a plan on (001) of a 2×2 array of four unit-cells. What is the lattice type? Identify the conventional unit-cell, state its lattice type and calculate its unit-cell dimensions. What are the coordinates of the atoms in the conventional unit-cell?

4.4 In a certain orthorhombic crystal sets of equivalent atoms have coordinates

$$x, y, z; \ x, \bar{y}, \bar{z}; \ x, \tfrac{1}{2}-y, \tfrac{1}{2}+z; \ x, \tfrac{1}{2}+y, \tfrac{1}{2}-z;$$

$$\bar{x}, \bar{y}, \bar{z}; \ \bar{x}, y, z; \ \bar{x}, \tfrac{1}{2}+y, \tfrac{1}{2}-z; \ \bar{x}, \tfrac{1}{2}-y, \tfrac{1}{2}+z;$$

What is the space group?

4.5 Gadolinium orthoferrite, $GdFeO_3$, is orthorhombic with $a = 5\cdot349$, $b = 5\cdot611$, $c = 7\cdot669$ Å and space group Pbnm. The coordinates of the atoms in the unit-cell are:

Gd^{3+} $\pm(x, \tfrac{1}{2}+y, \tfrac{1}{4}); \ \pm(\tfrac{1}{2}-x, y, \tfrac{1}{4})$ with $x = 0\cdot984$, $y = 0\cdot063$

Fe^{3+} $0, 0, 0; \ 0, 0, \tfrac{1}{2}; \ \tfrac{1}{2}, \tfrac{1}{2}, 0; \ \tfrac{1}{2}, \tfrac{1}{2}, \tfrac{1}{2}.$

O^{2-} $\pm(x, \tfrac{1}{2}+y, \tfrac{1}{4}); \ \pm(\tfrac{1}{2}-x, y, \tfrac{1}{4})$ with $x = 0\cdot101$, $y = 0\cdot467$

O^{2-} $\pm(x, \tfrac{1}{2}+y, z); \ \pm(x, \tfrac{1}{2}+y, \tfrac{1}{2}-z); \ \pm(\tfrac{1}{2}+x, \bar{y}, \bar{z}); \ \pm(\tfrac{1}{2}+x, \bar{y}, \tfrac{1}{2}+z)$

$$\text{with } x = 0\cdot696, \ y = 0\cdot302, \ z = 0\cdot051$$

Draw an accurate plan on (001) of one unit-cell of the structure using a scale of 20 mm to 1 Å. Mark on the plan all the symmetry elements present. The ferric cations are in six-fold coordination to oxygen anions, each Fe^{3+} having as its nearest neighbours six O^{2-} situated at the apices of a nearly regular octahedron. Identify the six nearest neighbours of the Fe^{3+} ion at $\tfrac{1}{2}, \tfrac{1}{2}, \tfrac{1}{2}$.
How many independent Fe–O distances are there?

4.6 Draw the two standard diagrams for each of the space groups P2/m, P2$_1$/m, C2/m, C2$_1$/m. Show that P2/m and P2$_1$/m are distinct space groups and that C2/m and C2$_1$/m are equivalent.

4.7 The space group of an orthorhombic crystal structure has been quoted as Ac2m. List all the equivalent space group symbols which arise when the x, y, and z axes are chosen in different ways, the axial system being always right-handed.

4.8 Pure Au and pure Cu are both cubic with atoms at 000, $0\tfrac{1}{2}\tfrac{1}{2}$, $\tfrac{1}{2}0\tfrac{1}{2}$, $\tfrac{1}{2}\tfrac{1}{2}0$. The compound Cu_3Au is cubic with an Au atom at 000 and Cu atoms at $0\tfrac{1}{2}\tfrac{1}{2}$, $\tfrac{1}{2}0\tfrac{1}{2}$, $\tfrac{1}{2}\tfrac{1}{2}0$. Determine the lattice type of Au, Cu, and Cu_3Au and in each case state the number of atoms associated with each lattice point.

4.9 The sulphide mineral enargite, Cu_3AsS_4, has a crystal structure which is derivative from an hexagonal structure with $a = 3\cdot72$, $c = 6\cdot18$ Å. The crystal structure of enargite is orthorhombic with space group Pnm2$_1$ and unit-cell dimensions $a = 6\cdot46$, $b = 7\cdot43$, $c = 6\cdot18$ Å. The approximate coordinates of the atoms in the orthorhombic unit-cell are:

Cu: $\frac{1}{3}\frac{1}{4}0$; $\frac{1}{3}\frac{3}{4}0$; $\frac{2}{3}\frac{1}{4}\frac{1}{2}$; $\frac{2}{3}\frac{3}{4}\frac{1}{2}$; $\frac{1}{6}0\frac{1}{2}$; $\frac{5}{6}\frac{1}{2}0$.

As: $\frac{5}{6}00$; $\frac{1}{6}\frac{1}{2}\frac{1}{2}$.

S: $\frac{5}{6}0\frac{3}{8}$; $\frac{1}{6}\frac{1}{2}\frac{7}{8}$; $\frac{1}{6}0\frac{7}{8}$; $\frac{5}{6}\frac{1}{2}\frac{3}{8}$; $\frac{1}{3}\frac{1}{4}\frac{3}{8}$; $\frac{1}{3}\frac{3}{4}\frac{3}{8}$; $\frac{2}{3}\frac{1}{4}\frac{7}{8}$; $\frac{2}{3}\frac{3}{4}\frac{7}{8}$.

On a scale of 20 mm to 1Å draw a plan of the structure on (001). Mark on the plan all the symmetry elements present. Determine the sets of general and special equivalent positions in space group $Pnm2_1$ occupied by the Cu, As, and S atoms. List the coordinates of the atoms in the unit-cell in the manner illustrated in Table 4.5.

4.10 A substance AX has been shown to have a cubic F lattice with one formula unit associated with each lattice point. Refer to *International Tables for X-ray Crystallography*, vol. 1 (or vol. A) to determine the only two possible structures for AX; in each case give the space group and list the atomic coordinates.

5
Interplanar and interzonal angles: some methods of calculation and transformation

In the study of crystalline solids it is frequently necessary to calculate the separation of lattice points, the spacing of sets of lattice planes, the angles between lattice directions, and the angles between the normals to sets of lattice planes. In this chapter we deal with the mathematical techniques available for making such calculations. Spherical trigonometry, with which we deal first, provides an elegant and efficient method for one-off calculations, requiring only the most simple of 'scientific' calculators. Analytical geometry and vector methods are more appropriate for repetitive calculations and are, whether one-off or repetitive, very easily performed with programmable calculators or computers; both of these methods usually require the evaluation of quite complicated expressions, which cannot be achieved without risk of error on a simple calculator but can be done quickly and reliably with the aid of a more sophisticated calculating device. After showing how vector methods can be applied to the calculation of the separation of lattice points and of angles between lattice directions, we introduce the concept of the *reciprocal lattice* to provide a vector method for the calculation of the spacing of sets of lattice planes and of the angles between the normals to sets of lattice planes. The reciprocal lattice will be reintroduced in chapter 6 and will be used there and in subsequent chapters for the interpretation of the diffraction of radiation by crystalline solids. Three-dimensional analytical geometry is the next topic to be dealt with in this chapter; its techniques are closely related to those of the vector methods described earlier. The Miller formulae, which describe the geometrical relationships between planes which lie in a zone, follow, being developed here by vector methods. The chapter ends with a discussion of the transformation of axes, unit-cell coordinates, Miller indices, lattice vectors (zone axes), and reciprocal lattice vectors.

Spherical trigonometry

The calculation of interplanar and interzonal angles can usually be most easily programmed for computer calculation in the language of solid geometry, but for calculation 'by hand', that is with trigonometric tables and logarithms or a simple calculator, spherical trigonometry provides a means of attack that is at once elegant, rapid, and instructive.

A *spherical triangle* is defined as that portion of the surface of a sphere bounded by the intersection of the sphere with a three-sided pyramid whose apex is at the centre of

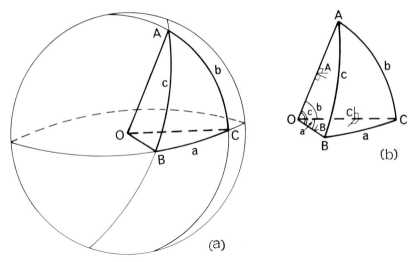

Fig 5.1 Spherical triangles. (a) shows the spherical triangle ABC as that portion of the surface of the sphere of centre O bounded by the great circles AB, BC, CA; the *angles* of the spherical triangle are denoted by A, B, C and its *sides* by a, b, c. (b) illustrates that the *angle* A is the angle between the normal to OA in the OAC plane and the normal to OA in the OAB plane, both normals being drawn from the same point on OA; and likewise for the *angles* B and C. The *sides*, a, b, c, of the spherical triangle ABC are equal to the lengths of the arcs BC, CA, AB respectively when the radius OA = OB = OC is unity.

the sphere. The sides of a spherical triangle are thus arcs of great circles. The *angles*, A, B, and C, of the spherical triangle ABC (Fig 5.1(a)) are the angles between the great circles whose planes are the faces of the pyramid OABC. The *sides*, a, b, and c, of the spherical triangle ABC are the angles between pairs of edges of the pyramid OABC and are therefore the angles subtended by the arcs, BC, CA, and AB, of the great circles at the centre, O, of the sphere. It follows that the sides, a, b, and c, are equal to the lengths of the arcs, BC, CA, and AB, when the sphere has unit radius (Fig 5.1(b)). Since the angles and sides of a spherical triangle are respectively the angles between faces and the angles between edges of a three-sided pyramid, every angle and side of a spherical triangle must be of magnitude less than π.

The general spherical triangle

In Appendix D we derive the eight fundamental relationships which relate the angles and sides of a general spherical triangle. There are three relationships between three sides and one angle,

$$\cos a = \cos b . \cos c + \sin b . \sin c . \cos A \tag{1}$$

$$\cos b = \cos c . \cos a + \sin c . \sin a . \cos B \tag{2}$$

$$\cos c = \cos a . \cos b + \sin a . \sin b . \cos C \tag{3}$$

There are three relationships of similar aspect between three angles and one side,

$$\cos A = -\cos B . \cos C + \sin B . \sin C . \cos a \tag{4}$$

$$\cos B = -\cos C . \cos A + \sin C . \sin A . \cos b \tag{5}$$

$$\cos C = -\cos A . \cos B + \sin A . \sin B . \cos c \tag{6}$$

And there is a set of relationships between angles and opposite sides,

$$\frac{\sin A}{\sin a} = \frac{\sin B}{\sin b} = \frac{\sin C}{\sin c} \tag{7}$$

These relationships are adequate for the solution of any problem in spherical trigonometry. They become greatly simplified if either one angle or one side of the spherical triangle is a right angle, the simplification being embodied in a set of rules formulated by John Napier (1550–1617). We shall not introduce Napier's Rules, but treat all problems in spherical trigonometry by use of the general relationships, which lead to quite easy calculations with a simple calculator.

Example (i)   To calculate the interplanar angle (100):(311) *in* BaSO₄ *which is orthorhombic with a* = 8·85 Å, *b* = 5·44 Å, *c* = 7·13 Å.

The first step is to draw a sketch stereogram showing all the zones relevant to the angle to be calculated (Fig 5.2(a)), that is the zones [(311),(100)], [(311),(010)], and [(311),(001)]. The required angle (100):(311) is a side of the spherical triangle (100), (311), (310), which is shown in isolation in Fig 5.2(b). It is self-evident that two zones are at right angles when the pole of the great circle representing one zone lies on the great circle representing the other zone (three cases are illustrated in Fig 5.3). In the case of the spherical triangle (100), (311), (310), the pole of the zone [(100),(310)] coincides with (001) and so lies in the zone [(311),(310)]: consequently the triangle is right-angled at (310). Since the normal to the plane (100) is parallel to the zone axis [100], the angle between the planes represented by the great circles [(100),(310)] and [(100),(311)] is the angle between the lines of intersection of these two zones with the plane normal

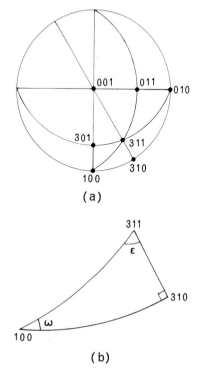

(a)

(b)

Fig 5.2   Calculation of the interfacial angle (100) : (311) in the orthorhombic system, the unit-cell dimensions being known. The stereogram (a) shows all the zones relevant to the calculation. The spherical triangle (100), (311), (310) is shown enlarged in (b): tan (100) : (311) = $a/3b$ and tan $\omega = b/c$.

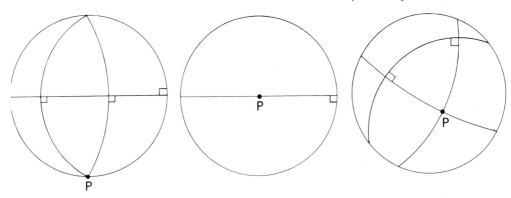

Fig 5.3 A right-angled spherical triangle arises when the pole of one great circle lies on another great circle. In the diagram at the left the pole P lies on the primitive and on two of the other great circles shown so that the horizontal diameter, of which P is the pole, intersects all three great circles orthogonally. In the middle diagram the primitive, whose pole is P, is perpendicular to all great circles passing through P. In the diagram at the right P lies at the intersection of two great circles which intersect the great circle whose pole is P orthogonally.

to their common direction [100]; so the angle ω in Fig 5.2(b) is equal to the angle between the normals to (011) and (010).

$$\omega = (011):(010)$$

Now in general in the orthorhombic system we have seen that

$$(010):(0kl) = \tan^{-1}\frac{b/k}{c/l}$$

Therefore $\quad \omega = \tan^{-1}\frac{b}{c} = \tan^{-1}\frac{5\cdot44}{7\cdot13} = 37\cdot35°$

Similarly $\quad (100):(hk0) = \tan^{-1}\frac{a/h}{b/k},$

so $\quad (100):(310) = \tan^{-1}\frac{8\cdot85}{3(5\cdot44)} = 28\cdot47°.$

Use of a relationship between three angles and one side, equations (4)–(6), gives

$$\cos\varepsilon = -\cos\omega.\cos 90° + \sin\omega.\sin 90°.\cos(100):(310)$$

$$= \sin 37\cdot35° \cos 28\cdot47°$$

therefore $\quad \varepsilon = 57\cdot77°.$

Application of equation (7) then yields

$$\frac{\sin 90°}{\sin(100):(311)} = \frac{\sin\varepsilon}{\sin(100):(310)}$$

whence the required angle $(100):(311) = \sin^{-1}\left\{\frac{\sin 28\cdot47°}{\sin 57\cdot77°}\right\} = 34\cdot30°.$

Alternatively the triangle (100), (301), (311) might have been used: here the angle between the zones [(100), (001)] and [(010), (301)] is 90°, the angle between the zones [(100), (001)] and [(100), (311)] is 90° − ω, and the angle (100):(301)

$= \tan^{-1}(a/3c)$, so the angle at (311) can be evaluated and thence equation (7) gives the required angle (100):(311).

Example (ii)   To calculate the interzonal angle $[100]:[0\bar{2}1]$ *in gypsum,* $CaSO_4 \cdot 2H_2O$, *which is monoclinic with* $a = 5\cdot68\,\text{Å}$, $b = 15\cdot18\,\text{Å}$, $c = 6\cdot29\,\text{Å}$, $\beta = 113\cdot83°$.

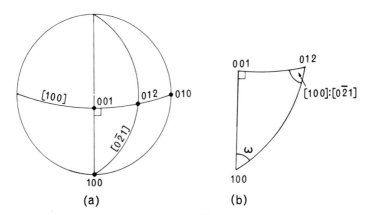

Fig 5.4 Calculation of the interzonal angle $[100]:[0\bar{2}1]$ in the monoclinic system, the unit-cell dimensions being known. The stereogram (a) shows all the zones relevant to the calculation. The significant spherical triangle (100), (001), (012) is shown in (b): (100):(001) $= 180° - \beta$ and $\tan\omega = c/2b$.

The first step is to draw a sketch stereogram (Fig 5.4(a)) to display the principal zones and the zone $[0\bar{2}1]$. The indices of planes common to any two zones can be found by application of the *zone equation*: the zone $[100]$ contains all planes with indices of the type $(0kl)$ and in particular (010) and (001); the zone $[0\bar{2}1]$ contains all planes with indices such that $2k = l$, e.g. (100), (012); therefore the face common to both zones has $h = 0$, $2k = l$ and must be (012). The triangle (100), (001), (012), shown in Fig 5.4(b), is right-angled at (001) since $[100] \perp [010]$ in the monoclinic system. It has already been shown that in the monoclinic system

$$(100):(001) = 180° - \beta$$

and $$\tan\omega = \frac{c}{2b}$$

Application of the relationship between three angles and one side of a spherical triangle then yields

$$\cos[100]:[0\bar{2}1] = -\cos 90° \cos\omega + \sin 90° \sin\omega \cos(100):(001)$$

$$= \sin\omega \cos(100):(001)$$

$$= \sin\tan^{-1}\frac{c}{2b} \cdot \cos(180° - \beta)$$

$$= \sin\tan^{-1}\frac{6\cdot29}{30\cdot36}\cos 66\cdot17°.$$

Hence $[100]:[0\bar{2}1] = 85\cdot30°$ or $94\cdot70°$.

Vector methods

The vector methods appropriate for calculating interplanar spacings and angles in crystallography are simple applications of the scalar and vector products of two vectors. Readers unfamiliar with these techniques are referred to any standard textbook of mathematics which includes a section on elementary vector analysis.[1] It will be our practice to show a vector in bold type and its magnitude or modulus either in italics or enclosed in the modulus sign as may be appropriate; thus the vector **p** has modulus p or $|\mathbf{p}|$.

It is often convenient to describe a vector in terms of its components along three non-coplanar axes; in crystallographic usage these axes will normally be the crystallographic reference axes which define the geometry of the unit-cell of the lattice. The base vectors are then the vectors which define the edges of the unit-cell, **a**, **b**, and **c**. The lattice vector $[UVW]$ is thus

$$\mathbf{t}_{[UVW]} = U\mathbf{a} + V\mathbf{b} + W\mathbf{c}.$$

Crystallographic base vectors will not necessarily be orthogonal.

Interzonal angles

The scalar product of the vectors **p** and **q** is a scalar quantity, written as $\mathbf{p}.\mathbf{q}$ and equal in magnitude to $pq\cos\theta$, where θ is the angle between the directions of **p** and **q**, the angle being within the range $0 \leqslant \theta \leqslant \pi$. So $\mathbf{p}.\mathbf{q} = pq\cos\theta$.

The vector product of the vectors **p** and **q** is a vector, written as $\mathbf{p} \wedge \mathbf{q}$, normal to the plane defined by **p** and **q** such that the vectors **p**, **q**, and $\mathbf{p} \wedge \mathbf{q}$ form a right-handed screw; its magnitude is $pq\sin\theta$, where θ is the angle between the directions **p** and **q**. So the magnitude of the vector product $\mathbf{p} \wedge \mathbf{q}$ is the area of the parallelogram bounded by **p** and **q** (Fig 5.5).

The magnitude of a vector is given by the scalar product of the vector with itself; thus $\mathbf{p}.\mathbf{p} = p^2$. So in three dimensions, with complete generality

$$t_{[UVW]}^2 = |U\mathbf{a} + V\mathbf{b} + W\mathbf{c}|^2. \tag{8}$$

It follows also from the argument above that the angle θ between the lattice vectors $[U_1V_1W_1]$ and $[U_2V_2W_2]$ is given by

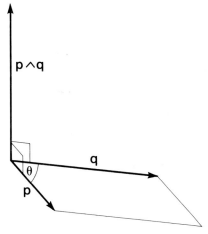

Fig 5.5 The vector product $\mathbf{p} \wedge \mathbf{q}$ is in the direction perpendicular to the plane defined by **p** and **q** and its magnitude is the area of the parallelogram bounded by **p** and **q**, $pq\sin\theta$. The vectors **p**, **q** and $\mathbf{p} \wedge \mathbf{q}$ form a right-handed screw.

[1] For example Riley (1974), Stephenson (1973), or Thomas and Finney (1984), details of which are given in the bibliography.

$$t_{[U_1V_1W_1]} \cdot t_{[U_2V_2W_2]} = |t_{[U_1V_1W_1]}| \, |t_{[U_2V_2W_2]}| \cos\theta,$$

i.e.
$$\cos\theta = \frac{(U_1\mathbf{a} + V_1\mathbf{b} + W_1\mathbf{c}) \cdot (U_2\mathbf{a} + V_2\mathbf{b} + W_2\mathbf{c})}{|U_1\mathbf{a} + V_1\mathbf{b} + W_1\mathbf{c}| \, |U_2\mathbf{a} + V_2\mathbf{b} + W_2\mathbf{c}|} \tag{9}$$

If the crystallographic reference axes are orthogonal, that is if the crystal is cubic, tetragonal, or orthorhombic, then the scalar product of one base vector with another, such as $\mathbf{a}.\mathbf{b}$, is zero and in consequence the general expressions given above reduce to

$$t_{[UVW]}^2 = U^2a^2 + V^2b^2 + W^2c^2 \tag{10}$$

and
$$\cos\theta = \frac{U_1U_2a^2 + V_1V_2b^2 + W_1W_2c^2}{\sqrt{(U_1^2a^2 + V_1^2b^2 + W_1^2c^2)}\sqrt{(U_2^2a^2 + V_2^2b^2 + W_2^2c^2)}} \tag{11}$$

Some simplification is achieved in the tetragonal system, where $a = b$, so that

$$t_{[UVW]}^2 = (U^2 + V^2)a^2 + W^2c^2$$

and
$$\cos\theta = \frac{U_1U_2 + V_1V_2 + W_1W_2c^2/a^2}{\sqrt{(U_1^2 + V_1^2 + W_1^2c^2/a^2)}\sqrt{(U_2^2 + V_2^2 + W_2^2c^2/a^2)}} \tag{12}$$

Further simplification is achieved in the cubic system, where $a = b = c$, so that

$$t_{[UVW]} = a\sqrt{(U^2 + V^2 + W^2)}$$

and
$$\cos\theta = \frac{U_1U_2 + V_1V_2 + W_1W_2}{\sqrt{(U_1^2 + V_1^2 + W_1^2)}\sqrt{(U_2^2 + V_2^2 + W_2^2)}} \tag{13}$$

By way of a simple example in the cubic system we evaluate the angle $[111]:[\bar{1}11]$,

$$\cos[111]:[\bar{1}11] = \frac{1\times(-1) + 1\times1 + 1\times1}{\sqrt{3}\ \sqrt{3}} = \frac{1}{3}$$

therefore $[111]:[\bar{1}11] = \cos^{-1}\frac{1}{3} = 70\cdot53°$

The evaluation of interzonal angles in systems with non-orthogonal axes is less easy. For instance in the *hexagonal* system $a = b \neq c$, $\alpha = \beta = 90°$, and $\gamma = 120°$. So $\mathbf{a}.\mathbf{b} = a^2\cos120° = -\frac{1}{2}a^2$ and $\mathbf{a}.\mathbf{c} = \mathbf{b}.\mathbf{c} = 0$ and in consequence

$$t_{[U_1V_1W_1]} \cdot t_{[U_2V_2W_2]} = U_1U_2\mathbf{a}.\mathbf{a} + V_1V_2\mathbf{b}.\mathbf{b} + U_1V_2\mathbf{a}.\mathbf{b} + U_2V_1\mathbf{a}.\mathbf{b} + W_1W_2\mathbf{c}.\mathbf{c}$$

$$= [U_1U_2 + V_1V_2 - \tfrac{1}{2}(U_1V_2 + U_2V_1)]a^2 + W_1W_2c^2 \tag{14}$$

all other scalar products being zero, and

$$t_{[UVW]} = \sqrt{U^2a^2 + V^2a^2 + W^2c^2 - UVa^2} = \sqrt{(U^2 + V^2 - UV)a^2 + W^2c^2} \tag{15}$$

In the *monoclinic* system $a \neq b \neq c$, $\alpha = \gamma = 90°$, and $\beta \neq 90°$ so that $\mathbf{b}.\mathbf{c} = \mathbf{a}.\mathbf{b} = 0$ and $\mathbf{a}.\mathbf{c} = ac\cos\beta$ and in consequence

$$t_{[U_1V_1W_1]} \cdot t_{[U_2V_2W_2]} = U_1U_2a^2 + V_1V_2b^2 + W_1W_2c^2 + (U_1W_2 + U_2W_1)ac\cos\beta \tag{16}$$

and
$$t_{[UVW]} = \sqrt{U^2a^2 + V^2b^2 + W^2c^2 + 2UWac\cos\beta} \tag{17}$$

The metric tensor

The expressions become more cumbersome in the *triclinic* system and it is there more convenient to use a matrix notation for scalar products. The matrix manipulations which will be required for this purpose are set out in Appendix E. Each crystal system is characterized by a symmetric 3×3 matrix, the elements of which are the scalar products of the base vectors of the lattice,

$$\begin{bmatrix} \mathbf{a}.\mathbf{a} & \mathbf{a}.\mathbf{b} & \mathbf{a}.\mathbf{c} \\ \mathbf{b}.\mathbf{a} & \mathbf{b}.\mathbf{b} & \mathbf{b}.\mathbf{c} \\ \mathbf{c}.\mathbf{a} & \mathbf{c}.\mathbf{b} & \mathbf{c}.\mathbf{c} \end{bmatrix}$$

Such a matrix is commonly called the *metric tensor*[2] of the system. The metric tensor for each crystal system is set down in Table 5.1.

Table 5.1
The metric tensor in the seven crystal systems

Triclinic	$\begin{bmatrix} a^2 & ab\cos\gamma & ac\cos\beta \\ ab\cos\gamma & b^2 & bc\cos\alpha \\ ac\cos\beta & bc\cos\alpha & c^2 \end{bmatrix}$
Monoclinic	$\begin{bmatrix} a^2 & 0 & ac\cos\beta \\ 0 & b^2 & 0 \\ ac\cos\beta & 0 & c^2 \end{bmatrix}$
Trigonal and hexagonal	$\begin{bmatrix} a^2 & -\tfrac{1}{2}a^2 & 0 \\ -\tfrac{1}{2}a^2 & a^2 & 0 \\ 0 & 0 & c^2 \end{bmatrix}$
Orthorhombic	$\begin{bmatrix} a^2 & 0 & 0 \\ 0 & b^2 & 0 \\ 0 & 0 & c^2 \end{bmatrix}$
Tetragonal	$\begin{bmatrix} a^2 & 0 & 0 \\ 0 & a^2 & 0 \\ 0 & 0 & c^2 \end{bmatrix}$
Cubic	$\begin{bmatrix} a^2 & 0 & 0 \\ 0 & a^2 & 0 \\ 0 & 0 & a^2 \end{bmatrix}$

The scalar product $\mathbf{t}_{[U_1 V_1 W_1]} \cdot \mathbf{t}_{[U_2 V_2 W_2]}$ can be expressed in terms of the metric tensor as

$$[U_1 V_1 W_1] \begin{bmatrix} \mathbf{a}.\mathbf{a} & \mathbf{a}.\mathbf{b} & \mathbf{a}.\mathbf{c} \\ \mathbf{b}.\mathbf{a} & \mathbf{b}.\mathbf{b} & \mathbf{b}.\mathbf{c} \\ \mathbf{c}.\mathbf{a} & \mathbf{c}.\mathbf{b} & \mathbf{c}.\mathbf{c} \end{bmatrix} \begin{bmatrix} U_2 \\ V_2 \\ W_2 \end{bmatrix} \tag{18}$$

The use of the metric tensor is exemplified by repeating the calculation of Example (ii) on p. 136.

Example (ii)′   Gypsum is monoclinic with $a = 5\cdot68\,\text{Å}$, $b = 15\cdot18\,\text{Å}$, $c = 6\cdot29\,\text{Å}$, $\beta = 113\cdot83°$. The monoclinic metric tensor is

---

[2] We shall make much use of *tensors* to relate an effect vector to a cause vector in our treatment of crystal physics in volume two. We introduce the term tensor here in advance of a complete definition in order to give this important matrix its common name.

$$\begin{bmatrix} a^2 & 0 & ac\cos\beta \\ 0 & b^2 & 0 \\ ac\cos\beta & 0 & c^2 \end{bmatrix}$$

which for gypsum becomes

$$\begin{bmatrix} 32\cdot262 & 0 & -14\cdot435 \\ 0 & 230\cdot432 & 0 \\ -14\cdot435 & 0 & 39\cdot564 \end{bmatrix}$$

So

$$\mathbf{t}_{[0\bar21]}\cdot\mathbf{t}_{[0\bar21]} = [0\bar21]\begin{bmatrix} 32\cdot262 & 0 & -14\cdot435 \\ 0 & 230\cdot432 & 0 \\ -14\cdot435 & 0 & 39\cdot564 \end{bmatrix}\begin{bmatrix} 0 \\ \bar2 \\ 1 \end{bmatrix}$$

$$= [0\bar21]\begin{bmatrix} -14\cdot435 \\ -460\cdot864 \\ 39\cdot564 \end{bmatrix}$$

$$= 921\cdot728 + 39\cdot564 = 961\cdot292\,\text{Å}^2$$

therefore

$$t_{[0\bar21]} = \sqrt{961\cdot292} = 31\cdot00\,\text{Å},$$

$$t_{[100]} = a = 5\cdot68\,\text{Å}$$

and

$$\mathbf{t}_{[100]}\cdot\mathbf{t}_{[0\bar21]} = [100]\begin{bmatrix} 32\cdot262 & 0 & -14\cdot435 \\ 0 & 230\cdot432 & 0 \\ -14\cdot435 & 0 & 39\cdot564 \end{bmatrix}\begin{bmatrix} 0 \\ \bar2 \\ 1 \end{bmatrix}$$

$$= [100]\begin{bmatrix} -14\cdot435 \\ -460\cdot864 \\ 39\cdot564 \end{bmatrix}$$

$$= -14\cdot435$$

The required angle $[100]:[0\bar21] = \cos^{-1}\left\{\dfrac{\mathbf{t}_{[100]}\cdot\mathbf{t}_{[0\bar21]}}{|\,\mathbf{t}_{[100]}\,|\,|\,\mathbf{t}_{[0\bar21]}\,|}\right\}$

$$= \cos^{-1}\left\{\frac{-14\cdot435}{5\cdot68\times31\cdot00}\right\}$$

$$= 180° - 85\cdot30° = 94\cdot70°$$

The angle between the two lattice directions, each taken in its positive sense, is 94·70°; the acute angle between the zones is 85·30°.

We give one more example of the use of vector methods for the calculation of interzonal angles:

Example (iii)   To calculate the angle between the polar edges of the cleavage rhombohedron {10$\bar{1}$4} *in calcite,* CaCO$_3$, *which is trigonal* ($\bar{3}m$) *with* a = 4·990 Å, c = 17·061 Å

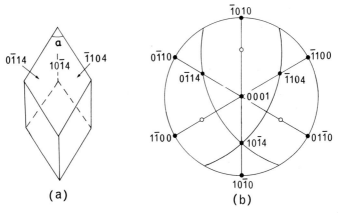

Fig 5.6 Calculation of the angle between the polar edges of the rhombohedron {10$\bar{1}$4} in the trigonal system, the unit-cell dimensions being known. The rhombohedron is shown in perspective in (a) and in stereographic projection in (b).

The *polar edges* of a rhombohedron are those edges that meet in the triad as shown for this case in the sketch Fig 5.6(a), where the required angle for the face (10$\bar{1}$4) is marked α. The polar edges of the face (10$\bar{1}$4) are the zone axes of the zone [(10$\bar{1}$4), (0$\bar{1}$14)] and [(10$\bar{1}$4), ($\bar{1}$104)] displayed on Fig 5.6(b). The zone axis [(10·4), (0$\bar{1}$·4)] is given by application of the Zone Equation as [$\bar{4}$41] and the zone axis [(10·4), ($\bar{1}$1·4)] is [$\bar{4}$81]. Therefore the required angle α is given by

$$\cos \alpha = \frac{t_{[\bar{4}41]} \cdot t_{[\bar{4}81]}}{|t_{[\bar{4}41]}| \, |t_{[\bar{4}81]}|}$$

The metric tensor in the trigonal system is

$$\begin{bmatrix} a^2 & -\tfrac{1}{2}a^2 & 0 \\ -\tfrac{1}{2}a^2 & a^2 & 0 \\ 0 & 0 & c^2 \end{bmatrix}$$

$$t_{[\bar{4}41]} \cdot t_{[\bar{4}81]} = [\bar{4}41] \begin{bmatrix} a^2 & -\tfrac{1}{2}a^2 & 0 \\ -\tfrac{1}{2}a^2 & a^2 & 0 \\ 0 & 0 & c^2 \end{bmatrix} \begin{bmatrix} \bar{4} \\ \bar{8} \\ 1 \end{bmatrix}$$

$$= [\bar{4}41] \begin{bmatrix} 0 \\ -6a^2 \\ c^2 \end{bmatrix}$$

$$= -24a^2 + c^2$$

$$t_{[\bar{4}41]} \cdot t_{[\bar{4}41]} = [\bar{4}41] \begin{bmatrix} a^2 & -\tfrac{1}{2}a^2 & 0 \\ -\tfrac{1}{2}a^2 & a^2 & 0 \\ 0 & 0 & c^2 \end{bmatrix} \begin{bmatrix} \bar{4} \\ 4 \\ 1 \end{bmatrix}$$

$$= [\bar{4}41] \begin{bmatrix} -6a^2 \\ 6a^2 \\ c^2 \end{bmatrix}$$

$$= 48a^2 + c^2$$

therefore $\quad |\mathbf{t}_{[\bar{4}41]}| = \sqrt{48a^2 + c^2}$

$$\mathbf{t}_{[\bar{4}\bar{8}1]} \cdot \mathbf{t}_{[\bar{4}\bar{8}1]} = [\bar{4}\bar{8}1] \begin{bmatrix} a^2 & -\frac{1}{2}a^2 & 0 \\ -\frac{1}{2}a^2 & a^2 & 0 \\ 0 & 0 & c^2 \end{bmatrix} \begin{bmatrix} \bar{4} \\ \bar{8} \\ 1 \end{bmatrix}$$

$$= [\bar{4}\bar{8}1] \begin{bmatrix} 0 \\ -6a^2 \\ c^2 \end{bmatrix}$$

$$= 48a^2 + c^2$$

therefore $\quad |\mathbf{t}_{[\bar{4}\bar{8}1]}| = \sqrt{48a^2 + c^2}$

which is consistent with the equivalence of $[\bar{4}41]$ and $[\bar{4}\bar{8}1]$ in the point group $\bar{3}m$, evident from Fig 5.6(b).
Therefore

$$\alpha = \cos^{-1}\left\{\frac{-24a^2 + c^2}{48a^2 + c^2}\right\}$$

$$= \cos^{-1}\left\{\frac{-306 \cdot 525}{1486 \cdot 283}\right\}$$

$$= \cos^{-1}(-0 \cdot 20624)$$

$$= 101 \cdot 90°$$

This sort of matrix method is particularly well suited to computer calculation.

Interplanar angles

It is customary to take as the angle between two sets of lattice planes the angle between the normals to the two sets of planes. That angle can be determined by means of the scalar product of the vectors parallel to the normals.

Consider the lattice plane of the set (hkl) which makes intercepts a/h, b/k, c/l on the reference axes x, y, z respectively (Fig 5.7). The vectors

$$\mathbf{p}_1 = \frac{1}{h}\mathbf{a} - \frac{1}{k}\mathbf{b}$$

$$\mathbf{p}_2 = \frac{1}{k}\mathbf{b} - \frac{1}{l}\mathbf{c}$$

$$\mathbf{p}_3 = \frac{1}{l}\mathbf{c} - \frac{1}{h}\mathbf{a}$$

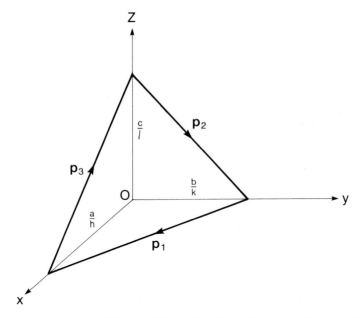

Fig 5.7 The first lattice plane of the set (*hkl*) out from the origin makes intercepts *a/h, b/k, c/l* on the reference axes and intersects the planes (001), (100), (010) through the origin in $\mathbf{p}_1$, $\mathbf{p}_2$, and $\mathbf{p}_3$ respectively.

are all three parallel to the plane (*hkl*) and therefore the direction of the normal to the plane is given by the vector products $\mathbf{p}_1 \wedge \mathbf{p}_2$, $\mathbf{p}_2 \wedge \mathbf{p}_3$, and $\mathbf{p}_3 \wedge \mathbf{p}_1$. Writing the normal to the plane (*hkl*) as $\mathbf{n}$ then

$$\mathbf{n} = \mathbf{p}_1 \wedge \mathbf{p}_2$$

$$= \left(\frac{1}{h}\mathbf{a} - \frac{1}{k}\mathbf{b}\right) \wedge \left(\frac{1}{k}\mathbf{b} - \frac{1}{l}\mathbf{c}\right)$$

$$= \frac{1}{hk}\mathbf{a} \wedge \mathbf{b} - \frac{1}{hl}\mathbf{a} \wedge \mathbf{c} + \frac{1}{kl}\mathbf{b} \wedge \mathbf{c}$$

$$= \frac{1}{hk}\mathbf{a} \wedge \mathbf{b} + \frac{1}{kl}\mathbf{b} \wedge \mathbf{c} + \frac{1}{lh}\mathbf{c} \wedge \mathbf{a}$$

Multiplication by *hkl* yields the result that the normal to (*hkl*) is parallel to $h\mathbf{b} \wedge \mathbf{c} + k\mathbf{c} \wedge \mathbf{a} + l\mathbf{a} \wedge \mathbf{b}$. This result can be used to calculate the vectors $\mathbf{n}_1$ normal to $(h_1k_1l_1)$ and $\mathbf{n}_2$ normal to $(h_2k_2l_2)$ and then after evaluation of $\mathbf{n}_1 . \mathbf{n}_2$ the interplanar angle $(h_1k_1l_1):(h_2k_2l_2)$ can be calculated in any crystal system. This method is valuable in principle; but, as we shall see in the ensuing paragraphs, the concept of the reciprocal lattice provides a very much easier means of calculating interplanar angles.

The reciprocal lattice
The reciprocal lattice concept provides a means of interpretation of diffraction patterns which is at once elegant, powerful and—once one has overcome the initial conceptual difficulty—easy to use. That is its principal use; but it has other applications in the interpretation of certain physical properties of crystalline solids and, which is why we introduce the reciprocal lattice at this point, it provides a neat

and simple means of calculating interplanar angles and interplanar spacings. We defer general discussion of the reciprocal lattice to chapter 6, where it will be applied immediately to the study of diffraction by crystalline solids; here we introduce the concept and restrict our development of it to its use as a powerful tool for crystallographic calculations. We have seen that the calculation of the magnitudes of lattice vectors and of the angles between lattice vectors by vector methods is very straightforward; we now proceed to show that the reciprocal lattice concept provides analogous and equally straightforward methods for the calculation of the spacings between lattice planes and of the angles between lattice planes.

It was shown in the previous section that the normal to the set of lattice planes (hkl) is parallel to the vector

$$h\mathbf{b} \wedge \mathbf{c} + k\mathbf{c} \wedge \mathbf{a} + l\mathbf{a} \wedge \mathbf{b}$$

Now $\mathbf{b} \wedge \mathbf{c}$ is a vector normal to y and z and is therefore normal to the lattice plane (100). Similarly $\mathbf{c} \wedge \mathbf{a}$ and $\mathbf{a} \wedge \mathbf{b}$ are normal respectively to the planes (010) and (001).

The volume of a unit-cell (Fig 5.8) is the product of the area of one face of the unit-cell, for example the (001) face, and the perpendicular distance between that face and the corresponding face on the other side of the unit-cell, which would then be $d_{(001)}$, the spacing of the (001) planes. The vector product $\mathbf{a} \wedge \mathbf{b}$ is a vector normal to (001) and of magnitude $ab\sin\gamma$ equal to the area of the (001) face of the unit-cell. The height of the unit-cell, that is the (001) interplanar spacing $d_{(001)} = \mathbf{c}.\hat{\mathbf{n}}$, where $\hat{\mathbf{n}}$ is a unit vector normal to the (001) face. So the volume of the unit-cell is $\mathbf{c}.\mathbf{a} \wedge \mathbf{b}$. But we could alternatively have considered the (100) or (010) faces and the corresponding interplanar spacings to give

$$V = \mathbf{a}.\mathbf{b} \wedge \mathbf{c} = \mathbf{b}.\mathbf{c} \wedge \mathbf{a} = \mathbf{c}.\mathbf{a} \wedge \mathbf{b} \tag{19}$$

Now the vector $\mathbf{a} \wedge \mathbf{b}/\mathbf{c}.\mathbf{a} \wedge \mathbf{b}$, denoted $\mathbf{c}^*$, is normal to (001) and of magnitude $1/\mathbf{c}.\hat{\mathbf{n}} = 1/d_{(001)}$. Similarly $\mathbf{a}^* = \mathbf{b} \wedge \mathbf{c}/\mathbf{a}.\mathbf{b} \wedge \mathbf{c}$ and $\mathbf{b}^* = \mathbf{c} \wedge \mathbf{a}/\mathbf{b}.\mathbf{c} \wedge \mathbf{a}$ are vectors normal to (100) and (010) respectively and of magnitude $1/d_{(100)}$ and $1/d_{(010)}$. We have already shown that $h\mathbf{b} \wedge \mathbf{c} + k\mathbf{c} \wedge \mathbf{a} + l\mathbf{a} \wedge \mathbf{b}$ is a vector normal to the (hkl) plane.

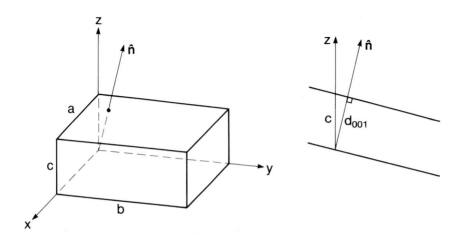

Fig. 5.8 The interplanar spacing $d_{(001)} = \mathbf{c}.\hat{\mathbf{n}}$ and the volume of a unit-cell can be expressed as $\mathbf{c}.\mathbf{a} \wedge \mathbf{b}$, where $|\mathbf{a} \wedge \mathbf{b}|$ is the area of the (001) face.

Dividing through by V and denoting the resulting vector as $\mathbf{d}^*_{hkl}$ we obtain

$$\mathbf{d}^*_{hkl} = \frac{h}{V}\mathbf{b} \wedge \mathbf{c} + \frac{k}{V}\mathbf{c} \wedge \mathbf{a} + \frac{l}{V}\mathbf{a} \wedge \mathbf{b}$$

Application of equations (19) yields

$$\mathbf{d}^*_{hkl} = h\frac{\mathbf{b} \wedge \mathbf{c}}{\mathbf{a}.\mathbf{b} \wedge \mathbf{c}} + k\frac{\mathbf{c} \wedge \mathbf{a}}{\mathbf{b}.\mathbf{c} \wedge \mathbf{a}} + l\frac{\mathbf{a} \wedge \mathbf{b}}{\mathbf{c}.\mathbf{a} \wedge \mathbf{b}}$$

i.e. $\mathbf{d}^*_{hkl} = h\mathbf{a}^* + k\mathbf{b}^* + l\mathbf{c}^*$ (20)

Since h, k, l are integers, the vectors $\mathbf{d}^*_{hkl}$ drawn from a common origin form a lattice, the reciprocal lattice, whose base vectors are $\mathbf{a}^*$, $\mathbf{b}^*$, and $\mathbf{c}^*$. The reference axes of the reciprocal lattice are denoted x^*, y^*, and z^* and its interaxial angles are $\alpha^* = \mathbf{b}^*:\mathbf{c}^* = (010):(001)$, $\beta^* = \mathbf{c}^*:\mathbf{a}^* = (001):(100)$, and $\gamma^* = \mathbf{a}^*:\mathbf{b}^* = (100):(010)$. Now $\mathbf{a}.\mathbf{a}^* = \mathbf{a}.\mathbf{b} \wedge \mathbf{c}/\mathbf{a}.\mathbf{b} \wedge \mathbf{c} = 1$ and likewise $\mathbf{b}.\mathbf{b}^* = 1$ and $\mathbf{c}.\mathbf{c}^* = 1$; also $\mathbf{b}.\mathbf{a}^* = \mathbf{b}.\mathbf{b} \wedge \mathbf{c}/\mathbf{a}.\mathbf{b} \wedge \mathbf{c} = 0$ and likewise $\mathbf{c}.\mathbf{a}^* = \mathbf{a}.\mathbf{b}^* = \mathbf{c}.\mathbf{b}^* = \mathbf{a}.\mathbf{c}^* = \mathbf{b}.\mathbf{c}^* = 0$. These nine equations are practically useful as we shall see immediately and in chapter 6.

We have seen that the reciprocal lattice vector $\mathbf{d}^*_{hkl} = h\mathbf{a}^* + k\mathbf{b}^* + l\mathbf{c}^*$ is normal to the plane (hkl) and that the vector $\mathbf{p}_1 = (1/h)\mathbf{a} - (1/k)\mathbf{b}$ and $\mathbf{p}_2 = (1/k)\mathbf{b} - (1/l)\mathbf{c}$ are parallel to the plane (hkl). Therefore

$$\mathbf{d}^*_{hkl}.\mathbf{p}_1 = (h\mathbf{a}^* + k\mathbf{b}^* + l\mathbf{c}^*).\left(\frac{1}{h}\mathbf{a} - \frac{1}{k}\mathbf{b}\right)$$

$$= \mathbf{a}^*.\mathbf{a} - \mathbf{b}^*.\mathbf{b}$$

$$= 0$$

and likewise $\mathbf{d}^*_{hkl}.\mathbf{p}_2 = 0$. The perpendicular distance from the origin of a plane whose intercepts on the reference axes x, y, z are a/h, b/k, c/l is given by $d_{(hkl)} = \mathbf{t}.\hat{\mathbf{n}}$, where $\mathbf{t}$ is the vector from the origin to any point in the plane and $\hat{\mathbf{n}}$ is a unit vector normal to the plane (Fig 5.9). Since $\hat{\mathbf{n}}$ is a unit vector parallel to $\mathbf{d}^*_{hkl}$,

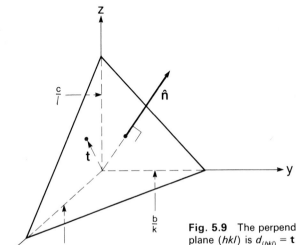

Fig. 5.9 The perpendicular distance from the origin to a plane (hkl) is $d_{(hkl)} = \mathbf{t}.\hat{\mathbf{n}}$, where $\mathbf{t}$ is the vector from the origin to any point in the plane and $\hat{\mathbf{n}}$ is a unit vector normal to the plane.

$$\hat{\mathbf{n}} = \frac{\mathbf{d}^*_{hkl}}{d^*_{hkl}}.$$

So $\quad \mathbf{t} . \dfrac{\mathbf{d}^*_{hkl}}{d^*_{hkl}} = d_{(hkl)}.$

Since $(1/h)\mathbf{a}$ is one possible value of $\mathbf{t}$

$$\frac{1}{h}\mathbf{a} . \mathbf{d}^*_{hkl} = \frac{1}{h}\mathbf{a} . (h\mathbf{a}^* + k\mathbf{b}^* + l\mathbf{c}^*)$$

$$= \frac{1}{h}\mathbf{a} . h\mathbf{a}^*$$

$$= 1$$

and therefore

$$\frac{1}{d^*_{hkl}} = d_{(hkl)}. \tag{21}$$

The vector $\mathbf{d}^*_{hkl}$ is thus normal to the planes (hkl) and its magnitude is equal to the reciprocal of the spacing of the (hkl) lattice planes. Thus each reciprocal lattice point hkl refers to the set of (hkl) lattice planes and lies at the end of the vector $h\mathbf{a}^* + k\mathbf{b}^* + l\mathbf{c}^*$ drawn from the origin, this vector lying in the direction of the normal to the set of lattice planes and having a magnitude equal to the reciprocal of the spacing of the planes of the set.

We now proceed to show that the reciprocal of the reciprocal lattice is the direct lattice. The reciprocal lattice point[3] hkl represents the set of planes (hkl) in real space, the normal to the planes being $\mathbf{d}^*_{hkl} = h\mathbf{a}^* + k\mathbf{b}^* + l\mathbf{c}^*$ and their interplanar spacing being $d_{(hkl)} = 1/d^*_{hkl}$. Correspondingly in the direct lattice the vector $\mathbf{t}_{[UVW]} = U\mathbf{a} + V\mathbf{b} + W\mathbf{c}$ is normal to a set of equally spaced planes of reciprocal lattice points whose spacing is $t^*_{[UVW]} = 1/t_{[UVW]}$. That this is so may quickly be seen by noting that the lattice vector $\mathbf{t}_{[UVW]} = U\mathbf{a} + V\mathbf{b} + W\mathbf{c}$ is the zone axis of the zone of planes (hkl) such that $hU + kV + lW = 0$. Since these planes are necessarily parallel to the zone axis, their normals, which are parallel to the reciprocal lattice vector $\mathbf{d}^*_{hkl} = h\mathbf{a}^* + k\mathbf{b}^* + l\mathbf{c}^*$, must be perpendicular to the zone axis. So

$$\mathbf{d}^*_{hkl} . \mathbf{t}_{[UVW]} = (h\mathbf{a}^* + k\mathbf{b}^* + l\mathbf{c}^*) . (U\mathbf{a} + V\mathbf{b} + W\mathbf{c})$$

$$= hU + kV + lW$$

$$= 0$$

The projection of a general reciprocal lattice point hkl on the zone axis $[UVW]$ is $\mathbf{d}^*_{hkl} . \hat{\mathbf{n}}$, where $\hat{\mathbf{n}}$ is a unit vector parallel to $\mathbf{t}_{[UVW]}$,

i.e. $\quad \hat{\mathbf{n}} = \dfrac{1}{t_{[UVW]}} \mathbf{t}_{[UVW]}.$

Thus $\quad \mathbf{d}^*_{hkl} . \hat{\mathbf{n}} = \dfrac{1}{t_{[UVW]}} (h\mathbf{a}^* + k\mathbf{b}^* + l\mathbf{c}^*) . (U\mathbf{a} + V\mathbf{b} + W\mathbf{c})$

[3] The indices of reciprocal lattice points are conventionally not enclosed in brackets: hkl denotes the reciprocal lattice point, (hkl) the corresponding lattice plane in real space.

$$= \frac{1}{t_{[UVW]}}(hU + kV + lW)$$

And, since h, k, l and U, V, W are by definition integers, the projection of any reciprocal lattice point hkl on $\mathbf{t}_{[UVW]}$ will be distant $m/t_{[UVW]}$ from the origin, where m may be a positive or negative integer or zero. Thus the lattice vector $\mathbf{t}_{[UVW]} = U\mathbf{a} + V\mathbf{b} + W\mathbf{c}$ is normal to a set of equally spaced planes of reciprocal lattice points whose interplanar spacing is $t^*_{[UVW]} = 1/t_{[UVW]}$.

It is left to the reader to particularize the argument and show that $\mathbf{a}$, $\mathbf{b}$, and $\mathbf{c}$ are related to $\mathbf{a}^*$, $\mathbf{b}^*$, and $\mathbf{c}^*$ in precisely the same way as we showed earlier that $\mathbf{a}^*$, $\mathbf{b}^*$ and $\mathbf{c}^*$ were related to $\mathbf{a}$, $\mathbf{b}$, and $\mathbf{c}$, that is to show that

$$\mathbf{a} = \frac{\mathbf{b}^* \wedge \mathbf{c}^*}{\mathbf{a}^* . \mathbf{b}^* \wedge \mathbf{c}^*}, \qquad \mathbf{b} = \frac{\mathbf{c}^* \wedge \mathbf{a}^*}{\mathbf{b}^* . \mathbf{c}^* \wedge \mathbf{a}^*}, \quad \text{and} \quad \mathbf{c} = \frac{\mathbf{a}^* \wedge \mathbf{b}^*}{\mathbf{c}^* . \mathbf{a}^* \wedge \mathbf{b}^*} \qquad (22)$$

It is obvious that the reciprocal lattice parameters, a^*, b^*, c^*, α^*, β^*, γ^*, are related to those of the direct lattice, a, b, c, α, β, γ. The relationship in the most general case of a triclinic crystal is best derived by spherical trigonometry (Fig 5.10). The reciprocal lattice axis x^* is normal to the plane (100) and is thus coincident with the pole of the great circle containing the y and z axes of the direct lattice. Similarly, y^* is normal to (010) and coincident with the pole of the great circle containing x and z, while z^* is normal to (001) and coincident with the pole of the great circle containing x and y. The interaxial angles of the reciprocal lattice are therefore $\alpha^* = (010):(001)$, $\beta^* = (100):(001)$, and $\gamma^* = (100):(010)$. The interaxial angle α of the direct lattice, being the angle between the y and z axes, is the angle between the zone [010], which contains (100) and (001), and the zone [001], which contains (100) and (010); α is thus

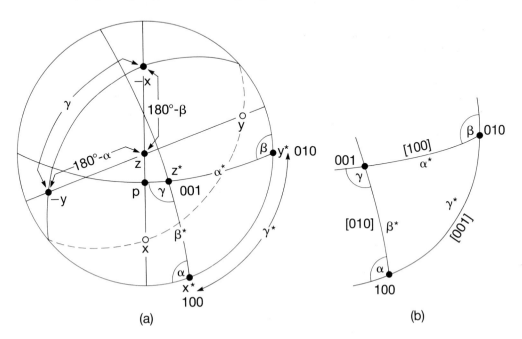

Fig 5.10 The direct and reciprocal axes and the principal zones of a triclinic crystal are shown in stereographic projection in (a). The triangle (100), (010), (001) whose angles are $\pi - \alpha$, $\pi - \beta$, $\pi - \gamma$ and whose sides are α^*, β^*, γ^* is shown in (b).

the external angle at (100) of the spherical triangle (100), (010), (001) shown in Fig 5.10(b). Similarly, $\beta = [100]:[001]$ is the external angle of the same triangle at (010) and $\gamma = [100]:[010]$ is the external angle at (001). The sides of this spherical triangle are the interaxial angles of the reciprocal lattice and the corner angles are the supplements of the interaxial angles of the direct lattice.

Equation (1) yields

$$\cos \alpha^* = \cos \beta^* \cos \gamma^* + \sin \beta^* \sin \gamma^* \cos (180° - \alpha)$$

i.e.

$$\cos \alpha = \frac{\cos \beta^* \cos \gamma^* - \cos \alpha^*}{\sin \beta^* \sin \gamma^*} \tag{23}$$

and equation (4) yields

$$\cos (180° - \alpha) = -\cos (180° - \beta) \cos (180° - \gamma) \\ + \sin (180° - \beta) \sin (180° - \gamma) \cos \alpha^*$$

i.e.

$$\cos \alpha^* = \frac{\cos \beta \cos \gamma - \cos \alpha}{\sin \beta \sin \gamma} \tag{24}$$

and similar expressions for β, γ, β^*, γ^* as shown in Table 5.2.

The relationships between direct and reciprocal lattice dimensions are derived from expressions such as $\mathbf{c} . \mathbf{c}^* = 1$. Since (010) is the pole of the great circle containing x and z (Fig 5.10) and since z^* makes an angle α^* with y^*, the projection of $\mathbf{c}^*$ on the xz plane is situated at p and is of length $c^* \cos (90° - \alpha^*) = c^* \sin \alpha^*$. Since x is the pole of the great circle $[(010), (001)]$, $x:p = 90°$ and $z:p = \beta - 90°$. The projection of p on z is thus $p \cos (\beta - 90°) = p \sin \beta$ and consequently the projection of c^* on z is $c^* \sin \alpha^* \sin \beta$. Since $\mathbf{c} . \mathbf{c}^* = 1$, $cc^* \sin \alpha^* \sin \beta = 1$ and therefore

$$c^* = \frac{1}{c \sin \alpha^* \sin \beta} \tag{25}$$

Analogous expressions for a^* and b^*, derived in precisely the same way, are listed in Table 5.2.

We have considered the relationships between direct and reciprocal lattice parameters in the most general case, the triclinic system. In all other systems there are simplifications. In the cubic, tetragonal, and orthorhombic systems x, y, and z are orthogonal and so x^*, y^*, and z^* are parallel to x, y, and z respectively as shown in Fig 6.29. In these systems the expression $\mathbf{a} . \mathbf{a}^* = 1$ yields $a^* = 1/a$ and likewise for the other axes.

In the trigonal and hexagonal systems (Fig 5.11) $\alpha = \beta = 90°$, $\gamma = 120°$, and $a = b \neq c$, so that x^* is the normal to $(10\bar{1}0)$ and $\mathbf{a} . \mathbf{a}^* = aa^* \cos 30° = 1$ and in consequence $a^* = 1/a \cos 30°$; similarly y^* is the normal to $(01\bar{1}0)$ and $b^* = 1/b \cos 30° = 1/a \cos 30°$ since $a = b$; but z^* is parallel to z and so $c^* = 1/c$. It is evident from Fig 5.11 that $\alpha^* = \beta^* = 90°$ and $\gamma^* = 120° - 2 \times 30° = 60°$. It should be noticed that the reciprocal lattice axis which derives from the u-axis in direct space is not used because it is not linearly independent, but that the fourth index is used in the indices of reciprocal lattice points simply to display symmetry relationships.

In the monoclinic system $\alpha = \gamma = 90°$, β is obtuse, and $a \neq b \neq c$; since y^* is parallel

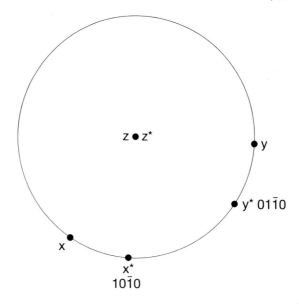

Fig 5.11   Stereogram showing the relationship between direct and reciprocal axis directions in the trigonal and hexagonal systems.

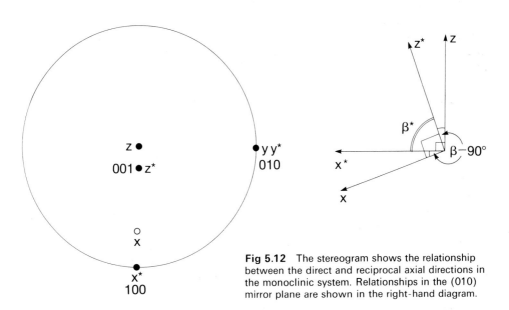

Fig 5.12   The stereogram shows the relationship between the direct and reciprocal axial directions in the monoclinic system. Relationships in the (010) mirror plane are shown in the right-hand diagram.

to y, $b^* = 1/b$ (Fig 5.12); and since x, z, x^*, z^* are coplanar with x^* normal to z and z^* normal to x, $\mathbf{a}.\mathbf{a}^* = aa^*\cos(\beta - 90°) = 1$ so that $a^* = 1/a\sin\beta$ and likewise $c^* = 1/c\sin\beta$. It is evident from Fig 5.12 that $\alpha^* = \gamma^* = 90°$ and $\beta^* = \beta - 2(\beta - 90°) = 180° - \beta$.

The reciprocal lattice parameters having been evaluated, interplanar spacings $d_{(hkl)}$ can simply be calculated by forming the scalar product of the reciprocal lattice vector $\mathbf{d}^*_{(hkl)}$ with itself

$$\frac{1}{d^2_{(hkl)}} = d^{*\,2}_{(hkl)} = \mathbf{d}^*_{(hkl)} \cdot \mathbf{d}^*_{(hkl)}$$

$$= (h\mathbf{a}^* + k\mathbf{b}^* + l\mathbf{c}^*) \cdot (h\mathbf{a}^* + k\mathbf{b}^* + l\mathbf{c}^*) \tag{26}$$

and the interplanar angle θ between the normals to $(h_1 k_1 l_1)$ and $(h_2 k_2 l_2)$ is given by

$$\mathbf{d}^*_{(h_1 k_1 l_1)} \cdot \mathbf{d}^*_{(h_2 k_2 l_2)} = d^*_{(h_1 k_1 l_1)} d^*_{(h_2 k_2 l_2)} \cos \theta \tag{27}$$

These expressions are very easy to use when the reference axes are orthogonal as in the cubic, tetragonal, and orthorhombic systems: there $\alpha^* = \beta^* = \gamma^* = 90°$ and in consequence $a^* = 1/a$, $b^* = 1/b$, $c^* = 1/c$ so that

Table 5.2
Relationships between direct and reciprocal lattice constants

Triclinic	$a^* = \dfrac{1}{a \sin \beta \sin \gamma^*} = \dfrac{1}{a \sin \beta^* \sin \gamma}$	
	$b^* = \dfrac{1}{b \sin \gamma \sin \alpha^*} = \dfrac{1}{b \sin \gamma^* \sin \alpha}$	
	$c^* = \dfrac{1}{c \sin \alpha \sin \beta^*} = \dfrac{1}{c \sin \alpha^* \sin \beta}$	
	$\cos \alpha^* = \dfrac{\cos \beta \cos \gamma - \cos \alpha}{\sin \beta \sin \gamma}$	$\cos \alpha = \dfrac{\cos \beta^* \cos \gamma^* - \cos \alpha^*}{\sin \beta^* \sin \gamma^*}$
	$\cos \beta^* = \dfrac{\cos \gamma \cos \alpha - \cos \beta}{\sin \gamma \sin \alpha}$	$\cos \beta = \dfrac{\cos \gamma^* \cos \alpha^* - \cos \beta^*}{\sin \gamma^* \sin \alpha^*}$
	$\cos \gamma^* = \dfrac{\cos \alpha \cos \beta - \cos \gamma}{\sin \alpha \sin \beta}$	$\cos \gamma = \dfrac{\cos \alpha^* \cos \beta^* - \cos \gamma^*}{\sin \alpha^* \sin \beta^*}$
Monoclinic	$a^* = \dfrac{1}{a \sin \beta}$	$\alpha^* = \alpha = 90°$
	$b^* = \dfrac{1}{b}$	$\beta^* = 180° - \beta$
	$c^* = \dfrac{1}{c \sin \beta}$	$\gamma^* = \gamma = 90°$
Orthorhombic $(a \neq b \neq c)$	$a^* = \dfrac{1}{a}$	$\alpha^* = \alpha = 90°$
Tetragonal $(a = b \neq c)$	$b^* = \dfrac{1}{b}$	$\beta^* = \beta = 90°$
Cubic $(a = b = c)$	$c^* = \dfrac{1}{c}$	$\gamma^* = \gamma = 90°$
Hexagonal $\quad$ $a = b \neq c$ $\quad$ $\gamma = 120°$	$a^* = \dfrac{1}{a \sin 60°}$	$\alpha^* = \alpha = 90°$
Trigonal	$b^* = \dfrac{1}{b \sin 60°}$	$\beta^* = \beta = 90°$
	$c^* = \dfrac{1}{c}$	$\gamma^* = (180° - \gamma) = 60°$

$$\frac{1}{d_{(hkl)}} = d^*_{(hkl)} = \sqrt{(h^2a^{*2} + k^2b^{*2} + l^2c^{*2})}$$

$$= \sqrt{\left(\frac{h^2}{a^2} + \frac{k^2}{b^2} + \frac{l^2}{c^2}\right)} \tag{28}$$

and

$$\cos(h_1k_1l_1):(h_2k_2l_2) = \frac{(h_1\mathbf{a}^* + k_1\mathbf{b}^* + l_1\mathbf{c}^*).(h_2\mathbf{a}^* + k_2\mathbf{b}^* + l_2\mathbf{c}^*)}{\sqrt{(h_1^2a^{*2} + k_1^2b^{*2} + l_1^2c^{*2})}\sqrt{(h_2^2a^{*2} + k_2^2b^{*2} + l_2^2c^{*2})}}$$

$$= \frac{h_1h_2a^{*2} + k_1k_2b^{*2} + l_1l_2c^{*2}}{\sqrt{(h_1^2a^{*2} + k_1^2b^{*2} + l_1^2c^{*2})}\sqrt{(h_2^2a^{*2} + k_2^2b^{*2} + l_2^2c^{*2})}}$$

$$= \frac{\dfrac{h_1h_2}{a^2} + \dfrac{k_1k_2}{b^2} + \dfrac{l_1l_2}{c^2}}{\sqrt{\left(\dfrac{h_1^2}{a^2} + \dfrac{k_1^2}{b^2} + \dfrac{l_1^2}{c^2}\right)}\sqrt{\left(\dfrac{h_2^2}{a^2} + \dfrac{k_2^2}{b^2} + \dfrac{l_2^2}{c^2}\right)}} \tag{29}$$

In the tetragonal system, where $a = b$, some degree of simplification is achieved,

$$\cos(h_1k_1l_1):(h_2k_2l_2) = \frac{h_1h_2 + k_1k_2 + (c^{*2}/a^{*2})l_1l_2}{\sqrt{(h_1^2 + k_1^2 + (c^{*2}/a^{*2})l_1^2)}\sqrt{(h_2^2 + k_2^2 + (c^{*2}/a^{*2})l_2^2)}} \tag{30}$$

$$= \frac{h_1h_2 + k_1k_2 + (a^2/c^2)l_1l_2}{\sqrt{(h_1^2 + k_1^2 + (a^2/c^2)l_1^2)}\sqrt{(h_2^2 + k_2^2 + (a^2/c^2)l_2^2)}} \tag{31}$$

which becomes particularly evident in the special case where $l_1 = l_2 = 0$,

$$\cos(h_1k_10):(h_2k_20) = \frac{h_1h_2 + k_1k_2}{\sqrt{(h_1^2 + k_1^2)}\sqrt{(h_2^2 + k_2^2)}} \tag{32}$$

In the cubic system, where $a = b = c$

$$\cos(h_1k_1l_1):(h_2k_2l_2) = \frac{h_1h_2 + k_1k_2 + l_1l_2}{\sqrt{(h_1^2 + k_1^2 + l_1^2)}\sqrt{(h_2^2 + k_2^2 + l_2^2)}} \tag{33}$$

Comparison of equations (13) and (33) makes it immediately apparent that in the cubic system the normal to the plane (pqr) is parallel to the zone axis $[pqr]$ and therefore the normals to planes in the zone $[pqr]$ are coplanar with (pqr). In the tetragonal system this is so only for $r = 0$ and in other systems only for more restrictive conditions. Figure 5.13 shows the poles of the (110) and (111) face normals together with the [110] and [111] zones and zone axes in cubic, tetragonal, and orthorhombic cases to emphasize this important point.

The reciprocal metric tensor

In crystal systems with non-orthogonal axes, the triclinic, monoclinic, trigonal, and hexagonal systems, equations (26) and (27) cannot conveniently be used directly. As we found when dealing with the calculation of angles between lattice directions, that is interzonal angles, the use of matrix methods is advantageous. The reciprocal metric tensor is a 3×3 symmetric matrix, the elements of which are the scalar products of the base vectors of the reciprocal lattice,

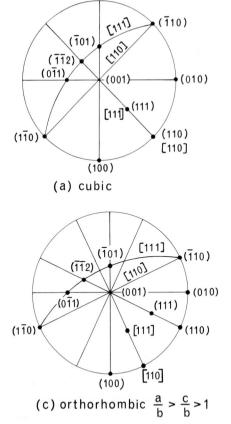

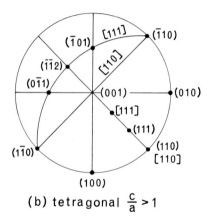

(a) cubic

(b) tetragonal $\frac{c}{a} > 1$

(c) orthorhombic $\frac{a}{b} > \frac{c}{b} > 1$

Fig. 5.13 The normal to the plane (pqr) is parallel to the zone axis $[pqr]$ generally in the cubic system (a), when $r = 0$ in the tetragonal system (b), and only for (100), (010), (001) and their opposites in the orthorhombic system (c).

$$\begin{bmatrix} a^* . a^* & a^* . b^* & a^* . c^* \\ b^* . a^* & b^* . b^* & b^* . c^* \\ c^* . a^* & c^* . b^* & c^* . c^* \end{bmatrix}$$

The reciprocal metric tensor for each of the seven crystal systems is shown in Table 5.3.

Table 5.3
The reciprocal metric tensor in the seven crystal systems

Triclinic	$\begin{bmatrix} a^{*2} & a^*b^* \cos \gamma^* & a^*c^* \cos \beta^* \\ a^*b^* \cos \gamma^* & b^{*2} & b^*c^* \cos \alpha^* \\ a^*c^* \cos \beta^* & b^*c^* \cos \alpha^* & c^{*2} \end{bmatrix}$		
Monoclinic	$\begin{bmatrix} a^{*2} & 0 & a^*c^* \cos \beta^* \\ 0 & b^{*2} & 0 \\ a^*c^* \cos \beta^* & 0 & c^{*2} \end{bmatrix}$		
Trigonal and hexagonal	$\begin{bmatrix} a^{*2} & \frac{1}{2}a^{*2} & 0 \\ \frac{1}{2}a^{*2} & a^{*2} & 0 \\ 0 & 0 & c^{*2} \end{bmatrix}$		

Orthorhombic
$$\begin{bmatrix} a^{*2} & 0 & 0 \\ 0 & b^{*2} & 0 \\ 0 & 0 & c^{*2} \end{bmatrix}$$

Tetragonal
$$\begin{bmatrix} a^{*2} & 0 & 0 \\ 0 & a^{*2} & 0 \\ 0 & 0 & c^{*2} \end{bmatrix}$$

Cubic
$$\begin{bmatrix} a^{*2} & 0 & 0 \\ 0 & a^{*2} & 0 \\ 0 & 0 & a^{*2} \end{bmatrix}$$

The use of the reciprocal lattice and metric tensor is illustrated by reworking Example (i) done by spherical trigonometry and by an example in the trigonal system.

Example (i)′   To calculate the interfacial angle (100):(311) *in* $BaSO_4$ *which is orthorhombic with* $a = 8\cdot85$ Å, $b = 5\cdot44$ Å, $c = 7\cdot13$ Å.
The angle (100):(311) is the angle between $\mathbf{d}^*_{(100)}$ and $\mathbf{d}^*_{(311)}$ and is given by

$$\mathbf{a}^* . (3\mathbf{a}^* + \mathbf{b}^* + \mathbf{c}^*) = a^* \,|\, 3\mathbf{a}^* + \mathbf{b}^* + \mathbf{c}^* \,|\, \cos(100):(311)$$

therefore

$$\cos(100):(311) = \frac{3a^{*2}}{a^*\sqrt{(9a^{*2} + b^{*2} + c^{*2})}}$$

$$= \frac{3a^*}{\sqrt{(9a^{*2} + b^{*2} + c^{*2})}}$$

$$= \frac{3(8\cdot85)^{-1}}{\sqrt{\{9(8\cdot85)^{-2} + (5\cdot44)^{-2} + (7\cdot13)^{-2}\}}}$$

therefore (100):(311) = 34·30°

Example (iv)   To calculate the interplanar angle (10$\bar{1}$4):($\bar{1}$104) *in a cleavage rhomb* {10$\bar{1}$4} *of calcite,* $CaCO_3$, *which is trigonal (class* $\bar{3}m$) *with* $a = 4\cdot990$ Å, $c = 17\cdot061$ Å.
When using the reciprocal lattice for calculation only three non-coplanar axes are required, so the i index is discarded. The required angle becomes (10·4):($\bar{1}$1·4). The reciprocal lattice parameters are (Table 5.2):

$$a^* = b^* = \frac{1}{4\cdot990\cos 30°} = 0\cdot23140 \text{ Å}^{-1}$$

$$c^* = \frac{1}{17\cdot061} = 0\cdot058613 \text{ Å}^{-1}$$

The reciprocal metric tensor is

$$\begin{bmatrix} 0\cdot053546 & 0\cdot026773 & 0 \\ 0\cdot026773 & 0\cdot053546 & 0 \\ 0 & 0 & 0\cdot0034354 \end{bmatrix}$$

$$\text{Therefore } d^{*2}_{(10\cdot4)} = [104] \begin{bmatrix} 0.053546 & 0.026773 & 0 \\ 0.026773 & 0.053546 & 0 \\ 0 & 0 & 0.0034354 \end{bmatrix} \begin{bmatrix} 1 \\ 0 \\ 4 \end{bmatrix}$$

$$= [104] \begin{bmatrix} 0.053546 \\ 0.026773 \\ 0.013742 \end{bmatrix}$$

$$= 0.108514$$

therefore $d^*_{(10\cdot4)} = 0.32941 \text{ Å}^{-1} = d^*_{(\bar{1}1\cdot4)}$ by symmetry. And

$$d^*_{(10\cdot4)} \cdot d^*_{(\bar{1}1\cdot4)} = [104] \begin{bmatrix} 0.053546 & 0.026773 & 0 \\ 0.026773 & 0.053546 & 0 \\ 0 & 0 & 0.0034354 \end{bmatrix} \begin{bmatrix} \bar{1} \\ 1 \\ 4 \end{bmatrix}$$

$$= [104] \begin{bmatrix} -0.026773 \\ 0.026773 \\ 0.013742 \end{bmatrix}$$

$$= 0.028195$$

therefore

$$\cos(10\cdot4):(\bar{1}1\cdot4) = \frac{0.028195}{0.108514}$$

therefore $(10\cdot4):(\bar{1}1\cdot4) = 74\cdot94°$.

The use of orthogonal axes in non-orthogonal systems

Frequently calculations in non-orthogonal systems can be performed more easily by contriving orthogonal reference axes. This is a particularly advantageous technique when computer programs are employed for the solution of problems in direct or reciprocal lattice geometry. We label the orthogonal reference axes X, Y, Z, take Z parallel to the crystallographic axis z and X parallel to the reciprocal axis x^*; Y is then positioned so as to make a right-handed set (Fig 5.14(a)). Examination of Fig 5.14(b) indicates that the projection of the unit-cell edge a on the XY plane is $a\cos(\beta - 90°) = a\sin\beta$ and, since y^* is perpendicular to the xz plane, the component of a on X is $a\sin\beta\sin\gamma^*$ and on Y is $-a\sin\beta\cos\gamma^*$; the component of a on Z is simply $a\cos\beta$. Similarly it is apparent that the components of the unit-cell edge b on X, Y, Z are respectively 0, $b\sin\alpha$, $b\cos\alpha$ and the components of c on X, Y, Z are respectively 0, 0, c. Hence the vector $t = U\mathbf{a} + V\mathbf{b} + W\mathbf{c}$ becomes $t = A\hat{\mathbf{i}} + B\hat{\mathbf{j}} + C\hat{\mathbf{k}}$ on the orthogonal reference axes X, Y, Z, where $\hat{\mathbf{i}}, \hat{\mathbf{j}}, \hat{\mathbf{k}}$ are unit vectors parallel to X, Y, Z respectively, and

$$\begin{bmatrix} A \\ B \\ C \end{bmatrix} = \begin{bmatrix} a\sin\beta\sin\gamma^* & 0 & 0 \\ -a\sin\beta\cos\gamma^* & b\sin\alpha & 0 \\ a\cos\beta & b\cos\alpha & c \end{bmatrix} \begin{bmatrix} U \\ V \\ W \end{bmatrix} \tag{34}$$

The reciprocal axes X^*, Y^*, Z^* are respectively parallel to X, Y, Z since the axial system is orthogonal. The reciprocal lattice axes x^* and y^* lie in the X^*Y^* plane while the projection of z^* on the YZ plane, lying at the intersection of that plane with the

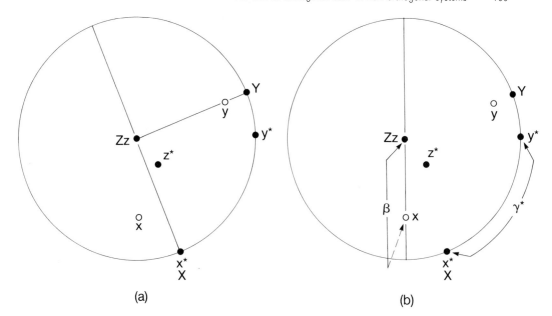

Fig 5.14 The stereogram (a) shows the relationship between the crystallographic axes x, y, z in a non-orthogonal system and orthogonal reference axes X, Y, Z such that Z ‖ z, X ‖ x*, and Y is positioned to make a right-angled set. The stereogram (b) illustrates the calculation of the components of the unit-cell edges a, b, c on X, Y, Z.

x^*z^* plane, is $z^* \sin \beta^*$. The components of a^* on X^*, Y^*, Z^* are therefore $a^*, 0, 0$, the components of b^* are $b^* \cos \gamma^*$, $b^* \sin \gamma^*$, 0, and the components of c^* are $c^* \cos \beta^*$, $c^* \sin \beta^* \sin (\alpha - 90°)$, $c^* \sin \beta^* \cos (\alpha - 90°)$, i.e. $c^* \cos \beta^*$, $-c^* \sin \beta^* \cos \alpha$, $c^* \sin \beta^* \sin \alpha$. Hence the reciprocal lattice vector $\mathbf{d}^* = h\mathbf{a}^* + k\mathbf{b}^* + l\mathbf{c}^*$ becomes $\mathbf{d}^* = H\hat{\mathbf{i}}^* + K\hat{\mathbf{j}}^* + L\hat{\mathbf{k}}^*$ on the orthogonal reciprocal axes X^*, Y^*, Z^* where $\hat{\mathbf{i}}^*$, $\hat{\mathbf{j}}^*$, $\hat{\mathbf{k}}^*$ are unit vectors parallel to X^*, Y^*, Z^* respectively, and

$$
\begin{bmatrix} H \\ K \\ L \end{bmatrix} = \begin{bmatrix} a^* & b^* \cos \gamma^* & c^* \cos \beta^* \\ 0 & b^* \sin \gamma^* & -c^* \sin \beta^* \cos \alpha \\ 0 & 0 & c^* \sin \beta^* \sin \alpha \end{bmatrix} \begin{bmatrix} h \\ k \\ l \end{bmatrix} \tag{35}
$$

It will be apparent to the reader that H, K, L are not necessarily integral.

The matrices required for transformation to orthogonal axes X, Y, Z and X^*, Y^*, Z^* in the four non-orthogonal crystal systems are shown in Table 5.4. The use of these matrices is illustrated by reworking Example (ii) done earlier by spherical trigonometry and again by use of the metric tensor.

Example (ii)″ To calculate the interzonal angle $[100]:[0\bar{2}1]$ *in gypsum.* $CaSO_4 . 2H_2O$, *which is monoclinic with* $a = 5·68$ Å, $b = 15·18$ Å, $c = 6·29$ Å, $\beta = 113·83°$
The matrix for transformation to orthogonal axes is

$$
\begin{bmatrix} 5·196 & 0 & 0 \\ 0 & 15·18 & 0 \\ -2·295 & 0 & 6·29 \end{bmatrix}
$$

The lattice vector [100] thus becomes $5 \cdot 196\hat{i} - 2 \cdot 295\hat{k}$ and [0$\bar{2}$1] becomes $-30 \cdot 36\hat{j} + 6 \cdot 29\hat{k}$, where $\hat{i}$, $\hat{j}$, $\hat{k}$ are unit vectors parallel to X, Y, Z respectively. Therefore

$$\cos[100]:[0\bar{2}1] = \frac{(5 \cdot 196\hat{i} - 2 \cdot 295\hat{k}) \cdot (-30 \cdot 36\hat{j} + 6 \cdot 29\hat{k})}{|5 \cdot 196\hat{i} - 2 \cdot 295\hat{k}| \, | -30 \cdot 36\hat{j} + 6 \cdot 29\hat{k}|}$$

$$= \frac{-2 \cdot 295 \times 6 \cdot 29}{\sqrt{\{(5 \cdot 196)^2 + (2 \cdot 295)^2\}} \sqrt{(30 \cdot 36)^2 + (6 \cdot 29)^2\}}}$$

$$= \frac{-14 \cdot 436}{5 \cdot 680 \times 31 \cdot 00}$$

$$= -0 \cdot 08198.$$

Therefore $[100]:[0\bar{2}1] = 94 \cdot 70°$.

Table 5.4
Matrices for the transformation of the base vectors of crystallographic axes to those of an orthogonal set of axes $Z \parallel z$, $X \parallel x^*$, Y to form a right-handed set.

	Direct space			Reciprocal space		
Triclinic	$a \sin \beta \sin \gamma^*$	0	0	a^*	$b^* \cos \gamma^*$	$c^* \cos \beta^*$
	$-a \sin \beta \cos \gamma^*$	$b \sin \alpha$	0	0	$b^* \sin \gamma^*$	$-c^* \sin \beta^* \cos \alpha$
	$a \cos \beta$	$b \cos \alpha$	c	0	0	$c^* \sin \beta^* \sin \alpha$
Monoclinic	$a \sin \beta$	0	0	a^*	0	$c^* \cos \beta^*$
	0	b	0	0	b^*	0
	$a \cos \beta$	0	c	0	0	$c^* \sin \beta^*$
Trigonal and Hexagonal	$a \sin 60°$	0	0	a^*	$b^* \cos 60°$	0
	$-a \cos 60°$	b	0	0	$b^* \sin 60°$	0
	0	0	c	0	0	c^*

Analytical geometry

Expressions similar to those developed in the preceding section for the calculation of angles between lattice planes and angles between lattice directions can be obtained by the methods of analytical geometry. We restrict our treatment here to orthogonal axes, that is to the orthorhombic, tetragonal, and cubic systems; extension to other systems by the use of orthogonal non-crystallographic reference axes presents no difficulty. For orthogonal axes the basic analytical equations (Appendix E) are the equation to a line whose direction cosines[4] are l, m, and n, where

$$l^2 + m^2 + n^2 = 1 \tag{36}$$

and the expression for the angle θ between two lines whose direction cosines are respectively $l_1 m_1 n_1$ and $l_2 m_2 n_2$

$$\cos \theta = l_1 l_2 + m_1 m_2 + n_1 n_2 \tag{37}$$

[4] The use of l to represent one of the direction cosines is common to many well-known texts of analytical geometry but the possibility of confusion in crystallography with the third Miller index is obvious.

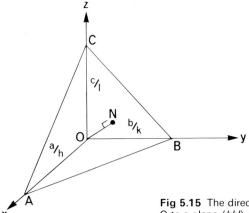

Fig 5.15 The direction cosines of the normal ON from the origin O to a plane (hkl) are $ON/(a/h)$, $ON/(b/k)$, $ON/(c/l)$.

Analytical expression for interplanar angles

The utmost generality that can be achieved with the restriction of orthogonality of axes is by considering an orthorhombic case. We take as mutually perpendicular reference axes x, y, and z springing from an origin at O (Fig 5.15). Let the plane (hkl) intercept the reference axes x, y, and z at A, B, and C so that $OA = a/h$, $OB = b/k$, and $OC = c/l$. If the normal from the origin to the plane (hkl) intersects the plane at N, then the direction cosines of ON will be $\cos \widehat{AON}$, $\cos \widehat{BON}$, $\cos \widehat{CON}$. It is evident from Fig 5.15 that $\cos \widehat{AON} = ON/(a/h)$, $\cos \widehat{BON} = ON/(b/k)$, $\cos \widehat{CON} = ON/(c/l)$, so that the direction cosines of the normal ON become $ON/(a/h)$, $ON/(b/k)$, $ON/(c/l)$. Therefore by equation (36)

$$\frac{h^2}{a^2} + \frac{k^2}{b^2} + \frac{l^2}{c^2} = \frac{1}{ON^2}$$

The direction cosines of ON can then be rewritten in terms of h/a, k/b, and l/c only as

$$\frac{h/a}{\sqrt{\left(\dfrac{h^2}{a^2} + \dfrac{k^2}{b^2} + \dfrac{l^2}{c^2}\right)}}, \quad \frac{k/b}{\sqrt{\left(\dfrac{h^2}{a^2} + \dfrac{k^2}{b^2} + \dfrac{l^2}{c^2}\right)}}, \quad \frac{l/c}{\sqrt{\left(\dfrac{h^2}{a^2} + \dfrac{k^2}{b^2} + \dfrac{l^2}{c^2}\right)}}$$

The angle between two planes ($h_1k_1l_1$) and ($h_2k_2l_2$) then follows from equation (37) as

$$\cos(h_1k_1l_1){:}(h_2k_2l_2) = \frac{\dfrac{h_1h_2}{a^2} + \dfrac{k_1k_2}{b^2} + \dfrac{l_1l_2}{c^2}}{\sqrt{\left(\dfrac{h_1^2}{a^2} + \dfrac{k_1^2}{b^2} + \dfrac{l_1^2}{c^2}\right)}\sqrt{\left(\dfrac{h_2^2}{a^2} + \dfrac{k_2^2}{b^2} + \dfrac{l_2^2}{c^2}\right)}} \tag{38}$$

Equation (38) is equivalent to equation (29) obtained earlier.

Analytical expression for interzonal angles

Again we restrict our treatment to orthogonal axes, taking as our example the orthorhombic system and considering a lattice direction or zone axis $[UVW]$ which is

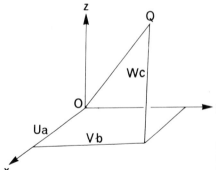

Fig 5.16   The zone axis [UVW] passes through the origin O and through a point Q with coordinates Ua, Vb, Wc.

parallel to the line OQ through the origin O and the point Q with coordinates Ua, Vb, Wc (Fig 5.16). Therefore

$$OQ^2 = U^2a^2 + V^2b^2 + W^2c^2$$

and the direction cosines of OQ, Ua/OQ, Vb/OQ, Wc/OQ can be expressed in terms of U, V, W, a, b, c only as

$$\frac{Ua}{\sqrt{(U^2a^2 + V^2b^2 + W^2c^2)}}, \frac{Vb}{\sqrt{(U^2a^2 + V^2b^2 + W^2c^2)}},$$
$$\frac{Wc}{\sqrt{(U^2a^2 + V^2b^2 + W^2c^2)}}$$

Application of equation (37) yields immediately the angle between the zone axes $[U_1V_1W_1]$ and $[U_2V_2W_2]$,

$$\cos[U_1V_1W_1]:[U_2V_2W_2] = \frac{U_1U_2a^2 + V_1V_2b^2 + W_1W_2c^2}{\sqrt{(U_1^2a^2 + V_1^2b^2 + W_1^2c^2)}\sqrt{(U_2^2a^2 + V_2^2b^2 + W_2^2c^2)}} \quad (39)$$

Equation (39) is the same as equation (11) obtained earlier; those familiar with vector methods will be aware that such equations are more easily derived than remembered.

Equations to the normal to a plane
Whether the reference axes x, y, z are orthogonal or not the plane (hkl) will make intercepts equal to a/h, b/k, and c/l on x, y, and z respectively. If the normal to the plane (hkl) through the origin O intersects the plane in N, then (Fig 5.15)

$$ON = \frac{a}{h}\cos x:(hkl) = \frac{b}{k}\cos y:(hkl) = \frac{c}{l}\cos z:(hkl) \quad (40)$$

where $x:(hkl)$ is the angle between the x-axis and the normal to (hkl) and so on. Equation (40) is of limited application in crystallography but may provide a rapid means of evaluating two such angles when the other is known. When the reference axes are orthogonal equation (40) can be rewritten as

$$\frac{a}{h}\cos(100):(hkl) = \frac{b}{k}\cos(010):(hkl) = \frac{c}{l}\cos(001):(hkl) \quad (41)$$

an expression that is occasionally quite useful.

Example (v)   *To calculate the interplanar angles* (010):(311) *and* (001):(311) *in* $BaSO_4$ *which is orthorhombic with* $a = 8{\cdot}85\,\text{Å}$, $b = 5{\cdot}44\,\text{Å}$, $c = 7{\cdot}13\,\text{Å}$.

The angle (100):(311) was evaluated by spherical trigonometry on p. 135 and therefrom the angles (010):(311) and (001):(311) can quickly be evaluated by the equations to the normal.

$$\cos(010){:}(hkl) = \frac{a}{b}\cdot\frac{k}{h}\cos(100){:}(hkl)$$

Since $(100){:}(311) = 34{\cdot}30°$

$$\cos(010){:}(311) = \frac{8{\cdot}85}{5{\cdot}44}\cdot\frac{1}{3}\cdot\cos 34{\cdot}30°$$

Therefore $(010){:}(311) = 63{\cdot}38°$

And $\cos(001){:}(hkl) = \frac{a}{c}\cdot\frac{l}{h}\cos(100){:}(hkl)$

Hence $\cos(001){:}(311) = \frac{8{\cdot}85}{7{\cdot}13}\cdot\frac{1}{3}\cdot\cos 34{\cdot}30°$

and so $(001){:}(311) = 70{\cdot}02°$

The sine ratio or Miller formulae

The Miller formulae relate the angles between four planes in a zone (i.e., four *tautozonal* planes) and are of general application in all systems.

Let the tautozonal planes P_1, P_2, P_3, and P_4 (Fig 5.17) have indices $(h_1k_1l_1)$, $(h_2k_2l_2)$, $(h_3k_3l_3)$, and $(h_4k_4l_4)$ respectively and denote the interplanar angles $\theta_{12} = P_1{:}P_2$, $\theta_{13} = P_1{:}P_3$ and so on.

Since the planes are tautozonal, their normals are coplanar and consequently the reciprocal lattice points $h_1k_1l_1$, $h_2k_2l_2$, $h_3k_3l_3$, and $h_4k_4l_4$ all lie in a plane through the origin of reciprocal space. Let $\mathbf{p}_1$, $\mathbf{p}_2$, $\mathbf{p}_3$, and $\mathbf{p}_4$ represent the reciprocal lattice vectors $h_1\mathbf{a}^* + k_1\mathbf{b}^* + l_1\mathbf{c}^*$, etc. (Fig 5.18). Then the vector product of any two of these reciprocal lattice vectors will be either parallel or anti-parallel to the zone axis. The vector product $\mathbf{p}_1 \wedge \mathbf{p}_2$ is given by

$$\mathbf{p}_1 \wedge \mathbf{p}_2 = (h_1\mathbf{a}^* + k_1\mathbf{b}^* + l_1\mathbf{c}^*) \wedge (h_2\mathbf{a}^* + k_2\mathbf{b}^* + l_2\mathbf{c}^*)$$

$$= (k_1l_2 - l_1k_2)\mathbf{b}^* \wedge \mathbf{c}^* + (l_1h_2 - h_1l_2)\,\mathbf{c}^* \wedge \mathbf{a}^* + (h_1k_2 - k_1h_2)\mathbf{a}^* \wedge \mathbf{b}^*$$

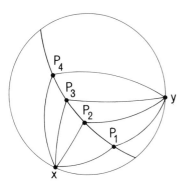

Fig 5.17   The Miller formulae: the four faces P_1, P_2, P_3, P_4 are tautozonal.

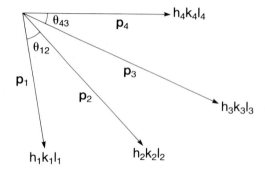

Fig 5.18 The angle θ_{12} is the angle between the reciprocal lattice vectors $\mathbf{p}_1 = h_1\mathbf{a}^* + k_1\mathbf{b}^* + l_1\mathbf{c}^*$ and $\mathbf{p}_2 = h_2\mathbf{a}^* + k_2\mathbf{b}^* + l_2\mathbf{c}^*$. The angle θ_{43} is similarly defined.

Now $\mathbf{b}^* \wedge \mathbf{c}^* = V^*\mathbf{a}$, where $V^* = \mathbf{a}^* \cdot \mathbf{b}^* \wedge \mathbf{c}^*$ is the volume of the reciprocal unit-cell, and similarly $\mathbf{c}^* \wedge \mathbf{a}^* = V^*\mathbf{b}$ and $\mathbf{a}^* \wedge \mathbf{b}^* = V^*\mathbf{c}$. Therefore

$$\mathbf{p}_1 \wedge \mathbf{p}_2 = V^*[(k_1 l_2 - l_1 k_2)\mathbf{a} + (l_1 h_2 - h_1 l_2)\mathbf{b} + (h_1 k_2 - k_1 h_2)\mathbf{c}]$$

which can be expressed more simply by putting $U_{12} = k_1 l_2 - l_1 k_2$ etc.,

$$\mathbf{p}_1 \wedge \mathbf{p}_2 = V^*(U_{12}\mathbf{a} + V_{12}\mathbf{b} + W_{12}\mathbf{c}).$$

Similarly

$$\mathbf{p}_1 \wedge \mathbf{p}_3 = V^*(U_{13}\mathbf{a} + V_{13}\mathbf{b} + W_{13}\mathbf{c}).$$

Since these two vectors must be parallel,

$$\frac{|\mathbf{p}_1 \wedge \mathbf{p}_2|}{|\mathbf{p}_1 \wedge \mathbf{p}_3|} = \frac{U_{12}}{U_{13}} = \frac{V_{12}}{V_{13}} = \frac{W_{12}}{W_{13}}$$

provided U_{12}, etc. are all non zero. Now $|\mathbf{p}_1 \wedge \mathbf{p}_2| = p_1 p_2 \sin \theta_{12}$ and $|\mathbf{p}_1 \wedge \mathbf{p}_3| = p_1 p_3 \sin \theta_{13}$ so that

$$\frac{U_{12}}{U_{13}} = \frac{V_{12}}{V_{13}} = \frac{W_{12}}{W_{13}} = \frac{p_2 \sin \theta_{12}}{p_3 \sin \theta_{13}} \tag{42}$$

In a similar way it can be shown that

$$\frac{U_{42}}{U_{43}} = \frac{V_{42}}{V_{43}} = \frac{W_{42}}{W_{43}} = \frac{p_2 \sin \theta_{42}}{p_3 \sin \theta_{43}} \tag{43}$$

Hence

$$\frac{\sin \theta_{12}/\sin \theta_{13}}{\sin \theta_{42}/\sin \theta_{43}} = \frac{U_{12}/U_{13}}{U_{42}/U_{43}} = \frac{V_{12}/V_{13}}{V_{42}/V_{43}} = \frac{W_{12}/W_{13}}{W_{42}/W_{43}} \tag{44}$$

These are the *Miller formulae*.

The Miller formulae, equation (44), fail only when two of the tautozonal planes are parallel; if any of the angles $\theta_{12}, \theta_{13}, \theta_{42}, \theta_{43}$, is 180° the division which produces (44) becomes indeterminate and if $\theta_{14} = 180°$ equations (42) and (43) become identical. If two or one of the digits, U, V, W, is zero two or one of the equations labelled (44) become indeterminate, but the remaining equation (or equations) holds.

Since U, V, and W are of necessity integers, $(\sin \theta_{12}/\sin \theta_{13})/(\sin \theta_{42}/\sin \theta_{43})$ must be rational. This is a property that can be used, as will be seen in the examples that follow, to determine the indices of one of the four tautozonal planes when those of the other three and all four interplanar angles are known.

The Miller formulae can be expressed in a variety of forms, some leading to greater ease of computation than others; the form, equation (44), in which they are set down here has in our experience the advantage of being the most symmetrical and consequently the easiest to memorize.

We consider two special cases of the Miller formulae which can be particularly useful. First the case when $\theta_{14} = 90°$, then $(\sin\theta_{12}/\sin\theta_{13})/(\sin\theta_{42}/\sin\theta_{43}) = \tan\theta_{12}/\tan\theta_{13}$ and the Miller formulae become

$$\frac{\tan\theta_{12}}{\tan\theta_{13}} = \frac{U_{12}/U_{13}}{U_{42}/U_{43}} = \frac{V_{12}/V_{13}}{V_{42}/V_{43}} = \frac{W_{12}/W_{13}}{W_{42}/W_{43}} \qquad (45)$$

If we now impose a further restriction by allowing $(h_1k_1l_1)$ to be, for instance, (100) and $(h_4k_4l_4)$ to be $(0k_4l_4)$ so that x is normal to (100), then the zone is of the type $[0VW]$.

$\dfrac{U_{12}/U_{13}}{U_{42}/U_{43}}$ is indeterminate,

$$\frac{V_{12}/V_{13}}{V_{42}/V_{43}} = \frac{(l_1h_2 - h_1l_2)/(l_1h_3 - h_1l_3)}{(l_4h_2 - h_4l_2)/(l_4h_3 - h_4l_3)} = \frac{l_2/l_3}{h_2/h_3}$$

and similarly

$$\frac{W_{12}/W_{13}}{W_{42}/W_{43}} = \frac{k_2/k_3}{h_2/h_3}.$$

Therefore

$$\frac{h_2}{h_3}\tan(100):(h_2k_2l_2) = \frac{k_2}{k_3}\tan(100):(h_3k_3l_3) = \frac{l_2}{l_3}\tan(100):(h_3k_3l_3) \quad (46)$$

For the zone $[(010),(h0l)]$ the corresponding equations are

$$\frac{k_2}{k_3}\tan(010):(h_2k_2l_2) = \frac{l_2}{l_3}\tan(010):(h_3k_3l_3) = \frac{h_2}{h_3}\tan(010):(h_3k_3l_3) \quad (47)$$

and for the zone $[(001),(hk0)]$

$$\frac{l_2}{l_3}\tan(001):(h_2k_2l_2) = \frac{h_2}{h_3}\tan(001):(h_3k_3l_3) = \frac{k_2}{k_3}\tan(001):(h_3k_3l_3) \quad (48)$$

The symmetry of equations (46)–(48) is apparent: the ratio of unique indices appears on the left-hand side while the central and right-hand expressions have the ratio of either of the other two indices, in each case for the planes (here P_2 and P_3) with all indices non-zero.

Example (vi)   To evaluate (hkl) given that $P_1(1\bar{1}0)$, $P_2(112)$, $P_3(hkl)$, $P_4(011)$ are tautozonal and that $\theta_{12} = 66·28°$, $\theta_{24} = 42·53°$, $\theta_{23} = 13·55°$ (*Fig 5.19*).

$$\frac{\sin\theta_{12}/\sin\theta_{13}}{\sin\theta_{42}/\sin\theta_{43}} = \frac{\sin 66·28°/\sin 79·83°}{\sin 42·53°/\sin 28·98°} = \frac{2}{3}$$

U_{12} etc. are obtained by cross-multiplication:

$$P_1, P_2$$

$$\begin{array}{c|c} 1 & \bar{1} \\ 1 & 1 \end{array} \times \begin{array}{c} 0 \\ 2 \end{array} \times \begin{array}{c} 1 \\ 1 \end{array} \times \begin{array}{c|c} \bar{1} & 0 \\ 1 & 2 \end{array}$$

$$[\bar{2} \quad \bar{2} \quad 2]$$

$$P_1, P_3$$

$$\begin{array}{c|c} 1 & \bar{1} \\ h & k \end{array} \times \begin{array}{c} 0 \\ l \end{array} \times \begin{array}{c} 1 \\ h \end{array} \times \begin{array}{c|c} \bar{1} & 0 \\ k & l \end{array}$$

$$[\bar{l} \quad \bar{l} \quad k+h]$$

$$P_4, P_2$$

$$\begin{array}{c|c} 0 & 1 \\ 1 & 1 \end{array} \times \begin{array}{c} 1 \\ 2 \end{array} \times \begin{array}{c} 0 \\ 1 \end{array} \times \begin{array}{c|c} 1 & 1 \\ 1 & 2 \end{array}$$

$$[1 \quad 1 \quad \bar{1}]$$

$$P_4, P_3$$

$$\begin{array}{c|c} 0 & 1 \\ h & k \end{array} \times \begin{array}{c} 1 \\ l \end{array} \times \begin{array}{c} 0 \\ h \end{array} \times \begin{array}{c|c} 1 & 1 \\ k & l \end{array}$$

$$[l-k \quad h \quad \bar{h}]$$

Application of equations (44) yields

$$\frac{U_{12}/U_{13}}{U_{42}/U_{43}} = \frac{-2/-l}{1/(l-k)} = \frac{2}{3} \quad \text{and} \quad \frac{V_{12}/V_{13}}{V_{42}/V_{43}} = \frac{-2/-l}{1/h} = \frac{2}{3}$$

i.e. $\dfrac{2(l-k)}{l} = \dfrac{2}{3}$ and $\dfrac{2h}{l} = \dfrac{2}{3}$

i.e. $2l = 3k$ and $3h = l$

Therefore (hkl) is identified as (123).
Check:

$$\frac{W_{12}/W_{13}}{W_{42}/W_{43}} = \frac{2/(k+h)}{-1/-h} = \frac{2/3}{-1/-1} = \frac{2}{3}$$

Example (vii)   *To calculate (100):(211) in* $BaSO_4$ *(barite) which is orthorhombic and has (100):(311)* = 34·30° *(cf. example (i), p. 135).*
The planes (100), (311), (211), (011) are tautozonal (Fig 5.20) with (100):(011) = 90°. Application of equation (46)

$$\frac{h_2}{h_3} \tan (100):(h_2 k_2 l_2) = \frac{k_2}{k_3} \tan (100):(h_3 k_3 l_3)$$

yields for P_2 (311) and P_3 (211)

$$\frac{3}{2} \tan (100):(311) = \tan (100):(211)$$

Therefore $(100):(211) = \tan^{-1} \left(\dfrac{3}{2} \tan 34·30° \right)$

$$= 45·65°$$

Transformation of axes

In the course of investigation of a crystalline solid it may be necessary, for one reason or another, to change the orientation of the reference axes. For instance the crystal morphology of a substance of class 422 may have been referred to reference axes x, and y, which are shown by subsequent X-ray study to be parallel to the diagonal diads of the smallest unit mesh in the (001) plane; a procedure for reindexing the crystal

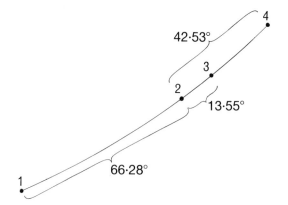

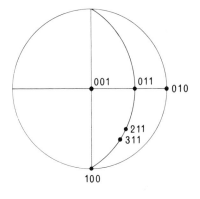

Fig 5.19 Determination of the indices of $P_3(hkl)$ by the Miller formulae, given that P_1 is $(1\bar{1}0)$, P_2 is (112), P_4 is (011) and that the angles θ_{12}, θ_{23}, and θ_{24} are known.

Fig 5.20 Calculation of (100) : (211) for an orthorhombic crystal, given (100) : (311), by the Miller formulae.

faces in terms of the true unit-cell will then be required. Axial transformations are quite commonly needed also in the comparison of related structures of different lattice type or crystal system; an example of this type of use appears at the end of this chapter.

In order to change from one set of reference axes x, y, z, (the 'old' axes) to another set x', y', z' (the 'new' axes), it is necessary first to express the vectors $\mathbf{a}'$, $\mathbf{b}'$, $\mathbf{c}'$, which represent the edges of the 'new' unit-cell, in terms of vector sums of the vectors $\mathbf{a}$, $\mathbf{b}$, $\mathbf{c}$, representing the edges of the 'old' unit-cell,

i.e.
$$\begin{aligned} \mathbf{a}' &= p_1\mathbf{a}+q_1\mathbf{b}+r_1\mathbf{c} \\ \mathbf{b}' &= p_2\mathbf{a}+q_2\mathbf{b}+r_2\mathbf{c} \\ \mathbf{c}' &= p_3\mathbf{a}+q_3\mathbf{b}+r_3\mathbf{c} \end{aligned} \quad \text{or} \quad \begin{bmatrix} \mathbf{a}' \\ \mathbf{b}' \\ \mathbf{c}' \end{bmatrix} = \begin{bmatrix} p_1 & q_1 & r_1 \\ p_2 & q_2 & r_2 \\ p_3 & q_3 & r_3 \end{bmatrix} \begin{bmatrix} \mathbf{a} \\ \mathbf{b} \\ \mathbf{c} \end{bmatrix}$$

where $p_1, q_1, r_1, p_2 \ldots r_3$ are integral or simple fractional coefficients. Such coefficients completely specify the relationship of the 'new' to the 'old' unit-cell vectors. The matrix

$$\begin{bmatrix} p_1 & q_1 & r_1 \\ p_2 & q_2 & r_2 \\ p_3 & q_3 & r_3 \end{bmatrix} \quad \text{or} \quad p_1q_1r_1/p_2q_2r_2/p_3q_3r_3$$

is called the *transformation matrix*. Transformation matrices are usually represented in print for obvious typographical reasons in the format shown on the right above.

In an exactly similar manner the 'old' unit-cell vectors can be expressed in terms of the 'new' unit-cell vectors, the transformation matrix $P_1R_1R_1/P_2Q_2R_2/P_3Q_3R_3$ relating $\mathbf{a}'$, $\mathbf{b}'$, $\mathbf{c}'$ to $\mathbf{a}$, $\mathbf{b}$, $\mathbf{c}$ by

$$\begin{bmatrix} \mathbf{a} \\ \mathbf{b} \\ \mathbf{c} \end{bmatrix} = \begin{bmatrix} P_1 & Q_1 & R_1 \\ P_2 & Q_2 & R_2 \\ P_3 & Q_3 & R_3 \end{bmatrix} \begin{bmatrix} \mathbf{a}' \\ \mathbf{b}' \\ \mathbf{c}' \end{bmatrix}$$

The transformation matrices $p_1 q_1 r_1 / p_2 q_2 r_2 / p_3 q_3 r_3$ and $P_1 Q_1 R_1 / P_2 Q_2 R_2 / P_3 Q_3 R_3$ which link the 'old' and the 'new' unit-cells are of course related to each other, the matrix for the transformation of 'new' to 'old' unit-cell vectors being the *inverse* of the matrix for the transformation of 'old' to 'new' unit-cell vectors. The relationship between a matrix and its inverse is fully explored by Buerger (1942, p. 22) and is discussed in all mathematical textbooks which deal with matrix algebra. In the crystallographical context, however, it is almost invariably just as easy to write down one matrix as the other and so we advocate writing down both matrices by inspection rather than just writing down one matrix and forming its inverse.

Transformation of zone axes and of coordinates

Here we make use of the obvious statement that the crystal structure and its lattice are independent of the choice of unit-cell. The lattice vector $\mathbf{t} = U\mathbf{a} + V\mathbf{b} + W\mathbf{c}$ referred to the 'old' axes becomes $\mathbf{t} = U'\mathbf{a}' + V'\mathbf{b}' + W'\mathbf{c}'$ when referred to the 'new' axes. So

$$\mathbf{t} = [UVW] \begin{bmatrix} \mathbf{a} \\ \mathbf{b} \\ \mathbf{c} \end{bmatrix} = [UVW] \begin{bmatrix} P_1 & Q_1 & R_1 \\ P_2 & Q_2 & R_2 \\ P_3 & Q_3 & R_3 \end{bmatrix} \begin{bmatrix} \mathbf{a}' \\ \mathbf{b}' \\ \mathbf{c}' \end{bmatrix}$$

and

$$\mathbf{t} = [U'V'W'] \begin{bmatrix} \mathbf{a}' \\ \mathbf{b}' \\ \mathbf{c}' \end{bmatrix}$$

Therefore

$$[U'V'W'] = [UVW] \begin{bmatrix} P_1 & Q_1 & R_1 \\ P_2 & Q_2 & R_2 \\ P_3 & Q_3 & R_3 \end{bmatrix}$$

and by transposing

$$\begin{bmatrix} U' \\ V' \\ W' \end{bmatrix} = \begin{bmatrix} P_1 & P_2 & P_3 \\ Q_1 & Q_2 & Q_3 \\ R_1 & R_2 & R_3 \end{bmatrix} \begin{bmatrix} U \\ V \\ W \end{bmatrix}$$

This transformation matrix for zone axes is the *transpose* of the reverse transformation ('new' to 'old' instead of 'old' to 'new') for reference axes; that is to say it is the matrix for the transformation from 'new' to 'old' axes with the columns and rows interchanged.

The distance of an atom from the origin of coordinates is $\mathbf{r} = x\mathbf{a} + y\mathbf{b} + z\mathbf{c}$, where x, y, z are the coordinates of the atom in the 'old' unit-cell. In terms of the 'new' unit-cell $\mathbf{r} = x'\mathbf{a}' + y'\mathbf{b}' + z'\mathbf{c}'$. These expressions are exactly analogous to those for zone axes: atomic coordinates therefore transform as zone axis symbols $[UVW]$.

Transformation of Miller indices

The scalar product of a given lattice vector $\mathbf{t}$ and a given reciprocal lattice vector $\mathbf{d}*$ must be independent of the choice of unit-cell.

Now $\mathbf{t} \cdot \mathbf{d}* = (U\mathbf{a} + V\mathbf{b} + W\mathbf{c}) \cdot (h\mathbf{a}* + k\mathbf{b}* + l\mathbf{c}*)$

$$= hU + kV + lW.$$

We can therefore write

$$\mathbf{t} . \mathbf{d}^* = [hkl] \begin{bmatrix} U \\ V \\ W \end{bmatrix} = [hkl] \begin{bmatrix} p_1 & p_2 & p_3 \\ q_1 & q_2 & q_3 \\ r_1 & r_2 & r_3 \end{bmatrix} \begin{bmatrix} U' \\ V' \\ W' \end{bmatrix}$$

But in terms of the 'new' unit-cell

$$\mathbf{t} . \mathbf{d}^* = [h'k'l'] \begin{bmatrix} U' \\ V' \\ W' \end{bmatrix}$$

Therefore

$$[h'k'l'] = [hkl] \begin{bmatrix} p_1 & p_2 & p_3 \\ q_1 & q_2 & q_3 \\ r_1 & r_2 & r_3 \end{bmatrix}$$

and transposing

$$\begin{bmatrix} h' \\ k' \\ l' \end{bmatrix} = \begin{bmatrix} p_1 & q_1 & r_1 \\ p_2 & q_2 & r_2 \\ p_3 & q_3 & r_3 \end{bmatrix} \begin{bmatrix} h \\ k \\ l \end{bmatrix}$$

Therefore Miller indices transform in the same way as the unit-cell or base vectors of the lattice.

We illustrate this important result by considering a set of lattice planes (hkl), which will divide the vectors $\mathbf{a}$ into h parts, $\mathbf{b}$ into k parts, and $\mathbf{c}$ into l parts; the general vector $p_1\mathbf{a}+q_1\mathbf{b}+r_1\mathbf{c}$ is therefore divided into $p_1h+q_1k+r_1l$ parts. That this is so is apparent from the two-dimensional case illustrated in Fig 5.21 where a set of lattice lines divides $\mathbf{a}$ into two parts and $\mathbf{b}$ into three parts (i.e. $h = 2, k = 3$ and the lines have indices (23)); the line $\mathbf{OM} = 2\mathbf{a}+\mathbf{b}$ (i.e. $p_1 = 2, q_1 = 1$) is seen to be divided into $2 \times 2 + 1 \times 3 = 7$ parts and the line $\mathbf{ON} = -\mathbf{a}+\mathbf{b}$ (i.e. $p_1 = -1, q_1 = 1$) into $-2+3 = 1$ part. Thus if new axes x', y' are taken respectively parallel to $\mathbf{OM}$ and $\mathbf{ON}$ the lattice lines, (23) on the old axes, become (71) when referred to these new axes. Formally the matrix for transformation from the old to the new axes is

$$\begin{bmatrix} 2 & 1 \\ \bar{1} & 1 \end{bmatrix}$$

hence $$\begin{bmatrix} h' \\ k' \end{bmatrix} = \begin{bmatrix} 2 & 1 \\ \bar{1} & 1 \end{bmatrix} \begin{bmatrix} 2 \\ 3 \end{bmatrix} = \begin{bmatrix} 7 \\ 1 \end{bmatrix}$$

Transformation of reciprocal lattice vectors

The reciprocal lattice vector $\mathbf{d}^*$ is independent of the choice of axes. Referred to 'old' axes

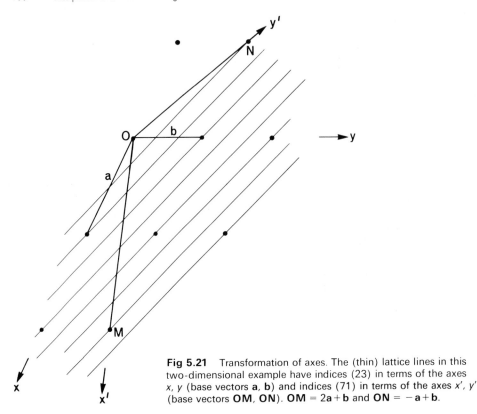

Fig 5.21 Transformation of axes. The (thin) lattice lines in this two-dimensional example have indices (23) in terms of the axes x, y (base vectors **a**, **b**) and indices (71) in terms of the axes x', y' (base vectors **OM**, **ON**). **OM** = 2**a**+**b** and **ON** = −**a**+**b**.

$$\mathbf{d}^* = h\mathbf{a}^* + k\mathbf{b}^* + l\mathbf{c}^*$$

$$= [\mathbf{a}^*\mathbf{b}^*\mathbf{c}^*] \begin{bmatrix} h \\ k \\ l \end{bmatrix}$$

$$= [\mathbf{a}^*\mathbf{b}^*\mathbf{c}^*] \begin{bmatrix} P_1 & Q_1 & R_1 \\ P_2 & Q_2 & R_2 \\ P_3 & Q_3 & R_3 \end{bmatrix} \begin{bmatrix} h' \\ k' \\ l' \end{bmatrix}$$

the inverse transformation matrix being used for the conversion from 'new' to 'old' Miller indices. Referred to 'new' axes

$$\mathbf{d}^* = [(\mathbf{a}')^*(\mathbf{b}')^*(\mathbf{c}')^*] \begin{bmatrix} h' \\ k' \\ l' \end{bmatrix}$$

Therefore

$$[(\mathbf{a}')^*(\mathbf{b}')^*(\mathbf{c}')^*] = [\mathbf{a}^*\mathbf{b}^*\mathbf{c}^*] \begin{bmatrix} P_1 & Q_1 & R_1 \\ P_2 & Q_2 & R_2 \\ P_3 & Q_3 & R_3 \end{bmatrix}$$

and transposing

$$\begin{bmatrix} (\mathbf{a}')^* \\ (\mathbf{b}')^* \\ (\mathbf{c}')^* \end{bmatrix} = \begin{bmatrix} P_1 & P_2 & P_3 \\ Q_1 & Q_2 & Q_3 \\ R_1 & R_2 & R_3 \end{bmatrix} \begin{bmatrix} \mathbf{a}^* \\ \mathbf{b}^* \\ \mathbf{c}^* \end{bmatrix}$$

So reciprocal lattice vectors transform by the transpose of the reverse transformation for reference axes, that is as zone axis symbols.

Summary and examples

The matrices applicable to the various types of transformation are shown in Table 5.5. We conclude with a word of advice: after setting up a transformation matrix and applying it, for example, to a given set of Miller indices, set up the matrix for the reverse transformation and transform back to check.

Table 5.5
Relationships between transformation matrices

$\begin{bmatrix} p_1 q_1 r_1 \\ p_2 q_2 r_2 \\ p_3 q_3 r_3 \end{bmatrix}$	$\begin{bmatrix} P_1 Q_1 R_1 \\ P_2 Q_2 R_2 \\ P_3 Q_3 R_3 \end{bmatrix}$
from old to new axes from old to new Miller indices	from new to old axes from new to old Miller indices
$\begin{bmatrix} P_1 P_2 P_3 \\ Q_1 Q_2 Q_3 \\ R_1 R_2 R_3 \end{bmatrix}$	$\begin{bmatrix} p_1 p_2 p_3 \\ q_1 q_2 q_3 \\ r_1 r_2 r_3 \end{bmatrix}$
from old to new coordinates from old to new zone axes from old to new reciprocal lattice vectors	from new to old coordinates from new to old zone axes from new to old reciprocal lattice vectors

Example (vii)   Relationships between three forms of $BaTiO_3$

$BaTiO_3$ exists in four structural modifications: below $-80\,°C$ it is trigonal, from $-80°$ to $5\,°C$ orthorhombic, from $5°$ to $120\,°C$ tetragonal, and above $120\,°C$ cubic. We first consider the relationship between the cubic and orthorhombic forms. The cubic modification has space group $Pm3m$, unit-cell $a \sim 4\,Å$, and one formula unit per unit-cell; its structure is identical with ideal *perovskite* and is illustrated in Fig 5.22(a). The orthorhombic modification has space group $Bmm2$, unit-cell dimensions, $a = 5·656\,Å$, $b = 3·986\,Å$, $c = 5·675\,Å$, and two formula units per unit-cell. That there is a simple approximate dimensional relationship between the cubic and orthorhombic unit-cells is evident from Fig 5.22(b); this relationship is expressed by the axial transformation matrices:

(I) from cubic to orthorhombic   (II) from orthorhombic to cubic

$$\begin{bmatrix} 1 & 0 & 1 \\ 0 & 1 & 0 \\ \bar{1} & 0 & 1 \end{bmatrix} \qquad \begin{bmatrix} \tfrac{1}{2} & 0 & \tfrac{\bar{1}}{2} \\ 0 & 1 & 0 \\ \tfrac{1}{2} & 0 & \tfrac{1}{2} \end{bmatrix}$$

The appearance of fractions in matrix II is because it refers to a transformation from a non-primitive to a primitive unit-cell. To determine the coordinates of the

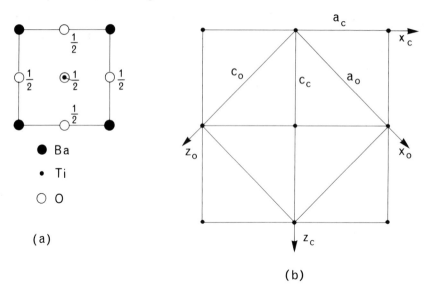

Fig 5.22 The $BaTiO_3$ structures. The plan (a) on (001) shows the *cubic* form of $BaTiO_3$, which has the ideal *perovskite* structure. The plan (b) shows the dimensional relationship between the unit-cells of the orthorhombic (a_o, b_o, c_o) and cubic (a_c, b_c, c_c) forms of $BaTiO_3$.

atoms in the orthorhombic B-cell our first step is to operate the appropriate matrix for transformation of coordinates on the coordinates of the atoms in one cubic unit-cell. Reference to Table 5.5 shows that the matrix for this purpose is matrix II above taken in columns, i.e. from cubic to orthorhombic coordinates

$$\begin{bmatrix} \tfrac{1}{2} & 0 & \tfrac{1}{2} \\[4pt] 0 & 1 & 0 \\[4pt] \bar{\tfrac{1}{2}} & 0 & \tfrac{1}{2} \end{bmatrix}$$

Our second step is to apply the B-centring translation $\tfrac{1}{2}, 0, \tfrac{1}{2}$, to obtain the coordinates of the other half of the atoms in the orthorhombic unit-cell. Thus Ti, which lies at $\tfrac{1}{2}, \tfrac{1}{2}, \tfrac{1}{2}$ in the cubic unit-cell, has coordinates $(\tfrac{1}{2}\cdot\tfrac{1}{2}+0\cdot\tfrac{1}{2}+\tfrac{1}{2}\cdot\tfrac{1}{2})$, $(0\cdot\tfrac{1}{2}+1\cdot\tfrac{1}{2}+0\cdot\tfrac{1}{2})$, $(-\tfrac{1}{2}\cdot\tfrac{1}{2}+0\cdot\tfrac{1}{2}+\tfrac{1}{2}\cdot\tfrac{1}{2})$, i.e. $\tfrac{1}{2}, \tfrac{1}{2}, 0$ and $(\tfrac{1}{2}+\tfrac{1}{2})$, $(\tfrac{1}{2}+0)$, $(0+\tfrac{1}{2})$, i.e. $0, \tfrac{1}{2}, \tfrac{1}{2}$ in the orthorhombic unit-cell. Table 5.6 lists the coordinates of all the atoms in the orthorhombic unit-cell derived in this way. These coordinates obviously refer to the cubic structure described on the orthorhombic unit-cell. The

Table 5.6
Derivation of atomic coordinates in orthorhombic $BaTiO_3$ and in trigonal $BaTiO_3$ from those in cubic $BaTiO_3$

	Cubic	Orthorhombic		Trigonal		
		x, y, z	$\tfrac{1}{2}+x, y, \tfrac{1}{2}+z$	x, y, z	$\tfrac{2}{3}+x, \tfrac{1}{3}+y, \tfrac{1}{3}+z$	$\tfrac{1}{3}+x, \tfrac{2}{3}+y, \tfrac{2}{3}+z$
Ba	$0, 0, 0$	$0, 0, 0$	$\tfrac{1}{2}, 0, \tfrac{1}{2}$	$0, 0, 0$	$\tfrac{2}{3}, \tfrac{1}{3}, \tfrac{1}{3}$	$\tfrac{1}{3}, \tfrac{2}{3}, \tfrac{2}{3}$
Ti	$\tfrac{1}{2}, \tfrac{1}{2}, \tfrac{1}{2}$	$\tfrac{1}{2}, \tfrac{1}{2}, 0$	$0, \tfrac{1}{2}, \tfrac{1}{2}$	$0, 0, \tfrac{1}{2}$	$\tfrac{2}{3}, \tfrac{1}{3}, \tfrac{5}{6}$	$\tfrac{1}{3}, \tfrac{2}{3}, \tfrac{1}{6}$
O_1	$\tfrac{1}{2}, \tfrac{1}{2}, 0$	$\tfrac{1}{4}, \tfrac{1}{2}, \tfrac{3}{4}$	$\tfrac{3}{4}, \tfrac{1}{2}, \tfrac{1}{4}$	$\tfrac{2}{3}, \tfrac{5}{6}, \tfrac{1}{3}$	$\tfrac{1}{3}, \tfrac{1}{6}, \tfrac{2}{3}$	$0, \tfrac{1}{2}, 0$
O_2	$\tfrac{1}{2}, 0, \tfrac{1}{2}$	$\tfrac{1}{2}, 0, 0$	$0, 0, \tfrac{1}{2}$	$\tfrac{5}{6}, \tfrac{1}{3}, \tfrac{1}{3}$	$\tfrac{1}{2}, \tfrac{2}{3}, \tfrac{2}{3}$	$\tfrac{1}{6}, 0, 0$
O_3	$0, \tfrac{1}{2}, \tfrac{1}{2}$	$\tfrac{1}{4}, \tfrac{1}{2}, \tfrac{1}{4}$	$\tfrac{3}{4}, \tfrac{1}{2}, \tfrac{3}{4}$	$\tfrac{1}{6}, \tfrac{5}{6}, \tfrac{1}{3}$	$\tfrac{5}{6}, \tfrac{1}{6}, \tfrac{2}{3}$	$\tfrac{1}{2}, \tfrac{1}{2}, 0$

actual orthorhombic structure is very similar, but has significant differences: the z coordinates of the barium atoms are not precisely 0 and $\frac{1}{2}$, nor are those of oxygen precisely $\frac{1}{4}$ and $\frac{3}{4}$; the effect of these small displacements is to lower the symmetry of the structure from $Pm3m$ (cubic) to $Bmm2$ (orthorhombic).

We turn now to the relationship between the trigonal and cubic forms of BaTiO$_3$. The triple hexagonal unit-cell of trigonal BaTiO$_3$ has $a_h = 5\cdot652$ Å, $c_h = 6\cdot945$ Å; the lattice type is rhombohedral. Its axes x_h, y_h, z_h are respectively parallel to the $[\bar{1}01]$, $[1\bar{1}0]$, and $[111]$ directions of the cubic polymorph (Fig 5.23) and $a_h = b_h \simeq a_c\sqrt{2}$, $c_h \simeq a_c\sqrt{3}$. Therefore

$$\mathbf{a}_h = -\mathbf{a}_c + \mathbf{c}_c$$

$$\mathbf{b}_h = \mathbf{a}_c - \mathbf{b}_c$$

$$\mathbf{c}_h = \mathbf{a}_c + \mathbf{b}_c + \mathbf{c}_c$$

and the axial transformation matrices are

(III) from cubic to trigonal

$$\begin{bmatrix} \bar{1} & 0 & 1 \\ 1 & \bar{1} & 0 \\ 1 & 1 & 1 \end{bmatrix}$$

(IV) from trigonal to cubic

$$\begin{bmatrix} \frac{\bar{1}}{3} & \frac{1}{3} & \frac{1}{3} \\ \frac{\bar{1}}{3} & \frac{2}{3} & \frac{1}{3} \\ \frac{2}{3} & \frac{1}{3} & \frac{1}{3} \end{bmatrix}$$

To obtain the coordinates of the atoms in the trigonal unit-cell it is necessary to operate the transpose of matrix (IV), i.e. $\frac{\bar{1}\bar{1}2}{333}/\frac{1\bar{2}1}{333}/\frac{111}{333}$, on the cubic coordinates to give the coordinates of the atoms associated with the lattice point at 000 and then to apply the rhombohedral lattice translations $\frac{211}{333}$ and $\frac{122}{333}$ to obtain the coordinates of the atoms associated with the other two lattice points of the triple hexagonal unit-cell.

We conclude by re-indexing the faces of a cube {100} of high-temperature cubic BaTiO$_3$ on the orthorhombic and trigonal unit-cells. To transform Miller indices from cubic to orthorhombic we operate matrix I $[101/010/\bar{1}01]$ on the faces (100), (010), (001) and their opposites to give $(10\bar{1})$, (010), (101) and their

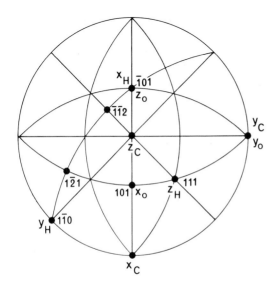

Fig 5.23 The stereogram shows the angular relationship between the axes of the trigonal (triple hexagonal) (x_H, y_H, z_H) orthorhombic (x_o, y_o, z_o) and cubic (x_c, y_c, z_c) unit-cells of three BaTiO$_3$ polymorphs. Indices shown refer to the cubic unit-cell.

opposites, a combination of the forms $\{010\}$ and $\{101\}$. To transform Miller indices from cubic to trigonal matrix III is operated on the same three cubic faces to give $(\bar{1}1\cdot1)$, $(0\bar{1}\cdot1)$, $(10\cdot1)$; conversion to Miller-Bravais indices gives $(\bar{1}101)$, $(0\bar{1}11)$, $(10\bar{1}1)$ and of course the opposites of these three faces so that it is evident that the cubic form $\{100\}$ becomes the trigonal form $\{10\bar{1}1\}$.

Problems

5.1 Evaluate $t_{[112]}$ for tungsten, which is cubic I with $a = 3\cdot17$ Å.

5.2 Determine d_{421} for KCl, which is cubic with $a = 6\cdot27$ Å.

5.3 Yoderite is monoclinic with $a = 8\cdot035$, $b = 5\cdot805$, $c = 7\cdot346$ Å, $\beta = 105\cdot63°$. Calculate the reciprocal lattice constants and the spacing of the (111) planes.

5.4 High-cordierite is hexagonal and the dimensions of its reciprocal cell are $a^* = 0\cdot1181$, $c^* = 0\cdot1068$ Å^{-1}. Determine the dimensions, a and c, of its unit-cell.

5.5 Calculate (100):(111) for (i) copper, which is cubic, (ii) tin, which is tetragonal with $a = 5\cdot82$, $c = 3\cdot17$ Å, (iii) zinc, which is hexagonal with $a = 2\cdot66$, $c = 4\cdot93$ Å.

5.6 Evaluate the coefficients of the metric tensor for augite, which is monoclinic with $a = 9\cdot73$, $b = 8\cdot91$, $c = 5\cdot29$ Å, $\beta = 105\cdot83°$. Hence determine the angle $[101]:[121]$.

5.7 Set up the matrices for transformation to orthogonal axes in direct and in reciprocal space for quartz, which is trigonal with $a = 4\cdot91$, $c = 5\cdot41$ Å. Hence evaluate the angles $[10\dagger0]:[11\dagger1]$ and $(10\bar{1}0):(11\bar{2}1)$.

5.8 In a cubic crystal a face (hkl) lies in the zone $[11\bar{1}]$ between the faces (101) and (011) at an angle of $19\cdot1°$ to (101). Use the Miller formulae to find the indices of this face, making use of the fact that the geometry of the cubic system implies a face (khl) at $19\cdot1°$ to (011).

5.9 A monoclinic substance has been described in terms of an unconventional I lattice with unit-cell dimensions $a = 9\cdot65$, $b = 8\cdot50$, $c = 5\cdot32$ Å, $\beta = 105\cdot32°$. It can alternatively be referred to a conventional C lattice with unit-cell dimensions $a = 5\cdot32$, $b = 8\cdot50$, $c = 9\cdot71$ Å, $\beta = 106\cdot58°$. Determine the matrices for transformation from the I-cell to the C-cell and *vice versa*. Write down the matrix for the transformation of atomic coordinates from the I-cell to the C-cell.

5.10 The low-temperature orthorhombic structure of praseodymium aluminate, PrAlO$_3$, is related to the ideal perovskite structure, ABO$_3$, which is cubic with atomic coordinates A: $\frac{1}{2}\frac{1}{2}\frac{1}{2}$; B: 000; O: $\frac{1}{2}$00, $0\frac{1}{2}0$, $00\frac{1}{2}$. The transformation matrix from the cubic unit-cell to the orthorhombic unit-cell is $101/020/\bar{1}01$. Deduce the matrix for the transformation from the orthorhombic to the cubic unit-cell. Assuming no distortion of the cubic structure write down the coordinates of the lattice points in the orthorhombic unit-cell and the coordinates of the atoms associated with one such lattice point.

In the low-temperature structure however the atoms are displaced from these ideal positions. The unit-cell dimensions are $a = 5\cdot34$, $b = 7\cdot48$, $c = 5\cdot32$ Å and the space group is $Icmm$. The atomic coordinates have been determined as:

$$(000; \tfrac{1}{2}\tfrac{1}{2}\tfrac{1}{2})+$$

Pr: $\pm(xy\tfrac{1}{2})$ with $x = -0\cdot002$, $y = \tfrac{1}{4}$

Al: $000; 0\tfrac{1}{2}0$

O(1): $\pm(\tfrac{1}{4}y\tfrac{1}{4}); \pm(\tfrac{1}{4}y\tfrac{3}{4})$ with $y = -0\cdot020$

O(2): $\pm(x\tfrac{1}{4}0)$ with $x = 0\cdot040$

On a scale of 20 mm to 1 Å draw a plan of the orthorhombic structure of $PrAlO_3$ projected on (001). Outline the projection of the ideal cubic unit-cell and correlate the actual atomic coordinates with those of the ideal cubic structure.

6
Diffraction of X-rays by crystals

X-radiation is the name given to that part of the electromagnetic spectrum in the wavelength range 0·1 to 500 Å, sandwiched between γ-radiation at shorter wavelengths and ultraviolet radiation on the high wavelength side. We are not concerned immediately either with the general theory of electromagnetic radiation, which crops up in a different context and another wavelength range in volume 2 where references to the general theory are given, or with details of the generation of X-rays (chapter 7). We assume that a source of X-rays, which for crystallographic purposes is usually restricted to the wavelength range 0·5–2·5 Å, is available and proceed to consider the nature of their interaction with matter in general and with crystalline solids in particular.

Interaction of X-rays with matter

When an X-ray beam passes through a material medium its intensity is reduced by the operation of a variety of effects which may be grouped under two general headings, *absorption* and *scattering*.

The *absorption* of X-ray photons by an atom leads among other effects to the ejection of electrons from the inner shells (K, L, or M) of the atom and consequent 'falling in' of electrons from higher energy levels (i.e. levels of less negative energy) to fill the vacancies so created. Such electronic transitions are accompanied by emission of X-rays of definite wavelength which is determined by the difference in energy between the initial and final state of the electron filling the vacancy in an inner shell. Such *fluorescent* X-rays may be reabsorbed by another atom to produce ejection of electrons from shells of higher energy, followed by 'falling in' of electrons from shells of even higher energy and emission of fluorescent X-rays of lower energy and therefore of longer wavelength. Absorption phenomena are utilized in various ways, including X-ray fluorescence analysis.

The scattering of X-rays by an atom may occur in either of two ways, both of which again involve interaction between X-radiation and extranuclear electrons. An X-ray photon passing close to an electron belonging to one of the constituent atoms of the material medium will be deflected by the electromagnetic field of the electron and will impart some of its energy to the electron as kinetic energy. The energy of the deflected X-ray photon will be correspondingly decreased and its wavelength increased. This is *incoherent* scattering which is not our immediate concern.

The second way in which X-rays are scattered by atoms is best considered by treating the incident X-ray beam as a plane wave-front. As the plane wave-front passes through an extranuclear electron belonging to an atom of the material medium it causes the electron to vibrate. The vibrating electron radiates X-rays of the same frequency as the incident beam. Such vibrating electrons act as secondary sources of X-radiation of fixed wavelength and give rise to interference effects. The interference phenomena associated with such *coherent* scattering are the basis of X-ray crystallography and will concern us in this and succeeding chapters.

That X-rays in the wavelength range $0.5–2.5\,\text{Å}$ incident on crystalline solids give rise to observable diffraction patterns is because the distances between adjacent atoms in crystalline solids are on the same scale and because the X-rays are scattered coherently from the extranuclear electrons of the constituent atoms of the solid substance. Simultaneously incoherent scattering, that is scattering with change of wavelength, occurs and that contributes to the background of the diffraction pattern. The study of the diffraction patterns produced by X-rays incident on crystalline materials, in particular single crystals, has made possible the determination of the size and shape of the relevant unit-cell and the coordinates of the atoms within the unit-cell. X-ray diffraction studies have proved to be the most powerful tool for the study of the internal structure of crystalline solids. The remainder of this chapter will be devoted to an elementary treatment of the diffraction of X-rays by single crystals.

Simplifying assumptions

The refractive index of most substances is less than unity by a very small amount, of the order of 10^{-6}, so that refraction of incident and scattered X-radiation at air/crystal interfaces can be neglected except when extremely precise measurement of unit-cell dimensions is required.

Absorption by the crystal of incident and scattered radiation affects the intensities and the directions of the scattered beams. Thermal vibration of the constituent atoms of the crystal will also modify the intensity of the scattered X-radiation. Corrections for both effects are necessary in structure determination and can quite simply be applied. Correction for absorption is necessary when the angular disposition of the scattered beams is to be measured very accurately. In the elementary treatment given in this chapter, it will be assumed that absorption is negligible and that all the constituent atoms of the crystal are at rest.

A scattered X-ray beam may, in the course of its travel through the crystal, be scattered a second time, but only in certain special circumstances is the resultant beam of appreciable intensity.

In general scattered X-ray beams have a phase difference π relative to the incident beam that generated them. Only when the wavelength of the X-radiation is close to an absorption edge (chapter 9) of one of the constituent elements of the crystalline substance is this not so. It will be assumed in this elementary treatment that scattering always introduces a phase change π and, since the phase change is the same for all scattered beams it can be ignored. The origin of the phase change on scattering is discussed later in this chapter.

Finally, it may be noticed that the effects of coherent scattering of X-rays by a crystal are going to be observed at distances from the crystal that are very large compared with X-ray wavelengths; that is to say we are dealing with an example of Fraunhofer, as distinct from Fresnel, diffraction.

Combination of X-rays

Consider a narrow pencil of monochromatic X-rays of wavelength λ, angular frequency ω, and amplitude a travelling in the x direction. A disturbance will be produced at a point x such that[1]

$$\psi = a\cos(\omega t + \phi) \tag{1}$$

where ψ is the displacement in the plane normal to the direction of propagation at time t, and ϕ is the phase of the wave when $t = 0$. It is evident from Fig 6.1 that a wave with phase ϕ is in *advance* of a wave with $\phi = 0$.

Suppose now that such a beam of monochromatic X-rays is divided into two beams, one with zero phase (this merely requires a suitable choice of the zero of the time scale) and the other with phase ϕ. It follows that the second beam must have travelled a shorter distance than the first by an amount equal to $\lambda\phi/2\pi$.

When a number, N, of such waves arrives simultaneously at a given point, the resultant wave is given by the *principle of superposition* as the sum of the individual waves. Thus the resultant ψ_R of the waves $\psi_1, \psi_2, \ldots \psi_{N-1}, \psi_N$ arriving simultaneously at the point is given, if the waves all have the same frequency, ω, by

$$\begin{aligned}
\psi_R &= \psi_1 + \psi_2 + \ldots + \psi_{N-1} + \psi_N \\
&= \sum_1^N \psi_n \\
&= \sum_1^N a_n \cos(\omega t + \phi_n) \\
&= \sum_1^N \{a_n \cos\omega t \cos\phi_n - a_n \sin\omega t \sin\phi_n\} \\
&= \cos\omega t \sum_1^N a_n \cos\phi_n - \sin\omega t \sum_1^N a_n \sin\phi_n
\end{aligned} \tag{2}$$

But the resultant wave motion is described by

$$\begin{aligned}
\psi_R &= a_R \cos(\omega t + \phi_R) \\
&= a_R \cos\omega t \cos\phi_R - a_R \sin\omega t \sin\phi_R
\end{aligned} \tag{3}$$

Therefore $\quad a_R \cos\phi_R = \sum_1^N a_n \cos\phi_n$

and $\quad a_R \sin\phi_R = \sum_1^N a_n \sin\phi_n.$

But, since $\cos^2\phi_R + \sin^2\phi_R = 1$

$$a_R^2 = \left(\sum_1^N a_n \cos\phi_n\right)^2 + \left(\sum_1^N a_n \sin\phi_n\right)^2 \tag{4}$$

and $\quad \tan\phi_R = \dfrac{\sum_1^N a_n \sin\phi_n}{\sum_1^N a_n \cos\phi_n}.$ $\tag{5}$

---

[1] The equation to a simple harmonic wave-motion can be written in various ways: $\psi = a\cos(\omega t + \phi)$ is preferred in X-ray crystallography, $y = A\sin(\omega t + \alpha)$ is one of several forms of the equation in common use in optics. It will be seen later in this chapter that $y = a\cos(\omega t + \phi)$ leads to the most convenient form of the expression for the structure factor.

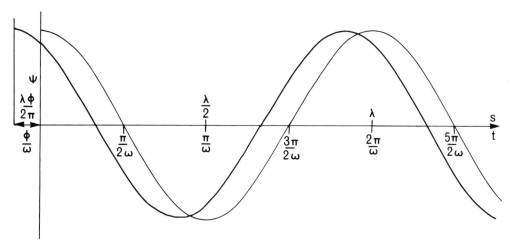

Fig 6.1   Two sinusoidal wave motions with the same frequency and velocity. Displacement ψ in the plane normal to the direction of propagation is plotted vertically while distance traversed s and time taken t are plotted horizontally. The wave motion represented by the thick line is in advance of that represented by the thin line by the phase difference ϕ, which corresponds to a time difference of ϕ/ω and a difference in distance traversed of $\lambda\phi/2\pi$.

Now, since there is no lens capable of refracting X-rays it is impossible to form an image of the crystal structure with the X-rays scattered by it and all that can be done in practice is to make measurements of the intensities and angular disposition of the scattered X-ray beams. Equation (4) provides an expression[2] for the resultant intensity I_R of N waves:

$$I_R = a_R^2 = \left(\sum_1^N a_n \cos \phi_n\right)^2 + \left(\sum_1^N a_n \sin \phi_n\right)^2 \tag{6}$$

To calculate the intensity of the resultant X-ray beam we thus need to know the amplitude and phase of the X-rays scattered by every electron in the crystal in the given direction.

Our ultimate goal is the calculation of the diffraction pattern produced by the interaction of an incident monochromatic X-ray beam with a single crystal and this is achieved by summing over all directions the waves scattered by every electron in the crystal. The summation is taken in steps, the first step being to consider the interaction of the X-ray beam with a single electron. The second step is to consider the scattering produced by all the electrons associated with each atomic species present in the crystal. The third step is to consider the scattering produced by all the atoms in a single unit-cell, having regard to their differing nature and to their spatial distribution within the unit-cell. The fourth and final step is to sum the scattering effect of one unit-cell over all the unit-cells in the crystal. In the course of this stepwise argument it is necessary to assume as a first approximation that each electron in the crystal can be assigned to a particular atom. This is not strictly true because valency electrons may be involved in bond formation and in consequence be shared between atoms; but the majority of electrons in the crystal will be core electrons, each belonging unambiguously to a particular atom, so that the assumption is valid as a first approximation at least.

The procedure outlined above is strictly logical but the steps are not progressive

[2] This form for the intensity expression is preferred at this stage because it gives emphasis to the way in which the amplitude of the scattered beam may be calculated. For the more elegant form of the expression in complex number rotation see later in this chapter.

in order of difficulty. In particular the final step is relatively easy, fundamental to crystallography, and immediately productive of useful results even before the earlier steps in the total summation have been taken. We shall therefore choose to take the final step first and consider the implications for the diffraction pattern of a crystal of a regular three-dimensional arrangement of unit-cells.

Laue Equations

Since a crystal is a regular three-dimensional arrangement of unit-cells it can be regarded as acting as a three-dimensional diffraction grating for X-rays. The effect of a grating is to limit the directions in which an observable diffracted beam occurs; in the diffraction of X-rays by crystals it will be shown that the directions of the diffracted X-ray beams depend on the dimensions of the unit-cell and their intensities on the nature and disposition of atoms within the unit-cell.

It is a familiar property of diffraction by gratings that diffracted intensity maxima occur only in directions for which the waves scattered by corresponding points in each grating element are in phase; in all other directions the scattered waves interfere destructively more or less. In other words the path difference in directions of intensity maxima between the waves scattered by corresponding points in different elements of the grating is, for all elements, an integral number of wavelengths (Fig 6.2).[3] By analogy with the corresponding points in the grating elements we can take corresponding points, one in each unit-cell of the crystal and so obtain an array of

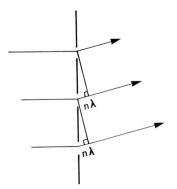

Fig 6.2 The path difference between waves scattered by corresponding points in each element of a diffraction grating is an integral number of wavelengths.

lattice points. Therefore if we are interested only in the angular disposition of intensity maxima of the X-radiation diffracted by a crystal, it is valid to make the assumption that the scattering is produced by the array of lattice points. This is a highly restrictive model that can give no information about the relative intensities of the local intensity maxima but it is adequate for our immediate purpose. Diffracted beams will occur in directions for which X-rays scattered by all lattice points are in phase and, because the lattice is a regular three-dimensional array of lattice points, this condition is satisfied if the X-rays scattered by pairs of adjacent lattice points lying on three non-coplanar rows are in phase.

We suppose a parallel beam of X-rays of wavelength λ to be incident on a row of lattice points of spacing t at an angle of incidence i (Fig 6.3) and consider a direction

---

[3] For thorough accounts of optical diffraction gratings see Jenkins and White (1976), Longhurst (1973) and Smith and Thomson (1971).

of scattering at an angle δ to the lattice row.[4] The path difference for X-rays scattered by adjacent lattice points is then, referring to Fig 6.3,

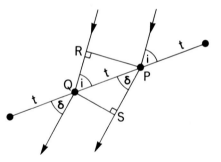

Fig 6.3   A parallel beam of X-rays of wavelength λ, represented by the wave front PR, is incident at an angle of incidence i on a row of lattice points, P, Q, . . . , of separation t. The condition for the X-rays scattered by adjacent lattice points to be in phase is $PS - RQ = n\lambda$, where n is an integer.

$$PS - RQ = PQ(\cos\delta - \cos i)$$
$$= t(\cos\delta - \cos i).$$

For a diffracted beam to occur the X-rays scattered by adjacent lattice points have to be in phase, that is to say their path difference must be an integral number of wavelengths. Therefore the condition for diffraction by a lattice point row is

$$t(\cos\delta - \cos i) = n\lambda,$$

where n is an integer. Figure 6.4 illustrates the result for four small values of n.

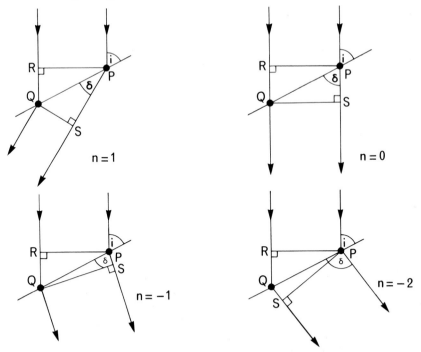

Fig 6.4   The condition for an intensity maximum in the X-radiation scattered by a lattice row of which P and Q are two adjacent lattice points. The condition $t(\cos\delta - \cos i) = n\lambda$ is illustrated for $n = 1, 0, -1, -2$.

[4] It is usual to measure i and δ with respect to opposite directions of the lattice row. We shall adopt the convention of taking i as the angle between the incident beam and the *negative* direction of the row and δ as the angle between the diffracted beam and the *positive* direction of the row.

Permissible directions of the diffracted beam are of course not confined to the plane defined by the incident beam and the lattice point row. The condition for diffraction, $t(\cos \delta - \cos i) = n\lambda$, is satisfied by any direction making the angle δ with the lattice row. Therefore for a given value of n the diffracted radiation is confined to the surface of a cone of semiangle δ; a set of cones coaxial about the lattice row represents solutions of the diffraction condition for $n = 0, \pm 1, \pm 2$, etc (Fig 6.5).

So far we have considered diffraction by a single point row, but a lattice is a regular three-dimensional array of points and as such is completely specified by the distance apart of adjacent lattice points in three non-coplanar directions. We label these three axes x, y, and z and take the separation of lattice points along each to be a, b, and c respectively. For a point row of separation a parallel to the x-axis we have seen that the diffraction condition is $a(\cos \delta_a - \cos i_a) = h\lambda$, where i_a and δ_a are the angles between the x-axis and the incident and diffracted beams respectively and h is an integer. If such a set of point rows is repeated successively with translation b parallel to the y-axis, each point row becomes a grating element and the diffraction condition

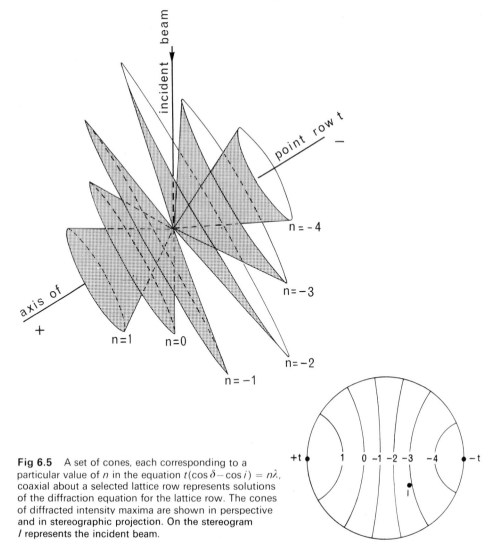

Fig 6.5   A set of cones, each corresponding to a particular value of n in the equation $t(\cos \delta - \cos i) = n\lambda$, coaxial about a selected lattice row represents solutions of the diffraction equation for the lattice row. The cones of diffracted intensity maxima are shown in perspective and in stereographic projection. On the stereogram I represents the incident beam.

for this grating is $b(\cos\delta_b - \cos i_b) = k\lambda$, where k is an integer. The condition for the production of an observable diffracted beam by all the lattice points in the xy plane is then $a(\cos\delta_a - \cos i_a) = h\lambda$ and $b(\cos\delta_b - \cos i_b) = k\lambda$ simultaneously. If such a set of lattice planes is repeated successively with translation c parallel to the z-axis, each plane constitutes an element of a third linear diffraction grating for which the diffraction condition is $c(\cos\delta_c - \cos i_c) = l\lambda$, where l is an integer. The simultaneous operation of all three conditions is necessary for the production of an observable diffracted beam by all the lattice points of the three-dimensional lattice. This statement implies that if X-rays scattered by any pair of adjacent lattice points in a three-dimensional lattice are to be in phase, then the X-rays scattered by adjacent lattice points along each of the reference axes must be in phase; this constitutes a necessary and sufficient condition for diffraction by a three-dimensional array of lattice points.

The diffraction condition for a three-dimensional lattice can thus be written as:

$$\left.\begin{array}{l} a(\cos\delta_a - \cos i_a) = h\lambda \\ b(\cos\delta_b - \cos i_b) = k\lambda \\ c(\cos\delta_c - \cos i_c) = l\lambda \end{array}\right\} \tag{7}$$

where the incident beam is inclined at angles i_a, i_b, i_c and the diffracted beam at angles δ_a, δ_b, δ_c with the x, y, z axes respectively, λ is the X-ray wavelength, and h, k, l are integers. These three equations (7) are known as the *Laue Equations* after Max von Laue, who in 1912, suggested that a crystal should act as a diffraction grating for X-rays.

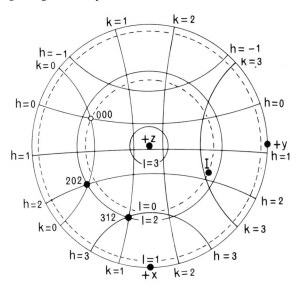

Fig 6.6 Solution of the Laue equations for a three-dimensional orthorhombic lattice for $h = k = l = 0$ (trivial), for $h = 2$, $k = 0$, $l = 2$ and for $h = 3$, $k = 1$, $l = 2$. The direction of the incident X-ray beam is represented in the stereographic projection by the pole I.

The first Laue Equation, $a(\cos\delta_a - \cos i_a) = h\lambda$, restricts the directions of observable diffracted beams to the surfaces of a set of cones coaxial about the x-axis and having semiangles δ_a consistent with the equation. This statement is illustrated in Fig 6.6 where, for a given direction I of the incident beam, the restriction placed by the first Laue Equation on the directions of diffracted beams is represented by a set of small circles of radius δ_a centred on the x-axis. The second Laue Equation, $b(\cos\delta_b - \cos i_b) = k\lambda$, further constrains the directions of observable diffracted beams to a set of small circles of radius δ_b centred on the y-axis. The simultaneous operation of these two Laue Equations thus restricts observable diffracted beams produced by X-rays incident in a particular direction to the intersections of pairs of small circles, the attitude of one small circle being dependent on h and of the other on k. Each pair of h, k values leads in general to two common directions. The third Laue Equation, $c(\cos \delta_c - \cos i_c) = l\lambda$, provides an additional restriction; for an observable diffracted beam to occur, a small circle of the set of radius δ_c centred on the z-axis must pass through the intersection of small circles of the x and y sets. An observable diffracted beam thus lies in a direction common to three cones, each coaxial with one of the reference axes and of semiangle consistent with the appropriate Laue Equation; such a direction is completely specified for a given lattice and for X-rays of given λ by the integers h, k, l. The orientation of the X-ray beam shown in Fig 6.6 leads to intersection of three small circles only for $h = k = l = 0$, for $h = 2$, $k = 0$, $l = 2$, and for $h = 3$, $k = 1$, $l = 2$. The first of these solutions of the Laue Equations is trivial, representing merely the forward direction of the incident beam.

It is worth noticing at this point that a three-dimensional grating, such as a crystal, differs from a one- or two-dimensional grating in the small number of diffracted beams produced by any particular orientation of the incident beam. Other diffracted intensity maxima can be observed only by changing the orientation of the incident beam relative to the crystallographic reference axes; in practical X-ray crystallography this is most conveniently done by rotating the crystal and keeping the attitude of the incident beam fixed.

Bragg Equation

Although the Laue Equations provide an elegant treatment of the diffraction of X-rays by crystals, they are difficult to manipulate and the diffraction condition is provided in what is, for most purposes, a more convenient form by the *Bragg Equation*. We now proceed to derive the Bragg Equation from the Laue Equations; later we shall show how the Bragg Equation can be obtained directly in a simpler, but less rigorous manner. It will be shown that the essential simplifying feature of the Bragg Equation is that a particular diffracted beam appears as a reflexion of the incident beam by a particular lattice plane, 'reflexion' occurring only at certain angles of incidence given by the equation.

We begin by considering[5] a general solution of the Laue Equations illustrated in Fig 6.7, where I represents the direction of the incident X-ray beam, D the direction of the diffracted X-ray beam for one solution of the Laue Equations, and z the positive direction of the z-axis of the lattice. Consistently with our previous usage I and D make angles $180° - i_c$ and δ_c respectively with the positive direction of the z-axis, which is placed in the centre of the stereogram. N is a direction in the plane defined by I and D and bisects the angle between I and D. The angle θ is defined such that $I:D = 180° - 2\theta$ and $I:N = N:D = 90° - \theta$. For the spherical triangle I N z equation (1) of chapter 5 yields

$$\cos(180° - i_c) = \cos z:N.\cos(90° - \theta) + \sin z:N.\sin(90° - \theta).\cos \omega$$

i.e. $$-\cos i_c = \cos z:N.\sin \theta + \sin z:N.\cos \theta.\cos \omega. \qquad (8)$$

Similarly for the spherical triangle DNz,

$$\cos \delta_c = \cos z:N.\cos(90° - \theta) + \sin z:N.\sin(90° - \theta).\cos(180° - \omega)$$

i.e. $$\cos \delta_c = \cos z:N.\sin \theta - \sin z:N.\cos \theta.\cos \omega. \qquad (9)$$

[5] We are indebted to Dr Helen D. Megaw for this simple ingenious argument.

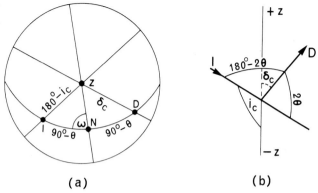

Fig 6.7   Derivation of the Bragg Equation from the Laue Equations. In the stereogram (a) and the perspective drawing (b) I represents the direction of the incident X-ray beam, D the diffracted beam and the pole z the positive direction of the z-axis; N is coplanar with I and D and bisects the angle ID.

Addition of equations (8) and (9) yields

$$\cos \delta_c - \cos i_c = 2 \cos z{:}N . \sin \theta. \tag{10}$$

But for an observable diffracted beam to occur the Laue Equations impose the restriction

$$c(\cos \delta_c - \cos i_c) = l\lambda, \tag{7}$$

which on substitution in equation (10) becomes

$$2 \cos z{:}N . \sin \theta = \frac{l\lambda}{c}$$

i.e. $\dfrac{c}{l} . \cos z{:}N = \dfrac{\lambda}{2 \sin \theta}.$

Similarly the other two Laue Equations can be rewritten in terms of the angle θ and the angles between the direction N and the x and y axes as

$$\frac{b}{k} . \cos y{:}N = \frac{\lambda}{2 \sin \theta}$$

and $\dfrac{a}{h} . \cos x{:}N = \dfrac{\lambda}{2 \sin \theta}$

so that $\dfrac{\lambda}{2 \sin \theta} = \dfrac{a}{h} . \cos x{:}N = \dfrac{b}{k} . \cos y{:}N = \dfrac{c}{l} . \cos z{:}N. \tag{11}$

Now, the length of the normal ON from the origin O to the plane (hkl) at N is given by equation (40) of chapter 5 as

$$ON = \frac{a}{h} . \cos x{:}(hkl) = \frac{b}{k} . \cos y{:}(hkl) = \frac{c}{l} . \cos z{:}(hkl).$$

where $x{:}(hkl)$ is the angle between ON and the x-axis and so on. Comparison with equation (11) indicates that the direction N shown on Fig 6.7 is the normal to the plane (hkl), where h, k, and l are the integral factors in the three Laue Equations, and

$$\frac{\lambda}{2 \sin \theta} = \text{ON}.$$

The perpendicular distance, ON, of the (hkl) plane from the origin is more conveniently described now as the spacing of the (hkl) planes of the lattice and written as d_{hkl}, so that

$$\lambda = 2d_{hkl} \sin \theta. \tag{12}$$

Equation (12), which was derived by W. L. Bragg in 1912, is known as the *Bragg Equation*. It simply states that for directions in which an observable diffracted beam can occur the incident and diffracted beams are coplanar with the normal from the origin to a set of lattice planes (hkl) and equally inclined at $90° - \theta$ to it and, further, that the angle θ (commonly called the *Bragg angle*) is related to the wavelength of the radiation λ and to the spacing d_{hkl} of the lattice planes by equation (12).

It is apparent from Fig 6.8 that the Bragg Equation provides an alternative way of looking at the diffraction of X-rays by a lattice: the diffracted beam can be regarded as a reflexion of the incident beam by the set of lattice planes (hkl), reflexion occurring only when the angle of inclination of the incident beam to the lattice planes, i.e. the Bragg angle θ, the spacing of the planes and the wavelength of the X-rays, satisfy the equation $\lambda = 2d_{hkl} \sin \theta$.

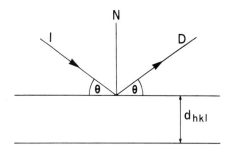

Fig 6.8   The Bragg Equation. The diffracted beam D can be regarded as a reflexion of the incident beam I by a set of lattice planes (hkl) when the angle of incidence θ satisfies the equation $\lambda = 2d_{hkl} \sin \theta$.

In his derivation of the Bragg Equation W. L. Bragg considered first how the X-rays scattered by all the lattice points in a plane (hkl) might be in phase and then established the condition for the X-rays scattered by the lattice points in all lattice planes parallel to (hkl) to be in phase. As the incident-plane wave-front passes over a set of points (which need not be lattice points or even regularly spaced) in a lattice plane, secondary wavelets build up a reflected wave-front according to the Huygens construction familiar in optics. A small part of the energy of the incident wave-motion is transferred to the reflected wave-motion, but most of it passes on. The condition for optical reflexion, that the angle of incidence is equal to the angle of reflexion, ensures that the waves scattered by all points in the lattice plane are in phase with one another (Fig 6.9(a)). In general the waves reflected from successive lattice planes will not be in phase. A plane wave-front incident on an adjacent pair of lattice planes is shown in Fig 6.9(b), where the lattice planes are shown as horizontal lines. The waves reflected from the upper plane have an optical path that is shorter than that for those reflected from the lower plane by $\text{PB} + \text{BQ} = 2d_{hkl} \sin \theta$. For reinforcement the path difference must be a whole number of wavelengths; therefore the condition for an observable diffracted beam is

$$n\lambda = 2d_{hkl} \sin \theta. \tag{13}$$

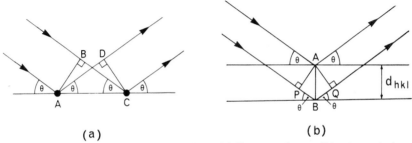

Fig 6.9 The Bragg Equation: the optical analogy. (a) illustrates the condition for optical reflexion, that the angle of incidence $A\hat{C}B = \theta$ is equal to the angle of reflexion $C\hat{A}D$ so that the waves scattered from all points in the plane AC are in phase with one another. (b) illustrates Bragg diffraction from successive lattice planes (*hkl*), AP and AQ being respectively incident and diffracted plane wave fronts; in general the waves reflected from successive lattice planes will not be in phase.

Equation (13) is a form of the *Bragg Equation* in common use in spectroscopy. In crystallographic usage it is rewritten as

$$\lambda = \frac{2d_{hkl}}{n} \sin \theta = 2d_{nh.nk.nl} \sin \theta$$

and the indices are divided through by the common factor n to give the form derived directly as equation (12), $\lambda = 2d_{hkl} \sin \theta$. In the derivation of equation (12) from the Laue Equations the only restriction placed on the values of h, k, and l was that they should be integral; they were not forbidden to have a common factor. However when the indices of planes (*hkl*) have a common factor n, the planes are not strictly lattice planes, but a set of planes parallel to the lattice planes (h/n, k/n, l/n) with a spacing of one nth of that of the lattice planes. This point is illustrated in Fig 6.10, where (210) lattice planes, (420) and (630) planes are shown in relation to an orthorhombic lattice. Every (210) lattice plane passes through lattice points and all planes are equivalent. However, only alternate planes of the (420) set pass through lattice points and in consequence not all planes of the set are equivalent. Likewise the planes (630) are not all equivalent, only one in every three passing through lattice points. In Fig 6.10 incident and diffracted beams that satisfy the Bragg Equation are shown for each set of planes and it is convenient at this point to comment on the statement, which has already been implied, that diffracted X-ray beams are commonly known as *X-ray reflexions*, it being understood that 'reflexion' implies a solution of the Bragg Equation and therefore a specific angle of incidence. The diffracted beams shown in Fig 6.10 would be described as the 210, 420, and 630 reflexions, the indices for reflexions being distinguished from those for planes by omission of brackets ( ). Reflexions are indexed so that the path difference for X-rays scattered by adjacent planes is λ, a point to which we shall return later. The statement that a particular diffracted beam is the *hkl* reflexion for X-rays of wavelength λ for a lattice of given dimensions completely specifies the angular relationship between incident and diffracted beam: the incident beam is required to lie on a cone of semi-axis $90° - \theta$ about the normal to the (*hkl*) plane and the diffracted beam lies on the same cone so as to be coplanar with the normal to the plane and the incident beam.

The Bragg Equation thus provides a simple and convenient statement of the geometry of the diffraction of X-radiation by crystals. It is the fundamental equation of X-ray crystallography. Its application in a variety of situations will be explored in the later parts of this chapter and in the two following chapters.

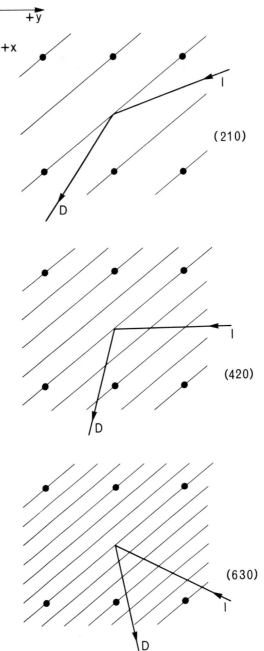

Fig 6.10 The Bragg Equation, $\lambda = 2d_{hkl} \sin \theta$. The top diagram shows an (001) projection of an orthorhombic P-lattice with lattice planes (210), and incident (I) and diffracted (D) X-ray beams. The central and lowermost diagrams show planes (420) and (630) for the same lattice and the directions of incident and diffracted X-ray beams. Since $d_{210} = 2d_{420} = 3d_{630}$, $\sin \theta_{210} = \frac{1}{2}\sin \theta_{420} = \frac{1}{3}\sin \theta_{630}$.

Intensities of X-ray reflexions

The regular arrangement of atoms in a crystal restricts diffracted intensity maxima to certain directions, which can be described either as directions that satisfy the Laue Equations or, alternatively, as directions in which reflexions from lattice planes satisfy the Bragg Equation. We need concern ourselves therefore only with the intensity of radiation scattered in the directions of X-ray reflexions, a much simpler problem than the investigation of the intensity of radiation scattered in any general direction.

Since the scattering produced by a crystal is almost entirely due to interaction between X-rays and electrons, the density of scattering matter at a point x, y, z in the unit-cell can be equated with the electron density[6] at that point. Strictly, electron density must be regarded as varying continuously throughout the unit-cell, but some simplifying assumptions can be made: since the electron density associated with an atom falls off rapidly with distance from the centre of the atom, it will be assumed that the electron cloud associated with any atom does not overlap that belonging to any other atom and is moreover spherically symmetric. All the atoms of a given element in the same state of ionization are thus taken to have identical distribution of electron density and to scatter X-rays similarly. We have deliberately ignored, at this first stage of approximation, valency electrons, which are involved in bond formation, because they are relatively few in number compared with the core electrons, each of which is unambiguously associated with one atomic nucleus; in short we are assuming that the core electrons are predominantly responsible for the scattering of X-rays and consequently that the radiation scattered by an atom will be independent of its environment. Thus the scattering produced by an atom of a given element in a given state of ionization will always be the same wherever it lies in the unit-cell and in all substances in which the element occurs.

In the foregoing we have referred qualitatively to the amount of radiation scattered by an atom; clearly we need to make this quantitative and a convenient unit is provided by the classical electrodynamic treatment of the scattering of X-radiation by a single free electric charge. When a wave falls on an electric charge it causes the charge to vibrate and to act as a secondary source of waves of the same frequency so that the charge can be regarded as scattering a small fraction of the radiation incident upon it. If unpolarized radiation of amplitude A and wavelength λ is incident on a free classical electron of charge e and mass m, it can be shown that the amplitude of the scattered radiation at a distance R from the electron, where $R \gg \lambda$, is

$$\frac{A}{R} \cdot \frac{e^2}{mc^2} \left\{ \frac{1 + \cos^2 2\theta}{2} \right\}^{\frac{1}{2}},$$

where c is the velocity of light and 2θ is the angle between the scattered beam and the forward direction of the incident beam.[7] The factor $\{(1 + \cos^2 2\theta)/2\}^{\frac{1}{2}}$ arises from the partial polarization of the scattered beam, which is out of phase with the incident beam by the amount π.

We now have to define the *atomic scattering factor*, f, of an atom as the ratio of the amplitude scattered in a particular direction by that atom to the amplitude scattered by a free classical electron in the same direction. The amplitude of the

---

[6] If ψ is the electronic wave function at a point, then the electron density at that point will be $|\psi|^2$.

[7] This result is proved in standard textbooks of electrodynamics and in James (1967) p. 29.

X-radiation scattered by an atom is thus

$$\frac{A}{R} \cdot \frac{e^2}{mc^2} f \left\{ \frac{1 + \cos^2 2\theta}{2} \right\}^{\frac{1}{2}}.$$

In the idealized, but impossible, case of a 'point atom', where all the electrons associated with the atomic nucleus are situated at a point, the atomic scattering factor of the atom would be equal to its atomic number,[8] Z. In reality the electrons associated with an atomic nucleus occupy a finite volume, the dimensions of which are of the same order of magnitude as the wavelength of X-rays. Each small element of volume ΔV within the electron cloud of the atom will give rise to scattered X-radiation of amplitude proportional to $\rho(r).\Delta V$, where $\rho(r)$ is the average electron density over the volume ΔV situated at a distance r from the centre of the atom. There will in general be a path difference between the X-rays scattered by any pair of volume elements within the electron cloud and this path difference will vary with scattering angle (Fig 6.11); the path difference will be zero for $2\theta = 0$ and will increase smoothly with increasing 2θ. We now suppose the radiation scattered by all volume elements of the electron cloud of the atom to be summed for a direction of scattering angle 2θ:

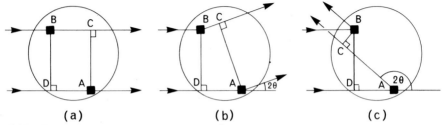

(a)   (b)   (c)

Fig 6.11   The dependence on scattering angle 2θ of the path difference between the X-rays scattered by two volume elements A and B within the electron cloud of the atom. The boundary surface of the electron cloud of the atom is represented by the large circle. (a) $2\theta = 0$; path difference $\delta = DA - BC = 0$. (b) 2θ small; path difference $\delta = DA - BC$. (c) 2θ large; path difference $\delta = DA + AC$.

for $2\theta = 0$ there can be no destructive interference and the atomic scattering factor, f, for the atom will be equal to its atomic number, Z; for small scattering angles all path differences will be small and there can be little destructive interference so that f will be only just less than Z; as the scattering angle increases the path difference for any pair of volume elements will increase and in general there will be a greater likelihood of destructive interference so that f will steadily decrease with increasing θ and at high θ, $f \ll Z$. The more tightly the electrons are bound to the nucleus, the greater will be the concentration of electron density towards the centre of the atom so that for given θ less destructive interference will be possible; for cations, especially for cations of high formal charge, the value of f will remain close to Z to higher θ values than for anions with a similar number of extranuclear electrons. In exploring the variation in magnitude of f we must also bear in mind that the phase difference consequent on a given path difference is dependent on the wavelength λ of the incident radiation: at given θ more pairs of volume elements will be able to produce destructive interference if λ is small. It is found that in general the atomic scattering factors of all elements and ions vary similarly with $\sin \theta / \lambda$, f decreasing from its value Z at $\theta = 0$ more slowly for cations with tightly bound electrons than for anions

[8] The atomic number is the charge on the nucleus which for a neutral atom is equal to the number of extranuclear electrons. It is the number of extranuclear electrons that concerns us here and we define this number as Z. For an ion Z is equal to the charge on the nucleus less the formal charge of the ion.

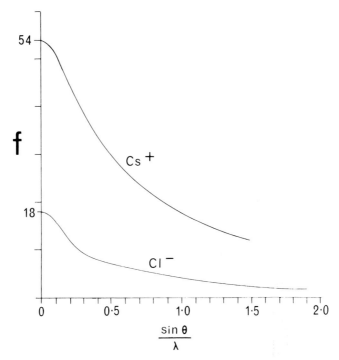

Fig 6.12   The dependence of atomic scattering factor f on $\sin \theta / \lambda$ for a cation Cs^+ and an anion Cl^-. The rate of fall-off for cations is generally less rapid than for anions.

with more diffuse electron clouds. The calculation of the electron density distribution within an atom is a problem in quantum mechanics; atomic scattering factors for X-radiation calculated by such methods are tabulated in *International Tables for X-ray Crystallography*, vol. III as functions of $\sin \theta / \lambda$. The dependence of f on $\sin \theta / \lambda$ is shown in Fig 6.12 for two examples, the cation Cs^+, in which the electrons are rather tightly bound, and the anion Cl^-, in which the electrons are relatively loosely bound.

Now although the amplitude of the X-rays scattered by each atom in a unit-cell is proportional to its atomic scattering factor, we cannot state the amplitude of the X-radiation scattered by the whole unit-cell until we know the phase of the X-rays scattered by each atom and that will depend on position in the unit-cell. We take the origin of the unit-cell as our reference point for phase and suppose that the X-rays scattered by an atom at the origin have phase $\phi = 0$. The phase of the X-rays scattered by an atom with coordinates x, y, z can then be deduced from the Bragg Equation by the following argument. The contributions to the hkl reflexion of X-rays scattered by atoms lying on the (hkl) plane through the origin will be in phase with one another and with the X-rays scattered by the atom at the origin, for which $\phi = 0$, because the path difference, Δ, between X-rays scattered by any point in such a plane is zero. Since the condition for an hkl reflexion is that the X-rays scattered from adjacent (hkl) lattice planes should have a path difference, $\Delta = \lambda$, which in phase terms amounts to saying a phase difference, $\phi = 2\pi$, the X-rays scattered from the first (hkl) plane out from the origin will have a path length less by λ than those scattered from the hkl plane through the origin and their phase will consequently be $\phi = 2\pi$ (Fig 6.13). Likewise the path length for X-rays scattered from the second plane out will be shorter again by λ and their phase will be 4π. In general for the nth plane out from the origin, the

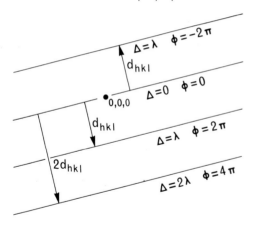

Fig 6.13 The phase difference ϕ between X-rays of wavelength λ scattered into the hkl reflexion by successive (hkl) lattice planes. The lattice planes (hkl) are perpendicular to the plane of the diagram.

path length of the scattered X-rays will be shorter by $n\lambda$ and their phase will be $2\pi n$. But the scattering is done by atoms, not by lattice planes, and it is unlikely except in very simple structures that all the atoms in the unit-cell will lie on the lattice planes corresponding to every observable reflexion. The phase of the X-rays scattered by an atom with general coordinates x, y, z can however easily be found by drawing a plane through x, y, z parallel to the (hkl) lattice planes and taking the perpendicular distance from the origin to that plane as D_{hkl} (Fig 6.14); then, if the separation of (hkl) lattice planes is d_{hkl}, the path difference between the X-rays scattered by the atom at x, y, z and an atom at the origin will be $(D_{hkl}/d_{hkl}).\lambda$ and the phase of the X-rays scattered by the atom at x, y, z will correspondingly be $2\pi(D_{hkl}/d_{hkl})$.

At this point it may be instructive to consider in some detail an example of phase calculation for a simple structure, caesium chloride (which is cubic with only two atoms in its unit-cell, Cs at $0,0,0$ and Cl at $\frac{1}{2},\frac{1}{2},\frac{1}{2}$ (Fig 6.15)). We consider first the 100 reflexion. The Cs atom, being at the origin, lies on the (100) lattice plane through the origin so that the phase of the X-rays scattered by it is zero. The amplitude of the X-rays scattered by the caesium atom will therefore simply be proportional to f_{Cs}, the atomic scattering factor of caesium; moreover this will be so for all X-ray reflexions because the Cs atom lies at the origin. The chlorine atom however does not lie on the (100) lattice plane through the origin, but at a perpendicular distance $\frac{1}{2}d_{100} = \frac{1}{2}a$ from that plane so that the X-rays scattered by the Cl atom have a path difference $\frac{1}{2}\lambda$ and a phase difference π relative to the X-rays scattered by the Cs atom. The amplitude of the X-rays scattered by the chlorine atom will of course be proportional to f_{Cl}, the atomic scattering factor of chlorine. The principle of superposition gives

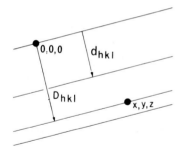

Fig 6.14 The phase of the X-rays scattered into the hkl reflexion by an atom at xyz is given by $2\pi(D_{hkl}/d_{hkl})$, where D_{hkl} is the perpendicular distance from the origin to the plane parallel to (hkl) passing through xyz. The plane of the diagram is perpendicular to (hkl) and does not necessarily contain the point xyz.

the intensity[9] of the X-rays scattered by one unit-cell in general as

$$I(hkl) = \left(\sum_1^N f_n \cos \phi_n\right)^2 + \left(\sum_1^N f_n \sin \phi_n\right)^2 \tag{6}$$

and for the 100 reflexion of CsCl as

$$I(100) = (f_{Cs} \cos 0 + f_{Cl} \cos \pi)^2 + (f_{Cs} \sin 0 + f_{Cl} \sin \pi)^2$$
$$= (f_{Cs} - f_{Cl})^2.$$

It should be noticed that we have taken our summation only up to $n = 2$, because there are only two atoms in the unit-cell, one Cs and one Cl; the other Cs atoms shown in Fig 6.15 belong to other unit-cells if the Cs atom at the origin is assigned wholly to the reference unit-cell.

We now turn to the 200 reflexion. The (200) lattice planes have an interplanar spacing $d_{200} = \frac{1}{2}d_{100}$ so that the Cl atom scatters X-rays of phase $2\pi(\frac{1}{2}a/\frac{1}{2}a) = 2\pi$ into the 200 reflexion. The phase of the X-rays scattered by Cs remains zero. The expression for the intensity of the 200 reflexion is thus

$$I(200) = (f_{Cs} \cos 0 + f_{Cl} \cos 2\pi)^2 + (f_{Cs} \sin 0 + f_{Cl} \sin 2\pi)^2$$
$$= (f_{Cs} + f_{Cl})^2.$$

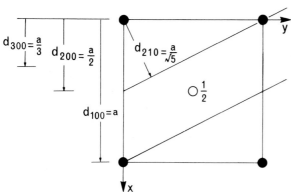

Fig 6.15 Projection of the structure of CsCl on (001) with Cs atoms represented as solid circles and Cl atoms as open circles. The spacings and orientation of (100), (200), (300) and (210) planes are indicated.

The Cl atom lies midway between (210) lattice planes with $D_{210} = \frac{3}{2}d_{210}$ so that the phase of the X-rays scattered by the Cl atom at $\frac{1}{2}, \frac{1}{2}, \frac{1}{2}$ into the 210 reflexion is $2\pi.\frac{3}{2} = 3\pi$. The intensity of the 210 reflexion is thus

$$I(210) = (f_{Cs} \cos 0 + f_{Cl} \cos 3\pi)^2 + (f_{Cs} \sin 0 + f_{Cl} \sin 3\pi)^2$$
$$= (f_{Cs} - f_{Cl})^2.$$

We have now expressed the intensities of the three reflexions 100, 200, and 210 of CsCl in terms of the atomic scattering factors of caesium and chlorine, f_{Cs} and f_{Cl}. But it must be borne in mind that atomic scattering factors are dependent on $\sin\theta/\lambda$ for the relevant expression and, by the Bragg Equation, $\sin\theta/\lambda = \frac{1}{2}d$. In Table 6.1 we give the cell edge, a, for CsCl; $\frac{1}{2}d$ for a selection of reflexions including 100, 200, and 210; f_{Cs} and f_{Cl} derived from the atomic scattering factor curves of Fig 6.12 for each reflexion; and the calculated intensity $I(hkl)$ of each reflexion.

[9] For practical purposes $I(hkl)$ is known as the 'intensity' of the hkl reflexion as will be explained later in this chapter.

Table 6.1
Intensities of some reflexions of CsCl

$a = 4 \cdot 123$ Å		Cs: 0, 0, 0			Cl: $\frac{1}{2}, \frac{1}{2}, \frac{1}{2}$	
hkl	d_{hkl}	$\dfrac{\sin\theta}{\lambda} = \dfrac{1}{2d_{hkl}}$	f_{Cs}	f_{Cl}	$I(hkl)$
100	a	$\dfrac{1}{2a} = 0 \cdot 12$	50	15	$(f_{Cs}-f_{Cl})^2 = (35)^2 = 1225$
200	$\dfrac{a}{2}$	$\dfrac{1}{a} = 0 \cdot 24$	42	11	$(f_{Cs}+f_{Cl})^2 = (53)^2 = 2809$
300	$\dfrac{a}{3}$	$\dfrac{3}{2a} = 0 \cdot 36$	36	8	$(f_{Cs}-f_{Cl})^2 = (28)^2 = 784$
210	$\dfrac{a}{\sqrt 5}$	$\dfrac{\sqrt 5}{2a} = 0 \cdot 27$	41	10	$(f_{Cs}-f_{Cl})^2 = (31)^2 = 961$
420	$\dfrac{a}{2\sqrt 5}$	$\dfrac{\sqrt 5}{a} = 0 \cdot 54$	28	7	$(f_{Cs}+f_{Cl})^2 = (35)^2 = 1225$

The amplitude of the X-rays scattered by a free electron is taken as unity. Atomic scattering factors for the ions Cs^+ and Cl^- are used rather than those for the neutral atoms because the structure is known to be ionic.

The procedure outlined above for finding the phase of the X-rays scattered by the various atoms of a crystal structure is impossibly cumbersome for all but the simplest structures and is then only convenient for reflexions with very simple indices. In general an analytical approach is more profitable and that will be explored in succeeding paragraphs.

We consider a general hkl reflexion and recall two points made previously: that the phase of the X-rays scattered by an atom lying on the (hkl) plane through the origin is taken to be zero and that the phase of the X-rays scattered by an atom lying on the first (hkl) plane out from the origin is 2π. Since the first (hkl) plane out from the origin makes an intercept a/h on the x-axis and corresponds to a phase change of 2π, we ean, by simple proportions, associate a translation ax along the x-axis with a phase change

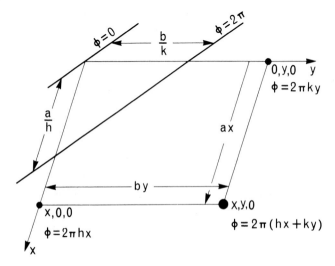

Fig 6.16 Atoms with coordinates $x00$, $0y0$, $xy0$ respectively scatter X-rays of phase $2\pi hx$, $2\pi ky$, $2\pi(hx+ky)$ into the hkl reflexion. The (hkl) plane through the origin, with phase difference $\phi = 0$, and the first (hkl) plane out from the origin, with $\phi = 2\pi$, are indicated by bold lines.

$(2\pi ax)/(a/h) = 2\pi hx$; thus an atom with coordinates x, 0, 0 will scatter X-rays of phase $2\pi hx$ into the hkl reflexion.

By analogy we can say that an atom with coordinates 0, y, 0 will scatter X-rays of phase $2\pi ky$ into the hkl reflexion. We can extend this idea to an atom with coordinates x, y, 0 (Fig 6.16): such an atom is associated with translations ax along the x-axis and by along the y-axis so that the phase of the X-rays scattered by it will be $[2\pi xa/(a/h)] + [2\pi yb/(b/k)] = 2\pi(hx + ky)$. Extension of the argument to three dimensions gives the phase of the X-rays scattered by an atom with coordinates x, y, z as $2\pi(hx + ky + lz)$, the amplitude of the scattered wave motion being proportional to the atomic scattering factor, f, of the atom at x, y, z. We can now apply the principle of superposition to a unit-cell containing N atoms of elements with atomic scattering factors f_1, $f_2, \ldots f_n$ and coordinates x_1, y_1, z_1; x_2, y_2, z_2; $\ldots x_N, y_N, z_N$. The intensity of the hkl reflexion will be given by

$$I(hkl) = \{f_1 \cos 2\pi(hx_1 + ky_1 + lz_1) + \ldots + f_N \cos 2\pi(hx_N + ky_N + lz_N)\}^2$$
$$+ \{f_1 \sin 2\pi(hx_1 + ky_1 + lz_1) + \ldots + f_N \sin 2\pi(hx_N + ky_N + lz_N)\}^2$$

Therefore $\quad I(hkl) = \left\{ \sum_1^N f_n \cos 2\pi(hx_n + ky_n + lz_n) \right\}^2$

$$+ \left\{ \sum_1^N f_n \sin 2\pi(hx_n + ky_n + lz_n) \right\}^2 \tag{14}$$

where the summation is taken over all the N atoms of the unit-cell.

From equation (14) the intensity of any X-ray reflexion can be calculated provided the coordinates of all atoms in the unit-cell and the relevant atomic scattering factors are known. Again taking CsCl as our example, we have two atoms in the unit-cell, Cs at 0, 0, 0 and Cl at $\frac{1}{2}, \frac{1}{2}, \frac{1}{2}$. The expression for the intensity of a general reflexion becomes

$$I(hkl) = \{f_{Cs} \cos 2\pi(h.0 + k.0 + l.0) + f_{Cl} \cos 2\pi(h.\tfrac{1}{2} + k.\tfrac{1}{2} + l.\tfrac{1}{2})\}^2$$
$$+ \{f_{Cs} \sin 2\pi(h.0 + k.0 + l.0) + f_{Cl} \sin 2\pi(h.\tfrac{1}{2} + k.\tfrac{1}{2} + l.\tfrac{1}{2})\}^2$$
$$= \left\{ f_{Cs} + f_{Cl} \cos 2\pi . \frac{h+k+l}{2} \right\}^2$$

The intensity of, for instance, the 420 reflexion is then given immediately by substitution as

$$I(420) = \{f_{Cs} + f_{Cl} \cos 2\pi . \tfrac{6}{2}\}^2$$
$$= \{f_{Cs} + f_{Cl}\}^2.$$

This expression is evaluated in Table 6.1.

It will have been noticed that in the case of CsCl the sine terms in the expression for $I(hkl)$ vanish because $\sin p\pi = 0$ for integral values of p. This is a rather special case of the general proposition that when the origin of the unit-cell of a centro-symmetric structure is taken at a centre of symmetry, the sine terms in the expression for $I(hkl)$ vanish. When the origin lies at a centre of symmetry the atoms in the unit-cell are related in pairs so that if an atom of a particular element has coordinates x_n, y_n, z_n an atom of the same element will lie at $\bar{x}_n, \bar{y}_n, \bar{z}_n$. Therefore for every term $f_n \sin 2\pi(hx_n + ky_n + lz_n)$ there will also be a term

$$f_n \sin 2\pi(-hx_n - ky_n - lz_n) = -f_n \sin 2\pi(kx_n + ky_n + lz_n)$$

and
$$\sum_{1}^{N} f_n \sin 2\pi(hx_n + ky_n + lz_n) = 0$$

The intensity expression then becomes,

$$I(hkl) = \left\{ \sum_{1}^{N} f_n \cos 2\pi(hx_n + ky_n + lz_n) \right\}^2 \tag{15}$$

In the case of CsCl both atoms lie on centres of symmetry so that there is no explicit pairing but the sine terms vanish immediately.

The trigonometric form of the intensity expression, equation (14), is obviously cumbersome; a more elegant and convenient formulation can be achieved by use of complex number notation. The expression

$$\psi = a \cos(\omega t + \phi) \tag{1}$$

for the disturbance produced at a given point by an incident wave-motion is the real part of the expression

$$\psi = |a| \exp\{i(\omega t + \phi)\}$$
$$= |a| \exp(i\phi) \exp(i\omega t).$$

Since we are concerned in diffraction with radiation of constant frequency, $\exp(i\omega t)$ is a common factor in all our expressions and need not be considered further. The complex amplitude $a = |a| \exp(i\phi)$ expresses both the amplitude $|a|$ and the phase ϕ of the disturbance; it is represented in the complex plane (Fig 6.17) by a line of length $|a|$ inclined at an angle ϕ to the real axis. Defining x as the real and y as the imaginary axis, then

$$a = x + iy$$
$$= |a| \cos \phi + i|a| \sin \phi.$$

The modulus $|a|$ of a thus represents the amplitude of the wave-motion,

$$|a| = \sqrt{(x^2 + y^2)},$$

and its intensity is given by aa^*, where $a^* = x - iy$ is the complex conjugate of a,

i.e. $aa^* = (x + iy)(x - iy) = x^2 + y^2 = |a|^2.$

The phase ϕ of the wave-motion is given by

$$\tan \phi = \frac{y}{x}.$$

Now the principle of superposition gives the resultant disturbance ψ_R at a point as the sum of the N disturbances, $\psi_1, \psi_2, \ldots \psi_N$, arriving simultaneously at the point. Therefore

$$a_R = \sum_{1}^{N} a_n$$
$$= \sum_{1}^{N} |a_n| \exp(i\phi_n)$$
$$= \sum_{1}^{N} |a_n| \cos \phi_n + i \sum_{1}^{N} |a_n| \sin \phi_n.$$

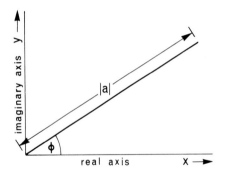

Fig 6.17 Representation of $a = |a|e^{i\phi}$ in the complex plane by a line of length $|a|$ inclined at the angle ϕ to the real axis.

We have already seen that the wave-motion scattered by an atom with coordinates x_n, y_n, z_n has an amplitude proportional to the atomic scattering factor f_n of the element concerned and phase $2\pi(hx_n + ky_n + lz_n)$, so that the wave-motion can be expressed as $f_n \exp 2\pi i(hx_n + ky_n + lz_n)$. The wave-motion scattered by all the atoms in one unit-cell into the *hkl* reflexion can then be represented as

$$\left. \begin{aligned} F(hkl) &= \sum_{1}^{N} f_n \exp 2\pi i(hx_n + ky_n + lz_n) \\ &= \sum_{1}^{N} f_n \cos 2\pi(hx_n + ky_n + lz_n) + i \sum_{1}^{N} f_n \sin 2\pi(hx_n + ky_n + lz_n) \end{aligned} \right\} \quad (16)$$

$F(hkl)$ being known as the *structure factor*. The amplitude of the resultant wave-motion is proportional to $|F(hkl)|$, which is known as the *structure amplitude*. Since the atomic scattering factor of an element is the ratio of the amplitude scattered by one atom of that element into the *hkl* reflexion to the amplitude scattered by a free classical electron, the structure amplitude $|F(hkl)|$ is therefore the ratio of the amplitude scattered into the *hkl* reflexion by the contents of one unit-cell to the amplitude scattered by a free classical electron in the same direction. Both quantities, f_n and $|F(hkl)|$, are thus pure numbers which represent the number of electrons that would have to be situated at a point, if that were possible, to produce a scattered wave of the same amplitude as that scattered by, in one case, an atom and, in the other, the whole contents of a unit-cell.

The intensity of the X-rays scattered by one unit-cell into the *hkl* reflexion is then

$$\begin{aligned} I(hkl) &= |F(hkl)|^2 \\ &= \left\{ \sum_{1}^{N} f_n \cos 2\pi(hx_n + ky_n + lz_n) \right\}^2 + \left\{ \sum_{1}^{N} f_n \sin 2\pi(hx_n + ky_n + lz_n) \right\}^2 \end{aligned}$$

The reader will observe that this expression is identical with equation (14) which was derived without recourse to complex numbers.

We have also seen that when the origin of the unit-cell of a centrosymmetric structure is taken at a centre of symmetry there will be atoms of the same element at x_n, y_n, z_n and $\bar{x}_n$, $\bar{y}_n$, $\bar{z}_n$; the expression for the structure factor then becomes

$$F(hkl) = \sum_{1}^{\frac{1}{2}N} f_n \{ \exp 2\pi i(hx_n + ky_n + lz_n) + \exp 2\pi i(-hx_n - ky_n - lz_n) \}$$

i.e.

$$F(hkl) = 2 \sum_{1}^{\frac{1}{2}N} f_n \cos 2\pi(hx_n + ky_n + lz_n)$$

i.e.
$$F(hkl) = \sum_1^N f_n \cos 2\pi(hx_n + ky_n + lz_n) \qquad (17)$$

The last form is preferred because it eliminates the risk of counting atoms situated at centres of symmetry twice. An important consequence of equation (17) is that for centrosymmetric structures the structure factor is always real and the phase of the X-rays scattered into the hkl reflexion can only take the values 0 or π.

In general the intensities of X-ray reflexions will be discussed in terms of structure factors $F(hkl)$ in preference to intensities $I(hkl)$, except when numerical values are required.

Systematic absences in non-primitive lattice types

In our discussion of the intensities of X-ray reflexions we have hitherto tacitly assumed that we have been dealing with a primitive unit-cell, that is a unit-cell that contains only one repeat unit. We have shown that diffracted intensity maxima occur only in directions for which the X-rays scattered by corresponding points in repeat units have path differences of integral numbers of wavelengths; such corresponding points constitute a lattice and our simple assumption that lattice points may be regarded as scatterers is justified. We have shown that the direction of a diffracted intensity maximum is a reflexion of the incident X-ray beam in a plane which must be parallel to a lattice plane but may belong to a set of planes with an interplanar spacing that is a simple sub-multiple of the interplanar spacing of the parallel set of lattice planes, reflexion being restricted to angles that satisfy the Bragg Equation, $\lambda = 2d_{hkl} \sin \theta$. The spacing d_{hkl} of the planes involved is independent of the choice of unit-cell, but the indices (hkl) of the set of planes does depend on the choice of unit-cell. The diffraction pattern produced by a crystal is obviously unaffected by any arbitrary choice of unit-cell; it must therefore always be referable to a primitive, if unconventional, unit-cell and we can say that all diffracted beams are reflexions from planes indexed on a primitive unit-cell.

Suppose that a C-cell, with axes x_c, y_c, z_c, is related to a primitive unit-cell, with axes x_p, y_p, z_p, by the axial transformation matrix

$$\begin{array}{ccc} 2 & 1 & 0 \\ 0 & 1 & 0 \\ 0 & 0 & 1 \end{array}$$

The transformation matrix from the C-cell to the P-cell is

$$\begin{array}{ccc} \frac{1}{2} & -\frac{1}{2} & 0 \\ 0 & 1 & 0 \\ 0 & 0 & 1 \end{array}$$

The lattice, with both unit-cells outlined, is shown in Fig 6.18. It follows from the axial transformation matrix that the indices $(h_p k_p l_p)$ of a set of planes referred to the P-cell are related to indices referred to the C-cell by the equations

$$h_p = \tfrac{1}{2}(h_c - k_c)$$
$$k_p = k_c$$
$$l_p = l_c$$

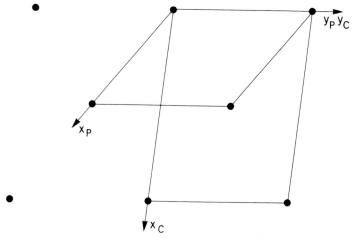

Fig 6.18   Relationship between a C-cell (x_c, y_c, z) and an unconventional P-cell (x_p, y_p, z) shown in projection on (001). Lattice points are represented by solid circles.

Since any diffracted beam produced by the crystal structure must be describable as a reflexion from a set of planes whose indices referred to the P-cell are integral, all X-ray reflexions must be describable as h_p, k_p, l_p reflexions. If h_p is to be integral, $h_c - k_c$ must be even; therefore $h_c + k_c$ must be even. When reflexions are indexed in terms of the C-cell, no reflexion can be observed for which $h_c + k_c$ is odd, i.e. the reflexions 100, 010, 120, etc will be systematically absent. A C-lattice is thus said to display a *systematic absence* or *extinction* for reflexions with $h + k = 2n + 1$, where n is an integer.

We can approach this sort of systematic absence, due to lattice type, in an alternative way. Let us assume that each lattice point scatters a wave equivalent to the resultant wave scattered by the atoms which it represents and consider the arrangement of lattice points in relation to lattice planes. Figure 6.19 shows the (001) projection of a C-lattice with the traces of (100), (200), and (120) planes outlined. Lattice points at the corners of the C-cell lie on (100) planes, but those at the centre of the (001) face of the unit-cell lie midway between (100) planes. The condition for occurrence of a 100 reflexion is that X-rays reflected from adjacent (100) lattice planes should have a phase difference of 2π; but the X-rays scattered by the (001)-face centring lattice points will then have a phase of π relative to the X-rays scattered by the lattice points at the corners of the unit-cell. Destructive interference will occur and, since in an effectively infinite lattice there will be equal numbers of both kinds of lattice point, the intensity of the 100 reflexion will be zero. From the traces of the (200) set of planes shown on the same lattice projection it is apparent that all lattice points lie on (200) planes and in consequence the X-rays scattered by all lattice points into the 200 reflexion are in phase. A 200 reflexion will therefore be observed unless, fortuitously, the resultant wave scattered in this direction by the atoms associated with a lattice point happens to be of zero amplitude. Figure 6.19 also shows the traces of the (120) set of planes on the same lattice projection. In this case the (001)-face centring lattice points lie midway between the planes of the set and scatter X-rays into the 120 reflexion with a phase $2\pi\frac{3}{2}$ relative to the X-rays scattered by the corner lattice points. Destructive interference takes place and the 120 reflexion has zero intensity. But, as in the case of the (200) planes, all lattice points lie on (240) planes, for which $d_{240} = \frac{1}{2}d_{120}$, and the 240 reflexion will, in general, have finite intensity.

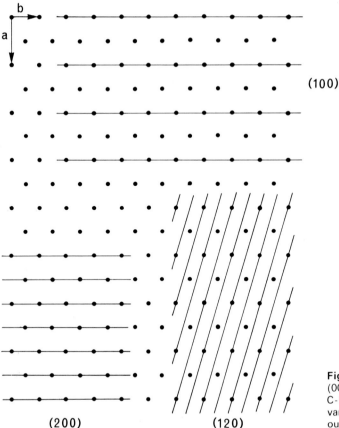

(100)

(200)   (120)

Fig 6.19 Projection on (001) of an orthorhombic C-lattice with the traces of various sets of lattice planes outlined.

Such arguments can be extended to general hkl reflexions: a reflexion hkl will be of non-zero intensity only if all lattice points lie on (hkl) planes. The equation (chapter 1) to a set of planes (hkl) is, in intercept form,

$$\frac{h}{a}x + \frac{k}{b}y + \frac{l}{c}z = n,$$

where n is an integer (positive, zero, or negative). The condition for a lattice point with coordinates x_n, y_n, z_n to lie on a plane of the (hkl) set is then

$$hx_n + ky_n + lz_n = n$$

If this equation is satisfied by all lattice points, then the hkl reflexion will have non-zero intensity; if the equation is not satisfied, a systematic absence will occur. In the case of a C-lattice there are lattice points with coordinates 0, 0, 0 and $\frac{1}{2}$, $\frac{1}{2}$, 0. The lattice point at the origin necessarily lies on all sets of planes; that at $\frac{1}{2}$, $\frac{1}{2}$, 0 lies only on planes for which $\frac{1}{2}h + \frac{1}{2}k$ is equal to an integer, that is on planes for which $h+k$ is even. Thus a systematic absence occurs for reflexions which have $h+k$ odd. A body-centred lattice has lattice points at 0, 0, 0 and $\frac{1}{2}$, $\frac{1}{2}$, $\frac{1}{2}$; the condition for the body-centring lattice points to lie on (hkl) planes is $\frac{1}{2}h + \frac{1}{2}k + \frac{1}{2}l = n$, where n is an integer, i.e. $h+k+l$ must be even for the hkl reflexion to be of non-zero intensity. An I-lattice thus gives rise to a systematic absence when $h+k+l$ is odd. All non-primitive

Table 6.2
Systematic absences displayed by conventional lattice types

	Coordinates of lattice points	Systematic absence
P	$0, 0, 0$	None
A	$0, 0, 0; \; 0, \frac{1}{2}, \frac{1}{2}$	$k + l = 2n + 1$
B	$0, 0, 0; \; \frac{1}{2}, 0, \frac{1}{2}$	$h + l = 2n + 1$
C	$0, 0, 0; \; \frac{1}{2}, \frac{1}{2}, 0$	$h + k = 2n + 1$
F	$0, 0, 0; \; 0, \frac{1}{2}, \frac{1}{2}; \; \frac{1}{2}, 0, \frac{1}{2}; \; \frac{1}{2}, \frac{1}{2}, 0$	h, k, l neither all odd nor all even
I	$0, 0, 0; \; \frac{1}{2}, \frac{1}{2}, \frac{1}{2}$	$h + k + l = 2n + 1$
R (hexagonal axes)	$0, 0, 0; \; \frac{2}{3}, \frac{1}{3}, \frac{1}{3}; \; \frac{1}{3}, \frac{2}{3}, \frac{2}{3}$	$-h + k + l = 3n \pm 1$
R (rhombohedral axes)	$0, 0, 0$	None

unit-cells exhibit such systematic absences; those arising from conventional non-primitive unit-cells are listed in Table 6.2.

The conditions that have to be fulfilled if a reflexion is to be of non-zero intensity when referred to a non-primitive unit-cell can be derived formally from the expression for the structure factor. If the lattice is referred to a C-cell, the N atoms in the unit-cell fall into two equivalent groups, each containing $M = \frac{1}{2}N$ atoms, related to each other by a translation $\frac{1}{2}a + \frac{1}{2}b$. For every atom with coordinates x_n, y_n, z_n there is an atom of the same element with coordinates $\frac{1}{2} + x_n, \; \frac{1}{2} + y_n, \; z_n$. The expression for the structure factor then becomes:

$$F(hkl) = \sum_{1}^{N} f_n \exp 2\pi i(hx_n + ky_n + lz_n)$$

$$= \sum_{1}^{M} f_n \left[\exp 2\pi i\{hx_n + ky_n + lz_n\} + \exp 2\pi i\{h(x_n + \tfrac{1}{2}) + k(y_n + \tfrac{1}{2}) + lz_n\} \right]$$

$$= \sum_{1}^{M} f_n \exp 2\pi i\{hx_n + ky_n + lz_n\} \cdot \left\{ 1 + \exp 2\pi i \frac{h + k}{2} \right\}$$

Now

$$\exp 2\pi i \frac{h + k}{2} = \cos \pi(h + k) + i \sin \pi(h + k)$$

$$= (-1)^{h+k}$$

since h and k are integers. Therefore

$$F(hkl) = \sum_{1}^{M} f_n \exp 2\pi i\{hx_n + ky_n + lz_n\} \cdot \{1 + (-1)^{h+k}\}$$

Therefore for $h + k = 2n$,

$$F(hkl) = 2 \sum_{1}^{M} f_n \exp 2\pi i\{hx_n + ky_n + lz_n\}$$

and for $h + k = 2n + 1$,

$$F(hkl) = 0$$

Thus there is a systematic absence when $h + k$ is odd and when $h + k$ is even the amplitude of the wave scattered by one unit-cell is twice the resultant amplitude of the wave scattered by the atoms associated with any lattice point.

An F-lattice has four lattice points so that an atom at x_n, y_n, z_n is accompanied by atoms of the same element at $x_n, \frac{1}{2}+y_n, \frac{1}{2}+z_n; \frac{1}{2}+x_n, y_n, \frac{1}{2}+z_n;$ and $\frac{1}{2}+x_n, \frac{1}{2}+y_n, z_n.$ If the number of atoms associated with each lattice point is $M = \frac{1}{4}N$, the structure factor is

$$F(hkl) = \sum_1^M f_n[\exp 2\pi i\{hx_n + ky_n + lz_n\}$$
$$+ \exp 2\pi i\{hx_n + k(\tfrac{1}{2}+y_n) + l(\tfrac{1}{2}+z_n)\}$$
$$+ \exp 2\pi i\{h(\tfrac{1}{2}+x_n) + ky_n + l(\tfrac{1}{2}+z_n)\}$$
$$+ \exp 2\pi i\{h(\tfrac{1}{2}+x_n) + k(\tfrac{1}{2}+y_n) + lz_n\}]$$
$$= \sum_1^M f_n \exp 2\pi i\{hx_n + ky_n + lz_n\}$$
$$\times \{1 + \exp \pi i(k+l) + \exp \pi i(h+l) + \exp \pi i(h+k)\}$$
$$= \sum_1^M f_n \exp 2\pi i\{hx_n + ky_n + lz_n\} \cdot \{1 + (-1)^{k+l} + (-1)^{h+l} + (-1)^{h+k}\}$$

If $h, k,$ and l are all odd or all even, $k+l, h+l,$ and $h+k$ will all be even and

$$F(hkl) = 4 \sum_1^M f_n \exp 2\pi i\{hx_n + ky_n + lz_n\}$$

But if $h, k,$ and l are not all odd or all even, two of the sums $k+l, h+l,$ and $h+k$ will be odd and the third even so that

$$F(hkl) = 0$$

Thus there is a systematic absence when $h, k,$ and l are neither all odd nor all even; but when $h, k,$ and l are all odd or all even the structure factor is four times that for the group of atoms associated with one lattice point.

Evidently then the labour of calculating structure amplitudes for non-primitive unit-cells will be appreciably lessened if all atoms that are related by lattice translations are first grouped together. By way of example we consider the cubic form of ZnS, the mineral *blende*, which has four formula units in the unit-cell and an F-lattice (Fig 6.20). One formula unit is necessarily associated with each lattice point and we choose to take as the repeat unit a zinc atom at the origin and a sulphur atom at $\frac{1}{4}, \frac{1}{4}, \frac{1}{4}$. Since we are dealing with an F-lattice reflexions of non-zero

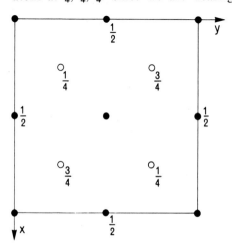

Fig 6.20 Projection of the structure of blende (ZnS) on (001). Solid circles represent Zn and open circles S atoms.

intensity will occur only when h, k, and l are all odd or all even, when the structure factor will be

$$F(hkl) = 4 \sum_{1}^{M} f_n \exp 2\pi i(hx_n + ky_n + lz_n)$$

Here $M = 2$, therefore

$$F(hkl) = 4\left\{ f_{Zn} \exp 2\pi i.0 + f_S \exp 2\pi i \frac{h+k+l}{4} \right\}$$

Therefore $\quad |F(hkl)|^2 = 16\left\{ \left(f_{Zn} \cos 0 + f_S \cos 2\pi \frac{h+k+l}{4} \right)^2 \right.$

$$\left. + \left(f_{Zn} \sin 0 + f_S \sin 2\pi \frac{h+k+l}{4} \right)^2 \right\}$$

$$= 16\left\{ \left(f_{Zn} + f_S \cos 2\pi \frac{h+k+l}{4} \right)^2 + \left(f_S \sin 2\pi \frac{h+k+l}{4} \right)^2 \right\}$$

Only three possibilities arise:

$$\begin{array}{lll} \text{if} & h+k+l = 4n & |F(hkl)|^2 = 16(f_{Zn} + f_S)^2 \\ \text{if} & h+k+l = 4n+2 & |F(hkl)|^2 = 16(f_{Zn} - f_S)^2 \\ \text{if} & h+k+l = 2n+1 & |F(hkl)|^2 = 16(f_{Zn}^2 + f_S^2) \end{array}$$

In the foregoing pages we have explored the systematic absences due to lattice type by examining the structure factors for the various lattice types. The resulting list (Table 6.2) of systematic absences provides the means of determining the lattice type of a crystalline substance by inspection of its diffraction pattern for systematically absent reflexions. Later we shall return to systematic absences to consider those due to the presence of translational symmetry elements.

Symmetry of X-ray diffraction patterns
Friedel's Law

We have seen that the intensities of X-ray reflexions are dependent on the positions of the atoms in the unit-cell and this would lead us to expect that the diffraction pattern produced by a crystal would exhibit a symmetry related to that of the disposition of the atoms in the unit-cell. Such a relationship would provide a means of determining the symmetry of the arrangement of atoms in the crystal by examination of its diffraction pattern. There is however a difficulty: there is no known means of determining the relative phases of X-ray reflexions. All that can be determined about an X-ray reflexion is its intensity and its angular position; the information about the symmetry of the atomic arrangement in a crystal that can be deduced from the symmetry of its diffraction pattern is in consequence limited.

We have already shown that the intensity of an X-ray reflexion is the product of the relevant structure factor and its complex conjugate,

i.e. $\quad F(hkl) = \sum_{1}^{N} f_n \cos 2\pi(hx_n + ky_n + lz_n) + i \sum_{1}^{N} f_n \sin 2\pi(hx_n + ky_n + lz_n)$

and $\quad F(hkl)^* = \sum_{1}^{N} f_n \cos 2\pi(hx_n + ky_n + lz_n) - i \sum_{1}^{N} f_n \sin 2\pi(hx_n + ky_n + lz_n)$

But $\qquad \cos(-\theta) = \cos\theta \quad$ and $\quad \sin(-\theta) = -\sin\theta$

Therefore $\quad F(hkl)^* = \sum_1^N f_n \cos 2\pi(-hx_n - ky_n - lz_n) + i\sum_1^N f_n \sin 2\pi(-hx_n - ky_n - lz_n)$

$$= F(\bar{h}\bar{k}\bar{l})$$

and similarly

$$F(hkl) = F(\bar{h}\bar{k}\bar{l})^*$$

Therefore $\quad I(hkl) = F(hkl)F(hkl)^* = F(\bar{h}\bar{k}\bar{l})^*F(\bar{h}\bar{k}\bar{l})$

$$= I(\bar{h}\bar{k}\bar{l}).$$

The intensities of the hkl and $\bar{h}\bar{k}\bar{l}$ reflexions are thus necessarily equal and in consequence an X-ray diffraction pattern always displays a centre of symmetry[10] or, in other words, the reflexions from either side of a set of (hkl) planes are invariably equal in intensity (Fig 6.21); this result is often called *Friedel's Law*.

If the crystal structure does not possess a centre of symmetry, although the intensities of the hkl and $\bar{h}\bar{k}\bar{l}$ reflexions must be equal their phases will be unequal. Consider the 111 and $\bar{1}\bar{1}\bar{1}$ reflexions of blende illustrated in Fig 6.21(a). The structure of this form of ZnS (Fig 6.20) is non-centrosymmetric, having a repeat unit consisting of Zn at $0, 0, 0$ and S at $\frac{1}{4}, \frac{1}{4}, \frac{1}{4}$. The wave scattered by zinc into all reflexions thus has zero phase and amplitude f_{Zn}, while that scattered by sulphur has phase $2\pi(h+k+l)/4$ and amplitude f_S. Since the (111) and ($\bar{1}\bar{1}\bar{1}$) planes are of the same set, the atomic scattering factors of Zn and S will be the same for both. It will be apparent from Fig 6.21(b) and (c) that the amplitude and consequently the intensity of the 111 and $\bar{1}\bar{1}\bar{1}$ reflexions will be equal and that their phases will be equal in magnitude but opposite in sign.

Friedel's Law holds only so long as the assumption that every atom in the structure scatters X-rays with a phase change of π remains valid. The assumption breaks down only when the X-ray wavelength is close to the wavelength of an absorption edge of one of the constituent atomic species of the crystal. Discussion of such *anomalous scattering* is outside the scope of this book; the reader is referred to James (1967), for a detailed treatment. It is important however to bear in mind that, in favourable circumstances, anomalous scattering may provide a means of establishing the absence of a centre of symmetry in a crystal structure (chapter 9).

Laue symmetry

If two or more planes are related by symmetry then the X-ray reflexions to which they give rise will have equal Bragg angles (but will in general require different orientations of the incident X-ray beam) and equal intensities. Therefore examination of the symmetry of the total diffraction pattern produced by a crystal should yield information about the point group of the crystal; but because diffraction patterns are, by Friedel's Law, necessarily centrosymmetric a unique determination of the point group is not possible. The point group symmetry of the diffraction pattern will be the same as that of the crystal if the crystal is centrosymmetric; but, if the crystal

---

[10] The reader should be aware that this statement refers to the whole diffraction pattern. Since a crystal is effectively a three-dimensional grating the whole diffraction pattern cannot be recorded with a single orientation of the crystal relative to the incident X-ray beam and consequently the symmetry of the whole diffraction pattern may not be evident in a single X-ray photograph. This point will be amplified in chapter 8.

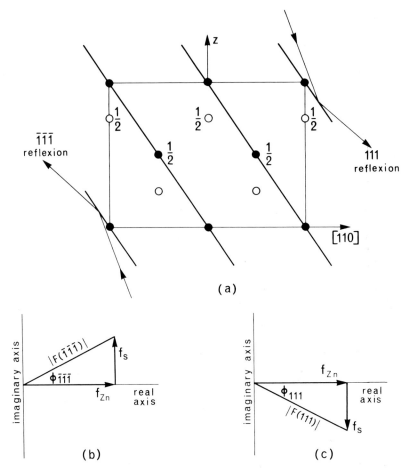

Fig 6.21 Friedel's Law. (a) is a projection of the structure of blende (Zn solid circles, S open circles) on the (1̄10) plane; atomic heights above that plane are shown as fractions of the lattice repeat $(a/\sqrt{2})$ in the [1̄10] direction. (b) and (c) are vector diagrams representing the amplitude and phase of the 1̄1̄1̄ and 111 reflexions.

is non-centrosymmetric, the point group of the diffraction pattern will be the point group obtained by adding a centre of symmetry to the point group of the crystal. For instance the symmetry of the diffraction pattern produced by a crystal of class 4 will be that of the point group $4/m$ (Fig 6.22). Conversely, if the diffraction symmetry of a crystal is identified as $4/m$, then the point group of the crystal may be $4/m$ or any point group that becomes $4/m$ when a centre of symmetry is added to it; reference to Fig 3.17 enables the non-centrosymmetric point groups to be identified as 4 and 4̄. Such a crystal is said to belong to the *Laue class* $4/m$, a statement which implies that the point group of the crystal is either 4 or 4̄ or $4/m$.

There are eleven centrosymmetric crystallographic point groups and it is evident from Fig 3.20 that addition of a centre of symmetry to the symmetry elements of any of the twenty-one non-centrosymmetric point groups yields one of these eleven. A centrosymmetric point group and all those non-centrosymmetric point groups which on addition of a centre of symmetry become identical with it constitute a *Laue class*. Each such group of point groups is assigned the symbol of its centrosymmetric point

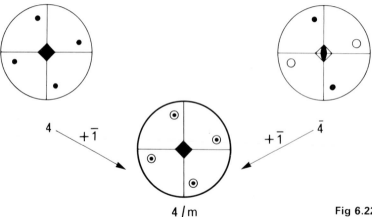

Fig 6.22 Laue class 4/m.

group; such a Laue class symbol is a statement of the point group symmetry of the diffraction pattern produced by a crystal of any point group in the Laue class. The point group symmetry of the diffraction pattern produced by a crystal is known as the *Laue symmetry* of the crystal. Examination of the symmetry of its diffraction pattern can only assign a crystal to a particular Laue class and cannot determine the point group of the crystal uniquely. The eleven Laue classes are listed in Table 6.3.

Table 6.3
The eleven Laue Classes

System	Laue class	Constituent point groups of the Laue group			
Triclinic	$\bar{1}$	1	$\bar{1}$		
Monoclinic	$2/m$	2	m	$2/m$	
Orthorhombic	mmm	222	$mm2$	mmm	
Trigonal	$\bar{3}$ $\bar{3}m$	3 32	$\bar{3}$ $3m$	$\bar{3}m$	
Tetragonal	$4/m$ $4/mmm$	4 422	$\bar{4}$ $4mm$	$4/m$ $\bar{4}2m$	$4/mmm$
Hexagonal	$6/m$ $6/mmm$	6 622	$\bar{6}$ $6mm$	$6/m$ $\bar{6}m2$	$6/mmm$
Cubic	$m3$ $m3m$	23 432	$m3$ $\bar{4}3m$	$m3m$	

Systematic absences due to translational elements

We have already seen that selection of a non-primitive unit-cell gives rise to systematically absent reflexions and that the rules governing such absences apply to all reflexions from the crystal. But systematic absences also arise from the presence of glide planes or screw axes in the space group of a crystal; such absences are restricted to one zone of planes, in the case of a glide plane, or to one set of planes, in the case of a screw axis.

The presence of an *a*-glide through the origin parallel to (001) causes an atom at x, y, z to be duplicated at $\frac{1}{2}+x$, y, $\bar{z}$. The expression for the structure factor then becomes

$$F(hkl) = \sum_{1}^{\frac{1}{2}N} f_n \{ \exp 2\pi i(hx_n + ky_n + lz_n) + \exp 2\pi i(\tfrac{1}{2}h + hx_n + ky_n - lz_n) \}$$

$$= \sum_{1}^{\frac{1}{2}N} f_n \exp 2\pi i(hx_n + ky_n) \{ \exp 2\pi i\, lz_n + \exp 2\pi i(\tfrac{1}{2}h - lz_n) \}$$

which becomes when $l = 0$,

$$F(hk0) = \sum_{1}^{\frac{1}{2}N} f_n \exp 2\pi i(hx_n + ky_n) \{ 1 + \exp \pi i\, h \}$$

$$= \sum_{1}^{\frac{1}{2}N} f_n \exp 2\pi i(hx_n + ky_n) \{ 1 + (-1)^h \}$$

$$= 0 \quad \text{for} \quad h = 2n + 1.$$

Thus an a-glide through the origin parallel to (001) produces a systematic absence in the $hk0$ reflexions when h is odd. The restriction that the glide plane should pass through the origin is not significant because it is evident that the phase of the X-rays scattered into the $hk0$ reflexions must be independent of the z coordinates of the atoms in the unit-cell, the phase difference between the X-rays scattered into any $hk0$ reflexion by two atoms related by an a-glide parallel to (001) being

$$2\pi(\tfrac{1}{2}h + hx_n + ky_n) - 2\pi(hx_n + ky_n) = \pi h;$$

wherever the glide plane intersects the z-axis the two atoms will scatter exactly out of

Table 6.4
Systematic absences produced by glide planes parallel to (001)

Type of glide	Translation	Systematic absences in $hk0$ reflexions
a	$\dfrac{a}{2}$	$h = 2n + 1$
b	$\dfrac{b}{2}$	$k = 2n + 1$
n	$\dfrac{a+b}{2}$	$h + k = 2n + 1$
d	$\dfrac{a \pm b}{4}$	$h + k = 4n + 2$ with $h = 2n$ and $k = 2n$

phase when h is odd to give a systematic absence and exactly in phase when h is even. Such a glide plane in effect halves the lattice spacing parallel to the x-axis where reflexions in the [001] zone are concerned.

All glide planes give rise to systematic absences in the reflexions of the zone whose axis is normal to the glide plane. The systematic absences produced by all possible types of (001) glide plane are listed in Table 6.4. Analogous systematic absences arise from glide planes parallel to (100) and (010). A complete list of the conventional glide planes and their associated systematic absences is to be found in *International Tables for X-ray Crystallography*, vol. I (1969), p. 54.

The presence of a screw diad through the origin parallel to the z-axis causes an atom at x, y, z to be duplicated at $\bar{x}$, $\bar{y}$, $\tfrac{1}{2}+z$. The expression for the structure factor then becomes

Table 6.5
Systematic absences produced by screw axes parallel to [001]

Screw axis	Translation	Systematic absences in $00l$ reflexions
2_1	$c/2$	$l = 2n+1$
4_1 and 4_3	$\pm c/4$	$l \neq 4n$
4_2	$c/2$	$l = 2n+1$
3_1 and 3_2	$\pm c/3$	$l \neq 3n$
6_1 and 6_5	$\pm c/6$	$l \neq 6n$
6_2 and 6_4	$\pm c/3$	$l \neq 3n$
6_3	$c/2$	$l = 2n+1$

$$F(hkl) = \sum_1^{\frac{1}{2}N} f_n\{\exp 2\pi i(hx_n + ky_n + lz_n) + \exp 2\pi i(-hx_n - ky_n + \tfrac{1}{2}l + lz_n)\}$$

$$= \sum_1^{\frac{1}{2}N} f_n \exp 2\pi i\, lz_n\{\exp 2\pi i(hx_n + ky_n) + \exp 2\pi i(-hx_n - ky_n + \tfrac{1}{2}l)\}$$

which becomes when $h = k = 0$,

$$F(00l) = \sum_1^{\frac{1}{2}N} f_n \exp 2\pi i\, lz_n\{1 + \exp \pi i\, l\}$$

$$= \sum_1^{\frac{1}{2}N} f_n \exp 2\pi i\, lz_n\{1 + (-1)^l\}$$

$$= 0 \quad \text{for} \quad l = 2n+1$$

Thus a screw diad parallel to z produces a systematic absence in the $00l$ reflexions when l is odd. The restriction that the 2_1 axis should pass through the origin is not significant because the phase of the X-rays scattered into the $00l$ reflexions must be independent of the x and y coordinates of the atoms in the unit-cell, the phase difference between the X-rays scattered into any $00l$ reflexion by two atoms related by a 2_1 axis parallel to z being $2\pi(\tfrac{1}{2}l + lz_n) - 2\pi lz_n = \pi l$; wherever the screw axis intersects the xy plane the two atoms will scatter exactly out of phase when l is odd and exactly in phase when l is even. Such a screw diad effectively halves the lattice spacing parallel to the z-axis for the $00l$ reflexions.

Screw axes parallel to other axes and with different translations give rise to analogous systematic absences. A list of systematic absences produced by all possible types of screw axis parallel to [001] is given in Table 6.5.

Diffraction symbols
We have already seen that observation of the symmetry of the diffraction pattern produced by a crystal enables the crystal to be assigned to a particular Laue class. Observation of systematic absences in the diffraction pattern enables the lattice type to be determined uniquely and the presence of translational symmetry elements to be detected. All this information can conveniently be expressed as the *diffraction symbol* of the crystal. Since non-translational symmetry elements do not give rise to systematic absences the diffraction symbol will represent a group of space groups; occasionally this will be a group of one and then the space group is uniquely determined.

A diffraction symbol consists of three parts: first, a statement of the Laue class; second, a statement of the lattice type; third, three spaces in which the symbols of any screw axes or glide-planes that have been detected are written in the same order as

Table 6.6
Diffraction symbols derived from
*mmm*P.*cn* by reorientation of axes

The original axes are denoted *x y z* and the
new axes *x' y' z'*.

x'	*y'*	*z'*	Diffraction symbol
x	*y*	*z*	*mmm* P.*cn*
z	*x*	*y*	*mmm* P*n*.*a*
y	*z*	*x*	*mmm* P*bn*.
x	*z̄*	*y*	*mmm* P.*nb*
y	*x*	*z̄*	*mmm* P*c*.*n*
z̄	*y*	*x*	*mmm* P*na*.

the corresponding non-translational symmetry element symbols appear in the relevant conventional point group symbol, absence of information about any of these three directions being indicated by a full stop. By way of example we take the diffraction symbol *mmm* P*bcn*; this means that the crystal belongs to Laue class *mmm*, has a primitive lattice, and has a *b*-glide parallel to (100), a *c*-glide parallel to (010), and an *n*-glide parallel to (001). In this case the space group is uniquely determined as P*bcn*. Suppose now that another crystal yields a diffraction pattern of symmetry *mmm* with systematic absences in *h0l* for $l = 2n+1$, *hk0* for $h+k = 2n+1$, *h00* for $h = 2n+1$, *0k0* for $k = 2n+1$, *00l* for $l = 2n+1$. The diffraction symbol of this crystal will be *mmm* P.*cn* and this implies that its point group must be either *mmm* or 2*mm*; the systematic absences observed in the *h00*, *0k0*, and *00l* reflexions are merely consequent on the more general conditions that apply to the *h0l* and *hk0* reflexions and do not necessarily imply the presence of screw diads parallel to *x*, *y*, and *z*. In this case however it can readily be seen by drawing out the space group diagrams for P.*cn* that this pair of glide-planes generates screw diads parallel to *x*. We know therefore that there is no (100) glide-plane and that there must be a [100] screw diad; there may or may not be a (100) mirror-plane. The determination of space group is not unique; there are two possibilities, P2₁*cn* (which belongs to point group 2*mm*) and P*mcn* (which belongs to point group *mmm* and is illustrated in Fig 4.18).

When the diffraction symbol of a crystal has been obtained it should be compared with the list of diffraction symbols of the space groups in *International Tables for X-ray Crystallography*, vol. I (1969), pp. 349–352; but a word of caution is necessary. The list has been drawn up in terms of certain arbitrary conventions so that our observed diffraction symbol may not appear in the list but be represented by the corresponding diffraction symbol for a different axial orientation; this difficulty, which is most likely to occur in the orthorhombic system, is simply resolved by transforming the observed diffraction symbol for all possible axial orientations and selecting the setting consistent with the *International Tables* convention. Alternative settings for *mmm* P.*cn* are listed in Table 6.6.

In the case of monoclinic diffraction symbols the first term, the Laue class symbol, is written in such a way as to indicate unambiguously the choice of unique axis: as 12/*m*1 if *y* is the unique axis and as 112/*m* if *z* is the unique axis. Also in this system three spaces are provided in the last term; but only one can be used, the second if *y* is unique and the third if *z* is unique. For instance consider a crystal with Laue symmetry 2/*m*, the diad being parallel to *y*, and systematic absences for *hkl* when $h+k = 2n+1$, for *h0l* when $h = 2n+1$, or $l = 2n+1$, for *0k0* when $k = 2n+1$. Its diffraction symbol is $12/m1C.c.$; the significant absences are in hkl for $h+k$ odd and in $h0l$ for l odd, the other observed absences being consequent on these. In the monoclinic system it is important to bear in mind that symmetry elements are associated with one axial direction only; if that is chosen as y, then systematic absences in the $0kl$ and $hk0$ reflexions can only be due to systematic absences in the general hkl reflexions and systematic absences in $h00$ and $00l$ reflexions may be derivative from systematic absences in either hkl or $h0l$ reflexions. Returning to our example, $12/m1C.c.$, two space groups are possible, Cc or $C2/c$. The former belongs to point group m. The latter belongs to point group $2/m$, has diads and screw diads parallel to y, and has an n-glide as well as a c-glide parallel to (010) (Fig 4.19); the diad does not give rise to any systematic absences, the systematic absences due to the screw diad, $0k0$ for $k = 2n+1$, are masked by those due to the C-lattice, hkl for $h+k = 2n+1$, and the systematic absences due to the n-glide, $h0l$ for $h+l = 2n+1$, are masked by the conditions for reflexion by a C-lattice and a c-glide, which require h and l to be even for a reflexion to be observed.

The reciprocal lattice and the reflecting sphere

The interpretation of X-ray diffraction patterns by direct application of the Laue Equations or the Bragg Equation can be highly tedious. The concepts of the reciprocal lattice and the reflecting sphere provide however a means of interpretation that is at once elegant, powerful and easy too use.

We take as our starting point the general expression for the complex amplitide a_R of the resultant disturbance ψ_R at a point when N disturbances $\psi_1, \psi_2, \ldots \psi_N$ arrive simultaneously at the point, $a_R = \sum_1^N a_n = \sum_1^N |a_n| \exp(i\phi_n)$, which was developed earlier in this chapter. We have already shown that this general expression leads in the case of scattering of radiation by a crystal to the expression for the structure factor

$$F(hkl) = \sum_1^N f_n \exp 2\pi i(hx_n + ky_n + lz_n) \tag{16}$$

This relationship in the form of equation (16) is appropriate for the calculation of structure factors; when rewritten in vector form it also provides an elegant way of demonstrating the orientational relationship of a crystal to the incident and diffracted X-ray beams, a subject which is our immediate concern.

The phase of the radiation scattered by a point P at a distance r from the origin O can, as we have shown earlier in this chapter, be obtained from the path difference

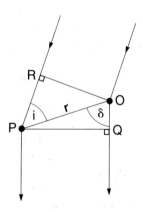

Fig 6.23 The path difference Δ between the waves scattered by the points P and O is given by $\Delta = OQ - RP = OP(\cos \delta - \cos i)$.

between the waves scattered by the points P and O. It is apparent from Fig 6.23 that the path difference Δ is given by

$$\Delta = OQ - RP = OP(\cos \delta - \cos i). \tag{17}$$

The path difference can be expressed as the difference between two scalar products when the direction of the incident beam is taken parallel to the vector $\mathbf{s}_o$, the direction of the diffracted beam is taken parallel to the vector $\mathbf{s}$ and the distance between the two scattering points OP is represented by the vector $\mathbf{r}$. Then the path difference $\Delta = OP(\cos \delta - \cos i) = (1/|\mathbf{s}|)\mathbf{r}.\mathbf{s} - (1/|\mathbf{s}_o|)\mathbf{r}.\mathbf{s}_o$. This expression can be simplified by putting $|\mathbf{s}| = |\mathbf{s}_o|$ so that $\Delta = (1/|\mathbf{s}|)\mathbf{r}.(\mathbf{s} - \mathbf{s}_o)$; further simplification can be achieved by introducing the vector $\mathbf{S} = \mathbf{s} - \mathbf{s}_o$ so that $\Delta = (1/|\mathbf{s}|)\mathbf{r}.\mathbf{S}$. We can now express the phase ϕ of the wave scattered from the point P relative to the wave scattered from the origin O as

$$\phi = \frac{2\pi}{\lambda}.\Delta$$

$$= \frac{2\pi}{\lambda}.\frac{1}{|\mathbf{s}|}\mathbf{r}.\mathbf{S}$$

and if we put $|\mathbf{s}| = \frac{1}{\lambda}$,

$$\phi = 2\pi\mathbf{r}.\mathbf{S} \tag{18}$$

and the expression for the structure factor (16) becomes

$$F(\mathbf{S}) = \sum_{1}^{N} f_n \exp 2\pi i \mathbf{r}.\mathbf{S} \tag{19}$$

The vector $\mathbf{S}$ is known as the *scattering vector* and provides a very useful statement of the relationship between the incident and the diffracted beams (Fig 6.24). Since we have put $|\mathbf{s}| = |\mathbf{s}_o| = 1/\lambda$, the triangle OTU is isosceles; and if the angle between the incident and diffracted beams, the scattering angle, is denoted as 2θ, then

$$|\mathbf{S}| = 2|\mathbf{s}|\sin\theta = \frac{2\sin\theta}{\lambda} \tag{20}$$

Moreover the incident and the diffracted beams are equally inclined at $90° - \theta$ to $\mathbf{S}$ and are of course coplanar with $\mathbf{S}$.

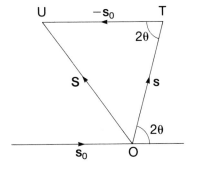

Fig 6.24 Definition of the scattering vector $\mathbf{S}$. Putting $|\mathbf{s}| = |\mathbf{s}_o| = 1/\lambda$ and denoting the scattering angle as 2θ, $|\mathbf{S}| = 2\sin\theta/\lambda$. The incident and the diffracted beams are equally inclined at $90° - \theta$ to $\mathbf{S}$ and are coplanar with $\mathbf{S}$.

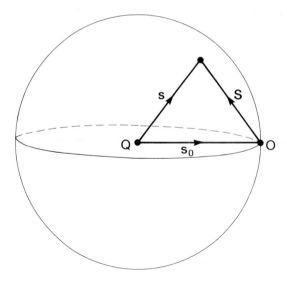

Fig 6.25 The reflecting sphere. The incident beam s_0 is represented by QO. $|s| = |s_0| = 1/\lambda$. The possible vectors s representing the diffracted beam define a sphere of centre Q and radius $1/\lambda$. The scattering vector $S = -s_0 + s$ has both its ends on the surface of the reflecting sphere.

The reflecting sphere

When the direction of the incident beam s_0 is fixed, a restriction is in consequence placed on the scattering vector S. The diffracted beam s may lie in any direction but its magnitude is fixed by $|s| = |s_0| = 1/\lambda$; therefore if all the possible vectors s are drawn from a common point Q, their ends will lie on the surface of a sphere of centre Q and radius $|s| = 1/\lambda$. If the incident beam is specified as $QO = s_0$, then the point O lies on the sphere (Fig. 6.25) and the scattering vector $S = -s_0 + s$ has both its ends, one being at O, on the surface of the sphere.

The moduli of the vectors s_0, s, and S are, as we have seen, respectively $1/\lambda$, $1/\lambda$, and $2\sin\theta/\lambda$; their dimensions are thus L^{-1} and they may be said to be vectors in reciprocal space. For any given direction of the incident beam the scattering vector S drawn from the origin O is, as we have seen, constrained to be a chord of the sphere of radius s_0 drawn on the incident beam as diameter so as to pass through the origin O. This sphere is called the *reflecting sphere* or the *Ewald sphere*[11]; its use in the interpretation of single crystal X-ray diffraction patterns will be developed in chapter 8 and in the interpretation of electron diffraction patterns in chapter 10.

The reciprocal lattice and the Laue Equations

It was shown earlier in this chapter that the diffraction condition for a three dimensional lattice is

$$\left.\begin{array}{l} a(\cos\delta_a - \cos i_a) = h\lambda \\ b(\cos\delta_b - \cos i_b) = k\lambda \\ c(\cos\delta_c - \cos i_c) = l\lambda \end{array}\right\} \tag{7}$$

where the incident beam is inclined at the angles i_a, i_b, i_c to the *negative* directions of the x, y, z axes respectively and the diffracted beam is inclined at the angles δ_a, δ_b, δ_c to the *positive* directions of the x, y, z axes respectively.

[11] P. P. Ewald was the first to apply reciprocal lattice theory to the diffraction of X-rays by crystals.

Rearrangement of the first Laue Equation as

$$\frac{a}{\lambda}(\cos\delta_a - \cos i_a) = h$$

followed by substitution of $\mathbf{s}_o$ for the incident beam and $\mathbf{s}$ for the diffracted beam, both being vectors of modulus $1/\lambda$, leads to the expression

$$\mathbf{a}.(\mathbf{s} - \mathbf{s}_o) = h$$

i.e. $$\mathbf{a}.\mathbf{S} = h$$

We can thus express the Laue Equations in vector form as

$$\left.\begin{array}{l} \mathbf{a}.\mathbf{S} = h \\ \mathbf{b}.\mathbf{S} = k \\ \mathbf{c}.\mathbf{S} = l \end{array}\right\} \tag{21}$$

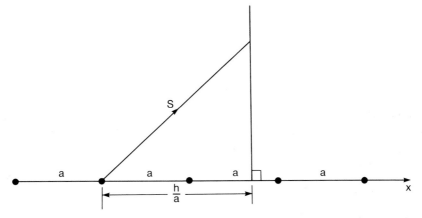

Fig 6.26 The first Laue Equation restricts the scattering vector **S** to directions for which its projection on the x-axis is h/a.

It is immediately apparent from equations (21) that the Laue Equations impose restrictions on the scattering vector. The first Laue Equation restricts **S** to such directions that its projection on the x axis $\mathbf{S}.\mathbf{a}/|\mathbf{a}| = h/a$ (Fig 6.26): when $h = 0$ **S** is normal to the x axis, when $h = 1$ the projection on the x axis is $1/a$, and so on (Fig 6.27). Thus when **S** is drawn from the origin the first Laue Equation implies that all possible scattering vectors must end on a set of parallel planes normal to the x axis and of interplanar spacing $1/a$. Likewise the second Laue Equation constrains **S** to

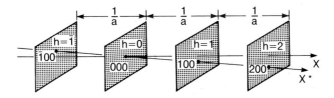

Fig 6.27 Reciprocal lattice points lie on planes perpendicular to the x-axis, each plane being characterized by a particular value of h and the interplanar spacing being $1/a$. Analogous planes with separation $1/b$, $1/c$ occur perpendicular to the y and z axes respectively.

end on a set of planes normal to the y axis and of interplanar spacing $1/b$, while the third Laue Equation constrains S to a set of planes normal to the z-axis and of spacing $1/c$. So all possible vectors S drawn from a common origin generate a lattice, the points of the lattice being situated at the mutual intersections of three sets of parallel planes.

It was shown in chapter 5 (p. 147) that the relationship between the direct and the reciprocal lattices is such that the reciprocal lattice points lie on planes normal to a zone axis $[UVW]$, the interplanar spacing being $1/t_{[UVW]}$. Since all possible values of S lie on planes normal to the x, y, and z axes and of spacing $1/a$, $1/b$, and $1/c$ respectively, then permissible scattering vectors S for a given set of hkl values are identical to the reciprocal lattice vector $\mathbf{d^*} = h\mathbf{a^*} + k\mathbf{b^*} + l\mathbf{c^*}$ for the same set of integers hkl. That this is so may be seen readily by substituting $S = h\mathbf{a^*} + k\mathbf{b^*} + l\mathbf{c^*}$ into each of the Laue Equations in turn:

$$\mathbf{a}.S = \mathbf{a}.(h\mathbf{a^*} + k\mathbf{b^*} + l\mathbf{c^*})$$

$$= h \text{ since } \mathbf{a}.\mathbf{a^*} = 1 \text{ and } \mathbf{a}.\mathbf{b^*} = \mathbf{a}.\mathbf{c^*} = 0,$$

$$\mathbf{b}.S = \mathbf{b}.(h\mathbf{a^*} + k\mathbf{b^*} + l\mathbf{c^*})$$

$$= k \text{ since } \mathbf{b}.\mathbf{b^*} = 1 \text{ and } \mathbf{b}.\mathbf{a^*} = \mathbf{b}.\mathbf{c^*} = 0,$$

$$\mathbf{c}.S = \mathbf{c}.(h\mathbf{a^*} + k\mathbf{b^*} + l\mathbf{c^*})$$

$$= l \text{ since } \mathbf{c}.\mathbf{c^*} = 1 \text{ and } \mathbf{c}.\mathbf{a^*} = \mathbf{c}.\mathbf{b^*} = 0.$$

The relationship between direct and reciprocal lattices is illustrated for the case of the (010) plane of a typical primitive monoclinic lattice in Fig 6.28. Lines of reciprocal

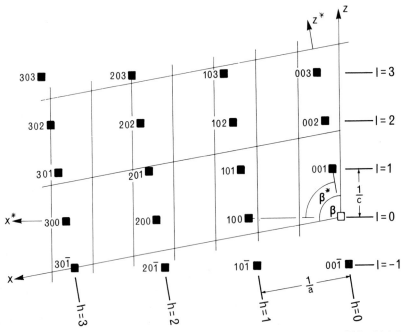

Fig 6.28 Relationship between reciprocal lattice points (solid squares except 000 which is an open square) and the direct lattice for the xz plane of a primitive monoclinic lattice. The thin lines represent the (100) and (001) planes of the direct lattice. The bold lines on the edges of the diagram indicate reciprocal lattice point rows of constant h or constant l in the $h0l$ reciprocal lattice plane. Reciprocal and direct axial directions are indicated.

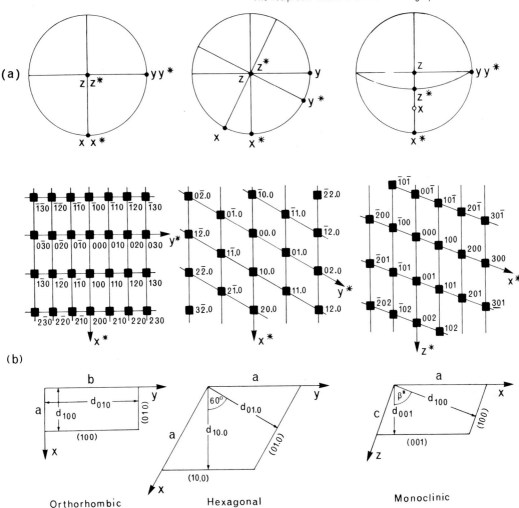

Fig 6.29   The angular relationships between reciprocal and direct axes in the orthorhombic, hexagonal and monoclinic systems are shown in the top row (a). Selected reciprocal lattice sections for these three systems are shown in (b), with diagrams to illustrate the relationship between direct and reciprocal lattice axial lengths in the bottom row.

lattice points $00l$, $10l$, $20l$, and $30l$ are normal to the x-axis and are spaced $1/a$ apart. Similarly lines of reciprocal lattice points $h0\bar{1}$, $h00$, $h01$, $h02$ and $h03$ are normal to the z axis and are spaced $1/c$ apart. Permitted scattering vectors **S** correspond to the reciprocal lattice vectors $h\mathbf{a}^* + l\mathbf{c}^*$, where $\mathbf{a}^*$ and $\mathbf{c}^*$ are the base vectors of the reciprocal lattice.

The interrelationships between direct lattice constants and reciprocal lattice constants were developed for the triclinic system in chapter 5 and are shown for all crystal systems in Table 5.2. In the solution of diffraction problems it is frequently necessary to draw a plane of reciprocal lattice points in its precise relationship to the direct lattice. This task presents no problem in the *cubic, tetragonal*, and *orthorhombic* systems where x^*, y^*, z^* are parallel to x, y, z and $a^* = 1/a$, $b^* = 1/b$, $c^* = 1/c$ (Fig 6.29). But in the hexagonal, trigonal, and monoclinic systems only the direct lattice axis which is perpendicular to the other two direct lattice axes is also parallel to the corresponding reciprocal lattice axis. Thus in the *hexagonal* and *trigonal* systems, also illustrated in Fig 6.29, z is normal to x and y and, since in general $\mathbf{c}^*.\mathbf{a} = \mathbf{c}^*.\mathbf{b} = 0$, z^* is parallel to z. The other two reciprocal lattice axes, x^* and y^*, are normal to z. The direction of x^* can be found by locating the direction normal to y and to z, that is the normal to the (10.0) plane. Since $\mathbf{a}.\mathbf{a}^* = 1$, $a^* = 1/a\cos 30°$; so the magnitude of a^* is the reciprocal of the spacing of the (10.0) planes. Likewise y^* is perpendicular to x and to z and consequently normal to the (01.0) plane; since $\mathbf{b}.\mathbf{b}^* = 1$ and $b = a$, $b^* = 1/a\cos 30° = 1/d_{(01.0)}$. In the *monoclinic* system y is normal to x and to z so that y^* is parallel to y and $b^* = 1/b$. The other two reciprocal lattice axes are perpendicular to y and so lie in the (010) plane as illustrated in Fig 6.29. Since x^* is perpendicular to z and $\mathbf{a}.\mathbf{a}^* = 1$ so that $a^* = 1/a\sin \beta$, x^* is parallel to the normal to (100) and a^* is the reciprocal of the spacing of the (100) planes. Similarly z^* is perpendicular to x and so normal to (001); $c^* = 1/c\sin \beta = 1/d_{(001)}$. In the *triclinic* system a reciprocal lattice axis is not in general coplanar with any pair of direct lattice axes and vice versa; in consequence each diffraction problem of the sort we have discussed generally in all other systems has to be treated as a special case.

The Laue Equations and the Bragg Equation
Rewriting the Laue Equations (21) in the form

$$\frac{1}{h}\mathbf{a}.\mathbf{S} = 1$$

$$\frac{1}{k}\mathbf{b}.\mathbf{S} = 1$$

$$\frac{1}{l}\mathbf{c}.\mathbf{S} = 1$$

it becomes apparent that they represent possible solutions of the equation $\mathbf{t}.\mathbf{S} = 1$, where $\mathbf{S}$ is normal to a plane and $\mathbf{t}$ is a vector from the origin to the plane. Moreover $\mathbf{S}$ is normal to the plane which makes intercepts a/h, b/k, c/l on the x, y, and z axes respectively (Fig 6.30), that is to say it is normal to the set of lattice planes (hkl). Since

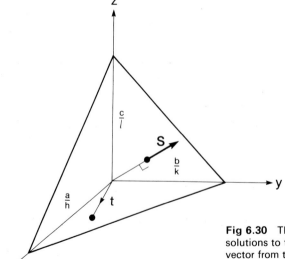

Fig 6.30 The Laue Equations are possible solutions to the equation $\mathbf{t}.\mathbf{S} = 1$, where $\mathbf{t}$ is a vector from the origin to a plane and $\mathbf{S}$ is normal to the plane, which makes intercepts a/h, b/k, c/l on the x, y, z axes respectively.

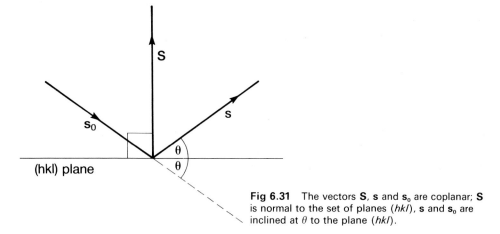

Fig 6.31 The vectors **S**, **s** and **s**$_0$ are coplanar; **S** is normal to the set of planes (*hkl*), **s** and **s**$_0$ are inclined at θ to the plane (*hkl*).

$\mathbf{S} = \mathbf{s} - \mathbf{s}_0$, **S**, **s**, and **s**$_0$ must be coplanar and, as we have already shown, **s** and **s**$_0$ make angles of $90° - \theta$ with **S**; therefore **s** and **s**$_0$ are both inclined at an angle θ to the plane (*hkl*) (Fig 6.31). Now the plane whose equation is $\mathbf{t} \cdot \mathbf{S} = 1$ will be distant $\mathbf{t} \cdot \hat{\mathbf{n}}$ from the origin, where $\hat{\mathbf{n}}$ is a unit vector normal to the plane. Therefore we can express the interplanar spacing of planes of the (*hkl*) set as

$$d_{(hkl)} = \frac{\mathbf{t} \cdot \mathbf{S}}{|\mathbf{S}|}$$

$$= \frac{1}{|\mathbf{S}|}$$

$$= \frac{\lambda}{2 \sin \theta} \text{ by equation (20),}$$

i.e. $\lambda = 2d_{(hkl)} \sin \theta,$

which is the Bragg Equation, equation (12). The vector **S** which corresponds to a solution of the Laue Equations for particular values of the integers *h*, *k*, and *l* is thus normal to the plane (*hkl*) and of modulus $1/d_{(hkl)}$.

For a diffracted beam to occur **S** must be a reciprocal lattice vector; but **S** is also a vector which depends on the direction of the incident beam **s**$_0$ in that it is a vector from the origin to a point on the surface of the reflecting sphere, which is a sphere of radius $1/\lambda$ with the incident beam as diameter. For a given direction of the incident beam the Laue Equations will thus only be satisfied for those reciprocal lattice vectors such as OP (Fig 6.32) which terminate on the surface of the reflecting sphere. The direction of the diffracted beam **s** is given by the radius of the reflecting sphere through the reciprocal lattice point P.

The concepts of the reciprocal lattice and the reflecting sphere thus provide a simple practical means of solving the Laue Equations or the Bragg Equation. If, for example, an X-ray beam were incident on an orthorhombic crystal in a direction in the *xy* plane inclined at 13° to *x* between $+x$ and $+y$ and we wished to discover which, if any, of the *hk*0 reflexions would occur, then we could simply draw out the *hk*0 reciprocal lattice plane and mark in the direction of the incident beam (Fig 6.33). The reflecting sphere passes through the origin O and has a radius of $1/\lambda$; its centre lies on the

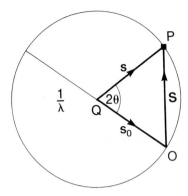

Fig 6.32 The incident beam $\mathbf{s_0}$ is represented by QO. The reciprocal lattice point P lies on the reflecting sphere and OP = **S**. The diffracted beam is $\mathbf{s}$ = QP. The radius of the reflecting sphere is $|\mathbf{s_0}| = |\mathbf{s}| = 1/\lambda$.

incoming incident beam at Q such that OQ = $1/\lambda$. Drawing the circle of intersection of the reflecting sphere with the $hk0$ reciprocal lattice plane quickly shows that only the reciprocal lattice points $3\bar{2}0$, $4\bar{1}0$ and 130 lie on the reflecting sphere; the only $hk0$ reflexions which will occur for this orientation of the incident beam are $3\bar{2}0$ $4\bar{1}0$ and 130. The directions of the diffracted beams are given by the radii through the corresponding reciprocal lattice points QE, QF, and QG.

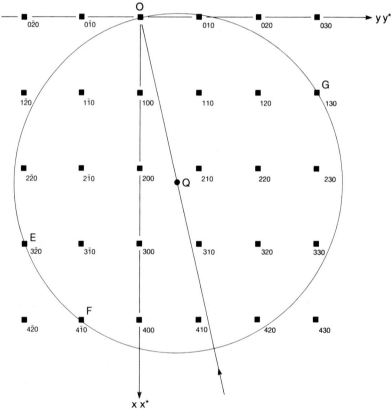

Fig 6.33 The $hk0$ reciprocal lattice plane of an orthorhombic crystal and the intersection of the reflecting sphere for an incident beam inclined at 13° to the x-axis in the xy plane between $+x$ and $+y$. The reciprocal lattice points $3\bar{2}0$, $4\bar{1}0$, and 130 lie on the reflecting sphere; the directions of the diffracted beams are respectively QE, QF, and QG.

The structure factor

We showed earlier in this chapter that the structure factor expression can be written in vector form as

$$F(\mathbf{S}) = \sum_1^N f_n \exp 2\pi i \mathbf{r}.\mathbf{S} \tag{19}$$

where $\mathbf{r}$ is the distance of the nth atom from the origin and $\mathbf{S}$ is the scattering vector for the hkl reflexion. Now $\mathbf{r} = x_n\mathbf{a} + y_n\mathbf{b} + z_n\mathbf{c}$ for the nth atom in the unit-cell, x_n, y_n, z_n being its coordinates expressed as fractions of the base vectors $\mathbf{a}, \mathbf{b}, \mathbf{c}$ of the lattice. In a crystal $\mathbf{S}$ is restricted to directions $h\mathbf{a}^* + k\mathbf{b}^* + l\mathbf{c}^*$. It follows immediately that

$$\mathbf{r}.\mathbf{S} = (x_n\mathbf{a} + y_n\mathbf{b} + z_n\mathbf{c}).(h\mathbf{a}^* + k\mathbf{b}^* + l\mathbf{c}^*)$$

$$= hx_n + ky_n + lz_n.$$

Thus the wave scattered by one unit-cell is given by

$$F(hkl) = \sum_1^N f_n \exp 2\pi i(hx_n + ky_n + lz_n),$$

an expression (equation 16) derived earlier by a geometrical argument.

Reciprocal lattice units

In the development of the concept of the reflecting sphere the expression for the phase of the X-ray beam scattered by a point distant $\mathbf{r}$ from the origin was written as $2\pi\mathbf{r}.\mathbf{S}$, where the scattering vector $\mathbf{S}$ is such that $|\mathbf{S}| = |\mathbf{s} - \mathbf{s}_o| = 2\sin\theta/\lambda$ and the moduli of the vectors representing the incident beam and the diffracted beam, $|\mathbf{s}_o|$ and $|\mathbf{s}|$ respectively, were taken to be equal to $1/\lambda$. It follows that the radius of the reflecting sphere is $|\mathbf{s}_o| = 1/\lambda$ and solutions of the Laue equations are possible only for reciprocal lattice points with $d^*_{hkl} = 1/d_{hkl}$. This approach is appropriate for the general theory of diffraction, but in some circumstances, such as the interpretation of single crystal X-ray diffraction patterns (chapter 8) and in the study of Brillouin zones (volume 2) it may be convenient to choose reciprocal lattice vectors with different moduli.

In general we may put $d^*_{hkl} = K/d_{hkl}$, implying that $|\mathbf{S}| = (2K\sin\theta)/\lambda$, that $|\mathbf{s}_o| = |\mathbf{s}| = K/\lambda$, and that $\mathbf{a}.\mathbf{a}^* = \mathbf{b}.\mathbf{b}^* = \mathbf{c}.\mathbf{c}^* = K$. If $K = \lambda$, then $d^*_{hkl} = \lambda/d_{hkl}$ and the radius of the reflecting sphere $|\mathbf{s}_o|$ becomes $\lambda/\lambda = 1$. That the radius of the reflecting sphere is then independent of the wavelength of the incident radiation provides a convenient simplification in the interpretation of single crystal X-ray diffraction patterns because it makes possible, as will be shown in chapter 8, the use of charts by means of which reciprocal lattice coordinates can be read directly from the photographic record of the diffraction pattern irrespective of the wavelength of the X-radiation.

When reciprocal lattice dimensions are quoted it is important to know what value of K has been used. The convention commonly adopted is: when $K = 1$, $d^*_{hkl} = 1/d_{hkl}$ and consequently if d_{hkl} is in Ångstrom units (Å), then d^*_{hkl} will be in Å^{-1}, but when $K = \lambda$, $d^*_{hkl} = \lambda/d_{hkl}$ and d^* is said to be in reciprocal units (r.u.) or in reciprocal lattice units (r.l.u.). Reciprocal lattice constants given in reciprocal or reciprocal lattice units are of course wavelength dependent.

Problems

6.1 By transforming to a primitive unit-cell derive the reflexion conditions for an I lattice of any system.

6.2 The crystal structure of blende, ZnS, is face-centred cubic with one formula unit per lattice point. For a Zn atom placed at the origin of the unit-cell and a S atom at $\frac{111}{444}$ write down and simplify the expression for the structure factor of blende. Evaluate the intensities of the 111, 200, and 220 reflexions assuming that atomic scattering factors are proportional to atomic number.

The crystal structure of silicon is related to that of blende, Si atoms being situated at all Zn and S sites. Deduce the intensities of the 111, 200, and 220 reflexions of silicon, again assuming that atomic scattering factor is proportional to atomic number. Show that the calculation of structure factors for silicon can be simplified by taking the origin at the centre of symmetry at $\frac{111}{888}$.

[Atomic numbers: Zn 30, S 16, Si 14].

6.3 The diffraction pattern of Li_3AlP_2, which is orthorhombic, displays the following systematic absences: hkl reflexions absent for $h+k+l = 2n+1$, $0kl$ reflexions absent for $k = 2n+1$, $h0l$ reflexions absent for $l = 2n+1$, $hk0$ reflexions absent for $h = 2n +1$. Determine the diffraction symbol and the space group.

6.4 The mineral struvite is orthorhombic and the only systematic absences shown on its diffraction pattern are $0kl$ reflexions for $k+l$ odd. Determine the diffraction symbol and the possible space groups.

6.5 LiF is cubic with $a = 4\cdot026$ Å. A crystal of LiF is mounted on a simple single crystal diffractometer with one of its tetrad axes, taken as [001], parallel to the axis of rotation and perpendicular to the plane defined by the incident X-ray beam and the axis of the detector. Initially the (100) face of the crystal is perpendicular to the incident beam. Through what angle must the crystal be rotated from this position to enable the 420 reflexion to be observed with CuKα radiation ($\lambda = 1\cdot5418$ Å)? At what angle will the detector then be inclined to the forward direction of the incident X-ray beam?

6.6 Calculate the intensities of the 331 and 420 reflexions of fluorite, CaF_2, which has a cubic F lattice with $a = 5\cdot459$ Å and one formula unit per lattice point. If the origin is taken at a Ca^{2+} cation, the coordinates of the fluoride ions associated with that lattice point are $\pm(\frac{111}{444})$. The variation of atomic scattering factors with θ may be expressed in the form $f(x) = \sum_{i=1}^{4} a_i \exp(-b_i x^2) + c$, where $x = \sin\theta/\lambda = \frac{1}{2}d^*$ and the parameters are:

for F⁻ $a_1 = 3\cdot63220$, $b_1 = 5\cdot27756$; $a_2 = 3\cdot51057$, $b_2 = 14\cdot7353$;
$\quad\quad\quad a_3 = 1\cdot26064$, $b_3 = 0\cdot442258$; $a_4 = 0\cdot940706$, $b_4 = 47\cdot3437$;
$\quad\quad\quad c = 0\cdot653396$.

for Ca^{2+} $a_1 = 15\cdot6348$, $b_1 = -0\cdot00740$; $a_2 = 7\cdot95180$, $b_2 = 0\cdot608900$;
$\quad\quad\quad a_3 = 8\cdot43720$, $b_3 = 10\cdot3116$; $a_4 = 0\cdot853700$, $b_4 = 25\cdot9905$;
$\quad\quad\quad c = -14\cdot875$.

6.7 A monochromatic beam of X-rays may be obtained by reflexion from the (0002) planes of graphite. Given that for graphite $c = 6.69$ Å, calculate the angle of incidence of the X-ray beam on the (0001) face of a graphite crystal for reflexion of (i) CuKα radiation for which $\lambda = 1.5418$ Å, and (ii) MoKα radiation for which $\lambda = 0.7071$ Å.

6.8 A cubic crystal with $a = 5.60$ Å is set initially with the incident beam travelling along the [$\bar{1}$00] direction. The crystal is then rotated clockwise about [001] through the angle ω until the 460 reflexion is first produced with CuKα radiation ($\lambda = 1.5418$ Å). Calculate the Bragg angle θ for the 460 reflexion and determine the angle ω. Determine the angles that the incident beam and the diffracted beam make with the crystallographic reference axes and hence show that all three Laue equations are satisfied.

6.9 Calculate the difference in Bragg angle θ for the α_1 and α_2 reflexions from the same set of lattice planes for CuKα_1 ($\lambda = 1.54050$ Å) and CuKα_2 ($\lambda = 1.54434$ Å) when the α_1 reflexion has (i) $\theta = 60°$, and (ii) $\theta = 85°$.

7
X-ray powder diffraction patterns

In this chapter and the next we discuss the production and interpretation of X-ray diffraction patterns, dealing first with those produced by crystalline powders and then moving on to single crystal diffraction patterns. One piece of groundwork remains to be done by way of preamble, that is the description of the spectrum emitted by an X-ray tube.

Emission spectrum of an X-ray tube

Our purpose in providing this account of the X-ray emission from an X-ray tube is to enable the reader to understand the general principles of the various types of X-ray goniometer as well as those features of diffraction patterns that are dependent on the wavelength distribution in the incident X-ray beam. Our treatment of this topic will be in outline only; for a detailed account the reader is referred to Klug and Alexander (1974).

A modern X-ray tube designed for crystallographic use (Fig 7.1) is in essence a permanently evacuated glass envelope into which is sealed a tungsten *filament* separated by about one centimetre from a *target* composed of a metallic element. The tungsten filament is heated by the passage of an electric current and emits electrons. A potential difference of the order of 50 kV applied between the filament, acting as cathode, and the target, as anode, accelerates the electrons emitted by the hot filament towards the anode so that a stream of high-energy electrons impinges on the anode. Most of the energy of the electron stream, about 98 per cent, is converted into heat so it is essential that the anode should be made of a material of high thermal conductivity and cooled from behind by a fast flowing stream of water. Anodes are usually made of copper; if a less well conducting element is required as target, either it is electroplated on to the copper anode or a small disc of it is soldered on. The filament is surrounded by a *focusing hood* which has a slot in its front face parallel to the length of the filament. The effect of the hood is to cause the electron stream to form a line focus ~1 cm × ~0·01 cm on the target; the dissipation of heat from such a line focus is relatively efficient so that the tube can be run at a higher electron current and so produce X-radiation of higher intensity. X-rays are emitted from the target in all possible directions, but only a narrow beam (in the angular range of highest intensity) making an angle from 3° to 6° with the face of the target is utilized by being allowed to pass out of the evacuated envelope through a *window* made of a

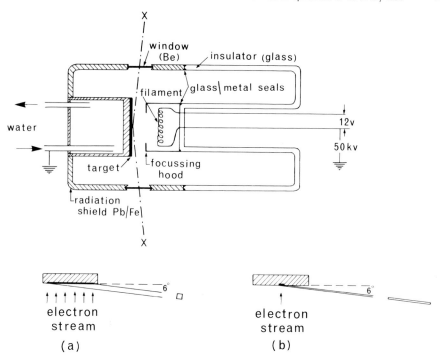

Fig 7.1   A crystallographic X-ray tube. The two lower diagrams are mutually perpendicular sections of a target, the thick black line representing the area irradiated by electrons, (a) in the plane containing the length of the filament and (b) in the plane perpendicular to the length of the filament. To the right of each diagram the cross-section of the emitted X-ray beam is shown; in (a) it is approximately square, a *spot focus*, and in (b) it is strongly elongated, a *line focus*.

substance with a very low absorption coefficient for X-rays; windows are usually made of beryllium. X-ray tubes are equipped with four windows, one situated on either side of the tube in line with the length of the line focus and two others at right-angles to these. Through the former the line focus appears as a nearly equidimensional spot and the X-ray beam that passes through the window is suitable for use with pin-hole collimators; but through the windows situated normal to the length of the line focus, the elongation of the focus is retained so that the X-ray beam transmitted has dimensions of about 1 cm × 0·01 cm and is suitable for use with a slit system of collimation.

The X-radiation emitted by the target is never monochromatic, but covers a considerable spectral range. In the X-ray spectrum it is convenient to distinguish between *white* (or *continuous*) radiation and *characteristic* radiation. We deal with the generation of white radiation first.

When an electron strikes the target it loses energy and part of the energy lost is converted into X-radiation of wavelength λ according to the equation $\lambda = hc/\Delta E$, where ΔE is the amount of energy lost by the electron, h is Planck's constant, and c is the velocity of light *in vacuo*. If the electron loses all the energy it has acquired by dropping through a potential V, then $\Delta E = eV$, where e is the charge on an electron, and the wavelength of the X-rays emitted will be

$$\lambda = \frac{hc}{eV} = \frac{12\cdot398}{V},\qquad(1)$$

where the wavelength is measured in Ångstrom units and V is in kilovolts; this will will be the shortest wavelength in the spectrum of the X-radiation emitted by the target. That an electron will lose all its kinetic energy in a single collision with a target atom is improbable; most will lose their kinetic energy in a series of collisions, each involving a loss of energy less than eV and resulting in the emission of X-radiation of wavelengths longer than that indicated by equation (1). The intensity of the X-radiation emitted by the target will vary continuously with wavelength and is of the general form shown in Fig 7.2. The intensity at a given wavelength and the variation of intensity with wavelength in such a *white radiation* spectrum depend on the operating voltage of the tube and on the nature of the target element. The kinetic energy of the electrons striking the target will increase with increasing applied potential difference between filament and target, so that there will be an overall increase in the intensity of the X-radiation emitted and moreover a movement to shorter wavelengths of both the maximum in the intensity distribution curve and the cut-off (or minimum wavelength). In general it can be said that the efficiency of the conversion of electron kinetic energy to X-radiation increases with the atomic number of the target element; thus a molybdenum ($Z = 42$) target produces more intense white radiation than a copper ($Z = 29$) target operated at the same voltage.

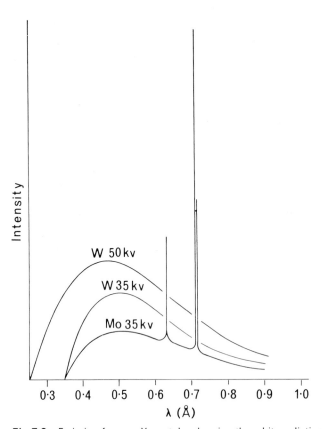

Fig 7.2   Emission from an X-ray tube showing the white radiation 'hump' and the characteristic α_1, α_2, and β lines for a molybdenum target operating at 35 kv; the characteristic lines of tungsten are at too high a wavelength to be shown.

We turn now to the generation of characteristic radiation. If an electron of sufficiently high energy strikes the target it may eject an electron from the K-shell of one of the atoms of the target element. Ejection of a K electron will be followed by the transfer of an electron from an electronic shell of higher energy to fill the vacant energy level; such a transfer will be accompanied by the emission of an X-ray photon whose energy is equal to the difference in energy between the two energy levels of the target atom. The X-rays emitted as a result of such a process will thus have a fixed wavelength characteristic of the target element and will constitute a line spectrum commonly described as the *characteristic* radiation of the target element. When the lower of the two energy levels concerned is in the K-shell of the target atom, the resulting spectral line is described as a K line.

K lines are classified in terms of the other energy level concerned. An electronic transition from the L-shell to the K-shell is said to give rise to a Kα line, while a transition from the M-shell to the K-shell gives rise to a Kβ line. Since the L-shell is of lower energy than the M-shell, the Kα line in the X-ray spectrum of a given element has a longer wavelength than the Kβ line of the same element. Moreover Kα lines are generally of higher intensity than the corresponding Kβ lines; in practice therefore Kα lines are invariably selected for isolation when monochromatic X-radiation is required. All elements give rise to two Kα lines, denoted $K\alpha_1$ and $K\alpha_2$, which have a very small difference in wavelength[1] and are resolved in diffraction only at high Bragg angle. The $K\alpha_1$ line has the shorter wavelength and is of about twice the intensity of the $K\alpha_2$ line; when the two lines are not resolved their wavelengths can be weighted to a fair approximation in the ratio 2:1 to give the wavelength of the unresolved doublet, usually written as Kα*, as $\lambda_{K\alpha^*} = \frac{2}{3}\lambda_{K\alpha_1} + \frac{1}{3}\lambda_{K\alpha_2}$. Of course the emission spectrum of an X-ray source includes characteristic spectral lines due to electronic transitions to the L-shell, Lα, Lβ lines, etc; such characteristic radiations are not commonly utilized crystallographically and need not be further discussed here.

The wavelengths of some commonly-used characteristic radiations are shown in Table 7.1. While the wavelength of characteristic radiation is dependent only on the nature of the target, its intensity is dependent on the magnitude of the voltage applied across the tube; in particular if the applied voltage is below a certain threshold value, none of the electrons incident on the target will have sufficient energy to eject a K-electron from a target atom and no K lines will be excited. The threshold voltage for excitation of K lines in any element is such as will impart to the electrons incident on the target kinetic energy equal to the photon wavelength of the K absorption edge of the target element. The operating voltage of an X-ray tube must therefore be greater than this threshold value and is chosen to give an optimum ratio of characteristic intensity to white radiation intensity.

When monochromatic radiation is required it is sufficient for most purposes merely to remove the Kβ line, the ratio of characteristic to white intensity being such that

[1] A Kα line is produced by an electronic transition from a $2p$ to a vacant $1s$ orbital. The reader familiar with X-ray spectroscopy will be aware that the $2p$ orbitals of an atom comprise two shells, denoted LII and LIII, of slightly different energy because the total orbital angular momentum and the spin angular momentum of the $2p$ electrons can be combined in two different ways. Electrons in the LII shell have slightly lower energy than those in the LIII shell and so electronic transitions $LII \rightarrow K$ give rise to the longer $K\alpha_2$ wavelength compared with electronic transitions $LIII \rightarrow K$ which give rise to the slightly shorter $K\alpha_1$ wavelength of X-radiation. The LII shell ($J = \frac{1}{2}$) contains two electrons whereas the LIII shell ($J = \frac{3}{2}$) contains four electrons so that the $K\alpha_2$ line is only half as strong as the $K\alpha_1$ line. Formally the $K\alpha_2$ line is produced by an electronic transition from a $2^2P_{\frac{1}{2}}$ to a $1^2S_{\frac{1}{2}}$ state whereas the $K\alpha_1$ line is produced by a transition from a $2^2P_{\frac{3}{2}}$ to a $1^2S_{\frac{1}{2}}$ state.

Table 7.1   Data for some common targets and filters

Element	Atomic Number	Line	Wavelength (Å)	Filter: Element	Atomic Number	Wavelength of K absorption edge (Å)
Mo	42	$K\alpha_1$	0·70926	Zr	40	0·6888
		$K\alpha_2$	0·71354			
		$K\beta$	0·63225			
Cu	29	$K\alpha_1$	1·54050	Ni	28	1·4869
		$K\alpha_2$	1·54434			
		$K\beta$	1·39217			
Co	27	$K\alpha_1$	1·78890	Fe	26	1·7429
		$K\alpha_2$	1·79279			
		$K\beta$	1·62073			
Fe	26	$K\alpha_1$	1·93597	Mn	25	1·8954
		$K\alpha_2$	1·93991			
		$K\beta$	1·75654			
Cr	24	$K\alpha_1$	2·28962	V	23	2·2676
		$K\alpha_2$	2·29352			
		$K\beta$	2·08479			

the latter can be ignored. The $K\beta$ line can simply be removed from the X-ray emission of a target by placing immediately outside the window a thin foil of an element with an absorption edge of wavelength just less than that of the required $K\alpha$ line. When the beam emitted from the X-ray tube passes through such a *filter* the intensity of the $K\alpha$ line is reduced by a small factor and the $K\beta$ line is reduced to negligible intensity. The appropriate filter for a target of an element of atomic number Z is an element of atomic number $Z-1$ usually, $Z-2$ in some cases, e.g. for $CuK\alpha$ the appropriate filter is nickel $(Z-1)$ whereas for $MoK\alpha$ zirconium $(Z-2)$ is employed. The dependence of mass absorption coefficient on wavelength is shown for one element in Fig 7.3. The optimum thickness of a filter varies from element to element; it should be adequate to place the intensity ratio $I(K\alpha)/I(K\beta)$ between 150 and 350, but no thicker. In the case of emission from a copper target, the former value of the $I(K\alpha)/I(K\beta)$ ratio corresponds of a reduction in $I(K\alpha)$ of 45 per cent and the latter of 60 per cent. Although the primary purpose of such a filter is to cut out the $K\beta$ line, white radiation of wavelength less than the absorption edge of the filter will also be drastically reduced in intensity, but white radiation with λ from just less than $\lambda_{K\alpha}$ to high wavelengths will pass the filter.

When strictly monochromatic radiation is required the X-ray beam emitted from the tube is reflected from a crystal face set at the appropriate Bragg angle for the $K\alpha$ line. Crystals with faces parallel to planes that yield very strong X-ray reflexions are employed, but even so the $K\alpha$ beam is much reduced in intensity. The reduction in intensity can be made less severe by bending the crystal plate to give a focusing effect; this was the usual way of achieving strictly monochromatic X-radiation until recent developments in the synthetic growth of perfect graphite crystals enabled flat crystals to be produced which give a much smaller reduction in intensity. Such *mono-chromators* are employed only when strictly monochromatic radiation is essential for the purpose in hand.

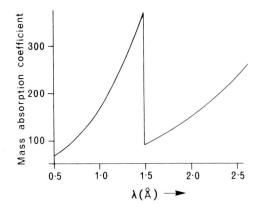

Fig 7.3 Dependence of the mass absorption coefficient of nickel on wavelength.

We conclude with a word of caution about the nature of the emission from commercial X-ray tubes. Tungsten slowly evaporates from the hot filament in use and eventually deposits as a thin film on the target and the windows. Then the radiation emitted by the target will be contaminated especially with the strongest line in the tungsten spectrum, $L\alpha_1$ ($\lambda = 1\cdot476$ Å), which represents an electronic transition from the M-shell to fill a vacancy in the L-shell. In old tubes also the intensity of emission is reduced by the presence of a strongly absorbent film of tungsten on the inner surface of the window.

Powder photographs

An essential difference between two and three-dimensional diffraction gratings is that the latter yield only a few diffracted beams for any one orientation of the grating relative to the incident beam. The number of diffracted beams that can be observed when X-radiation is diffracted by a crystalline substance can be increased in a variety of ways; several such ways that are applicable to a single crystal diffraction grating are discussed in the next chapter; here we remove the restriction that the grating be a single crystal and consider diffraction by a crystalline powder. It is assumed—and this is a practically valid assumption—that the powder has been so prepared that it consists of a very large number of minute crystal fragments in completely random orientation so that every possible lattice plane will be present in every possible orientation with respect to the incident X-ray beam. The Bragg Equation $\lambda = 2d \sin\theta$ will thus be satisfied for all planes (hkl) provided $d > \frac{1}{2}\lambda$. Since the only restriction placed by the Bragg Equation on the orientation of a reflecting plane is that it should make an angle θ with the incident X-ray beam, all planes with a given set of indices (hkl) whose normals lie on a cone of semiangle $90° - \theta$ about the direction of the incident beam will reflect. Figure 7.4 shows the normal N to a plane (hkl) with interplanar spacing d lying on a small circle of radius $90° - \theta$ about the direction X of the incident X-ray beam. Since the reflected beam R is required to be coplanar with X and N, R lies at the intersection of the great circle XN with a small circle of radius 2θ about the direction X' of the emergent direct beam. Planes with indices (hkl) will thus give rise to a cone of diffracted beams with semiangle 2θ. The total diffraction pattern produced by a crystalline powder is thus a set of cones, each cone corresponding to a solution of the Bragg Equation.

The diffraction pattern produced by a powder is commonly recorded on a narrow strip of photographic film in a cylindrical camera whose axis is coincident with the

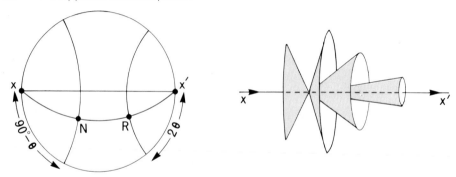

Fig 7.4 Generation of cones of diffracted X-rays by an X-ray beam incident on a powder specimen. The stereogram on the left shows the relationship between the normal N to a set of planes (hkl), the direction X of the incident X-rays, the direction R of the resulting diffracted beam and the Bragg angle θ. The drawing on the right shows cones of diffracted radiation emanating from a powder specimen for four solutions of the Bragg equation $\lambda = 2d \sin \theta$, the semi-angle of each cone being 2θ.

powder specimen. Each diffracted cone is recorded on the film strip as a pair of arcs, one on either side of the direction of the incident beam (Fig 7.5). The appearance of the diffraction pattern on the film strip after development depends on the way the film is mounted in the camera. There are three mountings in common use, which differ in the position of the free ends of the film relative to the incident beam. The *Bradley–Jay* mounting is such that the incident beam passes through the gap between the ends of the film and the undeviated beam leaves the camera through a hole punched in the centre of the film (Fig 7.5(a)). The *van Arkel* mounting simply has the positions of entry and exit reversed (Fig 7.5(b)). In the *Straumanis* mounting, both the incident and the undeviated beam pass through holes punched in the film, the gap between the ends of the film lying close to the radius of the camera normal to the incident beam (Fig 7.5(c)); most modern powder cameras use the Straumanis mounting.

Approximate Bragg angles can simply be determined by measuring the distance s between the midpoints of corresponding arcs on the film and assuming a value r equal to the radius of the camera for the radius of the film in the camera; then $\theta = s/4r$ if s is measured across the direction of the undeviated beam as in the Bradley–Jay mounting, $\frac{1}{2}\pi - \theta = s/4r$ if s is measured across the direction of the incident beam as in the van Arkel mounting. The Straumanis mounting yields concentric arcs about each of the two holes punched in the film so that measurements of s across one hole will give $\theta = s/4r$ and across the other $\frac{1}{2}\pi - \theta = s/4r$; the hole through which the incident beam enters the camera can easily be recognized by the splitting of the arcs closest to it into $\alpha_1 \alpha_2$ doublets. Accurate measurement of Bragg angles requires film shrinkage during development, fixing, washing, and drying to be taken into account as well as the deviation of the mean radius of the film in the camera from the radius of the camera due to the finite thickness of the film. In cameras that use the Bradley–Jay or the van Arkel mounting this is achieved by constructing the camera so that a knife-edge casts a shadow just short of each end of the film, the angle between the knife-edges, $4\phi_k$, being determined by calibrating the camera with a substance whose unit-cell dimensions are very accurately known; if the measured distance between the shadows cast by the knife-edges is s_k, then $\theta = \phi_k s/s_k$ for the Bradley–Jay mounting and $\frac{1}{2}\pi - \theta = \phi_k s/s_k$ for the van Arkel mounting. The Straumanis mounting is however self-calibrating provided the powder pattern

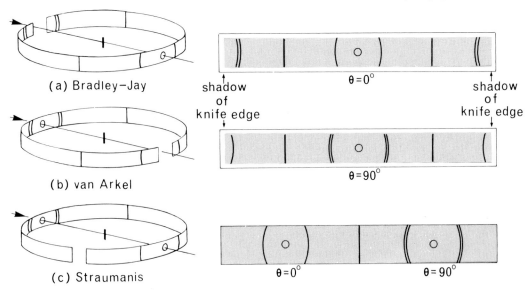

Fig 7.5   Film mountings for X-ray powder photography. The diagrams on the left show the arrangement of the film strip in the camera for three mountings that have been much used: the specimen is represented as a short thick line and the direction of the incident X-rays is arrowed. The diagrams on the right show the appearance of the film laid flat after development for each of the three mountings: exemplary low θ, moderate θ, and high θ powder rings are shown. Observation of $\alpha_1\alpha_2$ splitting serves to identify the high θ lines on the film.

concerned exhibits sharp high θ reflexions; then measurement of the positions of the midpoints of one pair of corresponding arcs, s_1 and s'_1, across the exit hole and of another pair, s_2 and s'_2, across the entrance hole yields a distance

$$s^* = \frac{s_1 + s'_1}{2} - \frac{s_2 + s'_2}{2}$$

equal to the distance between the centres of the two pairs; since one centre is at $\theta = 0$ and the other at $\theta = \frac{1}{2}\pi$, $s^* = \pi r$ so that $\theta = (\pi/4s^*)s$ if s is measured across the exit hole and $\theta = \frac{1}{2}\pi - (\pi/4s^*)s$ if s is measured across the entrance hole.

The Bradley–Jay mounting is suitable for recording powder patterns for comparative purposes and is capable of giving accurate d-spacings for low-θ lines; the van Arkel mounting is especially applicable to the accurate measurement of high angle lines; but both have been largely superseded in modern cameras by the Straumanis mounting, which combines the advantages of both with no intrinsic disadvantage.

It has already been said that high angle lines can be recognized immediately on inspection of a powder photograph because they display $\alpha_1\alpha_2$ splitting. That the Kα doublet will be resolved at high θ can simply be seen by differentiating the Bragg Equation

$$\lambda = 2d \sin \theta$$

Therefore $d\lambda = 2d \cos \theta \, d\theta$

whence $d\theta = \dfrac{\tan \theta}{\lambda} d\lambda$

As $\theta \to \frac{1}{2}\pi$, $\tan \theta \to \infty$ so that a small difference in λ will correspond to a relatively large difference in θ. For $CuK\alpha_1$ and $CuK\alpha_2$, for example, $\Delta\lambda = 0\cdot0038$ Å (Table 7.1) so that for $\theta = 80°$, $\Delta\theta = 0\cdot86°$ and for $\theta = 85°$, $\Delta\theta = 1\cdot72°$. In the doublet the $K\alpha_1$ line has the shorter wavelength and therefore the lower Bragg angle; its intensity is approximately twice that of the $K\alpha_2$ line.

Experimental procedure

For a full account of the practical details of taking X-ray powder photographs the reader is referred to Klug and Alexander (1974) or to Lipson and Steeple (1970). We present here an account in outline only.

The specimen is finely ground to a smooth powder, the constituent grains of which should have dimensions less than 45×10^{-3} mm. By mixing the powder with a small amount of gum tragacanth, moistening with water, and rolling between two microscope slides the powder specimen can be obtained in the form of a small cylinder of diameter 0·3 to 0·5 mm and about 1 cm in length. Alternatively the powder may be loaded into a thin-walled capillary tube made of borosilicate glass (which has a very low absorption coefficient for X-rays). The powder specimen so prepared is attached to a spindle that can be centred so that the length of the specimen can be brought into coincidence with the axis of the cylindrical camera (Fig 7.6). X-rays enter the camera through a *collimator* which is essentially a metal tube, of internal diameter $\sim 0\cdot5$ mm, extending to within a few mm of the centre of the camera; the hole in the collimator is widened to about 1 mm diameter at the exit end to form a *guard tube* that serves to trap the radiation scattered from the end of the fine hole. The undeviated beam is led out of the camera through a similar tube, of rather larger internal diameter, into a *beam trap*. The beam trap is so constructed that it incorporates a fluorescent screen which is useful for alignment of the camera along the beam emitted from the

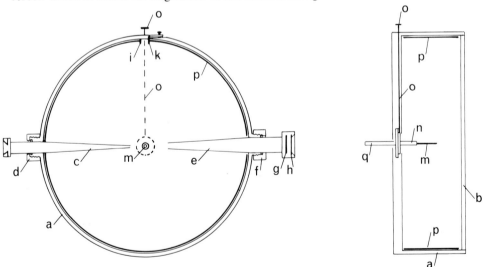

Fig 7.6 A type of powder camera of 57·3 mm radius shown about half-size in front and side elevation. a, cylindrical camera body. b, detachable lid. c, collimator. d, collimator locking screw. e, exit tube. f, exit tube locking screw. g, fluorescent screen. h, lead glass screen. i, fixed pin and k, movable pin with locking screw (movement of k away from i forces the film against the camera body; k is moved until resistance is felt and then locked). m, specimen. n, specimen holder. oo, centring plunger to move magnetic chuck (the spindle q is rotated manually and the plunger oo is applied until the specimen is centred throughout its rotation). p, film. q, drive spindle from detachable motor.

X-ray generator; the fluorescent screen is viewed through a lead-glass window. Centring of the specimen on the axis of the camera is achieved by removing the beam trap, illuminating the collimator with a light source, and adjusting the specimen mounting until the specimen is seen to remain stationary on rotation about its axis. The camera is in essence a light-tight box which must be loaded with film only in a darkroom. The various types of camera use different devices for holding the film strip firmly against a cylindrical metal former. Before loading the film is punched with one or two holes in the appropriate positions for the collimator and/or the beam trap to pass through. During exposure the specimen is rotated slowly about its axis by an electric motor; this serves greatly to increase the number of crystal orientations with respect to the incident beam that are present in the specimen. If the powder is too coarse or the specimen not rotated, complete randomness of orientation will not be achieved and the powder lines will appear 'spotty'.

Powder cameras of various diameters are available commercially, 57·3, 60, 114·6, 90, and 190 mm are diameters that have been used extensively. For collimation of the same quality a camera of large diameter will produce better resolution of the diffraction pattern; but this advantage is counterbalanced by the longer exposure required to produce lines of comparable intensity due to absorption and scattering of the diffracted X-rays by the air in the camera. Cameras of large radius are provided with a facility for evacuation, but this is usually only necessary when long wavelengths (e.g. CrKα) are being used or when the specimen is being heated. For general purposes a camera of diameter 114·6 mm is very suitable and has the added advantage for preliminary work that 1 mm measured on the film corresponds to 1° if film shrinkage is neglected: thus if the distance between the mid-points of corresponding arcs is measured across the exit hole with a ruler as 64·0 mm, the Bragg angle of this reflexion is immediately determined as 16·0° and this can simply be converted to a d-spacing by consulting tables giving $d - \theta$ relationships for commonly used radiations (e.g., Fang and Bloss, 1966). For accurate work of course this radius has no special advantage; measurement and calibration of the film will be required.

Powder diffractometry

In X-ray powder photography the whole diffraction pattern is recorded simultaneously on a photographic film; in X-ray powder diffractometry the diffraction pattern is scanned by a counter device which plots counter output against Bragg angle on a paper trace. The resolution obtainable in diffractometry under optimum working conditions is very much better than in photography and Bragg angles can be measured to much higher accuracy, but the apparatus is much more complex, very weak reflexions are difficult to distinguish from background noise, and a larger powder specimen is required. As will be exemplified later in this chapter the diffractometer is used mainly for problems that require highly accurate Bragg angles or high resolution, while the powder camera is used generally for identification.

A material is prepared for diffractometry by grinding to a smooth powder and sedimenting in a suitable volatile medium on to a microscope cover slip of diameter ～20 mm. It is usual to compress the sedimented specimen against a polished steel plate to ensure that its surface is flat. Randomness of orientation of crystallites in the powder specimen is further increased by rotating the specimen slowly during exposure about an axis normal to its plane. When crystallites tend to sediment with preferred orientation—that is when they are flakes or needles—special techniques of sample preparation have to be used.

The best operating systems currently in use make use of the parafocusing effect illustrated in Fig 7.7. The line focus of the X-ray tube and the entrance slit of the counter are constrained to lie on a circle, the *focusing circle*, so as to be equidistant from the specimen, the surface of which is tangential to the circle. This condition ensures that the diffracted X-rays to be measured are reflected from the surface of the specimen and so are effectively focused on the entrance slit of the counter. Theoretically the condition for the reflected X-rays to be focused is that the specimen should be an arc of the focusing circle, but it is practically more simple and found to be adequate if the specimen is tangential. The divergence of incident and diffracted beams parallel to the length of the line focus of the X-ray tube—that is, normal to the plane of Fig 7.7—is limited by passing each through *Soller slits*, a set of narrow slits formed by a pack of thin metal plates, each parallel to the plane of the diagram.

The diffraction pattern is scanned by rotating the counter at a steady speed about the centre of the surface of the specimen; the radius of this *scanning circle* varies from one make of instrument to another, but is often ~ 200 mm. In order to maintain the geometry of the focusing circle the specimen is geared to rotate about the same axis at half the speed of the counter. A scale provides direct measurement of the scattering angle 2θ. Scanning speeds vary from $\frac{1}{8}$ to 2 degrees per minute. The counter, which may be a Geiger, proportional, or scintillation counter, outputs through electronic circuits, that we shall not describe in detail, to a pen recorder which plots counts per second, as a measure of intensity, on a continuous paper chart. The chart moves at a steady speed so that distances parallel to its length provide a measure of differences in 2θ. Chart speeds vary from 200 to 1600 mm per hour. Scanning and chart speeds are independently adjustable to suit the nature of the problem under investigation. On the chart one degree of 2θ may be represented by as little as 1·67 mm or as much as 213 mm. For very accurate measurements counter and specimen can alternatively be moved in angular steps as small as 0·01° and the diffracted X-radiation counted at each step for a much longer time than would be possible in conditions of continuous scanning; the resultant counts are then plotted against 2θ to give a highly accurate peak profile.

The scanning range of diffractometers, except those built for special purposes, is limited to $\theta < 80°$ simply because at higher angles the counter would foul the X-ray tube; similarly they are limited to $\theta > 4°$ because at lower angles the intense undeviated beam would damage the counter. In practice these restrictions rarely matter.

It is worthy of note that since, as is evident from Fig 7.7, the area of the specimen irradiated by the incident beam varies with Bragg angle, it is important that the specimen should be homogeneous and large enough in area to catch the whole of the incident beam at the lowest Bragg angle to be used. If this second condition is not satisfied, peak heights at different Bragg angles will not be comparable.

It was pointed out at the beginning of this section that in powder photography the whole diffraction pattern is being recorded throughout the exposure whereas in diffractometry each part of the pattern is recorded at a different time. It becomes necessary therefore to stabilize the intensity of the incident X-ray beam; this is achieved by stabilizing the high voltage supply and the filament heating current or, less commonly, by continuous monitoring of the incident intensity.

A formal comparison of the advantages and drawbacks of powder photography and diffractometry would be misleading; the sort of use for which each technique is especially appropriate has already been pointed out in general terms and the reader

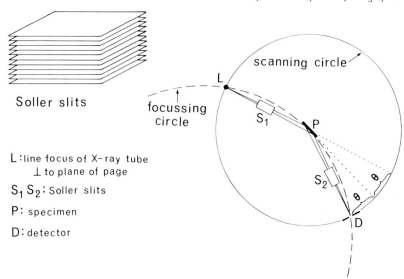

Fig 7.7 Geometry of a powder diffractometer utilizing the parafocusing effect. The construction of Soller slits is shown on a larger scale in the upper left-hand corner.

should draw his own conclusions from the examples discussed at the end of this chapter. It should be borne in mind that the area of overlap of the two methods is extensive. Most of the uses for which the diffractometer is superior stem from its better resolving power, as an indication of which we may compare the minimum Bragg angle at which the $CuK\alpha_1 - K\alpha_2$ doublet is resolved under optimum conditions, $\sim 20°$ by diffractometry compared with $\sim 55°$ by photography.

Interpretation of powder photographs

Measurement of a powder photograph provides a direct determination of the Bragg angle of every cone of diffracted radiation intersected by the film, that is of every *powder line* on the photograph. Solution of the Bragg Equation for each powder line yields the corresponding interplanar spacings, usually known as *d-spacings*. If the unit-cell dimensions of the substance are known it is then possible to determine the indices of the planes contributing to each powder line. If the unit-cell is unknown, it is generally possible with more or less certainty to find unit-cell dimensions consistent with the powder pattern and thence to index all the lines of the pattern. In either case it will be necessary to consider how many symmetry related planes contribute diffracted radiation to each powder line; we deal with this topic first.

Multiplicity Factors

All lattice planes of equal *d*-spacing of necessity give rise to reflexions at the same Bragg angle. Reflexions from each such plane will be independent of one another; consequently the intensity of a powder line is simply the sum of the intensities of all reflexions contributing to it. All lattice planes that are related by symmetry will have the same *d*-spacing so that the number of planes contributing to a powder line *hkl* will be the number of planes in the form {*hkl*} of the Laue class. Thus the line of longest *d*-spacing, that is of smallest Bragg angle, in a powder photograph of a primitive cubic substance of Laue class *m3m* is composed of reflexions from all the planes of the form {100}; the six planes (Fig 3.28) of this form (100), (010), (001),

Table 7.2   Multiplicity factors for cubic powder lines

Indices of powder line	$h00$	$hh0$	hhh	hhl	$hk0$	hkl
Laue group $m3m$						
Multiplicity factor	6	12	8	24	24	48
Line intensity	6 I($h00$)	12 I($hh0$)	8 I(hhh)	24 I(hhl)	24 I($hk0$)	48 I(hkl)
Laue group $m3$						
Coincident lines unrelated by symmetry	—	—	—	—	$hk0, kh0$	hkl, lkh
Multiplicity factor	6	12	8	24	12, 12	24, 24
Line intensity	6 I($h00$)	12 I($hh0$)	8 I(hhh)	24 I(hhl)	12 $[$I($hk0$) $+$I($kh0$)$]$	24 $[$I(hkl) $+$I(lkh)$]$

($\bar{1}00$), ($0\bar{1}0$), ($00\bar{1}$) all contribute equally to the powder line, the *multiplicity* of which is then said to be 6. Thus the intensity of the 100 reflexion will be $\frac{1}{6}$ of the intensity of the powder line containing the 100 reflexion. Likewise, as there are eight planes in the form $\{111\}$ in Laue class $m3m$, (111), ($\bar{1}11$), ($\bar{1}\bar{1}1$), ($1\bar{1}1$), and their opposites, the intensity of the 111 reflexion will be $\frac{1}{8}$ of that of the powder line which contains the 111 reflexion. In general the intensity of the powder line which contains the hkl reflexion will be mI(hkl) where I(hkl) is the intensity of the hkl reflexion and m, the *multiplicity* of the line, is equal to the number of planes in the form $\{hkl\}$ in the Laue class of the substance concerned. In short the multiplicity is the number of equivalent reflexions which contribute to the powder line. The lines on a powder pattern are conventionally indexed if possible with the indices of that contributing reflexion which has $h \geqslant k \geqslant l$; thus the powder line to which reflexions from all the planes of the form $\{100\}$ contribute is known as the 100 line, rather than as 010 or 001 or $00\bar{1}$.

Multiplicity factors for the cubic Laue group $m3m$ are shown in Table 7.2. The geometry of the cubic unit-cell is such that certain sets of lattice planes which are unrelated by symmetry have the same d-spacing. The spacing of the (hkl) planes in a cubic crystal (equation (28) of chapter 5) is given by $d_{(hkl)} = a/\sqrt{(h^2+k^2+l^2)}$. Certain values of $h^2+k^2+l^2$ can be obtained from different values of h, k, and l, and in consequence the corresponding powder lines coincide. Thus the $\{300\}$ and $\{221\}$ planes have the same d-spacing $\frac{1}{3}a$ so that the 300 and 221 powder lines are coincident. Measurement of the intensity of such a line provides information only about the combined intensities of the coincident lines 6I(300) + 24I(221); I(300) and I(221) can be separately evaluated only by study of the single crystal diffraction pattern.

In the other cubic Laue class $m3$ coincidence of powder lines due to planes unrelated by symmetry can arise in another way. Comparison of the relevant point group diagrams in Fig 3.28 makes it clear that whereas in the point group $m3m$ the form $\{210\}$ has 24 faces, in the point group $m3$ the form $\{210\}$ comprises only 12 faces while the distinct form $\{120\}$ has another 12 faces with the same d-spacing $a/\sqrt{5}$. The intensity of reflexion from planes of the two forms will be different so that the intensity of the composite line will be 12I(210) + 12I(120). This sort of coincidence affects all lines of the types $hk0$ and hkl in the powder patterns of substances of Laue class $m3$, but for all other types of reflexions multiplicity factors are the same as those for Laue class m3m. Multiplicity factors for Laue class m3 are shown alongside those for m3m in Table 7.2, where pairs of reflexions that are coincident but independent in the lower symmetry class are designated as hk0 and kh0, hkl and lkh.

In summary the intensity of a powder line hkl is related to the intensity of the Bragg reflexion hkl by the multiplicity factor for the form {hkl} in the appropriate Laue class provided no other form has the same interplanar spacing.

We have taken our examples of multiplicity factors entirely from the cubic system because we shall be mainly concerned with the interpretation of cubic powder patterns. What has been said here applies however in general to all other systems.

Interpretation when the unit-cell is known

From measurements of the Bragg angles of the lines of a powder pattern the d-spacings of the lines can simply be calculated by application of the Bragg Equation. It is convenient to arrange such a set of observed d-spacings in sequence of decreasing d. The observed d-spacings are then compared with the d-spacings of all planes that can give rise to reflexion calculated from the known unit-cell dimensions. In the cubic system the comparison invariably yields unambiguous indices for all low θ lines, but becomes progressively less certain as θ increases unless the unit-cell edge is very accurately known. The d-spacing of the highest θ line that can be unambiguously indexed is then used to recalculate a, from which a revised set of d-spacings is calculated; the indexing may then be carried to higher θ and the process of successive improvement of a continued. In systems of lower symmetry it is usual to calculate by computer from the unit-cell constants a set of d-spacings sorted in sequence of decreasing d. Ambiguity in the indexing of observed lines usually arises at a much lower Bragg angle than in the cubic system and it is advisable, whenever possible, to compare the intensities of observed powder lines with intensity data for reflexions obtained from single crystals of the substance, taking multiplicity factors into account. By comparison with single crystal intensity data it may be possible to index unambiguously a line that would otherwise have to be referred to two or more forms unrelated by symmetry but with equal, or nearly equal, d-spacings.

Interplanar spacings are most conveniently calculated from unit-cell constants by use of the reciprocal lattice. In chapter 5 we showed (equation (26)) that

$$\frac{1}{d_{(hkl)}^2} = d_{(hkl)}^{*2} = \mathbf{d}_{(hkl)}^* \cdot \mathbf{d}_{(hkl)}^*$$

$$= (h\mathbf{a}^* + k\mathbf{b}^* + l\mathbf{c}^*) \cdot (h\mathbf{a}^* + k\mathbf{b}^* + l\mathbf{c}^*).$$

In the triclinic system, the most general case, evaluation of this expression by use of the reciprocal metric tensor (Table 5.3) yields

$$\frac{1}{d_{(hkl)}^2} = d_{(hkl)}^{*2}$$

$$= h^2 a^{*2} + k^2 b^{*2} + l^2 c^{*2} + 2klb^* c^* \cos \alpha^* + 2hla^* c^* \cos \beta^*$$

$$+ 2hka^* b^* \cos \gamma^*$$

When the reference axes are orthogonal this reduces to equation (28) of chapter 5,

$$\frac{1}{d^2_{(hkl)}} = d^{*2}_{(hkl)} = h^2 a^{*2} + k^2 b^{*2} + l^2 c^{*2},$$

and there is a further simplification for the cubic system to

$$\frac{1}{d^2_{(hkl)}} = d^{*2}_{(hkl)} = a^{*2}(h^2 + k^2 + l^2)$$

i.e.

$$d_{(hkl)} = \frac{a}{\sqrt{(h^2 + k^2 + l^2)}}$$

Interpretation when the unit-cell is unknown

In the cubic system unambiguous indexing can usually be achieved, otherwise we are on much less certain ground especially in systems of low symmetry; even with elaborate computer programmes available for selecting and adjusting unit-cell constants to produce a perfect fit, within the limits of error of measurement, between calculated and observed d-spacings there can be no certainty that the interpretation is correct unless additional information is available from single crystal studies. An instructive example of misinterpretation is provided by Christophe-Michel-Lévy and Sandrea (1953), who indexed a powder pattern of 15 lines given by the mineral högbomite on a tetragonal unit-cell with a 8·34, c 7·96 Å; McKie (1963) employed single crystal data to show that the symmetry was hexagonal and the true unit-cell dimensions were a 5·72, c 23·0 Å. Indexing of powder photographs in these circumstances should only be attempted when crystals of a size suitable for single crystal X-ray study cannot be isolated and electron diffraction studies are for one reason or another impossible. In general it is true to say that as symmetry decreases from cubic to triclinic the number of lines in powder patterns taken with X-radiation of the same wavelength increases, the number of adjustable unit-cell parameters increases also, as does the number of correspondences available between observed and calculated d-spacings. Although there is always a risk of misinterpretation, plenty of interpretations have been shown by subsequent single crystal studies to be substantially correct. A clear account of the methods of indexing appropriate to each system and when the system is unknown is given by Lipson and Steeple (1970). We shall here confine ourselves mainly to cubic patterns which can always be indexed satisfactorily except when very few lines are present.

In the cubic system the expression for d_{hkl} may, as we have seen, simply be written as

$$d_{hkl} = \frac{a}{\sqrt{(h^2 + k^2 + l^2)}}.$$

Substitution in the Bragg Equation yields, for a cubic substance,

$$\sin^2 \theta = \frac{\lambda^2}{4a^2}(h^2 + k^2 + l^2)$$

which can conveniently for our present purpose be rewritten as

$$\sin^2 \theta = \frac{\lambda^2 N}{4a^2},$$

where N is an integer that can be expressed as the sum of three squares.

The equation at the end of the preceding paragraph provides a means of indexing the powder pattern of any substance that is known from other information or suspected from the simplicity of the pattern to be cubic. From each measured Bragg angle $\sin^2 \theta$ is calculated. The common factor of the set of $\sin^2 \theta$ values is found, either by inspection or graphically; the value of N for each measured line is then given by the ratio of $\sin^2 \theta$ for the line to the common factor. The graphical method is particularly suitable when θ has not been measured particularly accurately and no computing facilities are available. On a sheet of rectangular graph paper the measured values of $\sin^2 \theta$ are plotted along one axis and possible values of N, that is all integers that can be the sum of three squares, from 0 to at least 30 are plotted on the other axis. Lines are drawn through each $\sin^2 \theta$ value parallel to the N axis right across the sheet of graph paper. A ruler is laid along the $\sin^2 \theta$ axis and rotated slowly about the origin until all its intercepts with $\sin^2 \theta$ lines are at integral values of N; a line is drawn through the origin at this inclination and the value of N corresponding to each measured line is read from the graph (Fig 7.8).

Once a set of values of N has been obtained each line on the pattern can be indexed from $N = h^2 + k^2 + l^2$, applying the convention $h \geqslant k \geqslant l$. Some values of N can, as has been remarked earlier, correspond to more than one set of indices, for example $N = 9$ yields 221 and 300, both of which have $d = \frac{1}{3}a$. Certain other values of N cannot be expressed as the sum of three squares, e.g. 7, 15, 23, 28; such integers are given by $m^2(8n-1)$ where m and n are integers (a proof of this expression is given in Lipson and Steeple, 1970, Appendix 3). If the cubic substance has a non-primitive lattice, systematic absences will appear in the sequence of values of N: powder lines

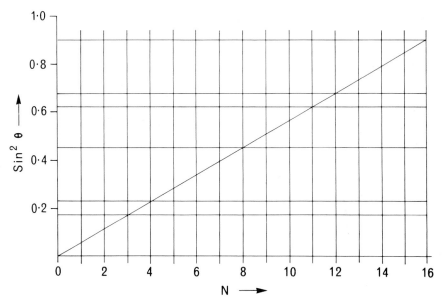

Fig 7.8 Indexing of a cubic powder pattern when the unit-cell dimension is not known, even approximately. Measured values of $\sin^2 \theta$ are plotted on the vertical axis and drawn out as horizontal lines. The sequence of integers from zero is plotted on the horizontal axis and each is drawn out as a vertical line. A ruler is rotated from the $\sin^2 \theta$ axis about the origin until all its intercepts with horizontal lines coincide with their intersections with the vertical set; a line is drawn through the origin at this inclination. For simplicity only values of N up to 16 are shown.

Table 7.3 Permissible reflexions for cubic lattice types
$N = h^2 + k^2 + l^2$

N	P	I	F	N	P	I	F
1	100	—	—	16	400	400	400
2	110	110	—	17	410, 322	—	—
3	111	—	111	18	330, 411	330, 411	—
4	200	200	200	19	331	—	331
5	210	—	—	20	420	420	420
6	211	211	—	21	421	—	—
7	—	—	—	22	332	332	—
8	220	220	220	23	—	—	—
9	300, 221	—	—	24	422	422	422
10	310	310	—	25	500, 430	—	—
11	311	—	311	26	510, 431	510, 431	—
12	222	222	222	27	333, 511	—	333, 511
13	320	—	—	28	—	—	—
14	321	321	—	29	520, 432	—	—
15	—	—	—	30	521	521	—

for which $h + k + l$ is odd will be absent when the lattice type is I, while lines for which h, k, l are neither all even nor all odd will be absent when the lattice type is F. The sequence of powder lines that may be observable in cubic substances with P, I, or F lattices is given in Table 7.3. Other lines may of course be absent due to the presence of translational symmetry elements or may be too weak to be observed due to the atomic arrangement associated with each lattice point. At first sight it might seem that, since N must always be even for an I lattice, P and I lattices would be indistinguishable, an I lattice yielding a spurious set of N values, each one half of the true N value; but the forbidden N values, 7, 15, 23, etc, serve to make the distinction, i.e. if N appears to be equal to any one of these forbidden numbers and the lattice appears to be P, then it is an I lattice with all the apparent values of N doubled. However, it is necessary to proceed cautiously because of the possibility of the lines with $N = 14$ and 30 in an I lattice being too weak to be observed; unless two or preferably three lines with forbidden values of N are actually absent it is not safe to conclude that the lattice is primitive and another powder photograph should be taken with X-rays of shorter wavelength to examine an extended range of N values.

Reference to Table 7.3 shows that, if there are no absences other than those due to lattice type, the ratio of the $\sin^2 \theta$ values for the three lines of lowest θ are 3:4:8 for a cubic F lattice and 1:2:3 for cubic P and I lattices. With experience these ratios become easy to recognize on inspection of the pattern; they should always be sought in the first instance when indexing either by inspection of the list of $\sin^2 \theta$ values or by the graphical method. Figure 7.9 shows the powder patterns of three cubic substances, one with a P, one with an I, and one with an F lattice.

Once the cubic powder pattern has been indexed the unit-cell edge a can be determined by application of the expression

$$a = \frac{\lambda}{2 \sin \theta} \sqrt{(h^2 + k^2 + l^2)}.$$

Maximum accuracy in the evaluation of a is obtained by using the measured value of θ for the line of highest θ as is evident from the following argument. Differentiation of the Bragg Equation $\lambda = 2d \sin \theta$ with respect to θ and d yields

$$2 \sin \theta . dd + 2d \cos \theta . d\theta = 0$$

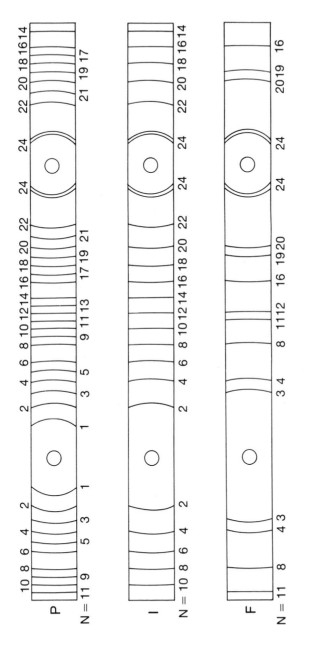

Fig 7.9 Diagrammatic representation of powder photographs, in the Straumanis setting, of three hypothetical cubic substances with $a = 3.86$ Å, taken with CuKα radiation ($\lambda = 1.542$ Å). The substance giving rise to the upper photograph has a primitive lattice so that the only absent lines are those for $N = (h^2 + k^2 + l^2) = 7, 15, 23$. In the middle photograph the substance has an I lattice so that there are systematic absences corresponding to odd values of $h + k + l$. In the lowermost photograph the substance has an F lattice so that reflexions are only present when h, k, l are all even or all odd.

Table 7.4   Indexing a cubic powder pattern

Procedure: (1) Measure film and evaluate θ for each line (column 3).
    (2) calculate $\sin^2 \theta$ for each line (column 4).
    (3) *either* plot $\sin^2 \theta$ for each line on graph (Fig 7.8) and draw a straight
      line through intersections of lines of calculated $\sin^2 \theta$ and lines of
      constant N; then read off the value of $N = h^2 + k^2 + l^2$ for each line
      (column 5).
      or determine highest common factor of $\sin^2 \theta$ values by inspection of
      column 4 and thence deduce values of N (column 5).
    (4) index each line from its N value (column 6).
    (5) determine the lattice type.
    (6) use the measured value of θ for the line of greatest N to calculate the
      unit-cell edge a.

line	intensity	$\theta°$	$\sin^2 \theta$	N	hkl	
a	w	24·3	0·169	3	111	Lattice type $= F$
b	m	28·3	0·225	4	200	
c	s	42·1	0·450	8	220	$a = \dfrac{\lambda\sqrt{N}}{2 \sin \theta}$
d	s	51·9	0·619	11	311	
e	vw	55·2	0·675	12	222	$= \dfrac{1\cdot542 \times 4}{2 \sin 71\cdot6°}$
f	m	71·6	0·900	16	400	

$(CuK\alpha) = 1\cdot542$ Å $= 3\cdot25$ Å

Intensity scale: $s > m > w > vw$

Therefore $\dfrac{dd}{d\theta} = -d \cot \theta.$

Therefore for a fixed error in θ, the error in d will be least when $\cot \theta$ has its smallest value, that is as $\theta \rightarrow 90°$.

  Examples of indexing and determination of a from cubic powder photographs are provided by Fig 7.8 and Table 7.4.

Some uses of powder methods

It is not proposed here to offer an exhaustive list of the uses to which X-ray powder diffraction studies have been put, but merely to indicate the scope of the methods by mention of a wide ranging variety of fruitful uses.

  Probably the best known use of powder methods, especially the powder photograph, is for identification. Over the past forty years a very extensive card file listing the d-spacings and relative intensities of the lines on the powder patterns of many thousands of substances has been built up; this is known as the Powder Diffraction File, formerly the ASTM Index (supplementary data are issued annually). Each compound on file has a card on which are given d and I for all lines in the powder pattern, chemical composition, unit-cell constants from single crystal data if available, density, and optical properties; in most cases powder lines are indexed. Accompanying the file is an index in which all the compounds on file are listed in order of the d-spacings of their strongest lines so that a particular substance will appear in the index under the d-spacing of each of its six strongest lines.[2] The technique of identification is to measure the d-spacings of all the strong lines on the powder

---

[2] This is the indexing system for inorganic compounds including minerals. For organic compounds there is a separate card file and a separate index in which a compound is listed under the d-spacing of each of its three strongest lines.

photograph of the unknown, if there are more than six to select the six strongest and make a visual estimate of their relative intensities, and to look up each strong line in order of decreasing intensity in the index until a satisfactory match is found. At this stage it is always advisable to obtain a powder photograph of a reliable specimen of the substance identified taken on a camera of the same radius and with the same radiation as the photograph of the unknown. Most laboratories concerned with identification maintain a collection of powder photographs of well authenticated substances in their field. To ease the task of reducing a powder pattern to a set of d-spacings, there are commercially available rulers that give a direct measurement of d, for a camera of given radius and radiation of given λ, to sufficient accuracy for use with the File. For purposes of identification, powder photography is generally preferred to diffractometry because the whole range of θ can be sampled more quickly and the pattern converted to a list of d-spacings more easily; nevertheless in particular circumstances diffractometry may be selected for identification. There are of course some potential snags to be borne in mind when using the File: the unknown may not be on file, or it may be a member of a solid solution series only the end members of which are on file, or it may be a mixture. In general however the Powder Diffraction File is a powerful tool for the identification of unknown substances.

Identification of the several constituents of a mixture by powder methods is never a straightforward task unless the substances that may be present are restricted in number and not more than three, or at most four, of them are present in the mixture. For reliable identification it is essential to compare the powder pattern of the mixture with the superimposed patterns of the suspected constituents. The limit of detection of any constituent depends very much on whether its powder pattern contains a very strong line that is clearly resolved from the lines of the other constituents; even when this criterion is satisfied it is rarely possible to detect a substance present to the extent of <5 per cent. For the detection of substances present in small concentrations diffractometry and photography are generally balanced: the former gives better resolution while the latter exhibits very weak lines with less ambiguity. The limit of detection by optical examination (volume 2) is generally very much lower, but identification of the impurity is less satisfactory than by powder methods; if the impurity is in the form of grains of manageable size, a sufficient number of grains to make a powder specimen can be picked out under the microscope and identification achieved by powder photography.

Powder diffractometry is particularly useful for accurate determination of the unit-cell dimensions of solid solutions. A simple example is provided by the Cu–Au system whose melts yield on rapid cooling (quenching) cubic solid solutions for which a varies linearly with atomic percentage from 3·608 Å for pure copper to 4·070 Å for pure gold. Measurement of the d-spacing of an indexed high-θ line by diffractometry will thus yield the composition of the quenched alloy. Precision can be greatly improved by admixture with a pure substance whose unit-cell dimensions are very precisely known and which serves as an internal standard of θ on the diffractometer trace; the standard is chosen to have a peak close to the peak whose d-spacing is to be determined and the proportions in the diffractometer sample are adjusted by trial and error until the two peaks are of comparable height so that uncertainties in scanning and chart speeds have minimal effect. An example of the use of internal standards is provided by the determination of composition of olivine in the binary solid solution series Mg_2SiO_4–Fe_2SiO_4 (Yoder and Sahama, 1957). For this purpose silicon is selected as internal standard because its 111 peak, at $2\theta = 28\cdot465°$ for CuKα

radiation, is strong, precisely known, and close to the 130 peak in olivine at 2θ ranging from $\sim 32\frac{1}{2}°$ for Mg_2SiO_4 to $\sim 31°$ for Fe_2SiO_4; the greatest length to be measured on the diffractometer trace is thus only $4°$ of 2θ. The diffractometer is run at least six times over the pair of peaks and the average of their separation used to calculate d_{130} for the olivine specimen. Thence the composition of the solid solution can be obtained by substitution in the expression

$$\text{Mol per cent } Mg_2SiO_4 = 4233\cdot 91 - 1494\cdot 59\, d_{130},$$

which is based on measurements of d_{130} for olivines of known composition. The resultant composition determined by diffractometry is in this case subject to an error that may be as high as 4 per cent mainly because most naturally occurring olivines are not strictly binary solid solutions but contain small amounts of other components such as Mn_2SiO_4 in solid solution. In developing such a method an essential preliminary is of course to obtain and index diffractometer traces of compositions close to the end numbers because not only will d_{hkl} vary with composition but so will I_{hkl}; it is necessary to choose a peak that remains strong throughout the compositional range and to choose a standard peak that is not interfered with by peaks of the solid solution in any compositional range.

The techniques described in the preceding two paragraphs are of particular use in synthetic studies and in the determination of phase diagrams.

The method of diffractometry with an internal standard can be applied to the accurate determination of the unit-cell dimensions of a substance of high symmetry whose crystal system is known from other evidence or in general when approximate unit-cell dimensions only are available, for one reason or another, from single crystal X-ray or electron diffraction photographs.

Powder photography and diffractometry provide a satisfactory means of determining coefficients of thermal expansion, which will be anisotropic (volume 2) for crystalline solids other than those belonging to the cubic system. Furnaces capable of heating a powder specimen in a camera or on a diffractometer to temperatures in excess of $2000\,°C$ are available; in high temperature work it is usually necessary to evacuate the camera or diffractometer space and the furnace must be split so that it does not interfere with incident or diffracted beams. Low temperature cameras in which the specimen is cooled by a stream of coolant, liquid air or rarely liquid helium, are available and can, with special precautions, be operated at temperatures as low as $2\,K$. In both high and low temperature cameras and diffractometers the temperature of the specimen is recorded by a thermocouple placed as close to the specimen as possible without interference with the incident or selected diffracted beams; it is usually necessary to make a calibration correction for the temperature difference between specimen and thermocouple by replacing the specimen with substances that have accurately known melting or transformation temperatures.

So far we have considered only uses of powder methods that depend primarily on measurements of d-spacings, but the relative intensities of powder lines can be measured and, after appropriate corrections, utilized in crystal structure determination. For this purpose intensity data collected from single crystals are preferable, not only are there more data, but the problem of coincident reflexions does not arise; nevertheless many structures of substances, such as alloys, for which it is difficult to obtain single crystals have been based on powder intensity data.

Powder methods have also been used in kinetic studies of polymorphic transformations. If the intensities of adjacent lines, one belonging to the reactant and the other to the product phase, are measured in a series of specimens subjected to isothermal heating for various lengths of time, a plot of their ratio against time enables the time to be evaluated for a certain fraction of the reactant phase to be transformed. Similar measurements for other temperatures then yield a determination of the activation energy of the process.

Special cameras

Many variants of the powder camera and a few variants of the powder diffractometer have been designed as instrumental aids specifically for the solution of particular problems. Some of these instruments have been designed to operate at high temperatures, others at low temperatures, and yet others over extended temperature ranges. Other specialized instruments have been designed for the study of layer structures with very large interlayer spacings, such as clay minerals. We confine our attention at the conclusion of this chapter to two such specialities which are in more general use than most. The Guinier type of camera has many variants—the Guinier–de Wolff camera is the one we describe—all of which make use of the parafocusing effect which we have already discussed in our treatment of powder diffractometry. The Gandolfi camera provides an ingenious and effective means of generating a powder photograph from a single crystal, which may be small or large; Gandolfi photography is widely applicable in almost all fields of solid state studies.

Guinier–de Wolff cameras

The most commonly used type of powder camera has a radius of 57·3 mm so that one degree of θ corresponds to 2 mm on the film. The photographic record of the diffraction pattern is so compressed that except for substances of high symmetry and small unit-cell dimensions overlap of powder lines is likely to occur even at quite low Bragg angles. Improved resolution of the powder pattern can be achieved by improved collimation of the incident beam, by evacuating the camera to eliminate air-scattering, and by increasing the radius of the camera, which will mean a proportionally longer exposure time. Cameras of the type generally described as 'Guinier cameras'—there are many variants, each being appropriate to a particular usage—provide a means of improving resolution without increasing the camera radius, but they do so at the expense of a restricted range of θ. We shall confine ourselves to the basic principles on which this type of camera depends and refer the reader for details to Klug and Alexander (1974) and the references quoted therein.

Because X-rays have refractive indices close to unity for all media, as we have already indicated on p. 173, it is not possible to focus an X-ray beam with lenses. However it is possible to reflect X-rays diverging from a point source by means of a crystal monochromator so that they converge on a point: the monochromatic X-ray beam is then effectively focused at that point. This result is fundamental to the design of all types of Guinier camera.

The point X in Fig 7.10 is ideally a point source from which the X-rays diverge. An approximation to the ideal is achieved in practice by placing a short narrow slit in front of the line focus of an X-ray tube to achieve a divergent beam of X-rays, the length of the slit being perpendicular to the length of the line focus. The divergent X-ray beam is incident on the surface of a crystal monochromator, the monochromator being bent so that its whole surface and the source X lie on a circle of radius r_M, the focusing circle. In consequence Bragg reflexions from all points on the surface of the monochromator crystal will converge on the point F, which also lies on the focusing

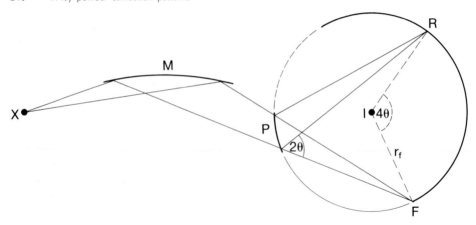

Fig 7.10 Guinier–de Wolff cameras. The divergent X-ray beam from the point source X is incident on a bent crystal monochromator M. The whole surface of the monochromator and X lie on the focusing circle of radius r_M. Bragg reflexions from all points on the surface of the monochromator converge on F, which also lies on the focusing circle. The circle of radius r_f represents the cylindrical film cassette; F lies on this circle. The powder specimen P is tangential to the cassette and lies between M and F. X-rays diffracted by (*hkl*) planes in the powder specimen are focused on the film at R which is distant $4r_f\theta$ from F.

circle. For efficient focusing it is necessary that the monochromator crystal should be bent so as to permit Bragg reflexion from the selected lattice planes at any point on its surface. It is usual for the surface of the monochromator crystal to be ground into a cylindrical shape, the radius of the cylinder being r_M: the source X and the whole surface of the monochromator will then lie on the focusing circle.

In the Guinier–de Wolff camera (Fig 7.10) the film is mounted in a cylindrical cassette of radius r_f, the cassette being so positioned that the point F, where the incident X-rays are focused, lies on the film. The powder specimen P is positioned tangential to the cylindrical camera so that X-rays incident from the crystal monochromator and travelling towards the focus F pass through the specimen. Some part of the incident X-radiation will be diffracted to give Bragg reflexions inclined at 2θ to the incident beam. Since the camera is designed so that the incident X-rays are focused at F and since for every point in the powder specimen the angle $\widehat{FPR} = 2\theta$, the X-rays diffracted by the lattice planes (*hkl*) will be focused at R. Since the arc RF subtends an angle 2θ at the circumference of the cylindrical camera, i.e. at P, it subtends an angle 4θ at the centre of the cylindrical camera. A distance s on the film thus corresponds to $4r_f\theta$. For a camera radius of 57·3 mm one degree of θ corresponds to a distance of 4 mm on the film; this is double the resolution obtained with a conventional powder camera. Moreover focusing not only increases the sharpness of the lines recorded on the film, but it also reduces the effect of the thickness of the specimen. Further improvement in resolution is achieved over a limited range of θ by the camera geometry because the $K\alpha_2$ line lies on one side of the $K\alpha_1$ line after reflexion by the monochromator but on the other side after reflexion by the crystalline specimen so that choice of the radius r_M of the focusing circle, of the radius r_f of the film cassette, and of the angle $\widehat{IPF}$ can be made to bring $K\alpha_1$ and $K\alpha_2$ lines for a particular value of θ into coincidence. Then the separation of the $K\alpha_1$ and $K\alpha_2$ lines will be reduced on either side of the θ angle for which they have been made to coincide, so improving the resolution in this θ range.

Powder lines will be relatively sharp in those parts of the film where the reflected X-rays are nearly normally incident. As the diffracted beams become more nearly tangential the sharpness of the powder lines decreases and some resolution is lost. In the Guinier–de Wolff camera powder lines with $2\theta < 90°$, i.e. $\theta < 45°$, can be recorded.

The powder specimens used in Guinier cameras are generally flat and quite large in area. Such specimens usually produce smooth powder lines, but provision is made in most Guinier–de Wolff cameras for the specimen to be moved backwards and forwards in its own plane to overcome lack of randomness in the powder sample.

There are many versions of the Guinier camera. In the Guinier–de Wolff camera the line focus produced by the crystal monochromator is utilized to take two or four photographs from different specimens simultaneously. Another variant is the Guinier–Lenne camera which has a heating facility combined with a controlled movement of the film cassette so that successive or continuous photographic records of the diffraction pattern can be obtained over a large temperature range.

The Gandolfi camera

The situation not infrequently arises where it is necessary to identify a substance from which a powder specimen cannot be prepared. This may be because insufficient material is available for the preparation of a powder specimen (e.g. an inclusion which has been extracted from a host crystal) or because grinding would be expected to change the structural state (e.g. the *wurtzite* polymorph of ZnS on grinding inverts to the *blende* polymorph), or because the single crystal is too valuable for any part of it to be lost (e.g. a gemstone), or for a variety of other reasons applicable to the problem in hand. In such circumstances what is required is a powder pattern produced by a single crystal; several cameras have been designed with that objective in mind, but only the camera described by Gandolfi (1967) is in general use.

If a single crystal is mounted in place of a powder specimen in a conventional powder camera, the lattice planes of the set (hkl) will be able to diffract only if $\mathbf{d}^*_{hkl}$ happens to be inclined at $90° - \theta_{hkl}$ to the incident X-ray beam; a stationary single crystal exposed to monochromatic radiation (Fig 7.11(a)) will thus give rise fortuitously to a very few reflexions and only some of those will be recorded on the narrow cylindrical film strip. If the crystal is rotated about the axis, A_1, normal to the incident beam just as the powder specimen would be rotated in conventional use of a powder camera, then the typical reciprocal lattice vector $\mathbf{d}^*_{hkl}$ will describe a small circle about the pole of A_1 and so intersect the small circle of radius $90° - \theta_{hkl}$ about the pole X, which represents the incident X-ray beam, twice in a complete revolution. The opposite reciprocal lattice vector $\mathbf{d}^*_{\overline{hkl}}$ will likewise make two intersections with the $90° - \theta_{hkl}$ small circle, once in the upper and once in the lower hemisphere. Each (hkl) plane and its opposite will thus give rise to four reflexions provided the angle $\phi = \mathbf{d}^*_{hkl} : A_1 > \theta_{hkl}$ (Fig 7.11(b)). The cylindrical film strip will record only a few of these reflexions.

In the Gandolfi camera the crystal is mounted on a spindle inclined at $45°$ to the camera axis A_1 ($A_1 \equiv q$ in Fig 7.6) and the spindle is rotated about its length, the rotation axis A_2, by irrational gearing to the primary rotation axis A_1. It is of course essential that after one complete rotation on A_1 there will not have been an integral number of rotations on A_2. In one complete rotation about A_2 a typical reciprocal lattice vector $\mathbf{d}^*_{hkl}$ will describe a small circle of radius ψ about the pole of A_2 where $\psi = \mathbf{d}^*_{hkl} : A_2$ and simultaneously the position of A_2 will have moved along the $45°$ small

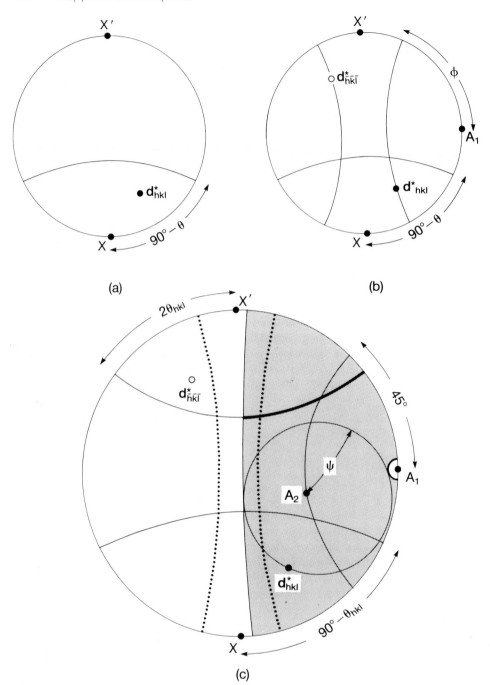

Fig 7.11 The Gandolfi camera. A stationary single crystal will give rise to an *hkl* reflexion only if $\mathbf{d}_{hkl}^{*}$ lies on the $90° - \theta_{hkl}$ small circle about the direction of the incident beam *XX'* as illustrated in (a). If the crystal is rotated about A_1, normal to *XX'*, the crystal will give rise to four *hkl* reflexions provided $\phi > \theta_{hkl}$ as illustrated in (b). In the Gandolfi camera, as shown in (c), the crystal is rotated about A_1 and A_2, where $A_1 : A_2 = 45°$, so that $\mathbf{d}_{hkl}^{*}$ occupies the whole of the shaded area during the exposure time and $\mathbf{d}_{\overline{hkl}}^{*}$ occupies a symmetry related area. Many complete and some incomplete powder lines will be recorded; there will usually be some absences. The dotted small circles represent the angular width of the film.

circle about A_1 (Fig 7.11(c)). After the many rotations that will have occurred during a normal exposure of the film for several hours the reciprocal lattice vector $\mathbf{d}^*_{hkl}$ will have occupied all points in the shaded area of the stereogram shown in Fig 7.11(c), the area covered being bounded by small circles about the pole A_1 of radius $45° + \psi$ and $45° - \psi$. Simultaneously an equivalent area of the stereogram symmetrical about XX', but not shown on the figure, will have been occupied by $\mathbf{d}^*_{hk\bar{l}}$. In the case exemplified in the figure an almost complete powder ring will be generated: segments of the powder line will appear at the top and bottom of the film strip, the central segment being absent. In general for $\psi = 45°$ a complete powder ring will be generated, for $45° < \psi < 90°$ there will be a complete powder ring with four lengths of overlap, and for $0° < \psi < 45°$ (the case illustrated) the powder ring will be incompletely developed. The resultant photographic record will thus show many complete powder lines (some of which will be of doubled intensity due to overlap extending beyond the limits of the record), a few lines terminating on the film, and there will be some absences. We have considered the least favourable case, a triclinic crystal where $\mathbf{d}^*_{hkl}$ is related only to $\mathbf{d}^*_{hk\bar{l}}$; in systems of higher symmetry other symmetry related reciprocal lattice vectors will contribute to the powder line and so the likelihood of absences will be diminished provided the crystal has not been mounted with a symmetry axis nearly coincident with A_2.

The Gandolfi camera is so constructed that the rotating spindle assemblage is mounted on the lid of the camera (b in Fig 7.6) rather than on the camera body or cassette (a in Fig 7.6) as in a conventional powder camera. This makes it possible after an initial exposure of several hours to remove the part of the camera on which the crystal is mounted from the film cassette in the dark-room, to remount the crystal at an orientation to the spindle axis A_2 close to 90° from its original orientation, to reassemble the camera in the dark-room without any disturbance to the film, and to superimpose on the first exposure a second exposure with the crystal in a very different orientation. In the second exposure a different set of reciprocal lattice vectors may be expected to be suitably oriented for the corresponding reflexions to be recorded on the film strip, which has an angular width of about 28°, shown on Fig 7.11(c) by dotted small circles about A_1.

If several small crystals are available, the second exposure may be obviated by mounting two or three crystals in different random orientations on the A_2 spindle.

One of the several advantages of the Gandolfi camera is that it is basically an adaptation of a standard Straumanis powder camera of radius 57·3 mm so that Gandolfi photographs can be directly compared with conventional powder photographs, which is a significant advantage when identification is the objective.

Problems

7.1 The X-ray powder photograph, taken with CuKα radiation, of the sodium tungsten bronze $Na_{0.8}WO_3$, which is cubic, has lines at the following Bragg angles:

11·60°, 15·52°, 20·38°, 23·71°, 26·71°, 29·50°, 34·65°, 37·09°, 39·49°, 41·81°, 44·13°, 46·45°, 48·78°, 53·52°, 55·98°, 58·52°, 61·19°.

Index these powder lines, determine the lattice type, and evaluate the unit-cell dimension.

7.2 Molybdenum is cubic and its X-ray powder photograph, taken with MoKα radiation ($\lambda = 0.7107$ Å), has lines in order of increasing Bragg angle at the following values of θ:

9·18°, 13·04°, 16·04°, 18·61°, 20·90°, 23·00°, 24·97°, 26·82°, 28·59°, 30·30°, 31·95°, 33·55°, 35·12°, 38·16°, 39·66°, ...

Index these lines, determine the lattice type, and evaluate the unit-cell dimension a.

7.3 Plutonium sulphide, PuS, is cubic and the eight lines of lowest Bragg angle on its X-ray powder photograph, taken with CuKα radiation ($\lambda = 1\cdot5418$ Å), have the following values of θ:

13·95°, 16·17°, 23·19°, 27·50°, 28·84°, 33·84°, 37·37°, 38·51°.

Index these lines, determine the lattice type, and evaluate the unit-cell dimension a.

7.4 Silicon is cubic and the seven lines of lowest Bragg angle on its X-ray powder photograph, taken with CuKα radiation ($\lambda = 1.5418$ Å), have the following values of θ:

14·23°, 23·67°, 28·08°, 34·59°, 38·22°, 44·05°, 47·51°.

Index these lines, determine the lattice type, and evaluate the unit-cell dimension a.

7.5 Magnetite, Fe_3O_4, is cubic and its X-ray powder photograph, taken with FeKα radiation ($\lambda = 1\cdot9373$ Å), has lines in order of increasing Bragg angle at the following values of θ:

11·67°, 19·29°, 22·74°, 23·86°, 27·70°, 34·60°, 37·09°, 41·04°, ...

Index these lines, determine the lattice type, and evaluate the unit-cell dimension. On this photograph the line of highest θ is a doublet. Given that $\lambda(FeK\alpha_1) = 1\cdot9360$ Å, $\lambda(FeK\alpha_2) = 1\cdot9399$ Å and that the Bragg angles for the doublet are $78\cdot02° \pm 0\cdot06°$ and $78\cdot51° \pm 0\cdot06°$, index the doublet and evaluate the unit-cell dimension accurately.

7.6 On cooling through 120°C the cubic structure of barium titanate, $BaTiO_3$, for which $a \simeq 4$ Å changes to a tetragonal structure, which persists down to 5°C. The tetragonal structure (T) is closely related to the cubic structure (C) so that a_T and c_T do not differ greatly from a_C. A powder photograph of $BaTiO_3$ taken at 20°C with CuKα radiation ($\lambda = 1\cdot5418$ Å) shows two closely spaced lines at $\theta = 22\cdot73°$ and $\theta = 22\cdot49°$, the former having about twice the intensity of the latter. Determine the unit-cell dimensions of tetragonal $BaTiO_3$.

7.7 The composition of the orthorhombic mineral olivine can be determined by X-ray powder diffractometry when it is restricted to the range Mg_2SiO_4 to Fe_2SiO_4. A convenient olivine reflexion is 130 which has limiting values of 2θ for CuKα radiation ($\lambda = 1\cdot5418$ Å) of 32·365° for Mg_2SiO_4 and 31·620° for Fe_2SiO_4. A convenient internal standard is provided by the 111 reflexion of pure Si, which is cubic with $a = 5\cdot4306$ Å. A diffractometer trace was taken of a specimen of olivine mixed with pure silicon over the 2θ range 27–33° using CuKα radiation. The diffractometer speed was $\frac{1}{8}°$ of 2θ per minute and the chart recorder was operating at 5 mm min^{-1}. The separation of the Si 111 and the olivine 130 peaks at half peak height was measured as 149 mm. Determine $2\theta_{130}$ for the olivine. Evaluate d_{130} for the specimen and for the limiting compositions and hence, assuming linear variation of unit-cell dimensions with molar composition, the composition of the specimen.

7.8 Periclase, MgO, is cubic and its powder photograph, taken with CuKα radiation ($\lambda = 1\cdot5418$ Å) has lines at the following Bragg angles θ:

$18\cdot38°$, $21\cdot25°$, $31\cdot00°$, $37\cdot00°$, $39\cdot25°$, $46\cdot25°$, $53\cdot00°$, $55\cdot12°$, $63\cdot75°$, $72\cdot00°$.

Index these lines, determine the lattice type, and evaluate the unit-cell dimension. The line of highest Bragg angle is an $\alpha_1\alpha_2$ doublet, which is clearly resolved so that θ for the α_1 line can be measured as $71\cdot95°$; given that $\lambda(\text{CuK}\alpha_1) = 1\cdot54050$ Å, obtain a better value for the unit-cell dimension a. Given that the density of periclase is $3\cdot56$ Mg m^{-3}, that the atomic weights of Mg and O are, respectively, $24\cdot32$ and $16\cdot00$, and that 1 a.m.u = $1\cdot66 \times 10^{-27}$ kg, determine the number of formula units of MgO in the unit-cell.

Draw plans on (001) of the two possible structures of MgO as determined in question 4·10 and for each write down the coordinates of the atoms associated with the lattice point at the origin. Note that one of the possible structures is centrosymmetric and that the other is non-centrosymmetric. Write down for each structure the expression for the structure factor F(hkl) of a general hkl reflexion which is not systematically absent due to the lattice type. Simplify the structure factor expressions for the 420 and 331 reflexions for each structure. Write down the multiplicities of the 420 and 331 powder lines. The atomic scattering factors at the $\sin\theta/\lambda$ value of the 420 reflexion are $f_{Mg} = 4\cdot6, f_o = 2\cdot2$ and at the $\sin\theta/\lambda$ value of the 331 reflexion are $f_{Mg} = 4\cdot8, f_o = 2\cdot2$. Calculate the ratio I(420)/I(331) for each of the two possible structures of periclase assuming that for lines which do not differ greatly in Bragg angle there will be no necessity to take into account a variable factor dependent on the geometry of the powder record.

Given that the observed ratio I(420)/I(331) derived from the periclase powder photograph was $\sim6\cdot5$, determine the structure of periclase.

7.9 The caesium chloride, CsCl, structure is cubic and has just one formula unit per unit-cell. If the Cs$^+$ cation is placed at the origin, the Cl$^-$ anion will be at $\frac{1}{2}\frac{1}{2}\frac{1}{2}$. Write down the expression for the structure factor F(hkl) and observe that its magnitude depends on whether $h+k+l$ is odd or even. Thence index the nine lines of lowest θ on the powder pattern of CsCl which have d-spacings 4·13(w), 2·92(s), 2·38(w), 2·07(s), 1·85(w), 1·69(s), 1·46(s), 1·38(w), 1·31(s) Å, where w = weak, s = strong.

Evaluate the unit-cell dimension of CsCl.

Caesium iodide has the same structure with a slightly larger unit-cell. The first five lines on its powder photograph have d-spacings 3·23, 2·29, 1·87, 1·62, 1·45 Å. Index these lines, evaluate the unit-cell dimension of CsI, and account for the difference between the powder patterns of CsCl and CsI. [Atomic numbers: Cl 17, Cs 55, I 53].

7.10 At low temperatures Cu$_3$Au is cubic with one formula unit per unit-cell. If the Au atom is placed at the origin, then the Cu atoms will be situated at $0\frac{1}{2}\frac{1}{2}$, $\frac{1}{2}0\frac{1}{2}$, and $\frac{1}{2}\frac{1}{2}0$. Write down the expression for the structure factor F(hkl). What will be the form of the structure factor (i) when h, k, l are all even or all odd, (ii) when h, k, l are mixed?

Index the nine lines of lowest Bragg angle in the powder pattern of ordered Cu_3Au and indicate whether each line is strong or weak within its θ-range.

At higher temperatures Cu_3Au becomes disordered and has a random distribution of Cu and Au atoms over the lattice points of a cubic F lattice. The unit-cell dimensions of ordered and disordered Cu_3Au are very similar. How do their powder patterns differ?

8
Single crystal X-ray diffraction patterns

When a parallel beam of monochromatic X-radiation falls on a stationary single crystal very few lattice planes will be oriented so as to satisfy the Bragg Equation and in consequence very few reflexions will be observable. There are two ways in which the number of reflexions can be increased. One is to allow the crystal to oscillate or rotate during its exposure to monochromatic radiation; the other is to allow the wavelength of the incident radiation to be variable by using the total emission of the X-ray tube while keeping the crystal stationary. The first approach is the more productive and is employed in all but one of the experimental methods to be described in this chapter; the second approach gives rise to Laue photography which, although important, is generally less informative.

The apparatus necessary for each single crystal technique is described in the appropriate section of this chapter. The one experimental feature common to all, the mounting of the crystal on *arcs*, we shall deal with at this point. A small crystal whose dimensions should ideally be within the range 0·5 to 0·05 mm is selected under a binocular microscope. The crystal is then glued to a thin glass fibre (about 15 mm in length and less than 0·5 mm in diameter) so that a simple zone axis is approximately parallel to the fibre. The method of locating the zone axis in the crystal depends on the nature of the substance under investigation. If the crystal has well developed faces it will be possible to locate prominent zone axes by direct morphological inspection under the microscope. For instance if the substance is known to be tetragonal and to have {100} prism faces commonly well developed, [001] will lie parallel to the faces of this form, while [100] and [010], which are equivalent, will be normal to the prism faces. If however the crystal has no well developed faces but is transparent and optically anisotropic, the polarized light techniques described in volume 2 may serve to locate one or more simple zone axes in the crystal. When both these approaches fail, trial and error X-ray methods have to be used.

For ease of handling the glass fibre, to which the crystal is to be attached, its opposite end is pushed into a pea-sized blob of plasticine. When the crystal has been firmly glued to the fibre, the plasticine serves to fasten the fibre quite rigidly to the arcs. Crystallographic arcs, illustrated in Fig 8.1, consist of a manually adjustable slide *a*, carrying a projection against which the plasticine blob is pressed, surmounting a pair of worm operated arcs. The upper arc *b* provides a movement of 30° in either a clockwise or anticlockwise sense along the circumference of a vertical circle parallel

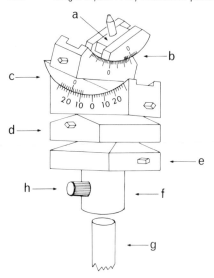

Fig 8.1   Crystallographic arcs. The labelling of the drawing is described in the text.

to the flat face of the arc. The lower arc *c* provides a similar movement in the vertical plane perpendicular to the plane of the upper arc *b*. Below the arcs are parallel worm operated slides *d, e*, to allow the crystal to be moved in a horizontal plane into coincidence with the axis of the spindle of the camera; the cap *f* at the base of the set of arcs fits over the camera spindle *g* and is clamped to it by tightening the knurled screw *h*.

The purpose of crystallographic arcs is to enable one to align the crystal so that it has a zone axis accurately parallel to a chosen direction in the X-ray camera. The techniques of crystal setting are described in Appendix F, which also provides details of crystal mounting techniques.

Oscillation photography

The oscillation camera, shown diagramatically in Fig 8.2, comprises a spindle *a* which is rigidly attached to the circular horizontal scale *b*, the arm *c* is clamped to the spindle and rests at its other extremity against a cam *d* driven by a synchronous electric motor *e* geared down to about one revolution per minute. Rotation of the cam causes the spindle, to which the arcs *f* are clamped, to oscillate through a definite angle; most cameras are equipped with alternative cams to give a choice of 5°, 10°, or 15° oscillation. A pin-hole collimator *g* is mounted so that its axis intersects the oscillation axis at right-angles; most cameras are supplied with fine and coarse collimators. The spindle can be raised or lowered so as to position the crystal in the X-ray beam. A cylindrical brass cassette whose axis is coincident with the oscillation axis constrains the photographic film *h*, which is in a light-tight envelope, to a cylindrical shape, the collimator protruding through the gap between the edges of the film; a typical film diameter is 60 mm. The undeviated X-ray beam passes through a hole in the cassette to be absorbed by a circular lead disc, the back-stop, attached to a removable cap (Fig 8.2(b)). By removing the cap it is a simple matter to test, with a fluorescent screen, whether the X-ray beam is passing correctly through the camera. Since the undeviated beam would produce serious fogging if it fell on the film a hole is punched in the film and a small tightly fitting brass collar is pressed through the hole in the film into the hole in the cassette to ensure correct alignment. The

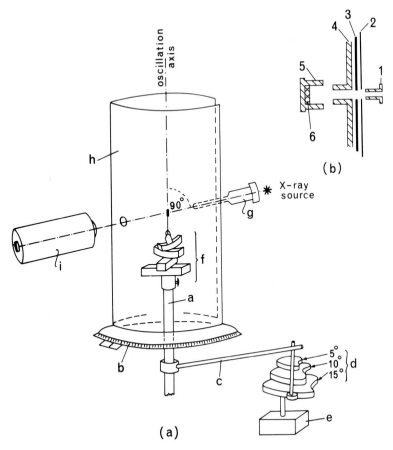

Fig 8.2 The oscillation camera. The essentials of the camera are shown in (a), the labelling of which is explained in the text. The detail of the backstop, or beam trap, is shown enlarged and exploded in (b): 1, collar; 2, opaque paper; 3, film; 4, wall of cassette; 5, cap; 6, lead disc.

oscillation camera is equipped with a telescope i whose axis is accurately aligned with that of the collimator to facilitate alignment and setting of the crystal. By swinging a lens into position in front of its objective the telescope i is simply converted into a microscope through which the crystal can be observed in a beam of light directed through the collimator; by focusing the microscope on the crystal and rotating the crystal, the crystal can be accurately centred on the oscillation axis of the camera and its height can be adjusted so that it lies precisely at the intersection of the oscillation axis with the incident beam from the collimator.

For oscillation photography the crystal is set so that a prominent zone axis is accurately parallel to the oscillation axis, that is the spindle axis, of the camera. Monochromatic X-radiation is incident on the crystal perpendicular to the oscillation axis. Suppose that the [001] axis of the crystal is parallel to the oscillation axis, then $i_c = 90°$ and the third Laue Equation reduces to

$$c \cos \delta_c = l\lambda.$$

Diffracted beams are consequently restricted to a series of cones, each cone being associated with a particular value of l, coaxial with the oscillation axis of the camera and with the z-axis of the crystal. The cones corresponding to $+l$ and $-l$ are symmetrical about the plane normal to the z-axis which corresponds to the solution of the third Laue Equation for $l = 0$, i.e. $\delta_c = 90°$ (Fig 8.3).

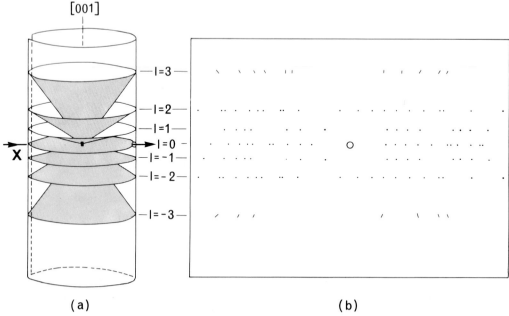

(a)   **(b)**

Fig 8.3   Oscillation photograph of a crystal with [001] as oscillation axis. (a) shows the coaxial cones generated by solutions of the third Laue Equation for successive values of l from -3 to $+3$ and the intersection of these cones with the coaxial cylindrical film. (b) shows the appearance of the film laid flat after development; the reflexions lying on any one layer line all have the same l index.

If the crystal is not oscillated, but kept stationary, very few diffraction maxima will occur because simultaneous solutions of the first and second Laue Equations,

$$a(\cos \delta_a - \cos i_a) = h\lambda$$

and $$b(\cos \delta_b - \cos i_b) = k\lambda,$$

will be extremely rare. The number of diffraction maxima is substantially increased by oscillating the crystal about its z-axis so that the angles i_a and i_b can vary to some definite extent while i_c remains fixed at 90°. All the resultant diffracted beams lie on the surfaces of the set of cones that constitute the solutions to the third Laue Equation for different values of l. The set of cones intersects the cylindrical film in a set of circles, which appears, when the film is laid flat after development, as a set of parallel straight lines. All the reflexions recorded on the film lie on these straight lines, which are known as *layer lines*. Each layer corresponds to a solution of the third Laue Equation for a particular value of l and the layer lines are symmetrically disposed about that one of their set corresponding to $l = 0$, the *zero layer line*. For the lth layer line it is evident from Fig 8.4 that δ_c can be calculated by measuring the height H_l of this layer line above the zero layer line since the camera radius r is known,

$$\cot \delta_c = \frac{H_l}{r}.$$

Combination of this equation with the third Laue Equation yields the generally useful equation

$$c = \frac{l\lambda}{\cos \cot^{-1} H_l/r}$$

which enables the lattice spacing parallel to the oscillation axis to be evaluated.

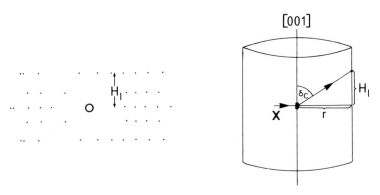

Fig 8.4 The geometry of layer line generation. The left-hand diagram is a portion of an oscillation photograph and the right-hand diagram illustrates the relationship between δ_c, H_l, and r.

Since the cotangent of δ_c is proportional to H_l, better accuracy would appear to be obtainable by measurement of the height of the highest observable layer line above the zero layer line, or better by measurement of the distance between the $+l$ and $-l$ layer lines for the greatest practicable value of l. However factors such as the divergence of the incident X-ray beam and the progressive reduction in angle between the diffracted beam and the surface of the film as l increases make the use of upper layer lines of no great advantage in improving the accuracy of δ_c. In general, measurement of layer line spacings yields unit-cell dimensions of accuracy no better than ± 1 per cent and can only be regarded as a preliminary means of determination of unit-cell dimensions; more sophisticated methods will be described later in this chapter.

The three unit-cell edges, a, b, and c, can be measured by mounting the crystal with its [100], [010], and [001] axes parallel in turn to the oscillation axis of the camera. If some other prominent zone axis can be located and the crystal mounted with this axis parallel to the oscillation axis of the camera, then the layer line spacing of the resultant oscillation photograph will provide a measurement of the spacing of lattice points in this direction by application of the general equation (p. 177) for $i = 90°$,

$$t \cos \delta = n\lambda$$

whence $$t = \frac{n\lambda}{\cos \cot^{-1} H_n/r}.$$

Thus for a monoclinic crystal measurement of the layer line spacing on a [110] oscillation photograph would serve to evaluate the lattice spacing parallel to the [110] zone axis and so to distinguish between a P lattice for which $t_{[110]} = \sqrt{(a^2 + b^2)}$ and a C lattice for which $t_{[110]} = \frac{1}{2}\sqrt{(a^2 + b^2)}$ if the magnitudes of a and b are already known. The determination of lattice type by this means is however fraught with danger because it is difficult to locate with certainty any but the most obvious zone axes even in a crystal with very well developed faces. In the early years of X-ray crystallography it was common practice to determine the β angle of monoclinic unit-cells by measuring the layer line spacing on $[101]$ oscillation photographs and then applying the relationship $t^2_{[101]} = a^2 + c^2 + 2ac \cos \beta$; but later work using more sophisticated methods has in many cases shown the selected oscillation axis to have been misidentified.

The indexing of specific reflexions on oscillation photographs is in general best achieved by application of the concepts of the reciprocal lattice and the reflecting sphere, but nevertheless some useful comments of a general nature can be made without recourse to these concepts. In an $[001]$ oscillation photograph the reflexions on the lth layer line will be reflexions from planes with that value of l, that is to say reflexions from $hk1$ planes will occur on the first layer line, reflexions from $hk2$ planes on the second layer line, and so on. That reflexions on the nth layer line of an oscillation photograph about a general axis $[UVW]$ will correspond to reflexions from planes (hkl) that satisfy the condition $hU + kV + lW = n$ can simply be seen by transforming the axes to a new set with $c' = Ua + Vb + Wc$, whence $l' = hU + kV + lW$.

The Bragg angles of reflexions on the zero layer line of an oscillation photograph can be determined directly in just the same way as for a powder photograph. For a cubic crystal the unit-cell dimension a can be determined from the layer line spacing of an oscillation photograph taken about an identifiable axis and the reflexions on the zero layer line can then be indexed by direct measurement of the Bragg angle θ and application of the expression

$$\lambda = \frac{2a \sin \theta}{\sqrt{(h^2 + k^2 + l^2)}}.$$

Consider for instance a $[100]$ oscillation photograph which will yield a value for a whatever the lattice type; the reflexions on the zero layer line will be $0kl$ reflexions for which $\lambda = (2a \sin \theta)/\sqrt{(k^2 + l^2)}$. If the oscillation axis is parallel to a zone axis along which the spacing of lattice points is such that the unit-cell dimension is not deducible unless the lattice type is known, care must be exercised. Consider for instance a $[111]$ oscillation photograph: if the lattice is P or F, $t_{[111]} = a\sqrt{3}$, but $t_{[111]} = \frac{1}{2}a\sqrt{3}$ for an I-lattice. In such a case as this, indexing of the zero layer reflexions on the $[111]$ oscillation photograph should not be attempted until a has been determined directly from a $[100]$ oscillation photograph.

Interpretation of oscillation photographs using the reciprocal lattice

The position of a reflexion on an oscillation photograph is specified by two coordinates (Fig 8.5), the height H of the reflexion above the zero layer line and M, which is defined as the distance of the reflexion measured along its layer line from the intersection of the plane containing the incident X-ray beam and the oscillation axis with the plane of the film. Our task now is to relate the measured coordinates of a reflexion to the coordinates of the reciprocal lattice point that gives rise to it on passing through the reflecting sphere. For this purpose it is convenient to define a system of cylindrical coordinates for the reciprocal lattice: ζ is defined as the perpendicular distance of the reciprocal lattice point from the plane which is perpendicular to the oscillation axis and contains the incident X-ray beam, and ξ is defined as the perpendicular distance of the reciprocal lattice point from the oscillation axis (Fig 8.6(a)). It is convenient, for reasons which will appear in due course, in

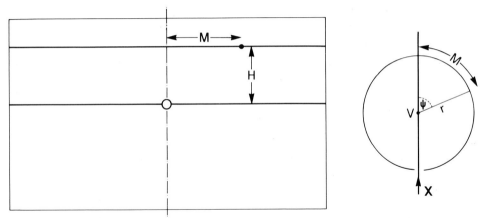

Fig 8.5   The coordinates H and M of a reflexion on an oscillation photograph. The selected reflexion is situated on the film at a perpendicular distance H from the zero layer line and at a distance M from the trace (shown as a dash-dot line) of the plane containing the incident beam and the oscillation axis. The right-hand diagram illustrates the relationship $M = r\psi$ of the film coordinate M to the angle ψ, which is the angle between two planes passing through the oscillation axis V, one containing the incident beam and the other the diffracted beam.

interpreting diffraction patterns to take the proportionality factor K in $d^* = K/d$ equal to the wavelength λ of the incident monochromatic radiation; this has the effect of making the radius of the reflecting sphere equal to one reciprocal lattice unit. One final preliminary statement remains to be made: that is the definition of the angle ψ as the angle made by the plane containing the incident X-ray beam and the oscillation axis with the plane containing the diffracted beam and the oscillation axis (Fig 8.6(b)); it is evident from Fig 8.5 that $\psi = M/r$.

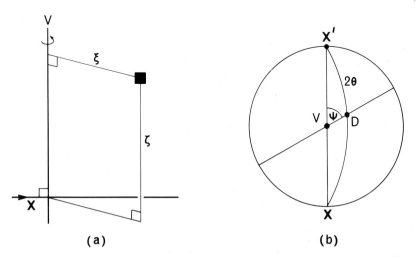

(a)   (b)

Fig 8.6   Cylindrical coordinates for the reciprocal lattice. (a) illustrates the definition of ζ as the perpendicular distance of the reciprocal lattice point (solid square) from the plane which is perpendicular to the oscillation axis V and contains the incident beam X, and ξ as the perpendicular distance of the reciprocal lattice point from the oscillation axis. (b) illustrates the definition of ψ as the angle between the plane containing the oscillation axis V and the incident beam XX' and the plane containing the oscillation axis V and the diffracted beam D (cf. Fig 8.5).

In the interpretation of oscillation photographs it is convenient to regard the three-dimensional reciprocal lattice as a stack of two-dimensional nets, each net being equidistant from those next above and below it. The separation of adjacent nets is λ/t where t is the spacing of lattice points along the oscillation axis; therefore for the nth layer line

$$\zeta = \frac{n\lambda}{t}.$$

It is apparent from Fig 8.7 that

$$\zeta = \cos \delta = \cos \cot^{-1} \frac{H}{r}$$

Therefore $t = \dfrac{n\lambda}{\cos \cot^{-1} \dfrac{H}{r}}.$

Thus t can be evaluated by measurement of H when r and λ are known.

Measurement of the reflexion coordinates M and H yields a direct determination of the reciprocal lattice coordinate ξ. It is evident from Fig 8.7 that the circular section of the reflecting sphere in a plane of constant ζ has radius $\sqrt{(1-\zeta^2)}$ and its centre is unit distance from the oscillation axis. For a reciprocal lattice point lying on the reflecting sphere (Fig 8.7(c))

$$\xi^2 = 2 - \zeta^2 - 2\sqrt{(1-\zeta^2)} . \cos \psi.$$

We have already seen that $\zeta = \cos \cot^{-1} H/r$, which can alternatively be expressed as

$$\zeta^2 = \frac{H^2}{H^2 + r^2}$$

and that $\psi = M/r$. Therefore

$$\xi^2 = 1 + \left(\frac{r^2}{H^2 + r^2}\right) - 2\left(\frac{r^2}{H^2 + r^2}\right)^{\frac{1}{2}} \cos \frac{M}{r},$$

which simplifies in the case of the zero layer line ($H = 0$) to

$$\xi^2 = 2 - 2 \cos \frac{M}{r}$$

$$= 4 \sin^2 \frac{M}{2r}$$

i.e. $\xi = 2 \sin \dfrac{M}{2r};$

but for the zero layer line $M = 2r\theta$ so that

$$\xi = 2 \sin \theta$$

and ξ becomes the distance $d*$ of the reciprocal lattice point from the origin (Fig 8.7(d)).

Since the calculation of ξ from measurements of M and H is rather awkward for upper layer lines, charts, known as *Bernal charts*, are commercially available with lines of constant ζ and curves of constant ξ drawn at 0.05 intervals (Fig 8.8) for the common camera radii. By superimposing a Bernal chart on an oscillation photograph

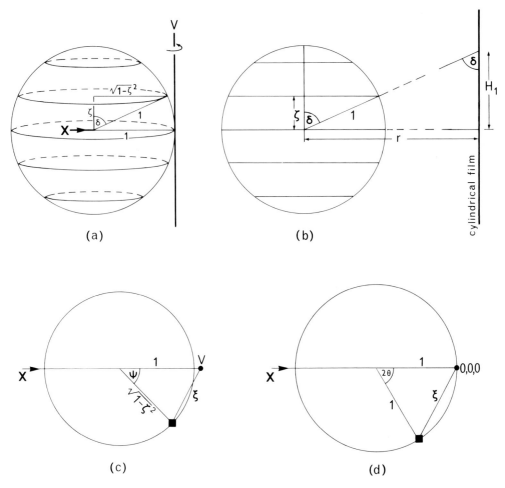

Fig 8.7 Interpretation of oscillation photographs in terms of the reciprocal lattice. (a) shows that a reciprocal lattice plane of constant ζ intersects the reflecting sphere in a circle of radius $\sqrt{(1-\zeta^2)}$ whose centre is unit distance from the oscillation axis V. (b) is a composite diagram in direct and reciprocal space to illustrate the relationship $\cot \delta = H_1/r$ for the first layer line on an oscillation photograph. (c) illustrates the relationship $\xi^2 = (1 - \zeta^2) + 1 - 2\sqrt{(1 - \zeta^2)} \cdot \cos \psi$ for a non-zero layer reciprocal lattice point lying on the reflecting sphere and (d) illustrates the simpler case of a zero layer reciprocal lattice point where $\xi = 2 \sin \theta$.

so that its line of zero ζ coincides with the zero layer line and the origin of the chart coincides with the centre of the punched hole through which the undeviated X-ray beam passed, the coordinates ξ and ζ of each reflexion on the film can be read directly. It is to facilitate the use of the Bernal chart that the proportionality constant K for the reciprocal lattice is commonly taken as equal to λ in practical work; the radius of the reflecting circle is then independent of the wavelength of the radiation used so that the curves of constant ξ and ζ shown on the chart for a camera of given radius are generally applicable. To avoid confusion about the magnitude of K it is usual to refer to reciprocal units when $K = 1$ as Å^{-1}, but when $K = \lambda$ no reciprocal unit is specified. As will be apparent from the argument that follows only rarely does ζ need to be known to higher accuracy than is obtainable by use of the Bernal chart.

As the crystal oscillates the reciprocal lattice net oscillates correspondingly about

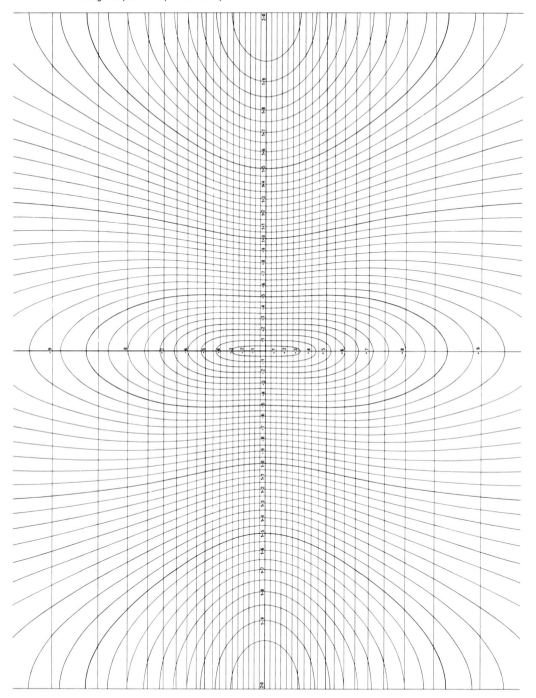

Fig 8.8 The Bernal chart. The parallel lines are lines of constant ζ in 0·05 intervals and the curves are curves of constant ξ in 0·05 intervals. The chart shown is approximately the size required for a camera of 30 mm radius. The chart is reproduced by courtesy of the Institute of Physics.

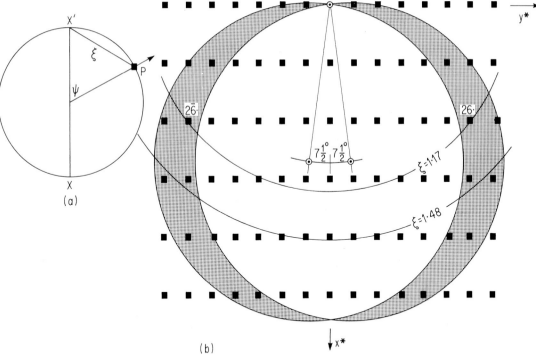

Fig 8.9   Interpretation of oscillation photographs in terms of the reciprocal lattice. (a) shows the general relationship between ξ and ψ for a reciprocal lattice point on the reflecting sphere. In (b) the x^*y^* reciprocal lattice net for a monoclinic crystal is taken to be stationary and the limits of oscillation of the reflecting sphere are shown as their circular sections of intersection with the net; the areas of the reciprocal lattice net intersected by the reflecting sphere are shown as shaded *lunes* for a 15° oscillation about x^*. The arc of $\xi = 1\cdot17$ passes through the reciprocal lattice points $2\bar{6}.$ and $26.,$ but the arc of $\xi = 1\cdot48$ does not pass through any reciprocal lattice points within the lunes; the resulting oscillation photograph may be expected to exhibit reflexions on the zero layer line with $\xi = 1\cdot17$, but no reflexions with $\xi = 1\cdot48$.

the vertical through its origin. Any reciprocal lattice point which passes through the reflecting sphere in the course of the oscillation gives rise to a diffracted beam in the direction of the radius of the reflecting sphere through the reciprocal lattice point[1] (Fig 8.9). Now and subsequently we shall treat the oscillation of the reciprocal lattice of the crystal relative to the reflecting sphere as a movement of the reflecting sphere through a stationary reciprocal lattice; to treat the relative motion in this way leads to some simplification of expression. Since the reflecting sphere oscillates about the vertical through the origin of the stationary reciprocal lattice, any reciprocal lattice point in the volume between the extreme positions of the reflecting sphere will give rise to a diffracted beam as the reflecting sphere passes through it. We illustrate this in Fig 8.9 for the zero layer of a monoclinic crystal in which the oscillation axis is parallel to [001]. The limiting positions of the reflecting sphere become for any one layer the limiting positions of the *reflecting circle* in which the sphere intersects the

---

[1] Since the origin of the reciprocal lattice lies on the oscillation axis, all diffracted beams pass through that point on the oscillation axis. But it is geometrically more convenient to suppose that all diffracted beams pass through the centre of the reflecting sphere when discussing the geometry of the oscillation photograph. It should be apparent to the reader that this geometrical simplification is generally acceptable even though it may occasionally make diagrams which combine reciprocal and direct space look inconsistent at first sight.

appropriate reciprocal lattice net; these are shown in the figure for a 15° (i.e. $\pm 7\frac{1}{2}°$) oscillation with the normal to (100) in the middle of the oscillation range. Only those reciprocal lattice points which lie within the *lunes* between the extreme circles give rise to reflexion; the reciprocal lattice points in the right-hand lune give rise to reflexions on the right-hand side of the film viewed in the direction of the incident X-ray beam while those in the left-hand lune give rise to reflexions on the left-hand side of the film.

Measurement of ζ and ξ for a reflexion on an oscillation photograph does not completely determine the coordinates of the corresponding reciprocal lattice point; it merely determines that the reciprocal lattice point lies in a plane of known ζ on an arc of constant ξ cut off at either end by the limiting positions of the reflecting sphere. However ζ and ξ are the only two reciprocal lattice coordinates determinable from an oscillation photograph and their measurement does enable many of the reflexions to be indexed unambiguously[2] provided the dimensions of the reciprocal lattice are known.

To index reflexions on the zero layer line ξ is estimated with the aid of a Bernal chart. On a plan of the reciprocal lattice net corresponding to the zero layer the limiting positions of the reflecting circle are drawn. An arc of radius ξ centred on the origin of the reciprocal lattice is drawn through the right-hand lune for every reflexion observed on the corresponding side of the photograph and likewise for the left-hand lune. Many of these arcs will pass clearly through one reciprocal lattice point and unambiguous indexing will have been achieved; but some arcs may pass close to more than one reciprocal lattice point so that the corresponding reflexions cannot be certainly indexed. If it is intended to determine the *diffraction symbol* of the substance the indices of reciprocal lattice points lying within the lunes but not giving rise to reflexion should be noted.

We take as our example a monoclinic crystal oscillating about its [001] axis with the incident X-ray beam normal to (100) in the middle of the 15° oscillation range. The zero layer line will contain only reflexions of the type $hk0$, the corresponding reciprocal lattice section will be the $a*b*$ plane, which has a rectangular mesh, and the limiting positions of the reflecting circle will be symmetrically disposed about the $x*$-axis. The procedure for indexing the zero layer line of the photograph is (i) draw out the reciprocal lattice net which passes through the origin and is normal to the oscillation axis [001], a convenient scale in most cases being 100 mm = 1 reciprocal unit, (ii) on this $a*b*$ net draw an arc of unit radius about the origin of the net, (iii) mark on the arc two points $7\frac{1}{2}°$ on either side of the intersection of the arc with the $x*$-axis, (iv) with these points as centres draw circles of unit radius to represent the limiting positions of the reflecting circle for the zero layer line, (v) draw an arc of radius ξ centred on the origin of the reciprocal lattice net for every observed reflexion to traverse the right-hand lune for reflexion on the right-hand side of the zero layer line (the film being viewed in the direction of the incident beam) and to traverse the left-hand lune for those on the left, (vi) passage of a ξ-arc through one reciprocal lattice point provides immediate indexing of the corresponding reflexion, (vii) note the indices of reciprocal lattice points lying within the lunes that do not give rise to reflexions.

In general the oscillation axis and the reciprocal axis corresponding to it will not be parallel so that in higher layers a reciprocal lattice point does not lie on the

---

[2] For unambiguous indexing of all the reflexions on a diffraction pattern recourse must be made to the moving film methods described later in this chapter.

oscillation axis. In the monoclinic example shown in Fig 8.10 the reciprocal axis z^* is inclined at $90° - \beta^*$ to $[001]$ in the plane normal to y^*. The nth reciprocal lattice layer intersects the oscillation axis at a distance $\zeta = n\lambda/c$ from the origin and contains reciprocal lattice points whose l-index is equal to n. The reciprocal lattice point $00n$ is distant $\zeta \cot \beta^* = nc^* \cos \beta^*$ (Fig 8.10(b)) from the oscillation axis along the x^*-axis, that is along the reciprocal lattice row $h0n$. To index an upper layer line of an $[001]$ oscillation photograph it is thus necessary to mark on the a^*b^* reciprocal lattice net the point of intersection O_n of the oscillation axis, which will lie for the nth layer at a distance $\zeta \cot \beta^*$ from the origin of the net, the reciprocal lattice point $00n$, in the $-x^*$ direction (i.e. between $00n$ and $\bar{h}0n$). The radius of the reflecting circle for this

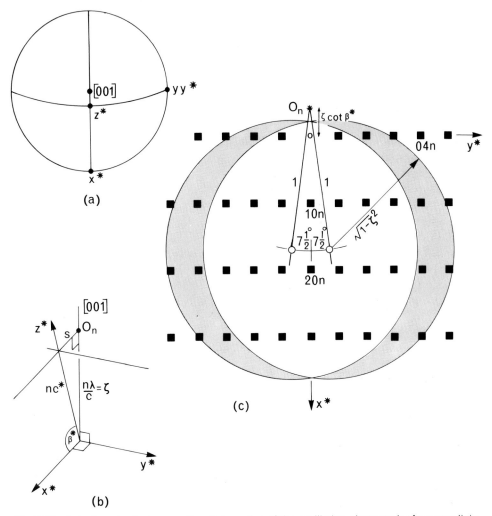

Fig 8.10 Indexing of reflexions on the nth layer line of the oscillation photograph of a monoclinic crystal whose oscillation axis is $[001]$. The stereogram (a) shows the relationship between direct and reciprocal lattice axes. The perspective drawing (b) shows the displacement $s = nc^* \cos \beta^* = \zeta \cot \beta^*$ of the point of intersection O_n of the oscillation axis with the a^*b^* reciprocal lattice net from the reciprocal lattice point $00n$; for n positive the displacement is in the direction $-x^*$.
(c) shows the a^*b^* reciprocal lattice net with circles of radius $\sqrt{(1-\zeta^2)}$ drawn with centres distant one reciprocal lattice point from O_n, which is distant $\zeta \cot \beta^*$ along $-x^*$ from the origin $00n$ of the net.

layer is $\sqrt{(1-\zeta^2)}$ and its centre lies at a distance of 1 reciprocal lattice unit from this intersection O_n in a direction parallel to the incident beam, in this case parallel to x^* for the middle of the oscillation range (Fig 8.10(c)). The procedure for indexing the nth layer of the photograph is (i) on the a^*b^* reciprocal lattice net draw an arc of unit radius about a point O_n along $-x^*$ at a distance $\zeta \cot \beta^*$ from the reciprocal lattice point $00n$, (ii) mark the two points $7\frac{1}{2}°$ on either side of the intersection of the arc with x^*, (iii) with these points as centres draw the limiting reflecting circles of radius $\sqrt{(1-\zeta^2)}$, (iv) then proceed as for the zero layer line (v)–(vii).

If the oscillation axis and the corresponding reciprocal axis are coincident, the reciprocal lattice points $00l$ will lie on the oscillation axis and therefore the reciprocal lattice net for the nth layer will be superimposed without displacement on that for the zero layer. Then O_n will coincide with the reciprocal lattice point $00n$ and the centre of the reflecting circle for the nth layer will be distant one reciprocal lattice unit from the $00n$ reciprocal lattice point. This simplified situation arises whenever the zone axis $[pqr]$ is normal to the face (pqr). This occurs for $[010]$ oscillation photographs of monoclinic crystals, $[100]$, $[010]$, and $[001]$ oscillation photographs of ortho-rhombic crystals, $[UV0]$ and $[001]$ oscillation photographs of tetragonal crystals, $[0001]$ oscillation photographs of hexagonal and trigonal crystals, and for any oscillation axis in a cubic crystal.

We have already considered in some detail a particular case in which the oscillation axis does not coincide with the corresponding reciprocal axis. We now consider a more general case by taking as our example the z-axis oscillation photograph of a triclinic crystal. We define the angle ε such that $z:z^* = \varepsilon$. The intersection of the oscillation axis with the nth reciprocal lattice net at a height ζ above the zero layer will thus lie at a distance $\zeta \tan \varepsilon$ from $00n$ (Fig 8.11(a)). The position of the intersection of the oscillation axis with the reciprocal lattice plane can be found conveniently by calculating the angle η between the plane containing the z and z^* axes and the plane containing the z and y^* axes (Figs 5.10(a) and 8.11(b)). The equations to a general spherical triangle (chapter 5) applied to Fig 8.11(b) yield by equation (1)

$$\cos \varepsilon = \cos 90° \cos \alpha^* + \sin 90° \sin \alpha^* \cos (\beta - 90°)$$

i.e. $$\cos \varepsilon = \sin \alpha^* \sin \beta$$

and by equation (2)

$$\cos \alpha^* = \cos \varepsilon \cos 90° + \sin \varepsilon \sin 90° \cos \eta$$

i.e. $$\cos \alpha^* = \sin \varepsilon \cos \eta$$

and by equation (7)

$$\frac{\sin \eta}{\sin \alpha^*} = \frac{\sin (\beta - 90°)}{\sin \varepsilon}$$

i.e. $$-\sin \alpha^* \cos \beta = \sin \varepsilon \sin \eta,$$

therefore

$$\tan \eta = -\tan \alpha^* \cos \beta$$

so that both η and ε can simply be calculated from known unit-cell dimensions. The a^*b^* net is then drawn out and the point O_n located on it by drawing a line inclined at η to y^* in the angle γ^* between $-x^*$ and $-y^*$ and measuring off on this line a distance $\zeta \tan \varepsilon$ from $00n$ (Fig 8.11(c)). When the oscillation axis is not a crystallo-

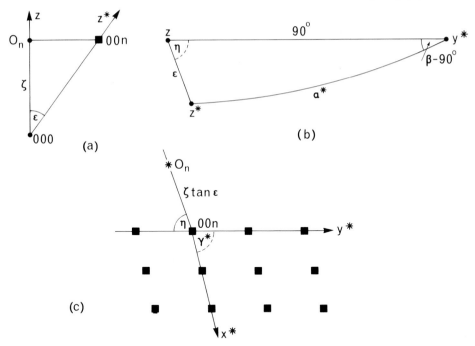

Fig 8.11   Indexing of a non-zero layer on the oscillation photograph of a triclinic crystal. (a) is a plane diagram showing the intersection of the nth reciprocal lattice net with z^* at $00n$ and with z (the oscillation axis) at O_n. The spherical triangle (b) illustrates the calculation of ε and η from α^* and β. The position of O_n on the a^*b^* reciprocal lattice net is shown in (c).

graphic reference axis, i.e. not [100], [010], or [001], then it is often simpler to choose a new set of reference axes with, say z', parallel to the oscillation axis, to index the crystal in terms of the new set of axes, x', y', z', and then to transform back to the original reference axes, x, y, z.

To conclude this section on oscillation photographs we draw the reader's attention to the observation that although it is always possible to determine one unit-cell dimension from an oscillation photograph, that parallel to the oscillation axis, reflexions can only be indexed when the reciprocal lattice geometry of the crystal is known. The determination of reciprocal lattice geometry is most easily achieved by the use of Weissenberg and precession photographs; oscillation photographs are practically useful only as a preliminary to more thorough investigation by moving film methods and for some specialized applications outside the scope of this textbook.

Rotation photography

The oscillation camera can simply be adapted to permit the crystal to rotate through $360°$ by removing the arm from the spindle to the cam and introducing a belt-drive from the camshaft to the spindle. The resultant rate of rotation of the crystal is of the order of one revolution per minute. The resultant rotation photograph is of course very similar to an oscillation photograph of the same substance taken about the same axis but displays many more reflexions. Measurement of layer line spacing enables the repeat distance between lattice points along the rotation axis to be determined in just the same way as from an oscillation photograph.

In the course of a complete rotation every lattice plane giving rise to a reflexion hkl will pass through an orientation which satisfies the Bragg Equation twice (Fig 8.12).

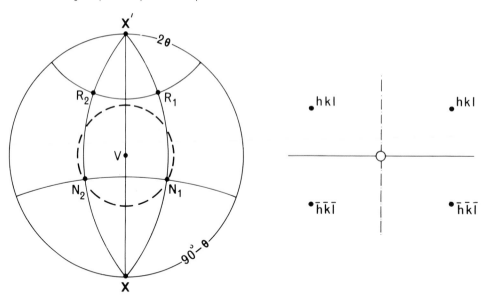

Fig 8.12 Rotation photography. On the stereogram the incident X-ray beam is denoted by X and X' and the rotation axis by V. If in the course of rotation of the crystal a plane (*hkl*) comes into the correct orientation for reflexion, when its normal is N₁, then during complete rotation it will also give rise to reflexion when its normal is N₂. The resultant reflexions, R₁ and R₂ on the stereogram, lie at the same height from the zero layer line (the primitive of the stereogram and the solid line on the drawing of a rotation photograph on the right) and are symmetrically disposed about the plane containing X, X' and V (shown as a dot-dash line in the right-hand diagram). The ($\bar{h}k\bar{l}$) plane will give rise to two identical reflexions which are related on the rotation photograph to the *hkl* reflexions by a line of symmetry coincident with the zero layer line.

The two reflexions produced will be symmetrically disposed about the vertical line representing the intersection of the film with the plane containing the incident X-ray beam and the rotation axis. Moreover the ($\bar{h}k\bar{l}$) lattice plane will give rise to two similarly disposed reflexions of equal intensity and these will be positioned on the film so that they are related to the *hkl* reflexions by a line of symmetry coincident with the zero layer line. Rotation photographs thus always have the symmetry of the two-dimensional point group *2mm*. In a case in which two or more lattice planes have the same *d*-spacing (and therefore identical Bragg angle) and give rise to reflexions on the same layer line these reflexions will be superimposed in the rotation photograph: for instance, on the [001] rotation photograph of a cubic crystal the 501, 051, $\bar{5}$01, and 0$\bar{5}$1 reflexions will all be superimposed at a point on the first layer line.

A rotation photograph is more useful than an oscillation photograph for the determination of the lattice type of a cubic or tetragonal crystal from an [001] photograph by measurement of θ for all reflexions on the zero layer line (the example discussed earlier) because it will display all reflexions of non-zero intensity for which θ is less than some angle little short of 90°. But exposure times for rotation photographs are very much longer than for oscillation photographs and, except for the sort of problem mentioned immediately above, rotation photographs are little used.

Interpretation of rotation photographs by use of the reciprocal lattice

As the crystal rotates about an axis, which for argument we take as [001], the reflecting sphere sweeps through the reciprocal lattice so that in the course of a complete 360°

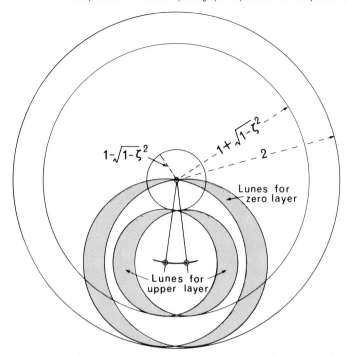

$1-\sqrt{1-\zeta^2}$

$1+\sqrt{1-\zeta^2}$

2

Lunes for zero layer

Lunes for upper layer

Fig 8.13 Rotation photography: the limiting torus. As the crystal describes a complete rotation the reflecting sphere sweeps through all reciprocal lattice points on the zero layer net within a circle of radius 2 reciprocal units and cuts an annular swathe with internal and external radii $1 \pm \sqrt{(1-\zeta^2)}$ through non-zero layer nets. The lunes for the zero layer and one non-zero layer of a 15° oscillation photograph are shown shaded.

rotation it passes through all reciprocal lattice points lying on the zero layer reciprocal lattice net within a circle of radius two reciprocal units about the origin of the net. For an upper layer the reflecting circle cuts an annular swathe through the relevant reciprocal lattice net with internal and external radii $1 \pm \sqrt{(1-\zeta^2)}$ (Fig 8.13). Thus for one setting of the crystal all the reciprocal lattice points within the torus produced by the rotation of the reflecting sphere about the rotation axis pass through the reflecting sphere. Moreover each such reciprocal lattice point passes through the reflecting sphere twice in the course of complete rotation of the crystal and so gives rise to a reflexion on the right and on the left-hand side of the incident X-ray beam. Reciprocal lattice points with the same value of ζ in the same reciprocal lattice net will of course give rise to superimposed reflexions.

Only very rarely, and then when the rotation axis is a symmetry axis of high order, is it profitable to attempt to index a rotation photograph. All the reciprocal space that can be sampled by a rotation photograph can be more informatively sampled by a series of 15° oscillation photographs taken at 15° intervals.

The *limiting torus* of a rotation photograph encloses all those reciprocal lattice points which may give rise to a reflexion in the course of a complete rotation and comprises all that volume of reciprocal space which can be studied with one rotation photograph or a series of oscillation photographs taken about the same axis. The only way in which reciprocal lattice points outside the limiting torus can be investigated is by changing the rotation or oscillation axis of the crystal. By use of a variety of rotation axes all the reciprocal lattice points within a sphere of radius equal to 2 reciprocal units centred on the origin of the reciprocal lattice can be brought into the reflecting position; this sphere is known as the *limiting sphere*. All reciprocal lattice points corresponding to $d^* < 2/\lambda$ lie within the limiting sphere and so the smaller the magnitude of λ the greater the number of reflexions that may be obtained; in other words, since $\sin\theta \leqslant 1$ it follows from the Bragg Equation that reflexion can only take place for lattice planes whose spacing $d \geqslant \frac{1}{2}\lambda$.

Moving film methods

We have already seen that the coordinates of a reciprocal lattice point are not completely determinable from an oscillation photograph because one cannot know at what stage of the oscillation the reflecting sphere passed through the reciprocal lattice point, or in other words at what inclination of the incident X-ray beam to its mean position the relevant lattice plane was so oriented as to satisfy the Bragg Equation. This problem can only be solved and unambiguous indexing of all reflexions achieved by selecting one reciprocal lattice net and by coupling a smooth movement of the film to the oscillatory motion of the crystal so that the reflexions produced by the reciprocal lattice points of the selected net are disposed over the whole area of the film. A separate film and different camera adjustments will be required to record each reciprocal lattice net. The two most effective means of achieving this objective are *Weissenberg* photography and *precession* photography. In the former a simple oscillatory translational motion is imparted to the film as the crystal oscillates about the camera axis; the mechanics of the camera are relatively simple but the resultant photograph is a distorted image of the reciprocal lattice net so that the indices of reflexions are not immediately obvious. In the latter the motion imparted to both film and crystal is more complicated and the camera more elaborate; but the resultant photograph is an undistorted image of the reciprocal lattice net and the reflexions are indexable by inspection. Some further comparative comments will be made at the end of the section on precession photography. For the explanation of both types of moving film photograph[3] we shall make use, of necessity, of the concepts of the reciprocal lattice and the reflecting sphere.

Weissenberg photography

The essential feature of the Weissenberg camera is that it selects one layer line of an oscillation photograph and distributes the reflexions of the layer line over the whole area of the film so that the coordinates of each reciprocal lattice point giving rise to a reflexion can be unambiguously determined. The selection is achieved by the use of screens and the layer line is spread over the area of the film by moving the film backwards and forwards parallel to the oscillation axis of the crystal.

We now proceed to describe the Weissenberg camera. The arcs carrying the crystal are attached to a spindle driven by a synchronous electric motor as in the oscillation camera except that the spindle is horizontal (Fig 8.14). A worm drive from the motor moves the cylindrical film cassette parallel to its axis, which is coincident with the oscillation axis of the crystal, so that for every $1°$ rotation of the crystal the cassette moves $0\cdot5$ mm; at either end of the traverse of the cassette the motor is reversed by a

[3] The account of the theory of moving film methods and the interpretation of Weissenberg and precession photographs provided here is necessarily confined to essentials. Of the many excellent textbooks and monographs to which the reader might turn for more detailed information we draw attention particularly to Henry, Lipson, and Wooster (1960), Buerger (1942), Buerger (1964), Nuffield (1966), and Jeffery (1971).

micro-switch. The cassette has an axial slit about 5 mm wide to accommodate the collimator. Immediately inside the cassette and mounted independently of it is a cylindrical *screen* opaque to X-rays; the screen, which likewise has a slit through which the collimator projects, is in two halves, one attached to each end of the camera. The gap between the two halves of the screen can be varied in width (usually 2–4 mm) and its mean position set to coincide with the cone of diffracted beams of a selected layer line. A back-stop to absorb the undeviated X-ray beam is mounted inside one half-screen on an adjustable slide so that it can be set opposite the collimator. The radius of the cylindrical film is usually 28·65 mm and that of the screens 23·02 mm. The oscillation range, usually set to cover just over 180°, can be varied by adjusting the positions of the micro-switches that reverse the movement of the film cassette. The camera can be rotated through a selected angle about a vertical axis which passes through the centre of the crystal; this facility is necessary for recording upper layer diffraction patterns. The film cassette consists of two parts, the cylindrical cassette itself which holds the film and a *carriage* which is moved backwards and forwards on rails parallel to the oscillation axis of the crystal by a reversible motor. The cassette can be locked to the carriage at different distances from a fixed point on the carriage; we shall see that this is a useful facility when taking upper layer photographs.

We consider first the formation of zero-layer photographs. The axis of the collimator is perpendicular to the oscillation axis; the crystal is situated at their intersection and oriented so that a prominent zone axis is parallel to the oscillation axis. The reflexions of the zero layer are generated by reciprocal lattice points situated on the net containing the origin and perpendicular to the oscillation axis. The resultant diffracted beams therefore lie in the plane containing the incident beam and perpendicular to the oscillation axis (Fig 8.15(a)); if the centre of the gap in the screens is set to coincide with this plane, diffracted beams generated by reciprocal lattice points not situated on this net will be absorbed by the screens.

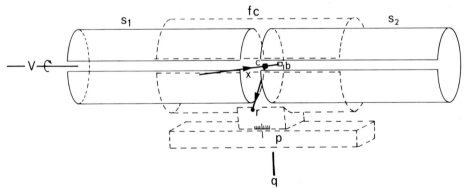

Fig 8.14 The Weissenberg camera. The perspective drawing shows the essential features of a Weissenberg camera. The crystal *c* is mounted on arcs on an oscillating spindle *V*. The split screens (*s₁*, *s₂*) are rigidly attached to the camera body in such a manner that the magnitude and position of their separation can be adjusted. The collimator, through which the incident X-ray beam, *X*, passes is rigidly mounted on the camera base. The back-stop, or beam trap, *b*, is attached to one screen. The film cassette (*fc*) is supported by the carriage *p* and can be set at a range of positions relative to a fiducial mark on the carriage. The carriage describes a linear motion parallel to the oscillation axis, its motion being geared to the oscillatory motion of the crystal. The bearings of the carriage are rigidly attached to the camera body. For zero layer photographs the oscillation axis is perpendicular to the incident X-ray beam; for non-zero layer photographs the camera body is inclined to the incident beam by moving the camera body through the appropriate angle about the vertical axis *q*, which passes through the intersection of the spindle axis *V* and the incident beam *X*. A diffracted beam passing through the gap between the screens is shown as *cr*.

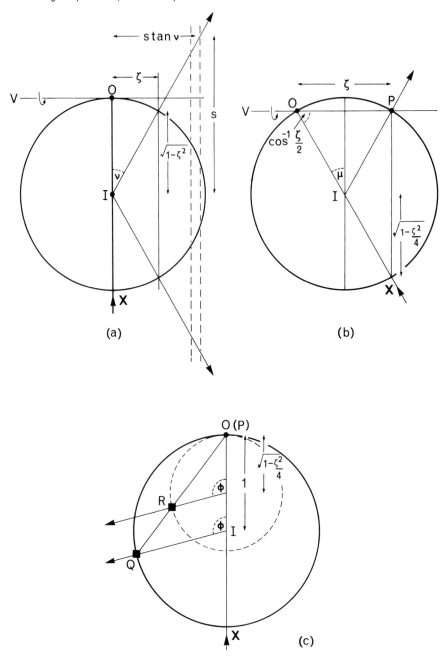

Fig 8.15 Weissenberg photography. (a) shows how a reciprocal lattice net normal to the oscillation axis V at a height ζ above the zero layer gives rise to a cone of diffracted rays of semi-angle $90° - \nu$ where $\nu = \sin^{-1} \zeta$ when the incident X-ray beam XO is perpendicular to the oscillation axis. For the zero layer, $\zeta = 0$ and the cone becomes a plane; for non-zero layers, reflexions for a layer of known ζ will pass through the gap in screens of radius s when the centre of the screen gap is separated by a distance $s \tan \nu$ from the plane normal to the oscillation axis and containing the incident beam (normal beam Weissenberg). (b) illustrates the essential geometry of the equi-inclination Weissenberg for non-zero layers: the incident beam XO is inclined to the oscillation axis V at the angle $\cos^{-1} \frac{1}{2} \zeta$. (c) illustrates the change of scale in non-zero layer equi-inclination Weissenbergs.

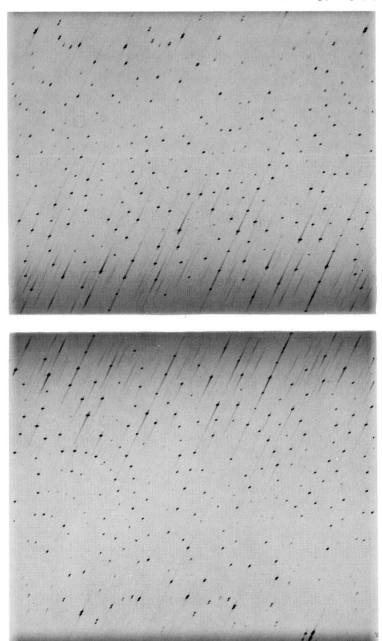

Fig 8.16   Zero layer *y*-axis Weissenberg photograph of the monoclinic mineral *latiumite* taken with CuKα radiation.

It is immediately apparent from inspection of a zero-layer Weissenberg photograph, such as that shown in Fig 8.16, that some reflexions lie on prominent straight lines running at a slant across the film. Another prominent feature of the photograph is the disposition of the reflexions on sets of curves. We now proceed to explain the geometry of a zero-layer photograph in general terms and then to discuss a specific example, a zero-layer photograph of a monoclinic crystal with [010] as oscillation axis.

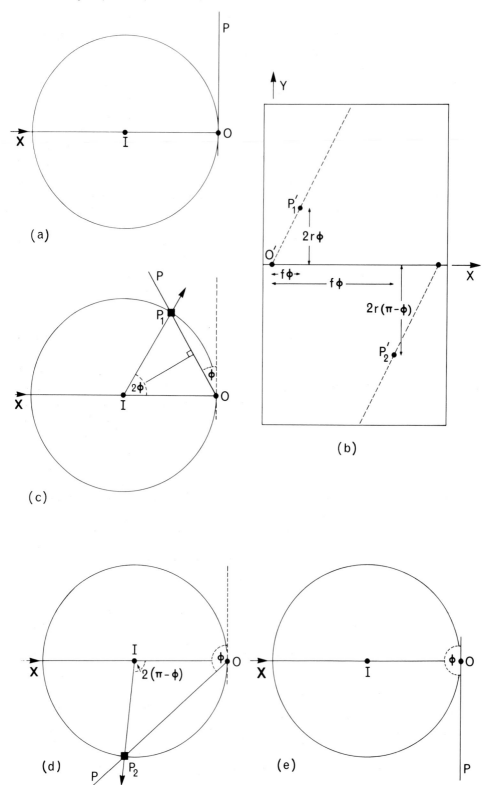

(a)

(b)

(c)

(d)

(e)

Consider a line OP in reciprocal space lying in the plane normal to the oscillation axis and passing through the origin O of reciprocal space. Suppose that initially the line OP is tangential to the reflecting circle (Fig 8.17(a)) and that for this orientation of the crystal the extreme left-hand side of the film (Fig 8.17(b)) is opposite the gap in the screen. For this position of the line OP any reciprocal lattice points that lie on OP cannot also lie on the reflecting circle and so cannot give rise to diffracted beams. The undeviated X-ray beam travelling along XO produced would pass through the gap in the screen to reach the film at O' were it not trapped by the back-stop attached to the screens. The origin of coordinates on the film is taken as O' and the reference axes X, Y are as shown in Fig 8.17(b). The X-axis is the median line of the film and corresponds to the locus of the point of intersection of the forward direction of the incident X-ray beam with the film as the crystal oscillates. The positive direction of the Y-axis is chosen so that the upper half of the film records diffracted beams produced by the passage of the reciprocal lattice net through the upper semicircle of the reflecting circle. As the line OP rotates anticlockwise the film moves from right to left so that the screen gap moves across the film from left to right. For a rotation of OP through $\phi°$ from its initial position the film moves through a distance $f\phi$, where f is for most instruments 0·5 mm per degree; at this stage (Fig 8.17(c)) OP intersects the reflecting circle in P_1 and if P_1 happens to be a reciprocal lattice point the reflected beam will make an angle 2ϕ with the forward direction of the incident beam. The reflexion P'_1 produced by the reciprocal lattice point P_1 will thus lie at a distance $2r\phi$ (for ϕ in radians, i.e. $\pi r\phi/90$ for ϕ in degrees) from the median line of the film, where r is the camera radius, so that the X, Y coordinates of this reflexion will be $f\phi$, $2r\phi$. Since both coordinates are proportional to ϕ, reflexions from reciprocal lattice points lying on OP will themselves lie on a straight line inclined at the angle $\eta = \tan^{-1}(\pi r/90 f)$ to the median line of the film. When the camera radius $r = 57\cdot3/2 \simeq 90/\pi$ mm and $f = 0\cdot5$ mm per degree, $\eta = \tan^{-1} 2 = 63\cdot43°$.

As the angle ϕ increases X and Y will increase until when ϕ is a little short of $90°$ Y will correspond to the edge of the film; reflexions with ϕ close to $90°$ are not recorded on the film but pass through the axial slit in the film cassette. For ϕ greater than $90°$ reflexions from reciprocal lattice points on the line OP lie on the lower half of the film (Fig 8.17(d)), the diffracted beam IP_2 making the angle $2(\pi - \phi)$ with the forward direction of the incident beam. The film coordinates of the reflexion P'_2 produced by the reciprocal lattice point P_2 will be $X = f\phi$, $Y = 2r(\phi - \pi)$. Reflexions due to lattice points on the line OP such that $\frac{1}{2}\pi < \phi < \pi$ will thus lie on a line of the same slope η as the line on the upper half of the film; this line will intersect the median line of the film at a point distant $180f$ mm from O' corresponding to $\phi = 180°$ (Fig 8.17(e)). When $\phi = 180°$ PO produced bears the same relationship to the reflecting circle as OP for $\phi = 0°$ (Fig 8.17(a)); further rotation allows reciprocal lattice points on PO produced to come into the reflecting position.

We turn now to consideration of a line QR parallel to OP in the plane normal to the oscillation axis and at a perpendicular distance q from OP. We choose QR such

Fig 8.17 The generation of lines of reflexions on a zero-layer Weissenberg photograph. Diagrams (a), (c), (d), and (e) show successive stages in the rotation of a line OP in reciprocal space normal to the oscillation axis and passing through the origin O. In (a) OP is tangential to the reflecting circle; at stage (c) it has rotated through the acute angle ϕ; at stage (d) ϕ has become obtuse; and in (e) ϕ is shown equal to $180°$. Diagrams (c) and (d) illustrate values of ϕ for which reciprocal lattice points, P_1 and P_2, lie on the reflecting circle and give rise to the reflexions P'_1 and P'_2 shown on the drawing (b) of the resulting Weissenberg photograph; the coordinates of P'_1 and P'_2 with respect to the film axes X and Y are indicated. (b) is drawn half-size for a camera with $r = 28\cdot65$ mm and $f = 0\cdot5$ mm per degree.

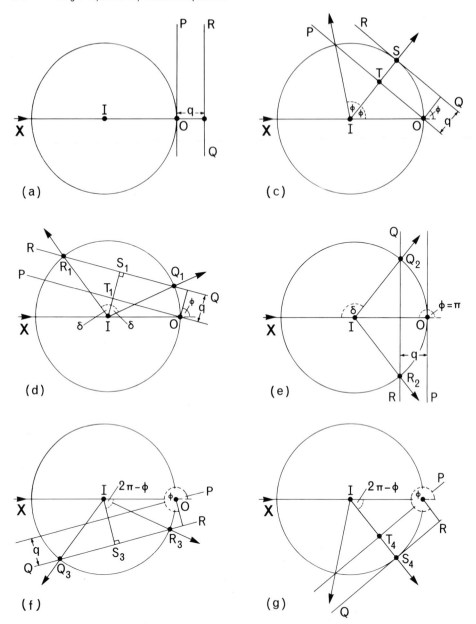

Fig 8.18 The generation of reflexions on a zero-layer Weissenberg photograph by a reciprocal lattice line which does not pass through the origin. The diagrams (a), (c), (d), (e), (f), (g) show successive stages of the rotation of the non-central line QR in the zero-layer reciprocal lattice net from $\phi = 0$ in (a) to $\phi = \pi$ in (e) and $\phi > \pi$ in (f) and (g). Diagram (b) is a drawing of the resultant Weissenberg photograph (half size for a camera with $r = 28 \cdot 65$ mm, $f = 0 \cdot 5$ mm per degree) with the hypothetical reflexions produced by a continuum of reciprocal lattice points on the line QR shown as broken curves.

that when OP is in its initial position tangential to the reflecting circle, QR does not intersect the reflecting circle (Fig 8.18(a)). As the crystal rotates anticlockwise from its initial position the line QR first touches the reflecting circle at a point S (Fig 8.18(c)) such that $\cos \phi = IT = IS - TS = 1 - q$, IO and IS being radii of the reflecting circle.

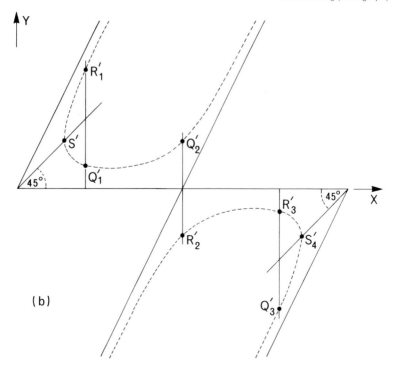

(b)

If there happens to be a reciprocal lattice point at S it will thus give rise to a reflexion S′ with film coordinates (Fig 8.18(b)) $X = f\phi$, $Y = r\phi$, where $\phi = \cos^{-1}(1-q)$.

Further increase in ϕ will cause QR to intersect the reflecting circle in two points Q_1 and R_1 (Fig 8.18(d)). The diffracted beams IQ_1 and IR_1 produced by reciprocal lattice points at Q_1 and R_1 respectively make equal angles δ with IS_1 and are thus inclined at angles $\phi \pm \delta$ to the forward direction of the incident X-ray beam. It is evident from the figure that $\cos\delta = IS_1 = IT_1 + T_1S_1 = \cos\phi + q$. The film coordinates of the reflexions Q_1' and R_1' produced by reciprocal lattice points at Q_1 and R_1 will thus be $X = f\phi$, $Y = r\phi - r\cos^{-1}(\cos\phi + q)$ and $X = f\phi$, $Y = r + r\cos^{-1}(\cos\phi + q)$ respectively. Reflexions corresponding to intersections of the line QR with the reflecting circle thus lie on the broken curve shown in Fig 8.18(b). Points on this curve, such as Q_1' and R_1', which correspond to simultaneous reflexion have their X-coordinates equal and the mean of their Y-coordinates lies on the line $X = f\phi$, $Y = r\phi$ which passes through O′ and S_1'; for $r = 57\cdot3/2$ mm and $f = 0\cdot5$ mm per degree, this line is inclined at $\tan^{-1}(\pi r/180f) = 45°$ to the median line of the film.

As ϕ increases from $\cos^{-1}(1-q)$ the Y-coordinate of the upper part of the curve increases until the top edge of the film is reached at Y just less than $r\pi$. Further increase in ϕ will cause the intersections of QR with the reflecting circle to be on opposite sides of the incident X-ray beam: a reciprocal lattice point which has crossed the incident beam will give rise to a reflexion whose coordinates are $X = f\phi$, $Y = r(\phi - 2\pi) + r\cos^{-1}(\cos\phi + q)$ so that the reflexion lies on the lower half of the film. When $\phi = \pi$ (Fig 8.18(e)) the two reflexions Q_2' and R_2' are equidistant from the median line of the film. When ϕ increases beyond π a stage will be reached at which both intersections of QR lie in the lower semicircle of the reflecting circle and then both the reflexions produced will be on the lower half of the film (Fig 8.18(f)).

Further rotation eventually brings QR into an attitude where it is tangential to the reflecting circle at S_4 (Fig 8.18(g)); the diffracted beam from a reciprocal lattice point at S_4 makes an angle $2\pi - \phi$ with the forward direction of the incident X-ray beam so that the film coordinates of the reflexion S_4' will be $X = f\phi$, $Y = r(\phi - 2\pi) = -r\cos^{-1}(1-q)$. In practice the range of movement of the film cassette and the standard size of X-ray film limit the oscillation range of the Weissenberg camera to about $200°$ and so it may not be possible experimentally to record the complete curves of constant q on one photograph.

Curves similar to those shown as broken lines on Fig 8.18(b) can be constructed for any value of $q < 2$ reciprocal units. In particular if two reference axes, x and y, in reciprocal space are chosen so that x is normal to OP and y is parallel to OP, we can construct a set of curves for reflexions from points on lines parallel to y with $x = 0$, ± 0.1, $\pm 0.2, \ldots \pm 1.9$ reciprocal units. By drawing a second set of identical curves displaced from the first set by $X = -90f$ mm along the median line of the film we have a set of curves of constant x with $y = 0$, ± 0.1, $\pm 0.2, \ldots \pm 1.9$ reciprocal units. A chart showing both sets of curves, known as a *Weissenberg chart*, is illustrated in Fig 8.19. The prominent straight lines on the chart are separated by $90f$ mm and correspond to reflexions from points along the reference axes x and y. The chart is usually 135 mm long so that it covers an oscillation range of as much as $270°$. By superimposing the base line of the chart (printed on transparent film) on the median line of the film, which is the line $Y = 0$, the rectangular coordinates of all reciprocal lattice points giving rise to reflexion can be read directly from the chart and the reciprocal lattice net can be plotted out on squared paper. It is immaterial where the chart is positioned relative to the film provided its base line is superimposed on the median line of the film; the coordinates of the reciprocal lattice points will of course depend on the positioning of the chart but when the coordinates are plotted on graph paper the effect of moving the base line of the chart along the median line of the film will be seen to correspond merely to a rotation of the zero layer reciprocal lattice net. It is always convenient to position the chart so that one prominent linear alignment of reflexions coincides with a diagonal line on the chart; then all reflexions will lie on one set of curves (the corresponding reciprocal lattice points will then lie on lines parallel to one of the reference axes) and, if the reciprocal lattice net is rectangular, at the intersections of the two sets. In the previous sentence the term 'set of curves' implies not only those curves actually drawn on the chart but also interpolated curves.

Before going on to consider upper layer Weissenberg photographs we illustrate the formation of a zero-layer photograph by showing how the $h0\bar{1}$, $h00$, and $h01$ reciprocal lattice points give rise to reflexions in the case of a monoclinic crystal oscillated about [010]. Figure 8.20(a) shows the orientation of the a^*c^* reciprocal lattice net when the gap in the screens exposes the point O' on the film (Fig 8.20(d)). In this orientation x^* is tangential to the reflecting circle. As the crystal rotates anticlockwise the reciprocal lattice point $20\bar{1}$ passes through the reflecting circle to give rise to a reflexion on the upper half of the photograph. Figures 8.20(b) and (c) show the successive orientations of the reciprocal lattice net as the 200 reciprocal lattice point passes into and out of the reflecting circle to give rise to reflexions respectively on the upper and lower halves of the photograph. Figures 8.20(e) and (f) show the successive orientations of the reciprocal lattice net as 301 passes into and out of the reflecting circle to give rise to reflexions respectively on the upper and lower halves of the film. The reflexions $h01$ lie on curves corresponding to $q = \lambda/c =$

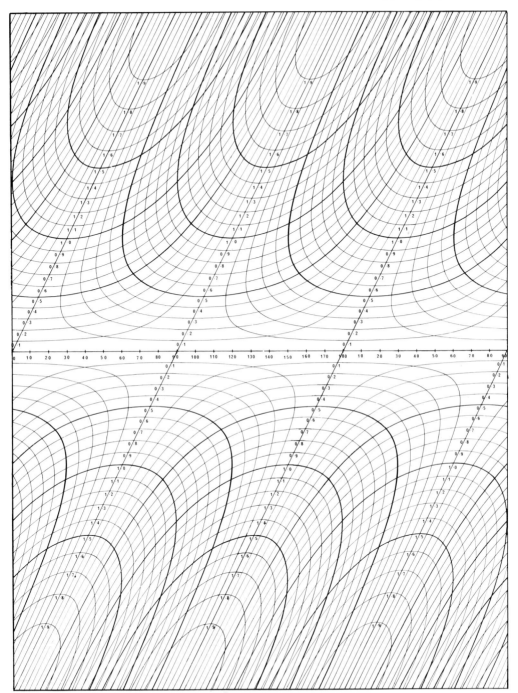

Fig 8.19   The Weissenberg chart. The chart shown is for a camera of diameter 57·3 mm, $f = 0·5$ mm per degree; it is reproduced by courtesy of the Institute of Physics.

$c^* \sin \beta^*$ and the $h0\bar{1}$ reflexions lie on curves corresponding to $q = -c^* \sin \beta^*$. The straight line through O′ and the reflexions $h00$ are the expression on the film of the axis x^* of the monoclinic reciprocal lattice. A parallel straight line through $00\bar{1}$, O″, and 001 on the film corresponds to the z^*-axis of the reciprocal lattice, the distance

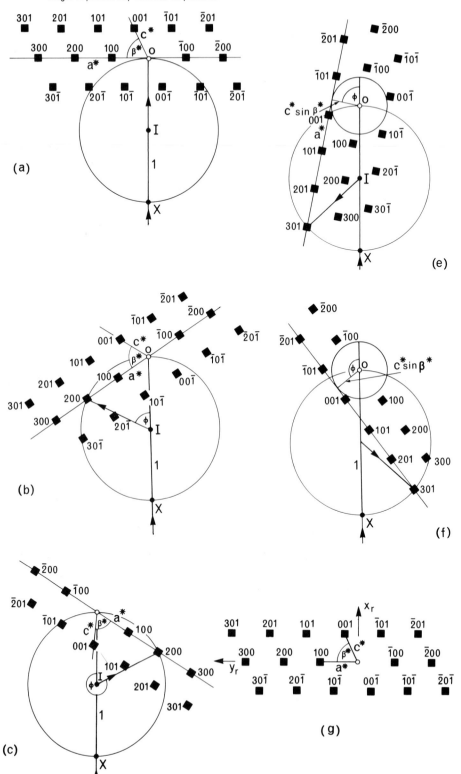

O'O'' on the film being equal to $f\beta^*$. On Fig 8.20(d) we have drawn curves of constant l; we could just as well have drawn curves of constant h, which would be similar in form, but displaced by the distance O'O'' from the set actually shown. To plot a reciprocal lattice net from a zero-layer Weissenberg photograph one of the straight lines on the Weissenberg chart should be superimposed on one of the prominent lines of reflexions, $h00$ or $00l$, and then the coordinates of each reflexion should be read off in rectangular reciprocal coordinates x_r, y_r (Fig 8.20(g)).

We now turn to upper-layer Weissenberg photographs, restricting our discussion to the most commonly used type, *equi-inclination* photographs. In an oscillation photograph—and a Weissenberg is only a specialized sort of oscillation photograph—a reciprocal lattice net normal to the oscillation axis at a height ζ above the zero layer gives rise to a cone of diffracted rays (Fig 8.15(a)) of semiangle $90° - v$, where $\sin v = \zeta$. Therefore if the screen is moved through a distance $s \tan v$, where s is the radius of the screen, the reflexions from this layer will pass through the gap in the screen to be recorded on the film. But the geometry of such a photograph (a *normal beam Weissenberg*) is rather inconvenient and it is better to rotate the incident beam relative to the crystal so that the incident beam lies on the surface of the cone of diffracted beams generated by the reciprocal lattice net (Fig 8.15(b)). When this is done the origin O of the reciprocal lattice lies a perpendicular distance $\frac{1}{2}\zeta$ below the plane passing through the centre of the reflecting sphere and normal to the oscillation axis so that incident and diffracted beams make equal angles $90° - \mu$, where $\mu = \sin^{-1}\frac{1}{2}\zeta$, with the oscillation axis. The selected reciprocal lattice net intersects the reflecting sphere in a circle of radius $\sqrt{(1 - \frac{1}{4}\zeta^2)}$ passing through the oscillation axis at the point P (Fig 8.15(b)). With such an *equi-inclination* arrangement it is possible to record all

Fig 8.20 Generation of the zero-layer Weissenberg photograph of a monoclinic crystal oscillated about [010]. (a) illustrates the orientation in which x^* is tangential to the reflecting circle. (b) and (c) illustrate the passage of the 200 reciprocal lattice point into and out of the reflecting circle as the crystal rotates anticlockwise. (e) and (f) illustrate the passage of the 301 reciprocal lattice point into and out of the reflecting circle. (d) shows the disposition of the reflexions produced on the resulting Weissenberg photograph by the reciprocal lattice points shown in (a)–(f); the rectangular film axes are labelled X and Y as in Fig 8.17 and the reference axes of the Weissenberg chart are labelled x_r and y_r. (g) shows the reciprocal lattice net plotted from (d) by reading off values of x_r and y_r for each observed reflexion.

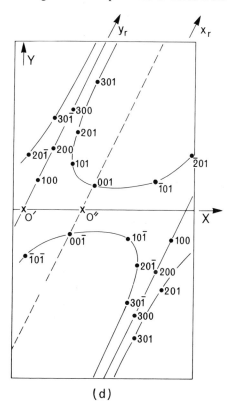

(d)

the reciprocal lattice points within a circle of radius $2\sqrt{(1-\frac{1}{4}\zeta^2)}$ centred on the oscillation axis and moreover the geometry of X-ray reflexion is identical with that for the zero-layer except that the radius of the reflecting circle is reduced from one to $\sqrt{(1-\frac{1}{4}\zeta^2)}$ reciprocal unit.

It is apparent from Fig 8.15(c) that a reciprocal lattice point Q lying in the plane through the origin perpendicular to the oscillation axis gives rise to a diffracted beam (which will be recorded as a reflexion on the zero-layer photograph) which makes an angle ϕ with the forward direction of the incident X-ray beam where the reciprocal lattice point is distant $OQ = 2\sin\frac{1}{2}\phi$ from the origin O. A reciprocal lattice point R in an upper layer giving rise to a diffracted beam inclined at the angle ϕ to the forward direction of the incident beam will lie at a distance $PR = 2\sqrt{(1-\frac{1}{4}\zeta^2)}.\sin\frac{1}{2}\phi$ from the intersection P of the oscillation axis with its reciprocal lattice net. Thus the Weissenberg chart constructed for the zero layer can be used to plot the reciprocal lattice net for an upper layer provided that all coordinates read from the chart are reduced by the factor $\sqrt{(1-\frac{1}{4}\zeta^2)}$. The reciprocal lattice net for an upper layer may not necessarily have a reciprocal lattice line passing through its origin and so the slanting lines which are so obvious on a zero layer photograph may be missing or, as in the case (Fig 8.10(a)) of a monoclinic crystal oscillating about [001], only one reciprocal lattice line (that parallel to x^* and containing the reciprocal lattice points $h0n$ in the nth layer) passes through the oscillation axis so that the nth layer equi-inclination Weissenberg photograph will show only $h0n$ reflexions lying on a straight diagonal line.

The essential requirement for an equi-inclination photograph, that the incident X-ray beam should lie on the cone of diffracted beams for the layer concerned, is achieved experimentally by turning the camera through the angle μ about the vertical axis through the crystal, where $\sin\mu = \frac{1}{2}\zeta$, and keeping the collimator, which is rigidly attached to the base of the camera, stationary. If the diffracted beams of the selected layer and no others are to pass through the gap in the screens, the centre of the gap must lie on the line PI of Fig 8.21. Therefore each screen must be moved in the same direction by an amount $s\tan\mu$ where s is the screen radius. If the cassette is locked on to the carriage in the same position as for the zero-layer photograph reflexions produced by the same orientation of the crystal will be displaced $r\tan\mu$ parallel to the median line of the photograph relative to those on the zero layer photographs. Such a translation of the reflexions on the film corresponds to a rotation of the reciprocal lattice net plotted from coordinates measured with the Weissenberg chart, the angle of rotation being $\{(r/f)\tan\mu\}°$. This rotation of successive reciprocal lattice nets is inconvenient and can simply be eliminated by moving the cassette relative to its carriage through $r\tan\mu$ before exposure starts and locking it in this position.

In order to illustrate one use of Weissenberg photographs we now discuss the determination of the unit-cell dimensions and diffraction symbol of a crystal which has been shown by optical examination to be biaxial (volume 2) so that it may be assumed to be orthorhombic, monoclinic, or triclinic. Suppose that a single crystal fragment of the substance has been mounted on a glass fibre and that a zone axis normal to a mirror plane has been located and set parallel to the oscillation axis. A 15° oscillation photograph will quickly yield a good approximate value for the spacing of lattice points normal to the mirror plane. The finding of one mirror plane of course immediately rules out the possibility that the substance is triclinic. If it is monoclinic, the oscillation axis must by convention be [010]; if it is orthorhombic,

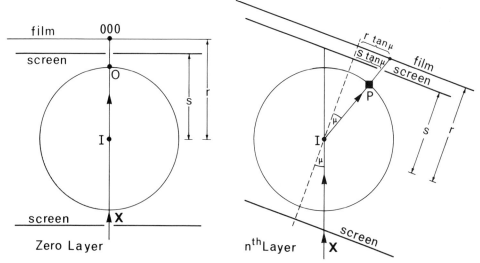

Fig 8.21 Upper layer equi-inclination Weissenberg photographs. The left-hand diagram shows the experimental arrangement for a zero-layer photograph contrasting with the arrangement for an upper-layer photograph shown on the right. For an *n*th layer photograph the camera axis is inclined relative to the fixed collimator through the angle $\iota = \sin^{-1}\left(\frac{1}{2}\zeta_n\right)$, the screen is translated through $s \tan \mu$ and the film carriage is translated through $r \tan \mu$, where s and r are screen and camera radii respectively.

the oscillation axis may be [100], [010], or [001]. We tentatively index the oscillation axis as [010], transfer the crystal on its arcs to the Weissenberg camera, and take zero, first, and second layer Weissenberg photographs with the presumed [010] as oscillation axis. These photographs will enable us to plot out the *h*0*l*, *h*1*l*, and *h*2*l* reciprocal lattice nets. If these three nets each exhibit symmetry 2*mm* with their mutually perpendicular lines of symmetry normal to reciprocal lattice point rows and if the intensities of the pairs of the reflexions related positionally by the lines of symmetry are approximately equal, then the substance is orthorhombic. The geometry of the Weissenberg camera is such that lines of symmetry in reciprocal lattice nets do not obviously appear as such on the photograph; a line of symmetry in the net generates one of the parallel slanting straight lines on the film and symmetry related reflexions of equal intensity will lie on either side of the line but will not have symmetry related film coordinates. If our crystal turned out to be orthorhombic x^* and z^* would be chosen as the directions normal to the two lines of symmetry on the reciprocal net plots. The reciprocal lattice dimensions a^* and c^* could then be evaluated and the dimensions a, b, c, of the unit-cell could be calculated from them and from the direct determination of b from the oscillation photograph. The next stage in this investigation would be to search the three reciprocal lattice nets for systematically absent reflexions. Adequate sampling of general *hkl* reflexions should have been achieved to determine whether the lattice type is A, B, C, P, I, or F. The reciprocal lattice net *h*0*l* would then be searched first for independent systematic absences which would indicate the presence of (010) glide planes translating $\frac{1}{2}a$ or $\frac{1}{2}c$ or $\frac{1}{2}(a+c)$ and secondly for independent systematic absences in the *h*00 or 00*l* reflexions which would indicate the presence of screw diads parallel to the x or z axes. Systematic absences in the 0*kl* and *hk*0 reflexions, which would imply the presence of b or c or n glides parallel to (100) or a or b or n glides parallel to (001), can be investigated by looking at the 00*l*, 01*l*, 02*l*, and *h*00, *h*10, *h*20 rows in the three nets, although it must be borne in mind that the amount of information available may not be adequate for a conclusive statement. What we particularly lack is information about systematic absences in the 0k0 reflexions because the 010 and 020 reflexions are certain to lie in the shadow cast by the back-stop. It is therefore necessary to remount the crystal so that it can oscillate about either its x or z axis; then a zero-layer Weissenberg photograph will supply the missing information. Inspection of the resulting 0kl or hk0 net will reinforce the conclusion reached earlier about systematic absences due to a (100) or (001) glide plane and the 0k0 row will indicate clearly whether there is a screw diad parallel to the y-axis. This completes the determination of the diffraction symbol.

If however the h0l, h1l, and h2l Weissenberg photographs do not exhibit any lines of symmetry, the crystal is monoclinic and it is most unlikely that the reciprocal lattice nets will be rectangular. Since the reciprocal lattice axes x^* and z^* in the monoclinic system are not determined by symmetry considerations, it is necessary to make an arbitrary choice of which prominent reciprocal lattice point rows should be taken as x^* and z^* in such a way as to satisfy the convention that β^* should be acute. Measurement of the reciprocal lattice nets will then yield values of a^*, c^*, β^* so that a, c, β can be calculated and we already have the magnitude of b from the oscillation photograph. One would then search the three photographs for systematic absences in general hkl reflexions to determine the lattice type, and for systematic absences in the h0l reflexions for evidence of an a or c or n glide plane parallel to (010). Again it would be necessary to remount the crystal and to take a zero-layer photograph about either [100] or [001] to investigate systematic absences in the 0k0 reflexions, which would indicate the presence of a screw diad parallel to [010]. The diffraction symbol will then be completely determined.

We conclude by noting that in both the orthorhombic and the monoclinic case we may have chosen the crystallographic reference axes unconventionally. If that is so, it may be necessary in the orthorhombic case to transform the axes so that they conform to the conventions laid down in the *International Tables for X-ray Crystallography*, or in the monoclinic case to choose alternative x and z axes. The reader will have noticed that in both cases it was necessary to remount the crystal; had a precession camera been available however, the crystal could have simply been transferred on its arcs from one camera to the other so that the supplementary data, the hk0 or 0kl layers, could be obtained without any necessity for remounting the crystal.

Precession photography

In essence the precession method differs from those previously described in that the movement of the crystal is not an axial oscillation but a precession. In consequence the camera motion is essentially three-dimensional and not easily described in terms of two-dimensional diagrams. The diffraction pattern is recorded on a plane film which provides an undistorted photograph of a selected reciprocal lattice plane; the provision of an undistorted representation of a reciprocal lattice plane coupled with uniform spot shape is the essential purpose of the method. We begin by discussing the geometry of zero-layer photographs in terms of the reciprocal lattice and reflecting sphere, pass on to consider upper layer photographs, and conclude with an outline description of the precession camera.

A crystal mounted on arcs, whose spindle axis is normal to the incident X-ray beam, is set initially so that a prominent zone axis is coincident with the X-ray beam and so perpendicular to the spindle axis. This contrasts with the requirement for oscillation and Weissenberg photography that a zone axis should be parallel to the spindle axis of the arcs and so perpendicular to the incident X-ray beam. In the initial setting of a crystal on the precession camera (Fig 8.22(a)) the reciprocal lattice layer through the origin and normal to the selected zone axis is thus tangential to the reflecting sphere; in consequence no zero-layer reflexions can occur in this orientation of the crystal. In order to record the zero-layer reflexions the zone axis is moved through the angle $\bar{\mu}$ so that the zero layer of the reciprocal lattice intersects the reflecting sphere in a small circle of radius $\sin \bar{\mu}$ (Fig 8.22(b)). The zone axis is then caused to *precess* about the direction of the incident X-ray beam so that it describes a conical surface whose apex is the centre of the crystal, whose axis is coincident with the incident X-ray beam, and whose semiangle is $\bar{\mu}$ (Fig 8.22(c)). The precession is achieved by coupling an oscillation of the spindle of the arcs about its own axis with an oscillation of the spindle about a second axis, which is normal to both the spindle and the incident X-ray beam and passes through their point of intersection (Fig 8.22(d)); this movement is such that the line in the crystal coincident with the spindle axis lies in the plane of the spindle axis and the incident X-ray beam throughout the precession.

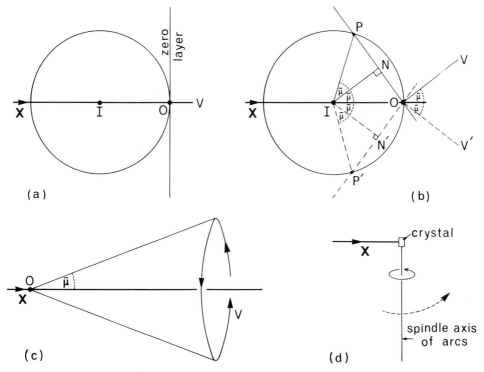

Fig 8.22   Geometry of zero-layer precession photography. (a) shows the initial setting of the crystal in which the zero-layer reciprocal lattice net normal to the selected zone axis OV is tangential to the reflecting sphere. In (b) the zone axis OV is inclined at the angle $\bar{\mu}$ to the incident X-ray beam so that the zero-layer reciprocal lattice net intersects the reflecting sphere in a small circle of radius NP $= \sin \bar{\mu}$; the two orientations in which the plane of the diagram contains the normals to the zero layer, IN and IN', are shown. (c) shows the precession of the selected zone axis OV about the incident X-ray beam on the surface of a cone of semi-angle $\bar{\mu}$. (d) illustrates the precession of the selected zone axis by coupled oscillations about two axes, both of which are perpendicular to the incident X-ray beam: one oscillation axis is the spindle axis of the arcs and the other is perpendicular to the plane of the diagram.

Figure 8.22(b) shows two positions in the course of the motion: IN and IN′ are directions parallel to the selected zone axis, which has a constant inclination $\bar{\mu}$ to the incident beam direction XIO; OP, and OP′ are the corresponding positions of the zero reciprocal lattice layer normal to the selected zone axis. The zero layer thus intersects the reflecting sphere at all stages of the motion in a small circle of radius $\sin \bar{\mu}$ which always passes through the origin O. In the course of a complete precession the small circle sweeps through all the reciprocal lattice points of the zero layer lying within a circle of radius $2 \sin \bar{\mu}$ about the origin of the reciprocal lattice. In Fig 8.23 the relative motion of the reciprocal lattice plane and the reflecting sphere is illustrated in terms of a stationary reciprocal lattice and a moving reflecting sphere. At any given time the reflecting sphere intersects the zero layer in a circle passing through the origin of reciprocal space. During the precession this circle of intersection effectively rolls about O: three successive positions of the circle of intersection are shown, two of which correspond to the positions shown in Fig 8.22.

The film is caused to precess in a manner identical to that of the crystal so that its centre remains stationary at a point distant F from the crystal in the forward direction of the incident beam and its plane is parallel to the zero layer at every stage of the motion.

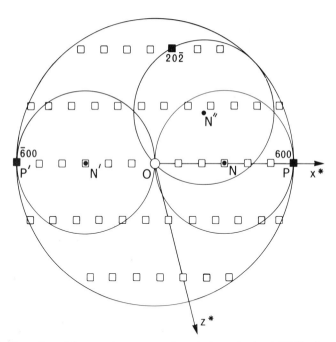

Fig 8.23 The formation of the zero-layer precession photograph about [010] of a monoclinic crystal, i.e. the *h0l* photograph. The reciprocal lattice net is taken to be stationary and the reflecting sphere rolls about O. Three instantaneous positions of the reflecting circle (radius = $\sin \bar{\mu}$) are shown; these circles (centres: N, N′, N″) give rise to the reflexions 600, $\bar{6}00$, and $20\bar{2}$, the relevant reciprocal lattice points being shown as solid squares. The reflecting circle sweeps out in the course of its motion a circle of radius $2 \sin \bar{\mu}$ about O.

In illustrating the production of precession photographs it is usually convenient to superpose diagrams illustrating the geometry in both direct and reciprocal space. We assume that the crystal is situated at the centre I of the reflecting sphere and choose a scale such that the crystal to film distance, F mm, in direct space measured

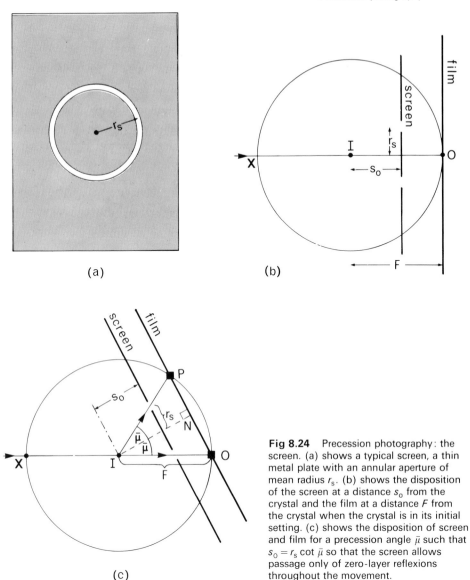

(a)

(b)

(c)

Fig 8.24 Precession photography: the screen. (a) shows a typical screen, a thin metal plate with an annular aperture of mean radius r_s. (b) shows the disposition of the screen at a distance s_0 from the crystal and the film at a distance F from the crystal when the crystal is in its initial setting. (c) shows the disposition of screen and film for a precession angle $\bar{\mu}$ such that $s_0 = r_s \cot \bar{\mu}$ so that the screen allows passage only of zero-layer reflexions throughout the movement.

in the direction of the undeviated X-ray beam is equal on the diagram to the radius of the reflecting sphere, one unit in reciprocal space. In Fig 8.24(c) then the centre of the film is at O and a reciprocal lattice point at P gives rise to a diffracted beam parallel to IP which intersects the film at P. The ratio $OP/IO = 2 \sin \bar{\mu}$ is the same in direct and in reciprocal space; in direct space $IO = F$ mm and in reciprocal space $IO = 1$ reciprocal unit. In direct space therefore $OP = Fd^*$, where d^* is equal to the distance represented by OP in reciprocal space, the distance of the reciprocal lattice point P from the origin O of reciprocal space. As the crystal precesses about IO the zero layer generates a cone of diffracted beams of semiangle $\bar{\mu}$ and this cone rolls about the line IO which remains on the surface of the cone throughout the motion. The film moves so that its plane is always parallel to the zero layer, the crystal to film distance is constant and equal to $IN = F \cos \bar{\mu}$, and the point O is fixed in position.

Thus the diffracted beams to which the zero layer gives rise strike the film so as to produce an undistorted image of the reciprocal lattice plane on a scale of F mm = 1 reciprocal unit.

If the film is to record only zero-layer reflexions it will be necessary to insert between the crystal and the film a *screen* so designed as to allow free passage of the cone of diffracted beams generated by the zero reciprocal lattice layer and to absorb those generated by all other reciprocal lattice layers. The form of the screen is a thin metal plate with an annular aperture of mean radius r_s (Fig 8.24(a)). When $\bar{\mu}$ is zero the centre of the annular aperture coincides with the forward direction of the incident X-ray beam (Fig 8.24(b)). The arm holding the screen is rigidly fixed to the spindle axis so that the movement of the screen follows precisely the movement of the crystal as the crystal precesses about the incident X-ray beam. If the crystal to screen distance is $r_s \cot \bar{\mu}$ the screen will isolate the zero-layer reflexions throughout the complete movement (Fig 8.24(c)). At all stages of the precession the annular aperture allows the undeviated X-ray beam to pass; the screen precesses round this direction while remaining parallel to the zero-layer reciprocal lattice net. Values of s_0 and r_s have to be selected so that the screen can move unimpeded within the restricted space between the crystal and the film.

We consider now the formation of an nth layer[4] precession photograph, the reflexions on which are generated by a net of reciprocal lattice points at a perpendicular distance $\zeta_n = n\lambda/t$ from the parallel zero-layer net, λ/t being the spacing of reciprocal lattice planes normal to the selected zone axis along which the spacing of lattice points is t in direct space. In Fig 8.25 $P_n Q_n$ and PO are respectively the

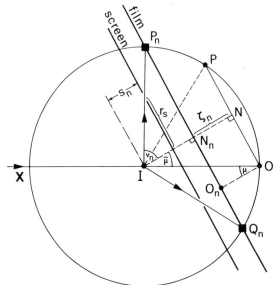

Fig 8.25   The formation of an nth layer precession photograph.

intersections of parallel nth and zero reciprocal lattice layers with the plane of the diagram. Again we imagine the crystal to be situated at the centre I of the reflecting sphere and the film to be coincident with the nth reciprocal lattice layer. IP_n and IQ_n lie on the surface of the cone of diffracted beams for the nth layer, the semiangle of this cone being υ_n such that $\cos \upsilon_n = IN_n = IN - N_n N = \cos \bar{\mu} - \zeta_n$. Just as for the

[4] For the nth layer to be an *upper* rather than a *lower* layer photograph the positive direction of $[UVW]$ must be directed back along the incident X-ray beam.

zero-layer photograph a screen has to be introduced between the crystal and the film to eliminate reflexions from all but the selected nth layer. It is apparent from Fig 8.25 that if the annular circular aperture in the screen is of mean radius r_s, then the screen has to be placed at a perpendicular distance s_n from the crystal such that $s_n = r_s \cot v_n = r_s \cot \cos^{-1}(\cos \bar{\mu} - \zeta_n)$. The magnitudes of r_s and s_n must be so chosen that the movement of the screen is not impeded by any other part of the camera. Since we imagine the film to be coincident with the nth reciprocal lattice layer it will have to be moved relative to its position for a zero-layer photograph by a distance $F\zeta_n$ so that the perpendicular crystal-to-film distance for an nth layer photograph is $F \cos v_n = F(\cos \bar{\mu} - \zeta_n)$ and the magnification factor remains equal to F. There is now no fixed point on the film; its centre O_n precesses about the forward direction of the incident X-ray beam in just the same way as does the intersection of the zone axis with the nth reciprocal lattice layer. In order to obtain the requisite movement of the film the film-cassette is mounted on an arm which can be brought forward in the direction normal to the plane of the film so that the cassette is displaced forwards relative to the mounting which controls the movement of the film; in terms of Fig 8.25 the film is positioned for recording the nth layer so that it moves about the stationary point O, which corresponds to the centre of the zero-layer film. As the crystal precesses the small circle (radius $\sin v_n$) in which the nth reciprocal lattice layer intersects the reflecting sphere sweeps through the reciprocal lattice plane in a circular area of radius $O_n P_n = O_n N_n + N_n P_n = ON + N_n P_n = \sin \bar{\mu} + \sin v_n$. Figure 8.26, which illustrates this point, shows the same instant intersection as Fig 8.25. It will be apparent from Fig 8.26 that nth layer photographs necessarily have a central blind spot of radius

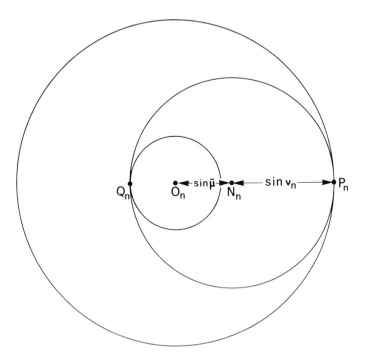

Fig 8.26 The central blind spot in nth layer precession photographs. The figure shows the same instant intersection of the reflecting circle with the nth reciprocal lattice net as Fig 8.25. Only reciprocal lattice points in the annular area centred on O_n between the limiting circles of radii $O_n Q_n$ and $O_n P_n$ give rise to reflexion. On the photograph $O_n Q_n = F(\sin v_n - \sin \bar{\mu})$ and $O_n P_n = F(\sin v_n + \sin \bar{\mu})$.

$O_nQ_n = N_nQ_n - N_nO_n = F(\sin v_n - \sin \bar\mu)$, corresponding to a circular area of the reciprocal layer which remains throughout the motion within the reflecting sphere.

In order to take an nth layer precession photograph it is necessary to know the magnitude of ζ. If ζ is not already known from prior study with oscillation photographs, it can very easily be determined by inserting a film in a light-tight envelope in the screen holder and taking a precession photograph at a known precession angle $\bar\mu$ to yield what is known as a *cone-axis photograph*. Reciprocal lattice layers normal to the zone axis will produce diffracted beams lying on coaxial cones; the semiangle for the cone produced by the nth layer is v_n and for the zero layer $v_0 = \bar\mu$. These cones of diffracted beams intersect the film in concentric circles of radius $r_n = s' \tan v_n = s' \tan \cos^{-1}(\cos \bar\mu - \zeta_n)$, where s' is the perpendicular distance from the crystal to the screen holder (Fig 8.27). By measurement of the radii of the circles so produced ζ can be evaluated with sufficient accuracy for first, second, etc layer photographs to be taken.

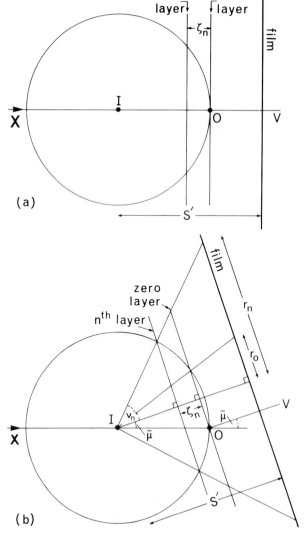

(a)

(b)

Fig 8.27 The formation of a cone-axis photograph. (a) and (b) show respectively the geometry for $\bar\mu = 0$ and for a non-zero value of $\bar\mu$. The film is positioned in the screen holder at a distance s' from the crystal at I. (b) shows an instant in the precession motion which causes the zero layer reciprocal lattice net to give rise to a cone of diffracted beams of semi-angle $\bar\mu$ and the nth layer net to give rise to a cone of semi-angle v_n.

Our description of the precession camera will be confined to an outline of the essentials of the instrument. Being fundamentally a three-dimensional rather than an axial apparatus it is difficult to describe in terms of two-dimensional diagrams (Fig 8.28), but easy enough to understand when seen. The motor drives a spindle a coincident with the incident X-ray beam. Rigidly attached to the spindle is a graduated arc b whose centre lies at the point of intersection of the forward direction of the incident beam with the film. A bearing at the centre of the film holder c maintains a rod perpendicular to the plane of the film throughout the motion; the other end of the rod engages the arc and is clamped in position at the chosen precession angle $\bar{\mu}$. The film holder and the spindle d, to which the arcs e carrying the crystal f are attached, are mounted on gimbals, i.e. free moving mutually perpendicular bearings. The motion of film and crystal are linked so that each precesses identically about the direction of the incident beam. The screen g is rigidly attached to the spindle d which

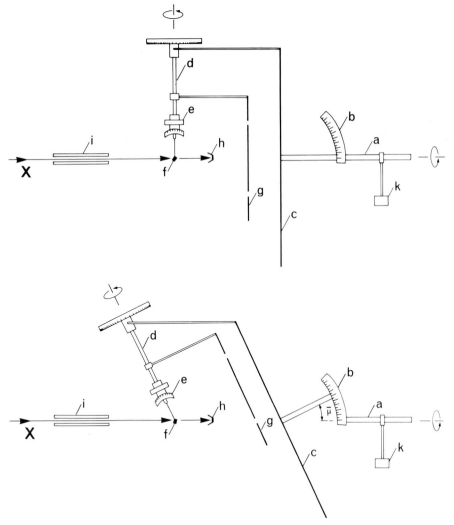

Fig 8.28 Schematic representation of a precession camera. The disposition of the parts of the camera are shown in the upper diagram for $\bar{\mu} = 0$ and in the lower diagram for a zero-layer photograph at $\bar{\mu} = 25°$. The labelling of the diagrams is explained in the text.

carries the arcs so that throughout its motion its plane is parallel to that of the film and its annulus is so placed as to transmit the selected cone of diffracted beams. A back-stop h which slips on to the collimator i has its absorbing cup situated immediately behind the crystal. Since smooth precession is essential if the intensities of the reflexions recorded on the film are to have significance, a counter weight, k, whose moment about the motor axis can be varied, is attached to the motor spindle to balance the moment of the various moving parts of the camera about this axis. For mechanical reasons the maximum attainable precession angle $\bar{\mu}$ is in most instruments not greater than 35°. The crystal-to-film distance is commonly fixed at 60 mm so that F is a constant equal to 60 mm; in some cameras the crystal-to-film distance is adjustable and then it can very simply be measured accurately by photographing a reciprocal lattice layer of a crystal whose unit-cell dimensions are known precisely. The back-plate of the film cassette is drilled with two pinholes which allow light to fall on the film and so serve to define the horizontal line through the centre of the film; the mid-point of the line joining the two black spots corresponds to the centre of the film.

There is very little to be said about the interpretation of precession photographs. The photograph is a direct representation of a reciprocal lattice net on the scale of F mm to one reciprocal lattice unit. Zero, first, and higher layer [010] photographs of a monoclinic crystal (Fig 8.29) will each display an array of spots that can be indexed on a unit mesh with axial repeats a^* and c^* and interaxial angle β^* (conventionally taken to be acute). Comparison of photographs of different layers taken about the same axis is straightforward, since there is no distortion and the magnification factor is constant, and may be achieved by direct superimposition of the films. Since the blank circle in the centre of upper layer photographs increases in radius as ζ_n increases, some important reflexions may be missing and that may be inconvenient. However a sufficiently large sample of hkl and $h0l$ reflexions should be obtainable from zero, first, and second layer [010] photographs to enable the systematic absences in these reflexions to be determinable. It will be necessary to supplement these photographs with a zero-layer [100] or [001] photograph to provide a reasonably accurate measurement of b and information about systematic absences in $0k0$ reflexions. Laue symmetry is always determinable by inspection of appropriately oriented precession photographs: for instance for monoclinic crystals (Laue class $2/m$) [010] photographs of any layer will display a central diad normal to the plane of the film while [100] and [001] photographs will have $2mm$ symmetry if zero layer and only lines of symmetry parallel to z^* and x^* respectively if upper layer.

The precession method is nicely balanced in its advantages and disadvantages relative to the Weissenberg method. Because it provides, when the crystal is correctly set and film and screen are properly adjusted, an undistorted photograph of a reciprocal lattice plane it enables interaxial angles in the plane to be very much more accurately measured than is possible on Weissenberg photographs. In the accuracy with which reciprocal cell edges can be measured there is little to choose between the two methods. Reflexions on a precession photograph tend to be uniform in shape so that accurate comparison of intensities is easier than on Weissenberg photographs where spot shape, especially on upper layer photographs, varies across the film. The unravelling of complicated orientational relationships in twins and intergrowths is simplified by the lack of distortion in precession photographs. Where exceptionally small or unstable crystals have to be used the shorter exposure time in which it is possible to obtain a satisfactory photograph by the precession method may be an

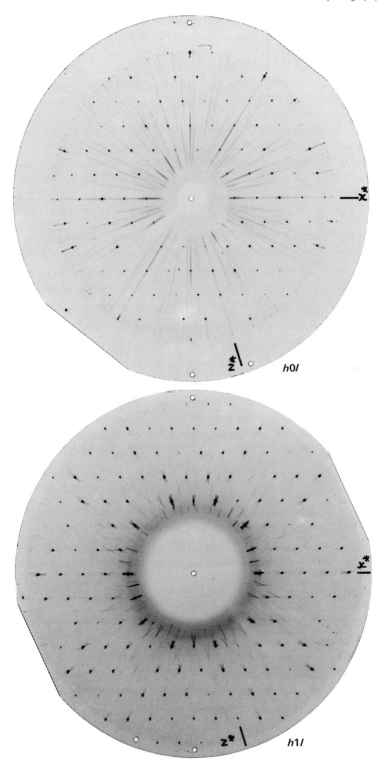

Fig 8.29 Zero and first layer [010] precession photographs of the monoclinic mineral *latiumite* taken with CuKα radiation, $\bar{\mu} = 25°$, $F = 60$ mm (three-quarters actual size).

important practical advantage. Where the precession method is inferior to the Weissenberg is in the smaller area of reciprocal space that, for reasons of camera geometry, can be sampled by a zero layer photograph. Usually the maximum precession angle that can conveniently be used is $30°$ and this allows a circle of radius $2 \sin \bar{\mu} = 1$ reciprocal unit about the origin of the zero layer net to be recorded; the area of nth layer nets that can be recorded is of course smaller. In the Weissenberg method however it is theoretically possible to record reflexions from a zero-layer net encompassed by a circle of radius two reciprocal units and on upper layer Weissenberg photographs, although the area that can be recorded decreases with increasing ζ_n, the area recorded is always substantially greater than on a precession photograph for the same value of ζ_n. Moreover it is possible with the Weissenberg camera to record layers of very much higher ζ value than with the precession camera; this is an important advantage when collecting intensity data for structural work and, moreover, occasionally symmetry elements that appear to be present when only reflexions of small Bragg angle are inspected may be seen to be absent when reflexions further out from the origin are considered.

In conclusion we draw attention to a useful facility provided by the precession camera: for one mounting of the crystal with a selected reciprocal lattice row parallel to the spindle axis of the arcs it is possible to record more than one zero-layer reciprocal lattice net and the corresponding upper layer nets. For example if x^* is parallel to the spindle axis of the arcs, [010] and [001] can be set parallel to the incident beam in turn so that the $h0l$, $h1l$, $h2l$, etc and the $hk0$, $hk1$, $hk2$, etc reciprocal lattice nets can be recorded simply by rotating the dial through α between the [010] and [001] exposures. It is just as easy to photograph reciprocal lattice nets normal to any zone axis $[OVW]$ by turning the dial through the appropriate angle. This is particularly useful when the reciprocal lattice row is parallel to a prominent zone axis and the arcs are interchangeable between oscillation, Weissenberg, and precession cameras; with a single mounting of the crystal oscillation, Weissenberg, and precession photographs can be taken and inter-related. For example suppose that an orthorhombic crystal has been mounted so that its [100] axis is parallel to the spindle axis of the arcs. This mounting would enable zero and upper layer [100] Weissenberg photographs and zero and upper layer [010] and [001] precession photographs to be taken by transferring the crystal on its arcs from one camera to the other.

In general the choice between Weissenberg and precession photography depends on the nature of the problem in hand and of the material under investigation. The principal advantage of the oscillation camera over the moving film cameras is that it provides a two-dimensional record of a truncated torus of reciprocal space, truncated because the film is of finite length, and this is the only way in which the regions between reciprocal lattice layers can conveniently be investigated; this facility is important in the study of phenomena, such as anti-phase domains, which give rise to diffracted intensity maxima that do not correspond to reciprocal lattice points.

Laue photography

For Laue photography the single crystal under investigation is maintained stationary in an incident beam containing a wide spectral range of X-ray wavelengths. As in the methods previously described for studying single crystal X-ray diffraction patterns the crystal is attached to a glass fibre mounted on arcs. Rigidly fixed to the spindle carrying the arcs is a circular scale, which permits the crystal to be rotated through a known angle between successive exposures; the incident X-ray beam is perpendicular to the spindle axis. In various circumstances it may be convenient to use either a cylindrical film coaxial with the spindle axis (Fig 8.30(a)) or a flat film with its plane perpendicular to the incident beam. Alternative positions for a flat film are in common use: in the *back-reflexion position* the film is situated between crystal and collimator (Fig 8.30(b)) to record only reflexions of high Bragg angle, whereas in the *front-reflexion position* the film is situated on the other side of the crystal (Fig 8.30(c)) and records only reflexions of low Bragg angle. The back-reflexion arrangement is particularly useful for crystals which absorb X-rays very strongly and for large crystals (especially metal crystals).

There is no specially designed camera for Laue photography: an oscillation or a Weissenberg camera can be used without adaptation for recording Laue patterns on cylindrical film, while a precession camera can be used directly for taking front-reflexion flat film photographs and an oscillation camera can be adapted for either front or back-reflexion photographs. The only camera conditions to be satisfied are that the crystal should remain stationary during exposure and that either the axis of a cylindrical film or the plane of a flat film should be perpendicular to the incident X-ray beam.

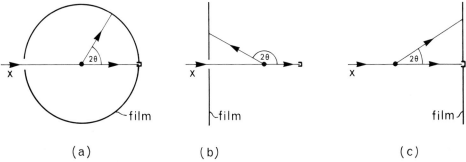

| (a) | (b) | (c) |

Fig 8.30 Laue photography. (a), (b), (c) show respectively the cylindrical film, back-reflexion flat film, and front-reflexion flat film arrangements. The crystal is shown as a solid circle and the beam trap as a section through a cup.

It was remarked at the beginning of this chapter that for a stationary crystal bathed in a parallel beam of monochromatic radiation few solutions of the Bragg Equation will occur. We have already explored ways in which the crystal can be moved in a regular manner while keeping the incident radiation monochromatic. In Laue photography we are concerned with increasing the number of solutions of the Bragg Equation for a stationary crystal and we do this by allowing the wavelength of the incident radiation to be variable. A broad X-ray spectrum in the incident beam is achieved by utilizing the unfiltered output of an X-ray tube; for this purpose the higher the atomic number of the target element the better, provided it is a sufficiently good thermal conductor to withstand a high current density, i.e. W is preferable to Mo, but Cu will do very well.

The reflexions on a Laue photograph cannot be easily indexed. The angle between the forward direction of the incident X-ray beam and the diffracted beam will be equal to 2θ so that θ is readily determinable. However a particular value of θ may correspond to more than one solution of the Bragg Equation $\lambda = 2d \sin \theta$ when λ is variable and d is unknown. For instance if the emission spectrum of the X-ray tube extends from $< \lambda'$ to $> 3\lambda'$, where λ' is a particular wavelength, the $2h, 2k, 2l$ and the $3h, 3k, 3l$ reflexions will have the same Bragg angle as the hkl reflexion and so all

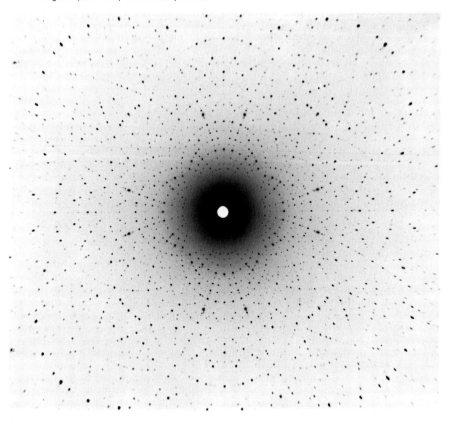

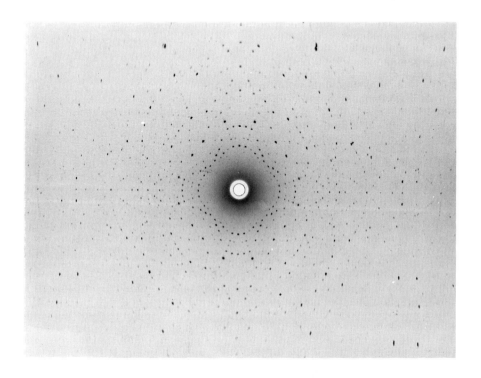

Fig 8.31 Laue photographs of a crystal of the tetragonal mineral *vesuvianite* (Laue class 4/*mmm*) taken with the incident beam parallel to the tetrad. The lower photograph was taken in a cylindrical camera; the upper photograph was taken with the front-reflexion flat film arrangement.

three reflexions will be superimposed. Indexing can be achieved by correlating the angular relationships between the normals to reflecting planes with the known axial ratios and interaxial angles. Indexed Laue photographs have some specialized uses, such as for the determination of the orientation of large single crystals of cubic metals, a topic which lies outside our scope. Except for such specialized applications it is rarely necessary to index Laue photographs; the commonest uses of the method do not involve indexing and are (i) the determination of the Laue symmetry of a crystal and (ii) the setting of a crystal with an identified zone axis in a particular direction relative to the camera geometry as a preliminary to oscillation or moving film photography.

We turn now to consider some of the general characteristics of Laue photographs. Simple inspection of a Laue photograph (Fig 8.31) reveals reflexions lying on curves such that each curve corresponds to the intersection with the film of a cone, the surface of which contains the forward direction of the incident beam. With flat films such intersections are conic sections, but with cylindrical films the nature of the curve is more complicated. Figure 8.32 illustrates the diffraction geometry for the generation of such a curve. Consider the plane represented by the great circle PZ whose pole is N. The reflected beam R will be coplanar with the incident X-ray beam XX' and with the normal N to the plane PZ; it will therefore lie on the great circle X'PN and will be so placed that RP = PX' = θ, where θ is the Bragg angle. Since the pole N of the great circle PZ lies on the great circle X'PR, these great circles intersect orthogonally so that $\widehat{ZPX'} = \widehat{ZPR} = 90°$. The spherical triangles ZPX' and ZPR are thus congruent (with common side ZP, right-angles at P, X'P = PR) so that ZR = ZX' = ψ. This result will be true for any plane which contains the direction Z and satisfies the Bragg Equation for this orientation of the incident X-ray beam. In

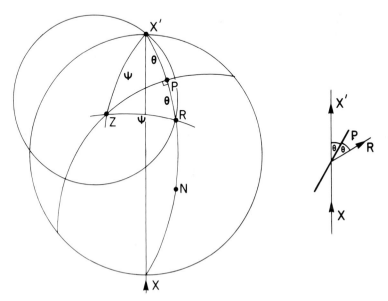

Fig 8.32 The diffraction geometry for the generation of a curve of reflexions on a Laue photograph. The diagram on the left is described in the text; that on the right illustrates the relationship of the incident and diffracted beams to the plane P.

particular if Z is a zone axis $[UVW]$ any plane containing the direction Z will lie in the zone $[UVW]$ and, if it is in the reflecting position, the direction of the diffracted beam arising from it will lie on the small circle whose stereographic centre is Z and which contains the forward direction of the incident X-ray beam. Reflexions generated by the planes of a zone $[UVW]$ will thus lie on a cone containing the incident direction, whose semiangle is the angle between the zone axis and the forward direction of the incident beam; on the film the cone will be represented by reflexions lying on a curve through the point of zero Bragg angle. If $[UVW]$ is a prominent zone axis in the crystal, there will be many reflexions on the curve and reflexions generated by planes of low indices will lie at the mutual intersection of several such curves, each related to a prominent zone axis. The shape of a zonal curve will depend very much on the angle ψ between the zone axis and the forward direction of the incident beam. For $\psi = 90°$ the cone of diffracted beams becomes a plane containing the incident X-ray beam; on a flat film this will be manifested as a radial line of reflexions through the $\theta = 0°$ (front-reflexion set-up) or $\theta = 90°$ (back-reflexion set-up) point. On a cylindrical film however the coplanar reflexions for which $\psi = 90°$ will only lie on a straight line on the film when the zone axis $[UVW]$ is either coaxial with the film cylinder or normal to the plane containing the incident beam and the axis of the cylindrical film. That is to say the only straight lines of reflexions on a cylindrical Laue photograph will be in the horizontal and vertical directions through the point on the film corresponding to $\theta = 0°$.

Another obvious feature of Laue photographs (Fig 8.31) that is worthy of comment is the absence of reflexions on the film over an area centred on the intersection of the forward direction of the incident beam with the film (this is of course not a feature of back-reflexion photographs). A reflexion close to the forward direction of the X-ray beam must have rather a small Bragg angle θ. Therefore for this reflexion $\lambda/2d$ must be small. As we have already seen (Fig 7.2) there is a sharp cut-off at the low wavelength end of the emission from an X-ray tube dependent on the operating voltage of the tube. Moreover the maximum value of d will be limited by the unit-cell dimensions of the crystal. For every radial direction about the $\theta = 0°$ point there will thus be a minimum value of θ below which reflexion is impossible. The size of the blank area on the film about the $\theta = 0°$ point will of course depend on the orientation of the crystal with respect to the incident beam; in general terms one can say that for an incident beam with a certain cut-off wavelength a substance with a small unit-cell will exhibit a larger blank area than a substance with one or more long unit-cell dimensions.

We turn now to the use of Laue photographs for assigning a crystalline substance to its Laue symmetry class. In what follows the statement that a crystal possesses a certain symmetry will refer to its Laue symmetry rather than to its point group or space group symmetry.

When X-rays are incident parallel to a symmetry axis of the crystal, the resultant Laue photograph will display the symmetry of that axis. Such axial symmetry is most clearly displayed on a flat film in the front-reflexion setting, but a photograph taken on a cylindrical film can be utilized although unambiguous determination of the nature of the axis is then rather more troublesome. Suppose, for instance, that the incident X-ray beam is coincident with the tetrad in a crystal of Laue class $4/m$ and that the plane (hkl) is so oriented that it will reflect X-radiation of wavelength λ. Then the symmetry related planes $(k\bar{h}l)$, $(\bar{h}\bar{k}l)$, and $(\bar{k}hl)$ will be similarly inclined to the direction of the incident beam and will also reflect X-rays of wavelength λ (Fig 8.33).

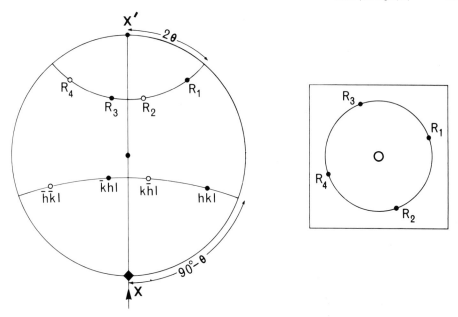

Fig 8.33 Laue photography of a crystal of Laue class 4/*m* with the incident X-ray beam parallel to the tetrad. The stereogram on the left shows the disposition of the normals to four planes related by the tetrad and the disposition of the resulting reflexions R_{1-4}. The portion of a flat film Laue photograph on the right shows the disposition of the four reflexions on a circle about the exit-hole in the centre of the film.

All four reflexions will be equal in intensity and will be disposed on the film about the point of intersection of the incident X-ray beam with the film in a manner consistent with tetragonal symmetry about that point. The Laue photograph taken as a whole will thus display tetragonal symmetry, every reflexion being equal in intensity to three others related to it spatially by successive rotation through 90° about the direction of the incident X-ray beam. Likewise when the incident beam is coincident with a hexad, triad, or diad axis in the crystal the corresponding symmetry will be apparent on a Laue photograph; and when the incident beam is coplanar with a mirror plane in the crystal, the resultant Laue photograph will display a line of symmetry parallel to the mirror plane and passing through the point of intersection of the incident beam with the film.

Since the Bragg Equation restricts θ to values between 0° and 90°, the X-ray beam must be incident on the same side of the (*hkl*) plane as the outward direction of the normal to the plane for the *hkl* reflexion to be produced; if the X-ray beam is incident, at the correct angle, on the other side of the (*hkl*) plane it will give rise to the $\bar{h}\bar{k}\bar{l}$ reflexion (Fig 8.34). It is thus impossible to record an *hkl* reflexion and a $\bar{h}\bar{k}\bar{l}$ reflexion without moving the crystal relative to the incident X-ray beam so that a Laue photograph never exhibits both *hkl* and $\bar{h}\bar{k}\bar{l}$ reflexions. For the tetragonal example that we have been considering this means that if reflexions are recorded from the planes (*hkl*), ($k\bar{h}l$), ($\bar{h}\bar{k}l$), ($\bar{k}hl$), reflexions will not be recorded from their opposites ($\bar{h}\bar{k}\bar{l}$), ($\bar{k}h\bar{l}$), ($hk\bar{l}$), ($k\bar{h}\bar{l}$). Of course if the crystal is rotated through 180° about an axis normal to the tetrad so as to bring the opposite sense of the tetrad into coincidence with the forward direction of the incident beam, then all four opposites will reflect and reflexions from (*hkl*), etc will be absent.

The general conclusion to be drawn from the tetragonal example discussed in the

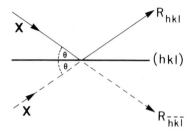

Fig 8.34   The diagram illustrates the impossibility of recording reflexions from a plane and from its opposite on the same Laue photograph.

preceding paragraphs is that the symmetry discernible on a Laue photograph is the symmetry about a direction in the crystal parallel to the incident X-ray beam; that is to say the symmetry of a Laue photograph must be assignable to one of the ten plane point groups.

In Table 8.1 the symmetry of Laue photographs of tetragonal crystals taken with the incident X-ray beam in a specified direction is listed for all possible directions for both tetragonal Laue classes (4/m and 4/mmm). The two tetragonal Laue classes are simply distinguished in practice by a Laue photograph taken with the incident X-ray beam parallel to [001]: the photograph for a 4/m crystal will have plane symmetry 4, whereas that for a 4/mmm crystal will have plane symmetry 4mm. If the crystal is already known to be tetragonal, its Laue class can be uniquely determined by taking just this one photograph. But if the possibility of the crystal being cubic has not been ruled out by other evidence, the observation that one Laue photograph has plane symmetry 4mm merely indicates that the Laue class of the crystal is either 4/mmm or m3m; Laue photographs in other orientations will have to be taken to distinguish between these two possibilities.

There is no standard procedure for determining the Laue symmetry of a crystal. The successive photographs necessary in a particular case will depend on the evidence provided by those already taken and on any reliable information that may happen to be available from prior study of certain physical properties of single crystals of the substance; for instance preliminary optical examination (volume 2) may have given a clear indication of crystal system. Quite commonly the Laue symmetry of a crystal is determined incidentally by observation of intensity relationships of reflexions in the course of the investigation of its reciprocal lattice geometry by one or other of the moving-film methods.

Table 8.1
Symmetry of Laue photographs of tetragonal crystals

Laue class	Direction of incident X-rays	Symmetry of Laue photograph
4/m	[001]	4
	$\langle UVO \rangle$	m
	$\langle UVW \rangle$	1
4/mmm	[001]	4mm
	$\langle 100 \rangle, \langle 110 \rangle$	2mm
	$\langle UVO \rangle, \langle UOW \rangle, \langle UUW \rangle$	m
	$\langle UVW \rangle$	1

The point groups of each tetragonal Laue class are:

4/m: 4, $\bar{4}$, 4/m

4/mmm: 422, 4mm, $\bar{4}2m$, 4/mmm

A difficulty commonly encountered in the course of determination of Laue symmetry by means of Laue photographs is that only when the crystal is very precisely set with its symmetry axis parallel to the incident beam will the resultant photograph clearly display the symmetry of the axis. An error in setting of as little as 5 minutes of arc may substantially affect the appearance of the photograph. Let us suppose that two symmetry related planes are inclined at angles $\theta + \delta\theta$ and $\theta - \delta\theta$ to the incident beam; then the two reflexions produced will be recorded on a flat film in the front-reflexion setting, at a perpendicular distance R from the crystal (Fig 8.35), at distances $R\tan 2(\theta + \delta\theta)$ and $R\tan 2(\theta - \delta\theta)$ from the centre of the film. Moreover the two planes will reflect different wavelengths $\lambda + \delta\lambda$ and $\lambda - \delta\lambda$, where $\delta\lambda = 2d\cos\theta.\delta\theta$ and, as we have seen earlier, intensity varies quite rapidly with wavelength in certain parts of the spectral range emitted by an X-ray tube. Thus the two reflexions may be markedly different in intensity as well as being noticeably asymmetrically disposed on the photograph even though the mis-setting of the crystal is slight. But it is difficult to generalize and the experienced crystallographer may be able to discern a suspicion of the presence of a symmetry axis at a considerable inclination to the incident beam. When inspection of a Laue photograph reveals a suspicion of the presence of a symmetry axis, the crystal should be adjusted to bring that direction into closer alignment with the incident beam and another Laue photograph should be taken.

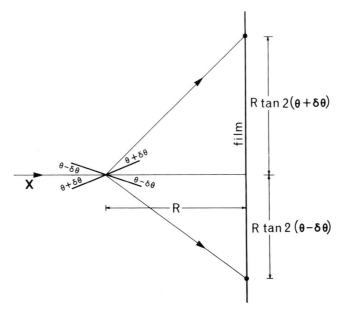

Fig 8.35 The figure illustrates the point that only when the incident X-ray beam is precisely parallel to a symmetry axis does the resultant Laue photograph clearly display the symmetry axis. The two lattice planes, shown as bold lines, are related by a diad axis inclined at the small angle $\delta\theta$, in the plane of the diagram, to the incident X-ray beam. The resultant reflexions will not be symmetrically disposed about the centre of the Laue photograph and will not be equal in intensity.

We deal generally with the setting of crystals on single crystal cameras in Appendix F, but it is appropriate to discuss here certain features of Laue photographs which are utilized for that purpose. A prominent zone in the crystal, because it contains a large number of lattice planes, will usually correspond to a curve with many closely spaced reflexions on the photograph and such curves are an obvious feature of most Laue photographs. The point of intersection of two or more such prominent zonal curves may be expected to correspond to the direction of the normal to a lattice plane of very simple indices. Symmetry axes are invariably parallel to such directions and it is, at least in principle, a simple matter to search for symmetry axes by bringing each such prominent zonal intersection in turn into coincidence with the incident beam.

It is always true to say that the presence of a mirror plane in a Laue class implies the presence of an axis of twofold or higher symmetry normal to it. Therefore if a mirror plane, which is necessarily perpendicular to a zone axis of simple indices, is located and the crystal is set with the mirror plane perpendicular to the spindle axis of the arcs on which the crystal is mounted, then a symmetry axis must be parallel to the spindle axis.

There is no generally applicable procedure for locating the crystallographic axes of a crystal of known or unknown symmetry. Each problem has to be tackled by the crystallographer in the light of what he knows at the start, or learns as he proceeds, about the Laue class of the crystal, in relation to the apparatus immediately available to him, and always bearing in mind the intensity of labour he is able to devote to the problem. The choice of procedure will depend very much on the experience and skill of the crystallographer. For a shapeless opaque crystal, the most difficult sort of subject, the present authors would usually choose to use Laue photographs taken on cylindrical film to locate symmetry directions; but other crystallographers might prefer to locate a principal zone by taking a series of precession photographs of small precession angle ($\bar{\mu} = 10°$) at appropriate intervals of rotation of the spindle axis. Either approach will succeed; one or the other may be more efficient in a particular case.

We turn now to consider the interpretation of Laue photographs in terms of the reciprocal lattice and the reflecting sphere: we have the choice of adopting either of two alternative approaches. We can either take the constant K to be equal to unity so that the dimensions of the reciprocal lattice will be independent of wavelength, but the radius of the reflecting sphere will be variable (Fig 8.36(a)); or we can take the constant K equal to λ so that the dimensions of the reciprocal lattice vary with wavelength, but the radius of the reflecting sphere is constant. If $K = 1$, all those reciprocal lattice points within the volume between the reflecting sphere of minimum radius (corresponding to maximum wavelength in the incident radiation capable of generating an observable reflexion) and the reflecting sphere of maximum radius (corresponding to the minimum wavelength, the cut-off wavelength, in the incident radiation) will be in the reflecting position for some wavelengths in the incident beam. Overlapping reflexions will occur if the reciprocal lattice points hkl; $2h, 2k, 2l$; $3h, 3k, 3l$; etc lie within this volume (Fig 8.36(b)). The second of the alternative approaches, with $K = \lambda$, is however more fruitful in general for the interpretation of Laue photographs. With $K = \lambda$, the reflecting sphere has radius equal to one reciprocal unit and each lattice plane is represented in reciprocal space by a radial streak; the end of the streak nearer the origin corresponds to the minimum wavelength emitted by the X-ray tube (the magnitude of λ_{min} depends on the nature of the target and on the applied voltage as indicated in Fig 7.2) and the end of the streak away from the origin fades away at a wavelength that is greater for planes which reflect strongly than for those which give rise only to weak reflexion.

For a particular wavelength the reciprocal lattice points corresponding to lattice planes lying in a zone are coplanar and this plane passes through the origin of reciprocal space. When the incident X-ray beam is polychromatic, as it is in Laue

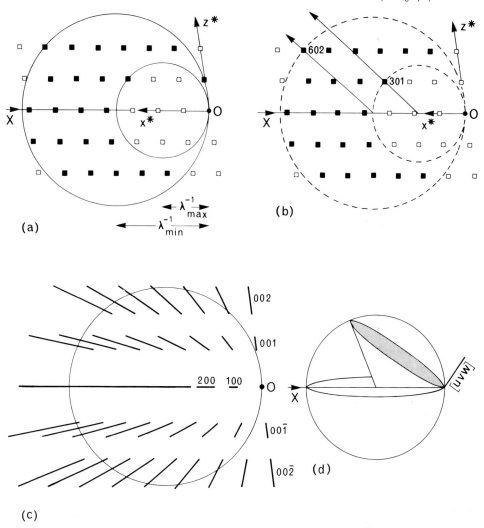

Fig 8.36 Reciprocal lattice interpretation of Laue photographs. In (a) and (b) $K = 1$ so that the reciprocal lattice dimensions are constant, but the radius of the reflecting sphere varies between λ_{min}^{-1} and λ_{max}^{-1} where λ_{min} is the cut-off wavelength of the incident spectrum and λ_{max} is arbitrarily taken at some wavelength where the intensity falls below a certain level; the ratio $\lambda_{max}/\lambda_{min}$ is taken as 2. (b) illustrates the parallelism of the reflected beams produced by the 301 and 602 reciprocal lattice points of a monoclinic crystal. In (c) and (d) $K = \lambda$ so that the reflecting sphere is of unit radius and each reciprocal lattice point becomes a streak radiating from the origin O as shown in (c); in (c) the ratio $\lambda_{max}/\lambda_{min}$ is taken at the lower value of 1·4 for clarity of the diagram and again a monoclinic lattice is exemplified. (d) serves to illustrate the point that the plane (shaded) in reciprocal space representing the zone $[UVW]$ intersects the reflecting sphere in a small circle; the Laue reflexions generated by such a zone lie on the surface of a cone whose apex is at the centre of the reflecting sphere and whose base is the small circle.

photography, each such reciprocal lattice point is replaced by a radial streak lying in the plane (Fig 8.36(c)). Moreover the streaks from the planes (hkl), $(2h, 2k, 2l)$, $(3h, 3k, 3l)$, etc will overlap if the range of wavelengths in the incident beam is sufficiently large. In general the plane in reciprocal space which represents the zone $[UVW]$ cuts the reflecting sphere in a small circle (Fig 8.36(d)) and the reflexions generated by such a zone lie in directions parallel to the radii of the reflecting sphere at the points of intersections of the streaks; all the reflexions generated by the zone of lattice planes thus lie on a cone whose semiangle is equal to the angle between the incident beam and the zone axis $[UVW]$. If the zone $[UVW]$ is a prominent zone, there will be a high density of reciprocal lattice streaks and consequently a large number of reflexions on the Laue photograph. The curve corresponding to the intersection of this cone with the film will thus stand out very clearly on the Laue photograph.

Since the closest approach of a reciprocal lattice streak to the origin is λ_{min}/d_{hkl}, where d_{hkl} is the spacing of the lattice planes giving rise to the streak and λ_{min} is the cut-off wavelength of the X-ray spectrum, it follows that there will be an irregularly shaped volume of reciprocal space about the origin totally devoid of reciprocal lattice streaks. This volume will extend in every direction to a distance of at least λ_{min}/d_{max} from the origin, where d_{max} is the greatest spacing of lattice planes in the crystal. Thus a Laue photograph will always exhibit an area devoid of reflexions close to the forward direction of the incident beam and the size of this area will be greater for crystals with small unit-cell dimensions.

The reflecting sphere has cylindrical symmetry about the incident X-ray beam so that the Laue photograph must have the symmetry of the crystal about that direction. That the symmetry of a Laue photograph may be lower than the symmetry of the reciprocal lattice about the direction of the incident beam we show by considering the incident beam to be parallel to a tetrad. The symmetry of the array of reciprocal lattice points about a tetrad is necessarily 4mm; but the crystal and the (flat film) Laue photograph generated by it will not necessarily show symmetry 4mm. If the Laue class of the crystal is 4/m, then the intensities of the hkl and $\bar{h}\bar{k}l$ reflexions will not necessarily be equal so that, when intensity as well as position of reflexions is considered, there will be no lines of symmetry on the Laue photograph. Only when the Laue class of the crystal is 4/mmm or m3m can the Laue photograph exhibit symmetry 4mm.

If the reflecting sphere passes through a point on the streak of an (hkl) lattice plane, it cannot intersect the streak of the $(\bar{h}\bar{k}\bar{l})$ plane for the same orientation of the crystal. In consequence a Laue photograph may lack a centre of symmetry even though the diffraction pattern as a whole may be centrosymmetric. The symmetry of a Laue photograph is always that of one of the ten two-dimensional crystallographic point groups.

In the treatment of Laue photographs as far as we have taken it in this chapter the use of the reciprocal lattice and the reflecting sphere are not essential; but they do provide, as we have sought to show in the preceding paragraphs, an elegant way of explaining the diffraction pattern produced.

Determination of accurate unit-cell dimensions

It is often necessary to be able to determine unit-cell dimensions very much more accurately than is possible by measurement of layer line spacings on oscillation photographs or by direct measurement of Weissenberg or precession photographs. Obvious uses for accurate unit-cell dimensions are in the determination of thermal expansion coefficients (volume 2) and for the conversion of the atomic coordinates which are the end result of a structure determination to accurate bond lengths and bond angles. Various methods for the accurate determination of unit-cell dimensions have been in general use over the past few decades; one obvious approach is to determine a^*, b^*, and c^* by very precise measurement of the Bragg angle for $h00$, $0k0$, and $00l$ reflexions. If the crystal is monoclinic, β^* can be measured directly on an [010] precession photograph and if it is triclinic the three interaxial angles α^*, β^*, and γ^* can be measured on the appropriate precession photographs. In these two systems, the only systems for which interaxial angles have to be measured, the form of the expressions for deriving unit-cell dimensions from reciprocal lattice dimensions (Table 6.7) are such that errors in the measurement of interaxial angles in reciprocal space may seriously affect the accuracy of both interaxial angles and unit-cell edges in direct space. We confine our discussion of the accurate measurement of unit-cell dimensions here to one very elegant, accurate, and generally applicable method, which makes use of the doublet splitting at high θ of reflexions on Weissenberg photographs.

As we have pointed out earlier (chapter 7) the characteristic X-radiation emitted from a crystallographic X-ray tube consists of a $K\beta$ line, which is filtered out, and the two closely spaced lines $K\alpha_1$ and $K\alpha_2$. The reflexions produced by the $K\alpha_1$ and $K\alpha_2$ wavelengths are resolved only at high Bragg angle. On a Weissenberg photograph all reflexions at low Bragg angles will appear to be single spots; but as θ increases the difference in Bragg angle for the $K\alpha_1$ and $K\alpha_2$ lines gradually increases until at high Bragg angle reflexions from a plane will be resolved into clearly separated pairs of reflexions, the inner and stronger of which is produced by the shorter α_1 wavelength in the incident beam and the outer by the longer, and weaker, α_2 wavelength.

The difference $\delta\lambda$ in wavelength between the α_1 and α_2 lines is accurately known for all X-ray sources in common use. Therefore the Bragg angle θ for the α_1 reflexion can be found by measuring the difference in Bragg angle $\delta\theta$ between pairs of reflexions produced by resolution of the α_1 and α_2 lines in the following manner. Suppose the wavelength of the α_1 radiation is λ, then for reflexion from a plane of spacing d,

$$\lambda = 2d\sin\theta \qquad \text{for the } K\alpha_1 \text{ wavelength}$$

and $\qquad \lambda + \delta\lambda = 2d\sin(\theta + \delta\theta) \qquad$ for the $K\alpha_2$ wavelength.

Therefore $\quad \sin(\theta + \delta\theta) - \sin\theta = \dfrac{\delta\lambda}{2d} = \dfrac{\delta\lambda}{\lambda}\sin\theta$

$$\sin\theta\cos\delta\theta + \cos\theta\sin\delta\theta - \sin\theta = \frac{\delta\lambda}{\lambda}\sin\theta$$

$$\cos\theta\sin\delta\theta - \sin\theta(1 - \cos\delta\theta) = \frac{\delta\lambda}{\lambda}\sin\theta$$

$$\cos\theta = \sin\theta\left(\frac{1 - \cos\delta\theta}{\sin\delta\theta} + \frac{\delta\lambda}{\lambda\sin\delta\theta}\right)$$

$$\cos\theta = \sin\theta\left(\tan\frac{\delta\theta}{2} + \frac{\delta\lambda}{\lambda\sin\delta\theta}\right)$$

$$1 - \sin^2\theta = \sin^2\theta\left(\tan\frac{\delta\theta}{2} + \frac{\delta\lambda}{\lambda\sin\delta\theta}\right)^2$$

and so $\qquad \sin^2\theta = \left\{1 + \left(\tan\dfrac{\delta\theta}{2} + \dfrac{\delta\lambda}{\lambda\sin\delta\theta}\right)^2\right\}^{-1}.$

For a zero-layer Weissenberg photograph, if the separation of the α_1 and α_2 reflexions from the same lattice plane is δs, measured perpendicular to the median line of the film, then $\delta\theta = \delta s/2r$, where r is the camera radius (Fig 8.37(a)). The magnitude of

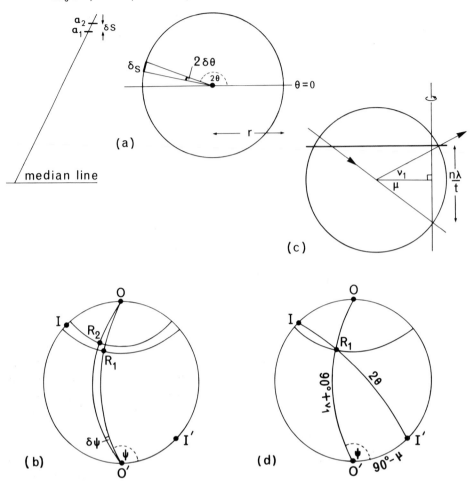

Fig 8.37 Accurate determination of unit-cell dimensions by measurement of $\alpha_1\alpha_2$ splitting. The measurement of the separation δs of the $\alpha_1\alpha_2$ doublet is made perpendicular to the median line of the zero-layer Weissenberg photograph, as shown on the left in (a), so that $\delta s = 2r \cdot \delta\theta$. The stereogram (b) illustrates the diffraction geometry for the α_1 reflexion R_1 and the α_2 reflexion R_2 in an nth layer Weissenberg photograph; OO' is the oscillation axis and II' is the incident beam direction so that $OI = 90° - \mu$, $OR_1 = 90° - v_1$, and $OR_2 = 90° - v_2$. (c) illustrates the generation of the α_1 reflexion in an nth layer equi-inclination Weissenberg; the intersection of the nth reciprocal lattice net with the plane of the diagram is shown as a bold line. The stereogram (d) serves to illustrate the relationship between θ, v_1, μ, and ψ for the α_1 reflexion R_1.

$\sin^2\theta$ can thus be calculated from measurements of δs for the pair of resolved $\alpha_1 - \alpha_2$ reflexions. The form of the relationship between δs and $\sin^2\theta$ is such that the value of $\sin^2\theta$ is rather insensitive to errors in measurement of δs.

For an nth layer equi-inclination Weissenberg photograph the relationship between the measured value of δs and $\sin^2\theta$ is more complicated. The Bragg angle θ of the α_1 reflexion cannot be deduced directly from measurement of δs. The distance s of a reflexion from the median line of the photograph gives the angle $\psi = s/r$ between the plane containing the forward direction of the incident X-rays and the oscillation axis and the plane containing the reflected beam and the oscillation axis; thus measurement of the separation of the resolved $\alpha_1 - \alpha_2$ doublet, δs, gives a value of $\delta\psi$

for the reflexion (Fig 8.37(b)). The equi-inclination angle μ for the nth layer is such that $\sin \mu = n\lambda/2t$, where t is the periodicity of lattice points along the zone axis parallel to the oscillation axis. The angle μ at which the camera is set cannot be right for both the α_1 and α_2 wavelengths. Reflexions will lie on two cones, one of semiangle $90° - v_1$ and the other of semiangle $90° - v_2$, where v_1 and v_2 are related to the camera angle μ. It is apparent from Fig 8.37(c) that for the α_1 reflexion

$$\frac{n\lambda}{t} = \sin v_1 + \sin \mu \tag{1}$$

and correspondingly for the α_2 reflexion of the doublet

$$\frac{n(\lambda + \delta\lambda)}{t} = \sin v_2 + \sin \mu. \tag{2}$$

We can obtain an expression for θ by use of the spherical triangle $O'I'R_1$ (Fig 8.37(d)):

$$\cos 2\theta = \cos (90° + v_1) \cos (90° - \mu) + \sin (90° + v_1) \sin (90° - \mu) \cos \psi \tag{3}$$

i.e. $$2 \sin^2 \theta = 1 + \sin v_1 \sin \mu - \cos v_1 \cos \mu \cos \psi \tag{4}$$

and correspondingly

$$2 \sin^2(\theta + \delta\theta) = 1 + \sin v_2 \sin \mu - \cos v_2 \cos \mu \cos (\psi + \delta\psi). \tag{5}$$

From these relations, numbered (1) to (5) above, coupled with the Bragg Equation an expression can be obtained for relating θ to the measured separation δs of an $\alpha_1 - \alpha_2$ doublet, the wavelengths λ and $\lambda + \delta\lambda$ of the relevant lines in the incident radiation, the camera angle μ, and the lattice spacing t. The lattice spacing t will be approximately known at the start and so each cycle of refinement in the computation will yield a more accurate value of t which is then used to calculate a better value of $\sin^2 \theta$.

We now illustrate the way in which cell dimensions can be calculated from a set of values of $\sin^2 \theta$ by considering a monoclinic example. The extension of the argument to the triclinic case is straightforward, but the expressions involved are of course more cumbersome. For a reciprocal lattice point in the monoclinic system $d*$ is given by

$$d*^2 = h^2 a*^2 + k^2 b*^2 + l^2 c*^2 + 2hla*c* \cos \beta*,$$

which becomes, on putting $q = d*$ and differentiating,

$$qdq = (h^2 a* + hlc* \cos \beta*)da* + k^2 b* db* + (l^2 c* + hla* \cos \beta*)dc*$$
$$- hla*c* \sin \beta*.d\beta*$$

i.e. $$qdq = A \, da* + B \, db* + C \, dc* - D \, d\beta*.$$

The coefficients A, B, C, D are calculated from the known approximate unit-cell dimensions; $q = \lambda/d = 2 \sin \theta$ is obtained from measurements of $\alpha_1 - \alpha_2$ doublet splitting in the way described earlier; and dq is taken to be the difference between this 'observed' value of q and that obtained from the known approximate unit-cell dimensions. Every measurement of an $\alpha_1 - \alpha_2$ doublet thus gives rise to one such linear equation in the four unknowns $da*$, $db*$, $dc*$, and $d\beta*$. The set of linear equations derived from all the measurements made is soluble by least-squares methods to yield the more accurate reciprocal lattice dimensions $a* + da*$, $b* + db*$, $c* + dc*$, and $\beta* + d\beta*$; from these more accurate unit-cell dimensions and thence more accurate values of $\sin^2 \theta$ for upper layer photographs can be calculated. A further cycle of refinement is then performed, and so on until by successive approximation unit-cell dimensions accurate to better than 1 in 10^3 are achieved.

The powerful technique outlined above, which was developed by Alcock and Sheldrick (1967) from a method restricted to zero-layer photographs (Main and Woolfson, 1963), has the advantages that all the measurements can be quickly and easily made with an ordinary travelling microscope (an accuracy of about 3 per cent in the measurements is all that is required) and that computer programmes for the least-squares calculations are readily available. The method is moreover found to be insensitive to even quite substantial errors in camera radius and in the equi-inclination angle μ. It is however observed that when data from Weissenberg photographs taken about only one oscillation axis are used, those cell dimensions which are not directly determinable from the zero-layer photograph are relatively less accurate. It is therefore advisable to include data from one or more Weissenberg photographs taken about a second axis. Of course one could merely use the three zero-layer photographs taken respectively about the x, y, and z axes; but much more data become available and higher accuracy is in consequence achieved if equi-inclination upper layer photographs about at least one axis are included.

We shall not discuss here other methods of determination of accurate unit-cell dimensions. For an excellent and thorough treatment of this topic the reader is referred to Woolfson (1970).

Problems

8.1 A crystal of gypsum is set up on an oscillation camera with [010] parallel to the oscillation axis. On an [010] rotation photograph taken with CuKα radiation ($\lambda = 1\cdot5418$ Å) the separation of the $h6l$ and $h\bar{6}l$ layer lines is found to be 46·12 mm, the radius of the cassette being 30·0 mm. Determine b for gypsum.

8.2 Rotation photographs of a crystal of rhodochrosite, $MnCO_3$, taken about three mutually perpendicular axes and Laue photographs taken with each of these axes parallel to the incident beam yielded the data given below. The rotation photographs were taken with CuKα radiation ($\lambda = 1\cdot5418$ Å) and the radius of the cylindrical cassette was 30·0 mm.

Axis	I	II	III
Layer line	8th	4th	2nd
Distance from zero layer (mm)	38·40	33·55	25·40
Symmetry of Laue photograph	$3m$	m	2

Determine the dimensions of the conventional unit-cell and the Laue class of rhodochrosite.

8.3 A cubic crystal is set on an oscillation camera with its [100], [110], and [111] axes successively parallel to the oscillation axis. Rotation photographs were taken with CuKα radiation ($\lambda = 1\cdot5418$ Å). The camera radius was 30·0 mm. The separation $2H_n$ of $+n^{th}$ and $-n^{th}$ layer lines was measured on each photograph with the following result:

Axis	[100]	[110]	[111]
n	3	5	3
$2H_n(mm)$	41·0	53·5	51·5

Determine the unit-cell dimension a and the lattice type.

8.4 On a rotation photograph of chalcopyrite, $CuFeS_2$, which is tetragonal, taken with $CoK\alpha$ radiation ($\lambda = 1·7902$ Å) using a cylindrical camera of radius 30·0 mm, the separation of the $hk4$ and $hk\bar{4}$ layer lines was measured as 58·05 mm. The ζ values of reflexions on the zero, first, and second layer lines were measured as

$\zeta = 0,$ $\xi = 0·68, 0·97, 1·08, 1·37, 1·53, 1·74, 1·93$

$\zeta = 0·174,$ $\xi = 0·34, 0·76, 1·02, 1·23, 1·41, 1·71, 1·84.$

$\zeta = 0·348,$ $\xi = 0·48, 0·68, 1·08, 1·37, 1·45, 1·53, 1·74.$

Determine the unit-cell dimensions a and c and the lattice type of chalcopyrite.

8.5 Rotation photographs of a monoclinic crystal, taken with $CuK\alpha$ radiation ($\lambda = 1·5418$ Å), about the three reference axes yielded the following information:

rotation axis	[100]	[010]	[001]
layer line	2nd	4th	3rd
ζ	$\zeta_2 = 0·62$	$\zeta_4 = 0·64$	$\zeta_3 = 0·60$

It is known that β is quite close to 90° and it is observed that on the second layer line of the [100] rotation photograph pairs of reflexions $h0l$, $h0\bar{l}$ occur close together with ξ values differing by 0·065. Determine the unit-cell parameters a, b, c, and β.

8.6 The $h0l$ Weissenberg photograph of the monoclinic mineral latiumite shown in Fig 8.16 is printed true to size. It was taken with $CuK\alpha$ radiation ($\lambda = 1·5418$ Å) in a camera of radius 28·65 mm. Use a Weissenberg chart to plot the undistorted $h0l$ reciprocal lattice net. Assuming no systematic absences in the $h0l$ reflexions determine a, c, and β. Given that $b = 5·08$ Å and the screen radius $= 23·02$ mm, evaluate μ, the screen shift, and the cassette shift required for the $h1l$ equi-inclination photograph to be taken with the same radiation.

8.7 Topaz is orthorhombic with unit-cell dimensions $a = 8·39$, $b = 8·80$, $c = 4.65$ Å. A crystal of topaz was set on a precession camera to record the $0kl$ reciprocal lattice net with $MoK\alpha$ radiation ($\lambda = 0·7107$ Å) and $\bar{\mu} = 25°$. Which of the $00l$ reflexions might be recorded on this photograph? In order to record the $2kl$ net through what distance will it be necessary to move the film cassette ($F = 60$ mm) and what will be the crystal to screen distance for $\bar{\mu} = 20°$ and a screen of radius $r_s = 20$ mm? Which $20l$ reflexions might be recorded on this photograph?

8.8 Blende, ZnS, is cubic with space group $F\bar{4}3m$ and unit-cell dimension $a = 5·413$ Å. A Laue photograph of a single crystal of blende taken on a flat plate perpendicular to the forward direction of the X-ray beam at a distance of 30 mm from the crystal displays symmetry $2mm$. In what direction was the X-ray beam incident on the crystal? The horizontal mirror line of the Laue photograph has four reflexions symmetrically disposed about the centre, two inner reflexions are 43·5 mm apart and the two outer reflexions are 77·5 mm apart. For each of these four reflexions determine the Bragg angle θ and the indices hkl. Calculate the wavelength of the X-rays diffracted by these four planes.

8.9 To which Laue class (or classes) would a crystal be assigned and in what direction would the incident X-ray beam be oriented if a Laue photograph exhibited symmetry (i) 6mm, (ii) 4mm, (iii) 3m, (iv) 2mm, (v) 6, (vi) 4, (vii) 3, (viii) 2?

9
Principles of structure determination

We turn now to a brief description of the determination of the structure of a crystal from its diffraction pattern. This provides a convenient opportunity to introduce an elementary account of the diffraction of neutrons by crystals in the course of which we shall emphasize the differences between the diffraction of X-rays and neutrons by crystals and point out the advantages and disadvantages of using one or another type of radiation to solve a particular problem. X-ray diffraction studies provide the primary means for the determination of crystal structures, as they have since the early days of the determination of the structures of such simple salts as NaCl and CsCl, soon after the discovery of the diffraction of X-rays by crystals. Structures of increasing complexity have become soluble as the techniques of X-ray structure determination have progressed until it is now possible to solve the crystal structures of the biologically very important proteins, the unit-cells of which contain several thousand atoms.

The techniques of crystal structure determination are outside the scope of this textbook. Here we are concerned simply with stating the essential problem and indicating, in general terms, the lines on which it may be soluble by X-ray methods with or without the assistance of complementary neutron diffraction studies. Electron diffraction, for reasons which will become apparent in chapter 10, is not generally useful for crystal structure determination, but is of immense value for the elucidation of the microstructures of crystals.

The first step in the determination of a crystal structure is the measurement of the unit-cell dimensions, the determination of the space group, and the determination of the atomic contents of the unit-cell. We have already described one method for the accurate determination of unit-cell dimensions using $\alpha_1\alpha_2$ splitting of high-angle reflexions on Weissenberg photographs in chapter 8 and we shall mention another method later in this chapter. In chapter 6 the determination of the diffraction symbol of a crystal from its X-ray diffraction pattern was discussed. However only 50 of the 230 space groups are uniquely determined by the diffraction symbol; for example, the diffraction symbol $mmmPban$ determines the space group uniquely as $Pban$, but the diffraction symbol $mmmPba-$ could be due to either of the space groups $Pba2$ or $Pbam$. Most of the remaining 180 space groups can, as in the example cited above, be uniquely determined if the point group as well as the diffraction symbol is known. We shall shortly turn our attention to point group determination.

For the determination of the atomic contents of the unit-cell the chemical composition of the compound, a determination of its density ρ, and an accurate determination of its unit-cell dimensions, from which the unit-cell volume $V = \mathbf{a} \cdot \mathbf{b} \wedge \mathbf{c}$ can be evaluated, are required. If the number of formula units in the unit-cell is Z and the weight of one formula unit in atomic mass units is M, then

$$\rho = \frac{ZM}{NV},$$

where $N = $ Avogadro's Number. So

$$Z = \frac{\rho NV}{M}$$

Point group determination

A complete description of a crystal structure has to include a statement of its space group. The statement of the space group gives a statement of the point group. If the space group is uniquely determined by the diffraction symbol, then no further experiments are required to determine the point group. But when the diffraction symbol is consistent with two or more possible space groups, then it becomes necessary to determine the point group, or at least to determine whether the point group is centrosymmetrical or non-centrosymmetrical.

If the substance is transparent and its crystals are not too small, optical examination (to be described in volume 2) will quickly show whether it is isotropic and therefore to be assigned to the cubic system; or uniaxial and therefore belonging to the trigonal, tetragonal or hexagonal systems; or biaxial and therefore triclinic, monoclinic or orthorhombic. Observations of the orientation of optical properties with respect to the morphology of the crystals may enable one tentatively (but no more than that) to assign the substance to a particular crystal system, other than the cubic system where the assignment is unambiguous. The next step is, usually, to make a single crystal X-ray diffraction study of the substance in order to determine its Laue symmetry and then, by observation of systematic absences on moving film photographs, to determine its diffraction symbol.

In general the restriction imposed by Friedel's Law prevents this sort of study going beyond the determination of the diffraction symbol. It is not usually possible to determine whether the substance is centrosymmetric or non-centrosymmetric. Some diffraction symbols do however lead to a unique determination of space group and thus of point group. For instance a substance with the diffraction symbol $mmmPbcn$ must be assigned to the point group mmm and to the space group $Pbcn$. But a substance whose diffraction symbol is determined as $mmmP.cn$ may have point group mmm or $2mm$ and space group $Pmcn$ or $P2_1cn$. In this second example the two possible space groups could be distinguished by determining whether or not the structure is centrosymmetric. If, in these circumstances, intensity data have been collected for a full structure analysis, then the most convenient way of determining whether the substance is centrosymmetric is by application of the $N(z)$ test, which we now describe in outline. Reflexions of similar Bragg angle are grouped together (for instance, the range of $\sin\theta$ from 0·3 to 0·5 might constitute one group, 0·4 to 0·6 the next, and so on). The assumption is then made that each atom in the structure scatters the same amplitude of X-radiation into every reflexion of a particular group. If the assumption is valid, it can be shown that the probability that the intensity of a given reflexion lies

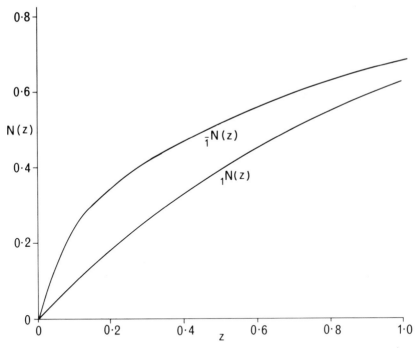

Fig 9.1 The $N(z)$ test. The theoretical curves for the centrosvmmetric case $_1N(z) = \text{erf}\sqrt{(\frac{1}{2}z)}$ and for the non-centrosymmetric case $_1N(z) = 1 - \exp(-z)$ are shown plotted against z.

between I and $I + \delta I$ is independent of the actual atomic positions but is dependent on the symmetry of the structure. We now define $N(z)$ as the percentage number of reflexions in a particular group with intensities less than or equal to $z\bar{I}$, where z is a fraction and $\bar{I}$ is the mean intensity of the group. The relationship between $N(z)$ and z has been calculated for centrosymmetric and non-centrosymmetric structures. Curves of $N(z)$ plotted against z are shown in Fig 9.1 for the two cases; it is noticeable that the two curves follow markedly different courses especially for z between zero and 0·4. The $N(z)$ test consists of plotting $N(z)$ averaged over all the selected groups of reflexions against z for the measured intensities of the unknown structure and comparing the resultant plot with the standard centro- and non-centrosymmetric curves; good agreement with either standard curve yields a clear determination of whether or not the structure is centrosymmetric.

When statistical methods are applied to an intensity data set in order to obtain information about symmetry or about the phases of diffracted X-ray beams it is generally preferable to work with a quantity which, unlike the observed intensity or the observed structure factor, is not dependent on the variation of atomic scattering factors with scattering angle. Such a quantity is the *normalized structure factor E*, which is defined as the ratio of the structure factor F to its root mean square value for the appropriate magnitude of $(\sin\theta)/\lambda$. Thus for a general reflexion hkl

$$E(hkl) = \frac{F(hkl)}{(\Sigma f_n^2)^{\frac{1}{2}}}, \tag{1}$$

where f_n is the scattering factor of the nth atom in the unit-cell at the $(\sin\theta)/\lambda$ value of the reflexion and the summation is over all the atoms in the unit-cell. Statistical analysis shows that the expected values of the mean of $|E|$ and of the mean of $|E|^2-1$ for centrosymmetrical and for non-centrosymmetrical crystals are:

	centrosymmetrical	non-centrosymmetrical
$\overline{|E|}$	0·798	0·886
$\overline{|E|^2-1}$	0·968	0·736

Although it is inherently dangerous to rely on just one or two numbers as the test for the presence of a centre of symmetry and generally safer to compare actual with theoretical distributions as one does in the $N(z)$ test, $\overline{|E|}$ and $\overline{|E|^2-1}$ are simply computed during the reduction of an intensity data set and will quite often provide a clear indication of whether or not the structure is centrosymmetric. The test will however sometimes fail; a structure in which the heavy atoms, the strong scatterers, are arranged in a centrosymmetric manner and the light atoms are disposed non-centrosymmetrically would be likely to yield intermediate values.

The same statistical tests can be applied to a zone of reflexions in order to determine whether there is a diad parallel to the zone axis. Suppose that a crystal has the diffraction symbol $2/mC$.. so that its possible space groups are C2, Cm, and C2/m. If a statistical test using all the reflexions indicates the presence of a centre of symmetry, then the space group is C2/m. If the absence of a centre of symmetry is indicated, application of the statistical test to the $h0l$ zone of reflexions will serve to distinguish between C2 and Cm because the $h0l$ zone corresponds to a projection down the y axis and, if there is a diad parallel to y, then the projection will be centrosymmetric. However tests on zones of reflexions do not invariably yield unambiguous results for the reason that there may well be overlapping atoms in projection so that the assumptions on which the statistical tests are based become invalid. In the course of a structure analysis it may be possible to detect, by methods which lie outside our scope here, the presence of non-translational symmetry elements and so, even in cases where statistical tests are inconclusive, obtain a unique determination of the space group.

We turn now to a property of X-ray diffraction which may be applied to point group determination in favourable circumstances; this is the property known as *anomalous scattering*. The X-radiation scattered by an atom is normally π out of phase relative to the incident beam. But if the incident X-rays are of a wavelength just short of the absorption edge of one of the constituent atomic species of the structure concerned, then the X-rays scattered by atoms of that element do not have a phase difference π and are said to be anomalously scattered. In general then it is necessary to express the atomic scattering factor f as $|f|\exp(i\delta)$, where δ is the phase relative to normal scattering. It is usual to write the atomic scatering factor as $f=f_0+\Delta f'+i\Delta f''$, where f_0 is the scattering factor of the atom for normal scattering of X-rays and $\Delta f'$ and $\Delta f''$ are the real and imaginary corrections for anomalous scattering. Both the corrections are dependent on λ, but only slightly on θ; that is because anomalous scattering is due to interaction of the more tightly bound electrons of the atom with the incident X-radiation so that the phase differences between the waves scattered by such a small volume will not be strongly dependent on the scattering angle. To a first approximation therefore $\Delta f'$ and $\Delta f''$ may be regarded as independent of $(\sin\theta)/\lambda$ for a given incident wavelength. It can be shown that, while $\Delta f'$ may be either positive of negative, $\Delta f''$ is necessarily positive. In consequence the phase of anomalously scattered X-radiation is always advanced relative to normally scattered radiation (Fig 9.2); the effective path of the anomalously scattered X-rays is thus less than it would be

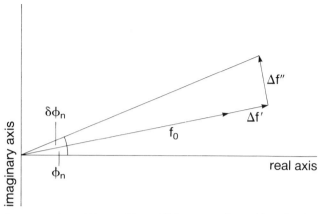

Fig 9.2 Anomalous scattering of X-rays. Since $\Delta f''$ is necessarily positive, the phase is always advanced relative to normally scattered radiation.

if the X-rays had been scattered normally, or, in other words, as if the anomalously scattering atoms were displaced away from the origin.

Such anomalous scattering leads to a breakdown of Friedel's Law, as may be seen by considering the 111 and $\overline{1}\overline{1}\overline{1}$ reflexions of blende (ZnS). It was shown in chapter 6 that although the blende structure is non-centrosymmetric, its 111 and $\overline{1}\overline{1}\overline{1}$ reflexions are of equal intensity (Fig 6.21). But when the incident X-radiation is $AuL\alpha_1$ which is scattered anomalously by Zn atoms, the diffraction pattern is modified in such a manner as though the Zn atoms were displaced from their actual positions so as to decrease the path travelled by the X-rays scattered by them. Thus for the 111 reflexion the zinc atoms behave as though they were displaced from their actual positions by a small distance in the AB direction and for the $\overline{1}\overline{1}\overline{1}$ reflexion as though they were displaced by an equal distance in the opposite direction BA (Fig 9.3(a)). Alternatively one might say that for the 111 reflexion the zinc atoms behave as though they were displaced away from the adjacent (111) plane of sulphur atoms, while for the $\overline{1}\overline{1}\overline{1}$ reflexion the zinc atoms appear to be displaced towards the adjacent plane of sulphur atoms. Formally one can write the structure factors as

$$F(111) = 4(f_{Zn} + f_S \cos 2\pi\tfrac{3}{4} + if_S \sin 2\pi\tfrac{3}{4})$$

$$= 4(f_{Zn} - if_S)$$

and $\qquad F(\overline{1}\overline{1}\overline{1}) = 4(f_{Zn} + if_S).$

Putting $f'_{Zn} = (f_0)_{Zn} + \Delta f'_{Zn}$ for the real part and $\Delta f''_{Zn}$ for the imaginary part of the scattering factor of Zn,

$$F(111) = 4[\, f'_{Zn} + i(\Delta f''_{Zn} - f_S)]$$

and $\qquad F(\overline{1}\overline{1}\overline{1}) = 4[f'_{Zn} + i(\Delta f''_{Zn} + f_S)].$

Thus the intensity of the $\overline{1}\overline{1}\overline{1}$ reflexion is greater than that of the 111 reflexion, as is apparent also from Fig 9.3(b), (c).

Such a breakdown of Friedel's Law may enable the point group of a substance to be determined. For instance the Laue group of blende is $m3m$, which embraces the crystal classes 432, $\overline{4}3m$, and $m3m$; the only one of the three classes in which (111) and $(\overline{1}\overline{1}\overline{1})$ are

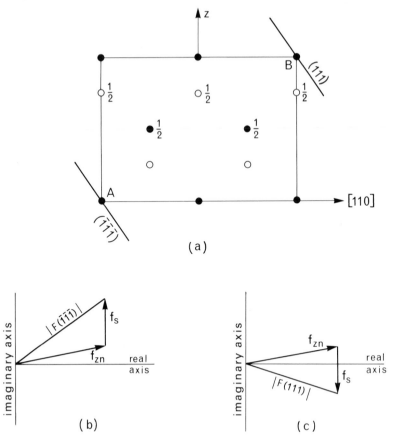

Fig 9.3 Anomalous scattering of AuLα_1 radiation by Zn atoms in blende (ZnS). (a) shows a projection of the blende structure on (1$\bar{1}$0) (cf. Fig 6.21); solid circles Zn, open circles S; the trace of the plane labelled (111) refers to the second plane out from the origin. (b) and (c) are respectively phase amplitude diagrams for the $\bar{1}\bar{1}\bar{1}$ and 111 reflexions; in each case the f_{Zn} vector makes an angle of $10\frac{1}{2}°$ with the real axis.

not symmetry related is $\bar{4}3m$. The observation of unequal 111 and $\bar{1}\bar{1}\bar{1}$ intensities with X-radiation of suitable wavelength thus determines the point group of blende as $\bar{4}3m$. Since such intensity differences are likely to be very small, it would be unwise to rely on the observation of the intensities of only one pair of reflexions for point group determination. The observation of several such inequalities is necessary for unambiguous point group determination; for instance in Laue group $m3m$ the observation of systematic inequality in pairs of hkl and $h\bar{l}k$ reflexions leads unambiguously to the determination of the point group of the substance as $\bar{4}3m$. The success of this approach depends in the first place on the availability of X-radiation of the appropriate wavelength and secondly on the observation of intensity differences which can clearly be attributed to the breakdown of Friedel's Law and cannot be due to some other cause such as the differential absorption of X-rays by a crystal of irregular shape.

While analysis of the X-ray diffraction pattern provides the most convenient and reliable means of determining the point group of a crystal when the objective is a structure determination, there are often circumstances where point group determination, or at least an answer to the question whether a crystalline substance is or is not centrosymmetric, is required when a crystal structure determination is not in prospect. Such circumstances often arise in the search for new materials with properties which depend on the absence of centrosymmetry. The need thus arises for a simple test for centrosymmetry based on an anisotropic physical property, that is to say a property which varies with direction in the crystal, the directional dependence of the property being controlled by the point group symmetry of the crystal.

The physical property which provides the most generally applicable and most sensitive test for non-centrosymmetry is the property of *second harmonic generation* (SHG). It is an optical property and so can only be utilized in transparent crystalline solids. We shall deal generally with the optical properties of anisotropic crystalline solids in volume 2; here we give a limited discussion of one specific optical property in the context of point group determination.

In accounting for the optical properties of crystals it is normally assumed that the electric polarization P of an atom is linearly proportional to the applied electric field E. In non-linear optics this is not so and the induced polarization is assumed to be given by the next approximation, which may be expressed as

$$P = a_1 E + a_2 E^2$$

where a_1 and a_2 are constants. The electric vector E associated with the incident light wave varies with time t as $E = E_0 \sin \omega t$, where ω is the angular frequency and thus E^2 can be expressed as

$$E^2 = E_0^2 \sin^2 \omega t$$
$$= \tfrac{1}{2} E_0^2 (1 - \cos 2\omega t)$$

so that the induced polarization P can be expressed as

$$P = \frac{a_2}{2} E_0^2 + a_1 E_0 \sin \omega t - \frac{a_2}{2} E_0^2 \cos 2\omega t \tag{2}$$

Thus the expression for the induced polarization contains a term which has twice the frequency ω of the incident wave. In consequence a wave with twice the frequency of the incident wave is generated by the crystal. The phenomenon of SHG is normally observed only when the light beam incident on the crystal is of very high intensity, such as a laser beam. If the observed intensity of the second harmonic beam is to approach its maximum possible value, the beams generated by different parts of the crystal must be in phase with each other. For this to be so the incident beam and the second harmonic beam, which are coherent at the point of origin of the latter, must travel with the same velocity in the direction of observation. If this condition is satisfied, then they are said to be phase-matched; if it is not satisfied, then destructive interference between the various second harmonic beams will occur.

For the induced polarization in the crystal to be anharmonic the crystal must lack a centre of symmetry, which would impose equivalence between opposite directions; further, as will be shown in volume 2, crystals in class 432 cannot exhibit SHG and crystals in classes 422 and 622 may produce only a second harmonic of anomalously low intensity. The electrons which interact most strongly with the electric field of the incident beam are the outer electrons of the most polarizable atoms, which in many mineral and inorganic structures will be oxygen atoms. SHG is thus particularly useful when the heavier atoms in the structure are arranged in a centrosymmetric manner and the light atoms have an acentric disposition; in such circumstances statistical analysis of single crystal X-ray intensity data often fails, as we have already indicated, to give an unambiguous answer. The experimental test for SHG should be exceptionally sensitive to small departures from centrosymmetry as even an ordinary laboratory photomultiplier is estimated to be capable of detecting just one photon of SHG produced from 10^{14} photons of incident laser light.

Dougherty and Kurtz (1976) have developed a sensitive second harmonic analyser to work with milligram amounts (typically 3–10 mg) of crystalline powder. The light source is a pulsed Nd^{3+}:glass laser which produces 300 μs long pulses of light of wavelength 10,600 Å. The second harmonic, with a wavelength of 5300 Å, is in the green region of the visible spectrum. The laser beam is focused to a spot of diameter 3 mm on the specimen. The powder specimen, prepared in the same way as for X-ray powder diffractometry, should be composed of crystallites with dimensions in the range 20–150 μm. The powder is sedimented on to a cover slip of diameter 15 mm. A few drops of a liquid of refractive index close to the mean refractive index of the crystalline specimen are added to reduce scattering of light and the whole is stirred to produce uniform particle distribution over an area of diameter about 8 mm. When the crystalline particles are platelets or needles it will be necessary to use, instead of a liquid, a viscous or solidifying medium in order to obtain random orientation of the particles. Since the light emitted by the specimen may arise from effects other than SHG, the detector system is designed to distinguish the more spectrally pure SHG from light produced by other effects. A crystalline quartz powder is used as a standard. Since the detection of SHG is so sensitive it is necessary to take precautions to ensure that the specimen is devoid of impurity and to make a null experiment using an empty cover slip or a powder of a centrosymmetrical substance such as NaCl.

Attenuation of the incident beam with filters has shown that with this apparatus a signal of strength 10^{-3} of the maximum obtainable from quartz can be recorded. Such a high level of sensitivity is thought to give a 99% probability of detecting non-centrosymmetry. Failure to observe SHG in a transparent crystal thus implies that the structure is very likely to be centric or to be in class 432.

The second property we consider is *piezoelectricity*. A crystal is said to be piezo-electric if it develops a dipole when subjected to an applied stress or, conversely, if it changes its shape when placed in an electric field. At equilibrium the applied stress will be centrosymmetric so that if the crystal is to develop charges of opposite sign at opposite ends of a line through its centre, it cannot have a centre of symmetry. Detailed analysis of the symmetry relations of the piezoelectric effect have shown that substances of all non-centrosymmetric crystal classes other than 432 may display piezoelectricity. Thus if a substance is shown to be piezoelectric, it must have a non-centrosymmetric point group. But the converse is not true because the magnitude of the piezoelectric effect may be below the limit of detection; failure to observe piezoelectricity in the crystal under examination does not necessarily imply that it belongs to one of the twelve point groups (the eleven centrosymmetric point groups and 432) which cannot display piezoelectricity.

The physical properties known as *pyroelectricity* (the development of an electric dipole when an unstressed crystal is uniformly heated or cooled) and *ferroelectricity* (the presence of a spontaneous electric dipole in a crystal) as well as piezoelectricity developed under hydrostatic pressure [1] are only observed when the symmetry of the crystal allows a resultant vector to occur. For instance in point group 2 the vector $U\mathbf{a} + V\mathbf{b} + W\mathbf{c}$ is related by the diad to the vector $-U\mathbf{a} + V\mathbf{b} - W\mathbf{c}$; the resultant of this pair of symmetry related vectors is $2V\mathbf{b}$ so that, if pyroelectricity is observed, the dipole will be parallel to $\mathbf{b}$. The symmetry of point group 2 allows a resultant vector along y and therefore y is said to be a *unique direction* in crystals of class 2. Clearly then a unique direction is simply a direction which is not repeated by the symmetry of the point group. Inspection of the chart of the 32 point groups (Fig 3.20) shows immediately that the rotation axes in the point groups 2, 2*mm*, 3, 3*m*, 4, 4*mm*, 6, and 6*mm* are unique directions, that all directions in point group 1 are unique, and that all directions parallel to the mirror plane in point group *m* are unique. Therefore if a crystal is found to be piezoelectric under hydrostatic stress or ferroelectric it must belong to one of these ten crystal classes which contain one or more unique directions and are known as the *polar classes*.[2] If piezoelectricity under hydrostatic stress or ferroelectricity is not observed in a crystal it is unsafe to conclude that the substance cannot belong to one of the ten polar classes; failure to observe either property may merely be due to its being too weak to be detected.

If a change in the temperature of a crystal causes an electric dipole to develop, the crystal is pyroelectric and in consequence must belong to one of the ten polar classes. But in practice it is difficult to be sure that the observed dipole moment is due to the pyroelectric effect rather than to a piezoelectric effect induced by the strains set up due to temperature gradients in the crystal during heating or cooling. For instance quartz, which belongs to class 32 and may therefore exhibit piezoelectricity under non-hydrostatic stress, commonly exhibits such 'false' pyroelectricity. In general it is wise to interpret the observation of pyroelectricity as indicating that the crystal belongs to one of the twenty crystal classes in which piezoelectricity may be observable, but does not necessarily belong to one of the ten polar classes in which true pyroelectricity may be observable.

Another physical property which is dependent on point group symmetry is *optical activity*.[3] A crystal is optically active if it rotates the plane of polarization of a beam of plane polarized light passing in certain directions through the crystal. For example crystals of quartz, class 32, are observed to rotate the plane of polarization in a clockwise sense in some crystals and anticlockwise in others. It can be shown that optical activity is restricted to crystals of those classes which contain no inversion axis of symmetry (including the mirror plane). These eleven *enantiomorphous* classes are listed in Table 9.1. In addition it can be shown that crystals of the classes *m*, *mm*, $\bar{4}$, and $\bar{4}2m$ may theoretically exhibit optical activity. However most crystals that are theoretically capable of exhibiting optical activity do not do so to any marked extent and moreover the observation of the property is difficult except for light travelling parallel to the principal axes of uniaxial crystals. Optical activity is thus not a property which can be utilized generally for point group determination.

---

[1] Piezoelectricity may be developed by hydrostatic, compressive, or torsional stress systems. We are concerned here only with piezoelectricity developed under hydrostatic stress, which imposes the most stringent symmetry constraints of the three types of stress system.

[2] The nomenclature is potentially confusing. *Polar directions* are directions whose opposite ends are not related by symmetry. Any non-centrosymmetric crystal has polar directions, but one polar direction may be related to other polar directions by symmetry. Thus while all unique directions are necessarily polar, polar directions are not necessarily unique.

[3] This property is discussed in volume 2.

The experimental methods of determination of the properties we have been concerned with in preceding paragraphs, other than SHG, are thoroughly discussed by Wooster and Breton (1970).

Table 9.1
Various groupings of non-centrosymmetric point groups

Crystal System	Piezoelectric and SHG classes	Polar classes	Enantiomorphous classes	Optically active classes
Triclinic	1	1	1	1
Monoclinic	2, m	2, m	2	2, m
Orthorhombic	222, $mm2$	$mm2$	222	222, $mm2$
Trigonal	3, $3m$, 32	3, $3m$	3, 32	3, 32
Tetragonal	4, $\bar{4}$, 422, $4mm$, $\bar{4}2m$	4, $4mm$	4, 422	4, $\bar{4}$, 422, $\bar{4}2m$
Hexagonal	6, $\bar{6}$, 622, $6mm$, $\bar{6}m2$	6, $6mm$	6, 622	6, 622
Cubic	23, $\bar{4}3m$	—	23, 432	23, 432

We conclude this brief survey of point group determination with some comments on the relevance of observations of crystal morphology. Before the advent of X-ray diffraction, morphological studies provided the principal means of point group determination, but such studies have for long been only of pedagogical and historical interest. Morphological studies are of course applicable only to substances which form well developed crystals. A well developed crystal of the substance under consideration is measured goniometrically and its faces are plotted on a stereogram. The stereogram is examined and the classes with which it is consistent are recorded. Other crystals of different habit, if available, are measured and the new faces added to the stereogram until as many faces as can be observed in crystals of the substance have been measured and plotted. In nineteen of the thirty-two crystal classes the general form is characteristic of the class; but in the remaining thirteen classes the general form is a special form[4] in one or more other classes so that, in these classes, the observation of one general form is insufficient to determine the point group symmetry. For example in the trigonal system the general form of class 32 is unique to that class; but the general form of class 3, a rhombohedron, is a special form in classes 32 and $\bar{3}m$ so that before a substance could be confidently assigned to class 3 crystals exhibiting other forms would have to be examined. Another commonly encountered difficulty is that the special forms of some point groups have higher symmetry than the point group: for instance the cube and the rhombic dodecahedron, which have symmetry $m3m$, are special forms in all the cubic point groups. Another potential source of error is that in non-holosymmetric point groups two forms of the same type may be developed so as to look as though they are a single form of higher symmetry. For example in classes 23 and $\bar{4}3m$ if the tetrahedra $\{111\}$ and $\{11\bar{1}\}$ are both present, they

---

[4] In this morphological context we define a *special form* as a form which bears some specialized relationship to the symmetry elements of the crystal class. The relationship may be that the normals to the faces of the form are parallel or perpendicular to a symmetry element (either an axis or a plane) or equally inclined to two symmetry axes. Special forms, defined in this way, are readily distinguishable by their appearance from the general form of the same crystal class. For example in class $\bar{4}$ the general form $\{hkl\}$ is a tetragonal sphenoid (that is a sort of tetrahedron elongated or shortened in the direction through the mid-points of one pair of opposite edges) while the special forms $\{hk0\}$ are tetragonal prisms. Likewise in class 23 the general form $\{hkl\}$ looks quite different from the special form $\{110\}$ although both have twelve faces; the former is a tetrahedral pentagonal dodecahedron and the latter is the familiar rhombic dodecahedron. In terms of the different definition of special form used in chapter 3 $\{hk0\}$ in class $\bar{4}$ and $\{110\}$ in class 23 would be general forms.

may present the appearance of an octahedron {111} which is a special form in all the other cubic classes. Another potential source of error arises from the possibility that crystals may be twinned in such a manner that the twinning is not readily discernible by morphological examination; such crystals will appear to be of higher symmetry than their true point group symmetry. With so many chances of going inadvertently wrong it is fair to say that it is quite remarkable that the early crystallographers succeeded in assigning so many substances to their correct point group.

The valuable, but all too often indecisive, information about the probable point group of a substance derived from morphological observations can in some cases be supplemented—and may then become decisive—by the study of *etch figures*. Etch figures are produced on the natural faces of a crystal by the brief application of an appropriate solvent. In the early stages of the interaction between the solvent and the crystal small pits appear randomly disposed on the crystal faces. Under correct experimental conditions—and it may require some experimental trial and error to achieve this—the etch pits have demonstrably plane faces indicative of symmetry control of rate of solution. The shape and orientation of the etch pits on a particular crystal face corresponds to the projection of the point group symmetry of the crystal structure on the plane of the face concerned and so must conform to one of the ten two-dimensional point groups. Caution must however be exercised in the interpretation of etch figures (that is, the shape and orientation of etch pits) because the rate of solution in directions that are not symmetry related may be fortuitously similar so that etch figures of apparently higher symmetry than is consistent with the point group symmetry of the crystal may be observed. It is therefore safe only to derive from the study of etch figures a statement of the maximum possible point group symmetry of the crystal. A thorough account of the utilization of etch figure studies in point group determination is given in Buerger (1956).

The Abbe theory of image formation

To illustrate the principle underlying the methods of crystal structure determination we turn to the familiar phenomenon of optical diffraction and, in particular, consider how a parallel beam of monochromatic light forms an image of a one-dimensional grating when a lens is inserted in the path of the light waves emergent from the grating. In Fig 9.4 a monochromatic light beam is incident on a grating G of transparent lines; the transmitted light waves then pass through a lens L so placed that a real image of the grating is formed in the plane I parallel to the plane of the grating G. The Abbe theory of image formation assumes that the formation of the image I takes place in two stages. As the incident plane wave impinges on the grating it is diffracted so as to give rise to sets of parallel beams corresponding to the zero, first, second, etc orders of diffraction. The lens L focuses these sets of diffracted beams in the plane D; the Fraunhofer diffraction pattern [5] of the grating is formed in the plane D, which is effectively at an infinite distance from the grating. The waves travel on in such a manner that the diffracted beams produced by a particular grating element cross in the plane I so that a real image of the diffraction grating is produced in the plane I. The image of a two-dimensional grating is produced in a precisely analogous manner. But for a three-dimensional grating the complete diffraction

---

[5] In Fraunhofer diffraction both the source and the observed diffraction pattern are effectively infinitely distant from the object. In Fresnel diffraction, in contrast, either the source, or the observed diffraction pattern, or both are not effectively at infinity.

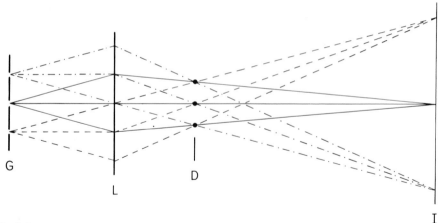

Fig 9.4   Abbe's theory of image formation. A parallel beam of monochromatic light is incident on the grating G. The diffracted rays are focused by the lens L to form a diffraction pattern in the plane D and a real image in the plane I.

pattern can only be recorded by allowing the angle of incidence of the light beam on the grating to be varied; it is thus impossible to focus the whole diffraction pattern simultaneously to form an image of the structure. With an optical lens only an image of a projection of the three-dimensional structure is obtainable.

The Abbe theory of image formation provides a means of calculating the form of the image produced by a grating of known size and shape. The direction, amplitude and phase of each diffracted beam can be calculated so as to give a complete description of the diffraction pattern formed in the plane D. The diffracted beams may then be imagined to be recombined so as to form the real image in the plane I and this can be calculated too. For a three-dimensional grating it is possible to calculate its complete diffraction pattern and to construct its image even through the image cannot be formed by an incident optical beam and a simple lens system.

Principles of crystal structure determination

If an experimental arrangement analogous to that used in optics could be used for X-rays it would then be possible, at least in principle, to form a real two-dimensional image of a crystal structure. But this cannot be done in practice because the refractive index of most solids for X-radiation differs so little from unity, by about 1 in 10^6, that it is impossible to construct a lens for X-rays. Without a lens there is no means of bending the diffracted X-rays to form an image of the crystal structure; but it remains possible to observe the diffraction pattern at a distance that is effectively infinite in comparison with the size of the grating elements of the crystal structure and from this Fraunhofer diffraction pattern it should be possible to calculate the image of the crystal structure.

We have already obtained the expression (equation (16) of chapter 6) necessary for the calculation of the Fraunhofer diffraction pattern of a crystal: it is the familiar structure factor expression

$$F(hkl) = \sum_{1}^{N} f_n \exp 2\pi i(hx_n + ky_n + lz_n), \tag{3}$$

which is one form of the expression for the Fourier transform[6] of an infinite single crystal. The image of the crystal may be calculated by taking the Fourier transform of its Fraunhofer diffraction pattern. The electron density $\rho(xyz)$ at the point with fractional coordinates x, y, z in the unit-cell is given (in electrons per $Å^3$) by the inverse Fourier transform as

$$\rho(xyz) = \frac{1}{V}\sum_h \sum_k \sum_l F(hkl) \exp\{-2\pi i(hx_n + ky_n + lz_n)\}, \tag{4}$$

where V is the volume of the unit-cell and the summations are taken from $-\infty$ to ∞. So it is possible from such a Fourier synthesis to calculate the electron density distribution over the unit-cell and to locate the positions of the atoms at the electron density maxima provided we have a set of accurate values of the structure factors $F(hkl)$. But we are thwarted by the practical difficulty that the structure factor $F(hkl) = |F(hkl)| \exp i\phi(hkl)$ cannot, except in certain special circumstances, be determined experimentally because there is no experimental means of determining the phase $\phi(hkl)$ of a diffracted X-ray beam. All that can be achieved directly by experiment is the orientation and structure amplitude of each diffracted X-ray beam. So the description of the diffraction pattern lacks information about phase and in consequence the determination of crystal structure cannot be achieved by straightforward calculation from the measurement of the diffraction pattern.

The basic problem of X-ray crystal structure analysis is the completion of the description of the diffraction pattern by finding some way of evaluating the phase of every observed reflexion. In a few special cases symmetry considerations limit the number of possible structures to so small a number that it becomes feasible to compare the observed structure amplitudes with those calculated for each of the possible structures and so to determine which is the correct structure. For example a cubic structure of Laue group $m3m$ with an F-lattice and four molecules of AX per unit-cell may have either the NaCl-type structure or the blende-type structure. From Table 9.2, in which the intensities of a few reflexions calculated for each of these possible structures are shown, it is apparent that comparison with the measured intensities of reflexions produced by the substance under consideration will rapidly show which of the two possible structures the substance actually has. In a simple case such as this a powder diffraction pattern would be adequate, care being taken to assign the appropriate multiplicity to each powder line and to take into account the several physical and geometrical factors which relate the measured intensity of any powder line to the intensity $I(hkl)$ of the corresponding diffracted beam produced by one unit-cell.

The problem of determining the phase of the structure factor $F(hkl)$ is central to the determination of crystal structures. We have already shown (equation (17) of chapter 6) that for centrosymmetric structures the phase of $F(hkl)$ can only have the values 0 or π, which is equivalent to saying that $F(hkl)$ is real and may be either positive or negative. For non-centrosymmetric structures however the phase $\phi(hkl)$ may have any value between the limits 0 and 2π.

At an early stage in the determination of a crystal structure it is necessary to deduce

---

[6] A function $F(S)$ is known as the *Fourier transform* of $f(X)$ if the two functions are related by $F(S) = \int_{-\infty}^{\infty} f(X) \exp(2\pi iSX)\, dX$. The reciprocal relationship $f(X) = \int_{-\infty}^{\infty} F(S) \exp(-2\pi iSX)\, dS$ necessarily follows. The Fourier transform in the context of X-ray diffraction is explored in many textbooks, but especially in Cowley (1981).

Table 9.2

Intensities of reflexions for AX structures of NaCl-type and blende-type.

Both structure types have cubic F-lattices with one formula unit per lattice point. The coordinates of the atoms associated with the lattice point at the origin are A 0, 0, 0; B 0, 0, $\frac{1}{2}$ for the NaCl structure and A 0, 0, 0; B $\frac{1}{4}, \frac{1}{4}, \frac{1}{4}$ for the blende (ZnS) structure. Intensities are given by

$$I(hkl) = 16(f_A + (-1)^l f_X)^2 \qquad \text{for NaCl}$$

and

$$I(hkl) = 16 \left(\left[f_A + f_X \cos 2\pi \frac{h+k+l}{4} \right]^2 + f_X^2 \sin^2 2\pi \frac{h+k+l}{4} \right) \quad \text{for ZnS}$$

Reflexion	$I(hkl)$ for NaCl		$I(hkl)$ for blende	
111	$16(f_A - f_X)^2$	w	$16(f_A^2 + f_X^2)$	m
200	$16(f_A + f_X)^2$	s	$16(f_A - f_X)^2$	w
220	$16(f_A + f_X)^2$	s	$16(f_A + f_X)^2$	s
311	$16(f_A - f_X)^2$	w	$16(f_A^2 + f_X^2)$	m
222	$16(f_A + f_X)^2$	s	$16(f_A - f_X)^2$	w

Relative intensities: s = strong, m = medium, w = weak

by any one of a variety of methods, to one of which we shall give some consideration later, a probable structure, which is termed the *trial structure*, and then to calculate structure factors $F(hkl)$ for the trial structure. If the trial structure is substantially correct, there will be quite good agreement between the structure amplitudes calculated from the trial structure and the measured structure amplitudes. The quality of the agreement is measured by the *reliability index*, or *R factor* defined as

$$R = \frac{\sum ||F_o(hkl)| - |F_c(hkl)||}{\sum |F_o(hkl)|} \tag{5}$$

where $|F_o|$ and $|F_c|$ are the observed and calculated structure amplitudes and the summations are over all hkl. Estimated values of R for a completely random trial structure are 83% in the case of a centrosymmetric structure and 59% for a non-centrosymmetric structure. For experimental data of high quality and an accurately determined crystal structure R would be expected to be no greater than 5% and there would be no significant discrepancies between $|F_o(hkl)|$ and $|F_c(hkl)|$ for any observed reflexion.

In calculating the structure factors for a trial structure it is essential to use the appropriate atomic scattering factors. The tabulated atomic scattering factors are calculated for atoms at rest, but in a crystal at room temperature the atoms will be vibrating about their mean positions with a frequency of the order of only 10^{-5} times that of the X-radiation. So as the X-ray beam passes through the crystal it will be diffracted by crystallographically equivalent atoms displaced from their mean positions in an irregular manner (it is assumed that the atomic vibrations are not correlated); the diffracted beams thus arise from the time-average of the positions of the vibrating atoms. The electron density of each atom is in effect smeared out over a volume which depends on the amplitude of thermal vibration. There will thus be a greater likelihood of destructive interference between the X-rays scattered from different volume elements of the atom than if the atom were at rest (Fig 6.11) and in consequence the atomic scattering factor will fall off more quickly with increasing $(\sin \theta)/\lambda$ than it does for a stationary atom. The effect of thermal vibration may be allowed for by writing for the atomic scattering factor

$$f = f_o \exp\left(-B \frac{\sin^2 \theta}{\lambda^2}\right) \tag{6}$$

where f is the atomic scattering factor at the temperature at which the intensity measurements were made, f_o is the atomic scattering factor calculated for the atom at rest, and B is known as the temperature factor. B is related to the thermal vibration of the atom by $B = 8\pi^2 \overline{u^2}$, where $\overline{u^2}$ is the mean square displacement of the atom from its equilibrium rest position. Since the temperature factor B depends on the amplitude of thermal vibration of the atom, it will in general vary from one atom to another in the structure even when they are of the same species, but not of course when their positions are symmetry related. Until quite a late stage in a structure determination it is adequate, and indeed wise, to assume that the thermal vibrations of the atoms are isotropic, in other words that the time-average of the electron density is spherical. When the structure determination has reached a stage at which the R factor is already quite small, it becomes necessary, in order to improve the fit between the observed and calculated structure amplitudes, to take account of the anisotropy of atomic thermal vibrations, that is that the time-average of the electron density at an atomic site is ellipsoidal rather than spherical. That it is necessary to lift the restriction of isotropy may be illustrated by considering the case of an atom which is strongly bonded only to one other atom: it is energetically more favourable for the atom to vibrate normal to the bond than along the bond since small changes in the bond angles at the other atom require the expenditure of less energy than does the changing of the bond length.

Once a viable trial structure has been obtained its parameters, that is the coordinates and temperature factors of all the atoms in the unit-cell, are adjusted to give the best possible fit between the observed and calculated structure amplitudes. In addition it is usually necessary to refine the scale factor which has to be applied to the observed structure amplitudes in order to place them on an absolute scale, the intensities of X-ray reflexions being measured usually on an arbitrary scale. If the coordinates of the atoms in the trial structure are substantially correct, least-squares methods may be used to refine the structure. The function which is minimized in the least-squares analysis is either

$$R_1 = \sum w(|sF_o(hkl)| - |F_c(hkl)|)^2,$$

or

$$R_2 = \sum w'(|sF_o(hkl)|^2 - |F_c(hkl)|^2)^2$$

where the summation is over all the measured reflexions, s is the scale factor and individual reflexions are weighted by the factor w, or w', representing an estimate of the relative accuracy of that particular structure amplitude. Since the relationship between the parameters to be determined and the measured structure amplitudes is non-linear, the least-squares refinement cannot lead to a direct determination of the parameters; each cycle of refinement calculates changes in the values of the parameters to produce better agreement between the observed and calculated structure amplitudes. Least-squares refinement is an automatic procedure which minimizes R_1, or R_2, by adjustment of the parameters of the trial structure.

If the trial structure contains a serious error, such as a grossly misplaced atom, then least-squares methods will not yield the correct position of that atom; least-squares refinement is effective only when the trial structure is substantially correct. It is however always possible at any stage of the refinement to compute an *electron density map*. If the coefficients $F(hkl)$ used in equation (4) are the calculated structure factors, the electron density map will be an image of the trial structure, modified because only structure factors corresponding to values of $(\sin \theta)/\lambda$ less than a certain value will have been used in the computation. The limiting values of $(\sin \theta)/\lambda$ are about 0.6 Å^{-1} and 1.3 Å^{-1} for data collected with $CuK\alpha$ and $MoK\alpha$ respectively. For a perfect image to be obtained the summation of equation (4) would have to be performed over the whole range of h, k, and l from $-\infty$ to ∞. If, alternatively, the coefficients used in equation (4) are the observed structure amplitudes combined with the phases calculated from the trial structure (which may be assumed to be approximately correct), then the electron density map will be an image of a structure which should be a closer approximation to the real structure than is the trial structure. Here too the quality of the image will be limited by the experimental restriction on the number of reflexions which can be used in the summation. If the structure is centrosymmetric, then $F(hkl) = \pm | F(hkl)|$ and it can be assumed that, unless $F(hkl)$ is close to zero, the sign of the observed structure factor is the same as that of the calculated structure factor; the computed electron density map should, in this case, be quite an accurate image of the structure. However, errors in the intensity measurements coupled with the inevitable curtailment of the number of terms in the summation have the consequence that, even if all the observed structure factors have been correctly signed, the calculated electron density map can never be a precise image of the electron density distribution in the unit-cell.

A particularly useful means of indicating in what respects a trial structure differs from the real structure is provided by the *difference synthesis*. The function $\Delta\rho(xyz)$ is defined by

$$\Delta\rho(xyz) = \frac{1}{V} \sum_h \sum_k \sum_l (| F_o(hkl)| - | F_c(hkl)|)$$

$$\times \exp i\phi_c(hkl) \exp \{-2\pi i(hx_n + ky_n + lz_n)\} \tag{7}$$

If the phases of the observed structure factors do not differ greatly from those calculated from the trial structure, then $\Delta\rho(xyz)$ will show how the trial structure differs from the real structure. Errors in atomic coordinates, in temperature factors, and in the scale factor can be estimated from a difference synthesis. Moreover a difference synthesis should show if an atom has been misplaced in the trial structure: there will be a negative peak at or near the coordinates given to the misplaced atom and a positive peak with coordinates that do not correspond to any atom in the trial structure may indicate the correct position of the misplaced atom. Difference syntheses can be used to refine a trial structure, but the procedure is less automatic and so more time-consuming than least-squares refinement. It is helpful to calculate a difference synthesis during refinement when least-squares methods are failing to achieve steady improvement of the R factor. The computation of a difference synthesis at the end of a least-squares refinement provides a valuable test of the accuracy of the refined structure: if the structure is effectively correct, then $\Delta\rho$ should not have a significant value at any point.

The deduction of the first trial structure is the stage in structure determination which makes the most demands on the skill, ingenuity, and experience of the investigator. That was universally so in the years before the advent of large, fast computers and is still so where inorganic and mineral structures and macromolecular organic structures are the subject of investigation. It is however now possible to make effective use of relationships between observed structure amplitudes to deduce the probable phases of the larger observed structure amplitudes. Such relationships, which are the basis of the so-called *direct methods* of structure determination arise from the premises that the electron density at any point in a unit-cell cannot be negative and that the electron density distribution over the unit-cell must correspond to an arrangement of atoms in the unit-cell. The electron density distribution over the unit-cell must thus comprise discrete and approximately spherical peaks, the number and relative weights of which correspond to the known atomic contents of the unit-cell. If the atoms are all identical and spherically symmetrical, then the structure factors are related by exact equations; but if the atoms are diverse and perhaps not all spherically symmetrical, then probability relationships, rather than exact equations, apply and can be used to deduce the probable phases of the structure factors corresponding to certain strong reflexions.

In centrosymmetric crystals, where the phase of a structure factor can only be 0 or π so that phase determination reduces to sign determination, it can be shown that if $|F(hkl)|$, $|F(h'k'l')|$, and $|F(h-h', k-k', l-l')|$ are all large, then

$$s(hkl)\, s(h'k'l')\, s(h-h', k-k', l-l') \approx +1 \tag{8}$$

where $s(hkl)$ represents the sign of $F(hkl)$ and $\approx$ means 'is probably equal to'. This probability is known as the *triple product sign relationship*. It can be simply illustrated by consideration of the 100, 101, and 001 structure factors of a primitive monoclinic crystal in point group $2/m$. For such a centrosymmetric crystal $F(hkl) = \sum f_n \cos 2\pi(hx_n + ky_n + lz_n)$. Therefore for $F(hkl)$ to be large and positive the cosine terms in the summation must in general be large and positive, which implies that the atoms must be close to *hkl* lattice planes (Fig 9.5). If $F(hkl)$ has a large negative magnitude, then the $\cos 2\pi(hx_n + ky_n + lz_n)$ terms must in general be close to -1,

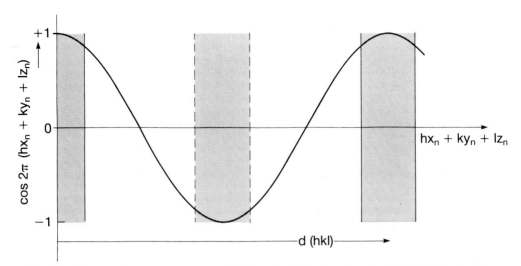

Fig 9.5 The smooth curve represents $\cos 2\pi(hx_n + ky_n + lz_n)$ plotted against $hx_n + ky_n + lz_n$. For the cosine to be large and positive the atom at x_n, y_n, z_n must be close to *hkl* lattice planes, that is within the shaded areas at the left and right of the figure. For the cosine to be large and negative the atom must lie close to an odd multiple of $\frac{1}{2}d(hkl)$, that is within the central shaded area.

which implies atoms situated close to the planes which interleave the (hkl) set at odd multiples of $\frac{1}{2}d(hkl)$. In this example the (100), (101), and (001) lattice planes are all parallel to the y axis and so are normal to (010). In projection on (010) the maximum value of $\cos 2\pi(hx_n + ky_n + lz_n)$ will correspond to the traces of $(h0l)$ lattice planes and its minimum value will correspond to the interleaving planes at odd multiples of $\frac{1}{2}d(h0l)$ as shown in Fig 9.6. The particular case of the (100), (101), and (001) lattice planes of our example is illustrated in Fig 9.7. If the 100, 101, and 001 structure amplitudes are large, then the atoms must be situated in the unit-cell close to positions where $\cos 2\pi(hx_n + ky_n + lz_n)$ has either its maximum magnitude of $+1$ or its minimum magnitude of -1 for each reflexion. The atoms must thus be close to the points A, B, C, or D so that either all three signs are positive or one sign is positive and two signs are negative: in either case $s(101)\, s(100)\, s(001) = 1$.

The unit-cell of a centrosymmetric crystal structure with a primitive lattice has eight centres of symmetry, at 000, $\frac{1}{2}00$, $0\frac{1}{2}0$, $00\frac{1}{2}$, $0\frac{1}{2}\frac{1}{2}$, $\frac{1}{2}0\frac{1}{2}$, $\frac{1}{2}\frac{1}{2}0$, and $\frac{1}{2}\frac{1}{2}\frac{1}{2}$, and in the least symmetrical case, $P\bar{1}$, the atomic environment of each of these centres of symmetry will be different; in primitive centrosymmetric space groups of higher symmetry some or all of the eight centres of symmetry will have different atomic environments. Since the phase of a diffracted beam is defined relative to a zero taken as the phase of the wave scattered by an atom at the origin of the unit-cell, the possibility of choosing the origin at as many as eight non-equivalent centres of symmetry leads to the possibility of producing as many as eight different sets of signs for the structure factors of the crystal. Suppose for instance that the origin was taken at A in Fig 9.7 and that the atoms in projection down the y axis were centred about D; then the signs of the 100 and 001 reflexions would be negative while the sign of the 101 reflexion would be positive. If however the origin of the unit-cell were to be taken at D, the signs of the 100, 001, and 101 reflexions would all be positive.

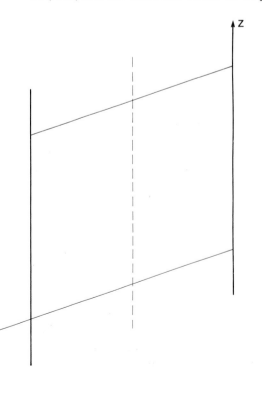

Fig 9.6 The maximum value of $\cos 2\pi hx_n$ corresponds to traces of $(h00)$ lattice planes, shown as bold lines in this projection on (010). The minimum value of $\cos 2\pi hx_n$ corresponds to interleaving planes at odd multiples of $\frac{1}{2}d(h00)$, the first of which is shown as a broken line.

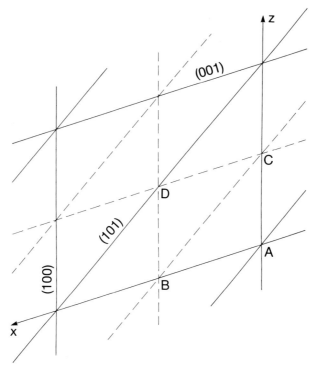

Fig. 9.7 The (100), (101), and (001) lattice planes are shown as solid lines and the interleaving planes at odd multiples of the appropriate $\frac{1}{2}d$ are shown as broken lines. If the three structure amplitudes are all large, atoms must be close to the points A, B, C or D. Either all three structure factors will be positive, or one will be positive and the other two negative; in either case $s(101)\, s(100)\, s(001) = 1$.

The origin of the unit-cell is effectively fixed on one of the eight centres of symmetry by arbitrarily giving signs to the structure factors for three appropriately chosen reflexions. Having made these arbitrary assignments, the three-dimensional intensity data set can be explored for sets of three reflexions to which the triple product sign relationship can be applied to determine the signs of the structure factors of more reflexions. In this manner it is usually possible to determine the signs of the structure factors of all reflexions with normalized structure amplitudes greater than 1.5. For some reflexions it may not be possible to determine the sign as $+$ or $-$, but only to show that certain reflexions have the same sign as, and other reflexions have the opposite sign to that of a particular reflexion, which may be given a *symbolic sign a*; the signs of all these reflexions will then be determined as a or $-a$, where a may be $+$ or $-$. One may expect to require quite a small number of such symbolic signs, a, b, c, d, ..., so that there will not be a large number of permutations of the possible values of the symbolic signs. Calculation of electron density maps, each of which uses all the fully signed reflexions and one permutation of the symbolic signs, enables the symbolic signs to be correctly determined because usually one of the maps shows an array of electron density peaks which can be correlated with the known atoms in the unit-cell so arranged as to have bond lengths of acceptable magnitudes. This array may then be taken as the trial structure for refinement by the methods described earlier. Should the refinement be unsuccessful it will be necessary to scrutinize the electron density maps again and to select a new trial structure.

In the case of non-centrosymmetric structures we are concerned with the phase $\phi(hkl)$ rather than with just the sign $s(hkl)$ of each structure factor. Here the triple product sign relationship is replaced by the expression

$$\phi(hkl) \approx \phi(h'k'l') + \phi(h-h', k-k', l-l') \qquad (9)$$

Similar methods to those already discussed for the centrosymmetric case are applied iteratively to determine the phases of a sufficiency of reflexions to yield an interpretable electron density map.

The methods of structure determination outlined in the preceding pages depend on the use of fast, efficient computers and on the various program packages, such as SHELX-76, which have been developed to perform the many lengthy calculations required.

Intensity data collection

A prerequisite for the determination of a crystal structure is the collection of a set of structure amplitudes $|F(hkl)|$ for as many independent reflexions (that is, reflexions which are not related by the space group symmetry) as possible together with an assessment of the accuracy of each measurement. This is achieved by measurement of the intensities of as many reflexions as possible from a small single crystal (usually $\not> 0.3$ mm in its greatest dimension) using either photographic methods or an automatic diffractometer. The intensity data collection—the raw data—is then processed by computer to give a set of structure amplitudes.

For photographic intensity data collection from inorganic, mineral, and small-molecule organic crystals the Weissenberg camera, described in chapter 8, is most commonly used. For photographic data collection from macromolecular organic crystals, such as proteins, special methods (which we shall not describe) are necessary because such crystals disintegrate rapidly on exposure to an intense X-ray beam so that it is imperative to collect as much data as possible in the short time available. In contrast, inorganic and mineral crystals are usually quite stable to irradiation by X-rays (there are some exceptions) so that in general there is no time limit imposed by crystal decay on the collection of intensity data.

Measurement of the optical density of the photographic record of a reflexion is performed by a microdensitometer. The optical density of the spot on the film is proportional to the intensity of the diffracted X-ray beam below a limiting level of intensity; it is therefore necessary to submit for microdensitometry a set of films such that every reflexion is recorded within the range for which the density is proportional to the intensity on at least one film. This is achieved by loading a pack of several—usually five—films into the Weissenberg cassette when collecting the intensity data for each reciprocal lattice layer. The weaker reflexions will be recorded within the measurable range on the innermost film of the pack; the stronger reflexions will yield spots of decreasing blackness on successive films from the innermost to the outermost so that, if the exposure time has been judiciously chosen they will be within the measurable range on a least one, but usually two or more, films of the pack. The computer-controlled microdensitometer measures the optical density of each reflexion on each film of the pack for each reciprocal lattice layer of each of the crystallographic axes about which data has been collected—it is usual to collect about two axes—and correlates the measurements from the different films to give a set of intensity data on a common arbitrary scale.

Whereas Weissenberg data collection requires the crystal to be precisely set with the first, and subsequently the second, selected zone-axis parallel to the camera axis and then proceeds layer by layer, the cassette having to be reloaded and the camera adjusted between successive layers, the other principal mode of data collection, automatic single-crystal diffractometry, involves the successive measurement of reflexions, one by one, over the whole, or some selected part, of accessible reciprocal space. The advantage of this latter, currently more common, method of intensity data collection is that once the parameters of the data collection have been entered into the controlling computer, the collection proceeds continuously and requires little or no human intervention. However it is always wise to take the preliminary precaution of recording some part of the diffraction pattern photographically before transferring the crystal to the diffractometer because it is very much easier both to determine whether the reflexions are well shaped (e.g., do not have tails, are not split) and to detect pseudo-symmetry by inspection of a photographic film than by examination of the numerical output from a diffractometer.

A diffractometer is an instrument for measuring the intensity of a diffracted beam by counting the number of X-ray photons arriving in a given time at an appropriately placed detector. The ionization spectrometer used by Bragg for collecting data for the early crystal structure determinations in 1913 was a rudimentary diffractometer, the use of which required great skill and was very slow. Photographic methods superseded the ionization spectrometer for the next half century and it was not until the 1960s that the development of stabilized X-ray tubes, the improvement of radiation detectors, and the advent of computer control made the single crystal diffractometer a viable and indeed advantageous instrument.

Most contemporary single-crystal diffractometers are of the four-circle type and are completely computer controlled. The essential geometry of a four-circle diffractometer is shown schematically in Fig 9.8(a). The incident X-ray beam and the diffracted beam when orientated for measurement lie in the horizontal plane which contains the crystal. The crystal can be rotated about three axes and the detector about a fourth axis; all four axes intersect at the crystal. The crystal is glued to a glass fibre, which is mounted on a head attached to a spindle. The head is provided with centring slides and a height adjustment so that the crystal can be precisely positioned at the intersection of the three axes about which it can be rotated. For four-circle single-crystal diffractometry in contrast to single-crystal diffraction photography it is desirable that the crystal should be randomly oriented and that is why the head is not provided with arcs. The spindle, on which the head is mounted, can be rotated about its axis, which is designated the ϕ axis, by the controlling computer. The spindle assembly, including the crystal, can be rotated about a perpendicular axis, designated the χ axis, by rotation of ring B inside the stationary ring A (Fig 9.8(a)). This χ-ring together with the spindle assembly and the crystal can be rotated about a vertical axis, designated the ω axis. The purpose of the three rotation axes, ϕ, χ and ω, is to enable the normal to a selected reflecting plane (hkl) to be brought into the horizontal plane, defined by the axes of the collimator C and the detector D, so that it is inclined to the incident beam at an angle $90° - \theta$. The detector D, usually a scintillation counter, is mounted on an arm which enables it to be rotated about a vertical axis, designated the 2θ axis, coincident with the ω axis so that it is inclined in the horizontal plane at 2θ to the forward direction of the incident beam; the detector is then correctly positioned to measure the intensity of the X-ray beam reflected from the selected (hkl) plane. In normal operation of the diffractometer the normal to the selected reflecting plane,

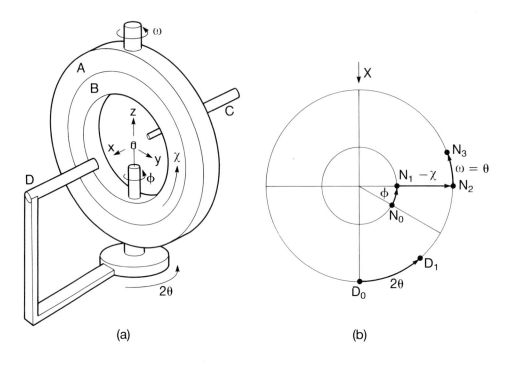

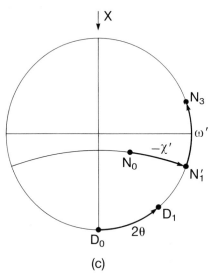

(c)

Fig 9.8 The four-circle single-crystal diffractometer. The essentials of the diffractometer are shown in (a). The rotation of a crystal in a random orientation about the ϕ-axis, the χ-axis, and the ω-axis to bring the normal to a set of reflecting planes from its initial orientation N_0 into the reflecting orientation N_3 is shown in (b). An alternative mode of achieving the same resultant orientation is shown in (c). In (b) and (c) the rotation of the detector through 2θ from D_0 to the collection position D_1 is shown.

initially in the arbitrary orientation N_0 (Fig 9.8(b)), is rotated about the ϕ axis to an orientation N_1 in the plane normal to the χ axis; it is then rotated about the χ axis to bring it to N_2 in the horizontal plane; finally rotation about the vertical ω axis brings it to the reflecting position N_3, where it is inclined in the horizontal plane at $90° - \theta$ to the incident beam.

That the crystal can be rotated about three axes leads to a degree of flexibility in the way in which it can be moved into the reflecting position for any given set of (hkl) planes. For instance the hkl reflexion could be measured by rotation of the crystal about the χ axis to bring N_0 into the horizontal plane at N_1', without any rotation about ϕ, followed by rotation about the ω axis into the reflecting position as shown in Fig 9.8(c). As we shall see, this flexibility can be put to good use.

Since the crystal is mounted at random, it is initially necessary to determine the orientation of its reciprocal lattice relative to the fixed reference axes of the diffractometer, X, Y, Z. These reference axes are orthogonal, right-handed, and such that X is directed along the forward direction of the incident X-ray beam while Z is vertical (Fig 9.8(a)). If the hkl reciprocal lattice vector, $\mathbf{d}_{hkl}^*$ where $|d_{hkl}^*| = \lambda/d_{hkl}$, has coordinates X, Y, Z when the scales of ϕ, χ, and ω are zero, then it is apparent from Fig 9.8(b) that in order to bring $\mathbf{d}_{hkl}^*$ into the reflecting position the rotation on ϕ must be $\tan^{-1} X/Y$, on $\chi \sin^{-1}(-Z/d_{hkl}^*)$, and on $\omega \sin^{-1}\frac{1}{2}d_{hkl}^*$, where $d_{hkl}^* = \sqrt{X^2 + Y^2 + Z^2}$. In order to be able to calculate these angles for a particular reflexion it is necessary to know X, Y, Z for the corresponding reciprocal lattice vector, which implies the necessity to know the orientation of the reciprocal lattice axes relative to the diffractometer reference axes. This can be achieved by searching a predetermined θ range for reflexions and then deducing the orientation of the conventional reciprocal unit-cell from the measured ϕ, χ, and θ of the recorded reflexions. If the components of the reciprocal lattice vector $\mathbf{a}^*$ along the diffractometer reference axes X, Y, Z are a_X^*, a_Y^*, a_Z^*, the components of $\mathbf{b}^*$ are b_X^*, b_Y^*, b_Z^*, and the components of $\mathbf{c}^*$ are c_X^*, c_Y^*, c_Z^*, then the coordinates X, Y, Z of the hkl reciprocal lattice point are given by the operation of what is known as the UB matrix on the indices hkl

$$\begin{bmatrix} X \\ Y \\ Z \end{bmatrix} = \begin{bmatrix} a_X^* & b_X^* & c_X^* \\ a_Y^* & b_Y^* & c_Y^* \\ a_Z^* & b_Z^* & c_Z^* \end{bmatrix} \begin{bmatrix} h \\ k \\ l \end{bmatrix} \tag{10}$$

In passing we note that careful measurement of the peak positions of reflexions with appropriate indices on both sides of the incident beam enables accurate unit-cell dimensions to be determined.

Since crystals usually reflect over a small range of θ and since it is the total intensity reaching the detector from the given reflexion that is required, the crystal is rotated through the reflecting position about the ω axis at constant speed. For accurate intensity measurements it is important both to set the limits of such a scan correctly for each reflexion and to maintain the intensity of the incident beam at a constant magnitude within quite narrow limits. The background intensity is estimated from measurements at the limits of the chosen ω scan and subtracted from the peak measurement to yield the *integrated intensity* of the reflexion.

The χ ring permits complete rotation of the crystal about the χ axis, but its construction involves quite sophisticated engineering, which is expensive, and moreover its thickness may inhibit the measurement of diffracted beams as well as restricting rotation on the ω axis by the necessity to avoid collision with the collimator. In the so-called *kappa diffractometer* an axis, designated κ, is inclined at 50° to the ω axis and can be rotated about the ω axis; the κ axis substitutes for the χ axis of the full four-circle diffractometer. In the kappa-diffractometer the ϕ axis is mounted on an arm attached to the κ axis so that it can be rotated on a small circle of radius 50° about the κ axis. The whole $\phi\kappa$ assembly can be rotated about the ω axis. The kappa diffractometer is in its engineering simpler than the full four-circle diffractometer and it has the advantage that the X-ray reflexion from any (*hkl*) plane can be recorded as the plane is rotated through 360° about its normal. Such an azimuthal scan may be utilized for making empirical absorption corrections (p. 331) to intensity measurements. With the full four-circle diffractometer azimuthal scans are also possible, but are limited by the necessity to avoid collision of the χ-ring with the collimator and by obstruction of the diffracted beam by the χ-ring.

Data reduction

The measured intensity $I(hkl)$ is related to the structure amplitude $|F(hkl)|$ of the reflexion by the expression

$$I(hkl) = s(Lp)|F(hkl)|^2 \tag{11}$$

The factor s is a scale factor, which may be estimated by use of intensity statistics and subsequently adjusted during refinement of the trial structure. The factor Lp denotes the Lorentz and polarization factors, which are always lumped together as both are dependent on the experimental conditions of the intensity measurement.

The polarization factor, which arises because X-rays are an electromagnetic radiation, has already been touched on in chapter 6. The intensity of the X-radiation scattered by a free classical electron when a plane-polarized beam of intensity I_0 is incident on it can be found by classical electrodynamics (see footnote 7 on p. 185); for our immediate purpose this result may be expressed as $I \propto I_0 \sin^2 \phi$, where ϕ is the angle between the direction of the scattered beam and the direction of polarization, that is the direction of the electric vector, of the incident beam. If the incident beam is unpolarized, it can be considered to have two component waves of equal intensity $\frac{1}{2}I_0$, one with its electric vector $E_\perp$ normal to the plane containing the incident beam XX' and the scattered beam D (Fig 9.9), and the other with its electric vector $E_\parallel$ lying in that plane. It is immediately apparent that $\phi_\perp = 90°$ and $\phi_\parallel = 90° - 2\theta$, where 2θ is the angle between the incident and scattered beams. Writing c for the proportionality factor we can now express the intensity of the components of the scattered radiation as

$$I_\perp = \frac{c}{2} I_0 \sin^2 90° = \frac{c}{2} I_0,$$

$$I_\parallel = \frac{c}{2} I_0 \sin^2 (90° - 2\theta) = \frac{c}{2} I_0 \cos^2 2\theta,$$

and the total scattered intensity as

$$I = I_\perp + I_\parallel$$

$$= \frac{c}{2} I_0 (1 + \cos^2 2\theta).$$

The factor $\frac{1}{2}(1+\cos^2 2\theta)$ is the *polarization factor* when the incident beam is unpolarized.

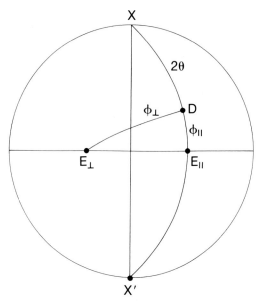

Fig 9.9 The polarization factor. The incident unpolarized beam $X'X$ can be resolved into two components, the electric vectors of which are $E_\perp$ and $E_\parallel$. Since $\phi_\perp = 90°$ and $\phi_\parallel = 90° - 2\theta$, $I = \frac{1}{2}c\,I_0(1+\cos^2 2\theta)$ and the polarization factor is $\frac{1}{2}(1+\cos^2 2\theta)$.

When, however, the incident beam has been monochromatized by reflexion from a crystal, as will often be the case in four-circle diffractometry, the intensity of the beam incident on the crystal being studied will be $(c/2)I_0(1+\cos^2 2\theta_m)$, where $2\theta_m$ is the angle between the incident and scattered beams in the monochromator. The polarization factor thus becomes

$$\frac{1+\cos^2 2\theta_m \cos^2 2\theta}{1+\cos^2 2\theta_m}$$

for the case in which the incident beam and the monochromatized beam lie in the horizontal plane.

The Lorentz factor is a geometrical factor concerned with the passage of the reciprocal lattice point through the reflecting sphere. We take as our example reflexions measured with a four-circle diffractometer, where the incident and the reflected beams are normal to the ω axis, the axis about which the crystal is rotated during intensity measurement. If the crystal is being moved at a constant angular velocity Ω, then a reciprocal lattice point P in the plane normal to the ω axis will have a linear velocity $|\mathbf{S}|\Omega$, where $\mathbf{S}$ is the reciprocal lattice vector from the origin to the reciprocal lattice point; the direction of motion of the reciprocal lattice point P will be normal to $\mathbf{S}$ (Fig 9.10). The speed with which the reciprocal lattice point passes through the reflecting sphere is given by the component of its velocity along the radius QP of the reflecting sphere. Since P is in the reflecting position $O\hat{Q}P = 2\theta$ and $Q\hat{O}P = Q\hat{P}O = 90° - \theta$; it follows that the velocity of the reciprocal lattice point P as it passes through the reflecting sphere is $|\mathbf{S}|\Omega \cos\theta$. But since $|\mathbf{S}| = (2\sin\theta)/\lambda$ (equation (20) of chapter 6), the speed with which the reciprocal lattice point passes through the reflecting sphere is

$$\frac{2\Omega \sin\theta \cos\theta}{\lambda} = \frac{\Omega \sin 2\theta}{\lambda}.$$

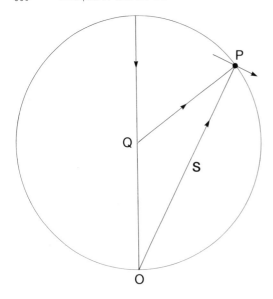

Fig 9.10 The Lorentz factor in four-circle diffractometry. The crystal moves at a constant angular velocity Ω about the ω axis, which is perpendicular to the plane of the diagram and passes through O; the linear velocity of the reciprocal lattice point P is thus $|S|\Omega$ along the normal to S. Since P is in the reflecting position, it follows that the component of its linear velocity along QP is $(\Omega \sin 2\theta)/\lambda$. The Lorentz factor for this geometry is thus $(\sin 2\theta)^{-1}$.

In practice a crystal reflects X-rays over a small angular range of θ for a variety of reasons so that the reciprocal lattice point becomes in effect a small volume element in reciprocal space. The time taken for such an expanded reciprocal lattice point to pass through the reflecting sphere is inversely proportional to the radial component of its velocity and therefore the measured intensity will be inversely proportional to the radial component of the velocity, that is the measured intensity will be proportional to $\lambda/(\Omega \sin 2\theta)$. Since λ and Ω will be the same for all reflexions the measured intensity of any reflexion in the data collection will be proportional to $(\sin 2\theta)^{-1}$. The variation in the speed with which the small element of reciprocal volume about each reciprocal lattice point passes through the reflecting sphere is allowed for by the *Lorentz factor*, which is in this case $(\sin 2\theta)^{-1}$. For a data collection with different geometry the Lorentz factor will have a different form.

Since the Lorentz and the polarization factors are dependent on θ, they are usually combined as the *Lorentz and polarization factor* (*the Lp factor*). When, as in the case we have considered, the original X-ray beam, the monochromatized beam, and the scattered beam are coplanar, the *Lp* factor will be

$$\frac{1 + \cos^2 2\theta_{\mathrm{m}} \cos^2 2\theta}{(1 + \cos^2 2\theta_{\mathrm{m}}) \sin 2\theta}$$

We have up to this point assumed that neither the incident nor the diffracted X-ray beam is absorbed by the crystal. That is not generally true in practice: the intensity I of an X-ray beam after it has travelled a distance x through a crystal is given by

$$I = I_0 \exp(-\mu x),$$

where I_0 is the intensity of the incident beam and μ is the linear absorption coefficient of the crystal. The linear absorption coefficient μ may be calculated from the tabulated mass absorption coefficients of the elements (*International Tables of X-ray Crystallography*, vol. III) provided the composition and density of the crystal and the wavelength of the incident radiation are known.

The observed intensity of a scattered X-ray beam will be less when absorption occurs, as it always does in a real crystal, than in the hypothetical case where there is no absorption. Equation (11) may now be rewritten as

$$I(hkl) = sA(hkl)(Lp)|F(hkl)|^2, \tag{12}$$

where $A(hkl)$ is known as the transmission factor. The calculation of the transmission factor for a particular reflexion requires some knowledge of the shape of the crystal because the path length within the crystal will be dependent on the location within the crystal of the point at which the diffracted beam originates. In practice, since absorption coefficients vary with wavelength, it is sensible to select a radiation for which μ is small for the crystal. But, if that is not possible, empirical absorption corrections may be calculated by analysing the variation of intensity of symmetry related reflexions as the crystal is rotated about the normal to the corresponding reflecting plane on a four-circle diffractometer; the intensity of a reflexion is measured after each successive rotation through a fixed angle about the normal to the reflecting plane and the procedure is repeated for the symmetry related reflexions. For photographic data collection the evaluation of absorption corrections is less straightforward and lies outside our scope here.

After Lp factors and, if necessary, absorption corrections have been applied to the raw intensity data symmetry related reflexions are merged to generate a set of symmetry independent $|F(hkl)|$ values for use in the determination of the crystal structure. For a comprehensive account of structure analysis the reader is referred to Dunitz (1979).

Synchrotron radiation

We conclude our account of crystal structure determination with a brief mention of a source of X-radiation which has recently become available and may well have a profound effect on the techniques of structure determination in the next decade. This source of X-radiation is *synchrotron radiation* which is becoming available for crystallographic use from an increasing number of electron storage rings in various parts of the world. The intensity is several orders of magnitude greater than the output from a conventional X-ray tube and the wavelength is tunable over the range 0·3–2·5 Å.

The X-radiation is produced by constraining a flux of electrons to move in a curved path; the electrons, being subject to a centripetal acceleration, emit electromagnetic radiation. The conditions in a storage ring are such that the emitted radiation is confined to a cone whose axis lies in the plane of the ring (Fig 9.11) and whose opening angle is approximately γ^{-1}, where the energy of the relativistic electrons is $\gamma m_0 c^2$. Synchrotron radiation is strongly polarized, the plane of polarization being parallel to the plane of the storage ring. Theoretically the polarization is complete for radiation travelling parallel to the plane of the ring; as the angle which the beam makes with that plane increases, the intensity of the component which is polarized perpendicular to the plane of the ring increases from zero. The intensity and the spectral range of the synchrotron radiation can be increased by incorporating into the storage ring at the appropriate point either an undulator or a wiggler magnet. These are devices which locally deflect the electron beam in alternate directions so as to produce no net change in the radius of curvature of the electron path.

The potential uses of synchrotron radiation are many and diverse, but almost all require the prior development of new apparatus appropriate to the greatly enhanced intensity and to the very high degree of polarization of this X-ray source. In the ensuing paragraphs we indicate a few of these applications.

Since the variation of intensity with wavelength in the beam of synchrotron radiation can be calculated, such 'white' radiation could be used with a powder sample to produce a powder pattern by allowing λ rather than θ to vary in the Bragg equation. Such an energy dispersive powder diffraction pattern would be observed at some convenient Bragg angle and the variation of intensity with wavelength in the scattered beam would be analysed with a solid state detector.

In conventional powder photography the very small natural divergence of the synchrotron radiation beam, less than 4×10^{-4} radians, should lead to greatly improved resolution. Moreover the very high intensity of the beam, making possible exposure times of only a few seconds, would enable phase transitions to be studied by continuous photography of the powder diffraction pattern.

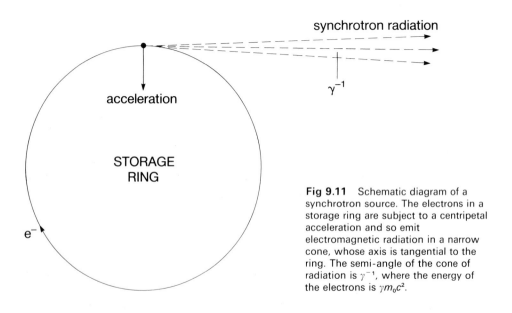

Fig 9.11 Schematic diagram of a synchrotron source. The electrons in a storage ring are subject to a centripetal acceleration and so emit electromagnetic radiation in a narrow cone, whose axis is tangential to the ring. The semi-angle of the cone of radiation is γ^{-1}, where the energy of the electrons is $\gamma m_0 c^2$.

Since the source of synchrotron radiation can, unlike a conventional X-ray source, be tuned, much more extensive use can be made of anomalous scattering by changing the wavelength of the incident radiation to a value just less than the absorption edge of one of the constituent atomic species. By this means it would become possible to overcome one of the intractable problems of X-ray structure determination, the inability to distinguish between atoms of adjacent atomic number, such as Cu and Zn, or between isoelectronic ions, such as Mg^{2+} and Al^{3+}, where the normal atomic scattering factors f_0 are almost identical. Such a technique has even wider applications in that by tuning the incident radiation to a succession of wavelengths at which various constituent atoms display significant anomalous scattering the variations in atomic scattering factors are such that the compound is effectively replaced by a series of isomorphous compounds; then, by analysing the differences in the intensities of the reflexions from one set of data to another, it should become possible to make significant deductions about the phases of the reflexions. Such a technique is likely to provide a new and very powerful method of phase determination.

Diffraction of neutrons by crystals

The interaction between a neutron beam and an atom gives rise to both coherent and incoherent scattering. This elementary treatment[7] is confined to coherent scattering, that is to scattering such that there is a definite phase relationship between incident and scattered beams so that beams scattered from different atoms in a crystal structure can interfere. Such scattering is analogous to the diffraction of X-rays.

The wavelength associated with a neutron is given by the de Broglie Equation $\lambda = h/mv$, where h is Planck's constant, m is the mass of the neutron, and v is its velocity. The range of velocities in the neutron beam extracted from a reactor is related to the temperature of the reactor by $\frac{1}{2}mv^2 = \frac{3}{2}kT$, where v is now the root mean square velocity, k is Boltzmann's constant, and T is in degrees Kelvin. In practice the temperature of a reactor is such that the emergent neutrons have root mean square velocities corresponding to wavelengths in the range 1·3 to 1·6 Å, an appropriate wavelength range for diffraction by crystals. In crystallography it is necessary to use a parallel neutron beam that is, at least approximately, monochromatic; this is achieved by collimating the neutron beam emergent from the reactor and allowing the collimated beam to be incident on a single crystal monochromator set at such an angle as to satisfy the Bragg Equation for reflexion of neutrons of the required wavelength from the face of the monochromator. However the neutron beam emergent from a reactor is generally much weaker than the X-ray beam emergent from an X-ray tube so that for diffraction studies it is necessary to use neutron beams of rather large cross-sectional area. Because of this, collimation cannot produce an accurately parallel beam and in consequence the monochromator does not produce a strictly monochromatic beam, but a beam with a wavelength spread of about 0·05 Å. Such a wavelength spread limits the resolution of the observable diffraction pattern. Since the cross-section of the neutron beam has to be large to provide adequate intensity, it is necessary for single crystal work to use crystals with linear dimensions about ten times larger than is usual in X-ray studies. No special problems arise in powder work; large powder specimens can be used because the absorption of neutrons is very much smaller than the absorption of X-rays by crystals.

Neutron diffraction is governed by the same physical principles as X-ray diffraction but the mechanism of scattering is different. We have earlier (chapter 6) shown that for X-ray diffraction, when an unpolarized plane wave of unit amplitude is incident on an atom of atomic scattering factor f, the amplitude of the scattered radiation at a distance R from the atom is given by

$$\frac{1}{R} \cdot \frac{e^2}{mc^2} \cdot f \left(\frac{1 + \cos^2 2\theta}{2} \right)^{\frac{1}{2}}$$

and there is a phase difference π between the scattered and incident X-radiation. The amplitude of the scattered radiation varies with direction because of the angular dependence of the polarization factor $\sqrt{(\frac{1}{2} + \frac{1}{2}\cos^2 2\theta)}$ and because the radius of the atom is comparable with the X-ray wavelength. Interference thus occurs between the X-radiation scattered from different parts of the atom so that the atomic scattering factor f decreases as $(\sin\theta)/\lambda$ increases. When neutrons are scattered by a non-magnetic atom however it is the nucleus that is solely responsible for the scattering,

---

[7] For a comprehensive account of neutron diffraction the reader is referred to Bacon (1975).

the electrons being too small to deflect the neutron beam. When a plane wave of unit amplitude is incident on the nucleus of an atom in a solid the nucleus is not free to recoil and the scattered neutron amplitude at a distance R from the nucleus is simply given by b/R, where b is a nuclear property known as *scattering length*. There is no polarization factor involved in neutron scattering and, since the radius of the nucleus is very small compared with the neutron wavelength, b is independent of scattering angle, for an atom at rest; b is also very nearly independent of neutron wavelength. The relationship however between scattering length for neutrons and atomic species is very much more complicated than the relationship between atomic scattering factor for X-rays and atomic species. The nuclei of different isotopes of the same element differ in their scattering lengths so that the coherent scattering length $\bar{b}$ for an element has to be taken as the average of $b \times$ (isotopic abundance) for all the isotopes of the element. Moreover if an isotope has non-zero nuclear spin, its scattering length has two possible values and this must be taken into account in calculating the coherent scattering length $\bar{b}$. Values of $\bar{b}$ cannot at present be calculated satisfactorily and so have to be determined experimentally. It is interesting to observe that $\bar{b}$ is of the same order of magnitude as $e^2 f/mc^2$ so that a crystal scatters neutrons and X-rays by about the same amount. It is simply because neutron beams of comparable intensity cannot be produced that it is necessary to use larger crystals for neutron diffraction than for X-ray diffraction studies.

Unlike atomic scattering factors, which vary regularly with atomic number, scattering lengths vary quite irregularly with atomic number. For a few atoms (e.g. hydrogen, titanium, Ni^{62}) the scattering length is negative, the sign of b corresponding to the phase change on scattering, π for b positive and zero for b negative. Most scattering lengths lie within quite a small range of magnitude, from $+1.0 \times 10^{-14}$ m to -1.0×10^{-14} m. Values of $\bar{b}$ are tabulated in *International Tables for X-ray Crystallography*, vol. IV.

We have so far confined our discussion of coherent neutron scattering to that produced by interaction with non-magnetic nuclei, but neutrons are also scattered by interaction of their magnetic moments with the permanent magnetic moments of atoms containing unpaired electrons. Here the scattering is due to neutron-electron interaction, so that the scattered neutron amplitude p, just like scattered X-ray amplitude, falls off with increasing $(\sin \theta)/\lambda$; moreover, since the unpaired electrons will not be in the innermost orbitals of the atom, the scattered neutron amplitude falls off more rapidly with $(\sin \theta)/\lambda$ than does the atomic scattering factor for X-rays for the same element. The magnitude of p at $(\sin \theta)/\lambda = 0$ is comparable with that of the scattering length b for the same atom, for example Fe^{2+} has $b = 0.96 . 10^{-14}$ m and p decreasing from $1.08 . 10^{-14}$ m at $(\sin \theta)/\lambda = 0$ to $0.45 . 10^{-14}$ m at $(\sin \theta)/\lambda = 0.25$ Å^{-1}.

In paramagnetic substances the magnetic moments of the atoms are randomly oriented so that the neutron magnetic scattering is incoherent and merely contributes to the background scattering. It is only for ferromagnetic, anti-ferromagnetic, and ferrimagnetic substances that neutron magnetic scattering is coherent. When the incident neutron beam is *unpolarized*, that is when it can be resolved into two equal and independent components of opposite spin states, the nuclear and the magnetic scattering contribute independently to the diffracted intensity so that $I(hkl) = I_n(hkl) + I_m(hkl)$, where $I_n(hkl)$ and $I_m(hkl)$ are respectively the nuclear and the magnetic scattered intensities. When, however, the incident neutron beam is *polarized*, that is when it contains neutrons in only one of the two possible spin states, the nuclear and the magnetic scattering are coherent and so interact with each other. It then becomes necessary when calculating the structure factors for neutron scattering by such magnetic substances to take into account the spatial relationship between the direction of the scattering vector $\mathbf{S}$ and the directions of the magnetic moments of the constituent atoms in the unit-cell; this is a matter to which we shall return in the course of dealing with the physical properties of magnetic crystals in volume 2. A comprehensive account of the scattering of neutrons by ferromagnetic, anti-ferromagnetic, and ferrimagnetic substances is provided by Bacon (1975). We note here that, since the magnetic moment of an atom has no effect on its scattering of X-rays, neutron diffraction studies provide an effective means of determining the orientation of the magnetic moments of the atoms in a ferromagnetic, anti-ferromagnetic, or ferrimagnetic substance.

Neutron diffraction is usually used to complement previous X-ray diffraction studies rather than for the primary determination of crystal structures. The reasons why are straightforward: neutron sources are much less widely available than X-ray tubes, the experimental techniques of neutron diffraction are more difficult than those of X-ray diffraction, and the resolution of neutron diffraction patterns is in general not as good as that of X-ray diffraction patterns. It is advantageous therefore, before starting a neutron diffraction study of a substance, to collect all the information that can readily be obtained by X-ray diffraction studies. The positions of all the atoms in the unit-cell are usually determinable by X-ray methods, with the possible exception of atoms of low atomic number where the atomic scattering factors for X-rays will be very small; atoms of hydrogen, the extreme case, may be difficult to locate by X-ray methods. In certain special circumstances, to which reference will be made later, crystal structure refinement using neutron diffraction data may elucidate more detail than is possible using X-ray data. If single crystal neutron intensity data are used the refinement techniques are essentially the same as those for X-ray structure refinement, with the exception that more attention must be given to the effects of extinction.

Extinction is the term applied to certain diffraction effects which are not encompassed by the simple theory of diffraction described in chapter 6. There it was implicit that the intensity of the incident beam would not be reduced significantly during its passage through the crystal and that no interaction would take place between the incident and the diffracted beams. If these assumptions were true, then the intensity of a reflexion would always be proportional to $|F(hkl)|^2$. In X-ray diffraction it is commonly observed that for very strong reflexions the observed intensity is systematically smaller than the intensity calculated from the atomic positions; in the later stages of the structure refinement allowance must be made for this extinction effect. In neutron diffraction extinction poses a more severe problem: neutrons, unlike X-rays, are scarcely absorbed by the crystal, so the attenuation of the neutron beam during its passage through the crystal is almost wholly due to the generation of diffracted beams. It follows that if large crystals (linear dimensions ~ 1 mm) are used for single crystal neutron diffraction, then the intensities of strong reflexions will in general not be proportional to $|F(hkl)|^2$.

The difficulties inherent in growing sufficiently large crystals for single crystal neutron diffraction work coupled with the problem of extinction has led to the development of methods of structure refinement based on neutron powder diffraction data. In the Rietveld method of profile refinement (Rietveld, 1969) use is made of the observation that in a neutron powder diffraction pattern recorded with a fixed wavelength the shape of each reflexion is almost precisely Gaussian. The intensity y_i of

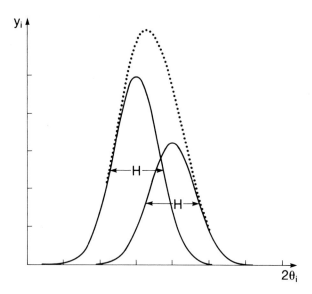

Fig 9.12 Overlapping peaks in a neutron powder diffraction pattern. Since the scattering angle $2\theta_i$ is similar for the two peaks shown, each will have effectively the same half-width H. The resultant (observed) peak is shown dotted.

a powder line at a scattering angle $2\theta_i$ can then be represented by

$$y_i = \frac{cmLI}{H} \exp\{-4\ln 2(2\theta_i - 2\theta_B)^2/H^2\}, \tag{13}$$

where c is a constant for a given sample and diffraction geometry, m is the multiplicity (p. 229) of the powder line, L is the Lorentz factor for the reflexion, I is the intensity of the hkl reflexion which will be equal to $|F(hkl)|^2$ for non-magnetic substances, θ_B is the Bragg angle for the reflexion, and H is the full width of the peak in degrees of 2θ at half peak-height (H is slightly confusingly known as the 'half-width'). The half-width, H, of the peak varies smoothly with θ, its variation being given by

$$H^2 = U\tan^2\theta_B + V\tan\theta_B + W, \tag{14}$$

where U, V and W are constants which have to be determined for the sample under investigation.

In a powder diffraction pattern peaks may overlap so that the resultant diffracted intensity at a scattering angle 2θ will be the sum of the intensities arising from each of the contributory peaks (Fig 9.12). The profile of the powder pattern can be calculated from equation (13) provided the unit-cell parameters, the crystal structure, and the parameters which define the shapes and positions of the peaks are known. If the crystal structure is not known accurately it can be refined using least-squares methods by minimizing the function $\sum_i w_i\{y_i(\text{obs}) - (1/s)y_i(\text{calc})\}^2$, where y_i is the intensity of the profile at $2\theta_i$, w_i is the weight of the ith term in the summation, s is the scale factor necessary to bring the observed intensities on to the same scale as the calculated intensities, and the summation is over all observations. In the Rietveld method of profile refinement two sets of parameters, that is structural parameters (atomic coordinates, temperature factors) and profile parameters (parameters defining the shape of the peak and any asymmetry, together with the precise position of $2\theta_B = 0$ and the precise value of θ_B for each reflexion), are refined together in the least-squares calculation. Two-stage refinements, in which the profile parameters are refined to give integrated intensities of Bragg reflexions, which are then used to refine the structural parameters, have also been developed; such methods give statistically better estimates of the standard deviations of the structural parameters.

Profile refinement methods have been conspicuously successful in neutron powder diffraction. Initial studies indicate the prospect of such methods becoming widely applicable to structure refinement using X-ray synchrotron powder data. The application of profile analysis methods to data collected by conventional X-ray powder diffractometry has however so far been very limited because of the difficulty of defining peak shape.

In non-magnetic substances neutrons are scattered by atomic nuclei so that a structure determination based on neutron intensity data yields the positions of the atomic nuclei in the unit-cell. In contrast a structure determination based on X-ray intensity data yields the distribution of electron density within the unit-cell. The two results are the same only if the atomic nuclei lie at the centroids of the electron density concentrations. It is however unlikely that in a crystal structure the electron density distribution will be spherically symmetric about all atomic nuclei; there will be concentrations of electron density between pairs of nuclei where bond formation occurs. In crystal structure refinements based on X-ray data such a departure of the electron density distribution from spherical symmetry will usually be accommodated in the anisotropic temperature factors. Temperature factors derived from neutron diffraction data are therefore likely to give a more accurate statement of the thermal vibrations of the atom. The asymmetry of the electron density distribution about an atomic nucleus is likely to be most pronounced for the valency electrons and therefore most likely to be observed in the case of light atoms, where the valency electrons are not heavily outnumbered by the core electrons. It is for this reason that the positions of hydrogen atoms determined from X-ray data may not correspond to those determined from neutron data: neutron diffraction yields the position of the hydrogen nucleus whereas X-ray diffraction yields the centroid of the associated electron distribution which will be displaced towards the atom to which the hydrogen is bonded.

We now conclude this brief account of neutron diffraction with two examples in which neutron studies have provided solutions to structural problems that could not be solved by X-rays alone.

A consequence of the irregular variation of scattering length with atomic number is that certain light elements scatter neutrons as effectively as some heavy elements, whereas for X-rays the atomic scattering factor increases regularly with atomic number so that the light elements consistently scatter least. It is thus often impossible to locate very light atoms in the unit-cell by X-ray diffraction because their scattering effects are swamped by the much stronger scattering produced by any heavy elements that may be present; in some cases the irregularity of neutron scattering lengths provides a means of locating such light atoms. For example in NaH X-ray studies show that the sodium atoms are in face-centred-cubic array but fail to locate the positions of the hydrogen atoms because their contribution to the total scattering is so slight. The scattering lengths of hydrogen and sodium are however similar in magnitude but opposite in sign: -0.37×10^{-14} m for H, 0.35×10^{-14} m for Na. Examination of the neutron powder diffraction pattern readily enables the hydrogen atoms to be located in the unit-cell and the structure of NaH to be determined as NaCl-type. This example illustrates one of the more important uses of neutron diffraction, the location of hydrogen atoms in crystal structures. Unfortunately hydrogen atoms yield a high intensity of incoherently scattered neutrons which merely increases the intensity of the background scattering. It is therefore normal practice to replace hydrogen atoms by deuterium atoms, for which the scattering length is large and positive ($b = 0.67 \times 10^{-14}$ m); deuterium produces little background scattering. We shall comment at length in volume 2 on the location of the positions of the deuterium atoms in heavy ice, D_2O, at $-50°C$.

Another way in which neutron diffraction data may be very useful is in distinguishing between the structural sites occupied by atoms of different elements whose atomic numbers are rather close, provided that their neutron scattering lengths are not, as may be the case, also similar. It is for example difficult to distinguish between the sites occupied by Mg and Al atoms in complex oxide structures by the use of X-ray methods because the number of extranuclear electrons in Mg^{2+} and Al^{3+} is the same; but the neutron scattering lengths for Mg 0.52×10^{-14} m and for Al 0.35×10^{-14} m are sufficiently different to make a clear distinction in neutron diffraction patterns between the sites occupied by these two elements. For instance neutron diffraction shows unambiguously that $MgAl_2O_4$ has the normal spinel structure (volume 2). In a similar way neutron diffraction can be applied to the study of ordering in alloys such as CuZn (see volume 2), the atomic numbers of Cu and Zn being respectively 29 and 30 so that X-ray diffraction cannot easily distinguish between Cu and Zn occupied sites. Contrariwise X-ray diffraction is better for the study of ordering in Cu–Au alloys since the atomic numbers of Cu and Au, respectively 29 and 79 are very far apart while their neutron scattering lengths are both equal to 0.76×10^{-14} m.

Problems

9.1 Many compounds with formulae of the type ABO_3 are modifications of the perovskite structure. The ideal perovskite structure is cubic and the atomic coordinates are: A 000; B $\frac{1}{2}\frac{1}{2}\frac{1}{2}$; O $0\frac{1}{2}\frac{1}{2}$, $\frac{1}{2}0\frac{1}{2}$, $\frac{1}{2}\frac{1}{2}0$. The structure of $PbZrO_3$ is such a compound; it is orthorhombic, space group $Pba2$, and the edges of its unit-cell, a_0, b_0, c_0, are related to those of the ideal cubic structure, a_c, b_c, c_c by the matrix

$$\begin{bmatrix} a_0 \\ b_0 \\ c_0 \end{bmatrix} = \begin{bmatrix} 1 & \bar{1} & 0 \\ 2 & 2 & 0 \\ 0 & 0 & 2 \end{bmatrix} \begin{bmatrix} a_c \\ b_c \\ c_c \end{bmatrix}$$

Draw a plan of a 3×3 array of nine unit-cells of the cubic structure on (001) and outline the (001) projection of the orthorhombic unit-cell. Mark on the plan the symmetry planes of the cubic structure which are normal to the x and y axes of the orthorhombic unit-cell. For each set list the possible types of symmetry planes to which the cubic symmetry planes may be reduced in the orthorhombic unit-cell. Compare these possibilities with the symmetry planes present in $Pba2$. Draw a projection on (001) of one unit-cell of the orthorhombic structure showing all the symmetry elements of $Pba2$ and the idealized positions of the atoms. Hence deduce the independent positional parameters required to describe the structure of $PbZrO_3$.

9.2 Sodium chlorate, $NaClO_3$, is cubic, space group $P2_13$, and is optically active. The structure is derivative from the NaCl type structure with pyramidal ClO_3^- ions substituting for Cl^-. A dextro-rotatory crystal was found to yield the intensity ratio

$I(511)/I(\bar{5}1\bar{1}) = 0.67$ for CuKα radiation. The coordinates of the atoms in the laevo-rotatory structure are related to those in the dextro-rotatory structure by inversion through the origin. The structure factor $F(511) = A + iB$ has been calculated for anomalous dispersion being absent; in these circumstances $A = 4.02$ and $B = 7.11$. Given that Cl scatters CuKα radiation anomalously with $\Delta f' = 0.348$, $\Delta f'' = 0.70$, and that the coordinates of the Cl atoms in the structure are xxx; $\frac{1}{2}+x, \frac{1}{2}-x, \bar{x}$; $\bar{x}, \frac{1}{2}+x, \frac{1}{2}-x; \frac{1}{2}-x, \bar{x}, \frac{1}{2}+x$ where $x = 0.417$, calculate the intensity ratio $I(511)/I(\bar{5}1\bar{1})$ for the enantiomorph whose Cl coordinates are given above. Is this a dextro- or a laevo-rotatory crystal?

9.3 The mineral *groutite*, α-MnOOH, is orthorhombic with $a = 4.560$, $b = 10.700$, $c = 2.870$ Å. Its density is 4.144 Mg m^{-3}. The only systematic absences on its X-ray diffraction pattern are $0kl$ for k odd, $h0l$ for $h+l$ odd, $h00$ for h odd, $0k0$ for k odd, $00l$ for l odd. The $N(z)$ test indicates that the structure is centrosymmetric. Determine the space group and the number of formula units in the unit-cell. Given the ionic radii Mn^{3+} 0.66 Å and O^{2-} 1.40 Å, deduce as much as you can about the coordinates of the Mn^{3+} and O^{2-} ions in groutite. How many parameters are required to describe the positions of the atoms in one unit-cell?

9.4 The crystal structure of CsCl may be described in eight different ways by taking one of the eight centres of symmetry in the unit-cell as the origin. The coordinates of the atoms for each of these eight possibilities are:

	(1)	(2)	(3)	(4)	(5)	(6)	(7)	(8)
Cs	000	$\frac{1}{2}$00	$0\frac{1}{2}0$	$00\frac{1}{2}$	$\frac{1}{2}\frac{1}{2}\frac{1}{2}$	$0\frac{1}{2}\frac{1}{2}$	$\frac{1}{2}0\frac{1}{2}$	$\frac{1}{2}\frac{1}{2}0$
Cl	$\frac{1}{2}\frac{1}{2}\frac{1}{2}$	$0\frac{1}{2}\frac{1}{2}$	$\frac{1}{2}0\frac{1}{2}$	$\frac{1}{2}\frac{1}{2}0$	000	$\frac{1}{2}$00	$0\frac{1}{2}0$	$00\frac{1}{2}$

For each possibility write down the structure factor $F(hkl)$ and evaluate it for the reflexions 100, 010, 001, 111, and 200. Notice that change of origin from one centre of symmetry to another produces only a change in the sign, but not in the modulus of the structure factor. Notice also that only by taking the origin at a Cs or a Cl atom can the symmetry which demands the equivalence of 100, 010, and 001 be made apparent.

9.5 The following reflexions from a substance with space group $P\bar{1}$ have quite large normalized structure factors $|E|$. Three of the structure factors have been arbitrarily signed by choice of origin. Apply the triple product sign relationship to sign as many as possible of the remaining structure factors.

hkl	507	$2\bar{3}2$	$7\bar{1}6$	$\bar{2}11$	413	241
$|E|$	2.4	2.6	3.1	2.5	2.0	2.6
s	+	+	+			

9.6 A crystal of pyrochlore (cubic, F-lattice, $a \simeq 10.4$ Å) was set up on a four-circle single crystal automatic diffractometer. A preliminary survey of its diffraction pattern yielded a unit-cell with dimensions $a = 7\cdot37$, $b = 7\cdot36$, $c = 7\cdot37$ Å, $\cos\alpha = 0\cdot50$, $\cos\beta = -0\cdot50$, $\cos\gamma = -0\cdot50$. Determine the transformation matrix required to convert this 'machine' cell to the true cubic unit-cell with $a = b = c \simeq 10\cdot4$ Å, $\alpha = \beta = \gamma = 90°$.

9.7 A crystal of γ-FeO(OH) was mounted in a random orientation on an automatic single crystal diffractometer. After a 'peak hunt' the diffractometer printed out the dimensions of a primitive unit-cell in reciprocal space: $a* = b* = 0\cdot239$ r.l.u., $c* = 0\cdot184$ r.l.u., $\alpha* = \beta* = 90°$, $\gamma* = 152\cdot4°$ (for MoKα radiation, $\lambda = 0\cdot7107$ Å). It is necessary then for the operator to input the transformation matrix from this machine cell to the conventional unit-cell, which has $a = 3\cdot06$, $b = 12\cdot50$, $c = 3\cdot86$ Å, space group $Cmcm$. Derive the matrix for the transformation from the machine cell to the conventional cell in (i) reciprocal space, (ii) direct space.

9.8 A beam of thermal neutrons from a reactor has a wavelength spread from $\sim 0\cdot50$ Å to $> 2\cdot5$ Å and the divergence of the collimated beam is $\pm 0\cdot3°$. The beam is monochromatized with a single crystal lead monochromator (cubic, $a = 4\cdot95$ Å) set to give a 111 reflexion. At what angle should the crystal be set to produce a neutron beam of wavelength $1\cdot10$ Å? Estimate the wavelength spread in the monochromatized beam. What other wavelengths (if any) will be present to a significant extent in the monochromatized beam? If a single crystal copper monochromator (cubic, $a = 3\cdot61$ Å) is used instead and it is not possible to change the angle of incidence of the neutron beam, what will be the wavelength of the monochromatized beam?

9.9 The alkali metal hydrides RH are all cubic, lattice type F, and have one formula unit per lattice point. Assuming that the structure is of the sodium chloride type, putting R at 000 and H at $\frac{1}{2}\frac{1}{2}\frac{1}{2}$, write down and simplify the expression for the structure factor. Evaluate $F(111)$ and $F(200)$ for LiH ($a = 4\cdot085$ Å) and for NaH ($a = 4\cdot880$ Å) using the atomic scattering factors given below. Evaluate also $F(111)$ and $F(200)$ for lithium and sodium hydrides and deuterides using the neutron scattering lengths given below. Repeat this procedure for the zinc sulphide (blende) structure type (cubic F lattice, one RH per lattice point, R at 000, H at $\frac{1}{4}\frac{1}{4}\frac{1}{4}$). Which type of radiation would be appropriate for confirming the NaCl structure in the sodium and in the lithium case? If neutrons are preferred, would it be better to use the hydride or the deuteride?

$\dfrac{\sin\theta}{\lambda} =$	0	0·1	0·2	0·3	0·4	0·5
H⁻ $f =$	2·00	1·20	0·52	0·24	0·12	0·06
Li⁺ $f =$	2·00	1·94	1·76	1·52	1·27	1·02
Na⁺ $f =$	10·00	9·55	8·37	6·89	5·47	4·29

	H	D	Li	Na
$b(10^{-4}$ m)	$-0\cdot37$	$0\cdot67$	$-0\cdot19$	$0\cdot35$

10
Electron diffraction and microscopy

One of the most important tools for the study of solids is the electron microscope. A beam of electrons travelling with velocity v has, by the de Broglie postulate, a wavelength $\lambda = h/mv$, where h is Planck's constant and m is the mass of the electron. An electron beam will, as we shall see, have a wavelength of the order of 0·05 Å and so be capable of diffraction by atoms. Since the direction of an electron beam can be deflected by an applied electric or magnetic field, it becomes possible to use electrostatic or, more commonly, electromagnetic lenses to form an image of an object with electrons. Since electromagnetic lenses are subject to spherical and chromatic aberrations, it is necessary to restrict the numerical aperture of the electron microscope to about 0·1° with the consequence that the theoretical resolving power for an accelerating potential of 100 kV is about 8 Å. It is thus not quite possible to resolve atoms and, as we shall see, the small numerical aperture restricts the detail observable in the image.

When an electron beam accelerated through a large potential difference is incident on a crystal it will interact with the crystal in a variety of ways (Fig 10.1). The electrons may be scattered elastically, that is without a change in energy, or inelastically, that is with a change in energy. In inelastic scattering the energy lost by the electron is transferred to electrons in the atoms of the crystal and may result in ejection of these electrons from their atomic orbitals. Alternatively the inelastic interaction may result in a change in the lattice vibrations of the atoms in the crystal and a concomitant increase or decrease in the energy of the scattered electrons. More important in electron microscopy is the change in energy of the electron beam produced when some of its energy is lost to valency, or conduction, electrons in the solid specimen which are thereby caused to oscillate with quantized frequencies. The energy lost by the incident beam is small, of the order of 10 eV,[1] and the electrons are scattered through angles of the order of 0·005°. Since electrons travelling through a crystal may suffer more than one scattering event, all diffracted beams, that is elastically scattered beams with a fixed phase relationship to the incident beam, will be surrounded by a narrow cone of inelastically scattered electrons and in consequence the quality of the image will be impoverished. Any interaction which involves a

---

[1] The electron volt (eV) is not an SI unit, but is commonly in use in the field of particle physics. It is defined as the energy required to change the potential of an electron by one volt. $1\,\mathrm{eV} = 1\cdot6021 \times 10^{-19}\,\mathrm{J}$.

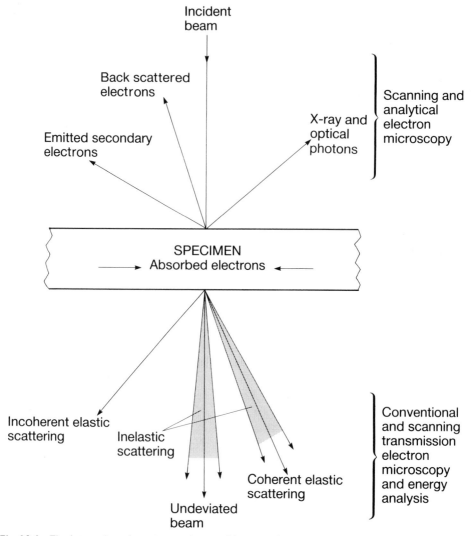

Fig 10.1 The interaction of an electron beam with a crystal.

positive transfer of energy from the incident electron beam to the crystal specimen will in general produce an increase in the temperature of the crystal; such electron beam heating may lead, especially in non-conducting crystals, to chemical decomposition, which may also be facilitated by the low working pressure of about 10^{-4} torr[2] required for efficient operation of an electron microscope. Contamination of the crystal surface by the deposition of products of interaction of the electron beam with residual gases in the vacuum system may give rise to difficulties in the interpretation of both diffraction patterns and images.

In this chapter we shall be concerned mainly with the use of coherent elastic scattering of electrons for the study of crystals, that is the use of scattered electrons which have the same wavelength as the incident electron beam and a constant phase relationship with it. Some other types of electron-crystal interaction do however have

[2] The torr is not an SI unit, but is commonly employed in any experimental field where low pressure apparatus is used. 1 torr = 1 mmHg = $133 \cdot 322 \, \mathrm{Nm}^{-2}$.

important applications, for example in electron microprobe analysis and in the study of crystal surfaces.

It is essential to appreciate at the outset of this introduction to electron microscopy and diffraction that an electron may be scattered more than once in the course of its passage through the crystal; in this respect electrons differ from both X-rays and neutrons. In the *kinematical theory*, which we have used earlier to interpret X-ray diffraction it is assumed that a beam of radiation travelling through a crystal will be scattered either not at all or only once in the course of its passage through the crystal and that the intensity of the incident beam will not be reduced significantly by its passage through the crystal. These assumptions are not valid for electron diffraction. The full theoretical treatment of electron diffraction lies outside the scope of this book: it makes use of the *dynamical theory* of diffraction, which takes account of multiple scattering (where a diffracted beam is scattered more than once), of the intensity of the diffracted beam being sufficiently high to cause a significant reduction in the intensity of the incident beam, and of the possibility of interaction between the incident and diffracted beams. It is however possible to obtain some insight into the formation of images in the electron microscope with the simpler kinematical approach provided one restricts the argument to thin crystals, that is to crystals of the order of $1000\,\text{Å}$ in thickness, so that it will be safe to assume that scattering does not significantly reduce the intensity of the incident beam in the course of its passage through the crystal. The assumption will be safe provided the intensity of the diffracted beams is small compared with that of the incident beam, a condition which implies a thin crystal or a crystal which is not oriented so as to give rise to a strong diffracted beam. This condition is more likely to be satisfied when the scattering factors of the constituent atoms are small.

We shall confine our treatment of electron microscopy and diffraction to the theory and application of the conventional transmission electron microscope. We shall start with a general description of the microscope, pass on to a discussion of electron diffraction and conclude with an account of the production of images either by amplitude (or diffraction) contrast or by phase contrast, that is the formation of an image by the incident beam and one or more diffracted beams. A fuller treatment together with an account of the application of electron microscopy to the study of materials is to be found in Grundy and Jones (1976) and in Thomas and Goringe (1979). For a comprehensive account of the theory of the electron microscopy of thin crystals the reader is referred to Hirsch, Howie, Nicholson, Pashley and Whelan (1977).

The electron microscope

The conventional transmission electron microscope has an electron-optical arrangement which is in essence the same as the optical arrangement in a light microscope. But in the electron microscope the magnification scheme is more elaborate and of course the electron source, the specimen, and the whole assembly of lenses are in an evacuated column.

The electrons emitted from a hot tungsten filament (the electron 'gun') are accelerated through a large potential difference V, of the order of 10^5 volts, and so acquire kinetic energy eV, where e is the electron charge. The velocities of the electrons so accelerated are not negligible in comparison with the velocity of light c so that a relativistic correction has to be made. The wavelength of an electron is thus given by

$$\lambda = h \left\{ 2m_0 e V \left(1 + \frac{eV}{2m_0c^2} \right) \right\}^{-\frac{1}{2}}$$

where m_0 is the rest mass of the electron. For $V = 1 \times 10^5$ volts, $\lambda = 0.037$ Å. For $V < 10^5$ volts the relativistic correction is insignificant so that the approximation $\lambda = h(2m_0 eV)^{-\frac{1}{2}}$ can be used.

The arrangement of the parts of the microscope are shown schematically in Fig 10.2. The ray paths for an electron beam with normal incidence on the specimen are shown with the microscope in the imaging mode in Fig 10.2(a) and set up for selected area diffraction in Fig 10.2(b). The electrons emitted from the electron gun are immediately controlled by a sequence of two electromagnetic lenses, the *condenser* lenses, which determine the area of the specimen illuminated by the electron beam and the divergence of the electron beam incident on the specimen. The gun and condenser system will include provision for imparting a small inclination to the electron beam so that its angle of incidence on the specimen can be varied from 90°. Since the specimen crystal is necessarily very thin, of the order of 1000 Å, it must be supported in a special holder and, since the microscope interior is at high vacuum, the holder is introduced through an airlock. The *specimen stage*, in which the holder fits, is situated within the bore of the next electromagnetic lens, the objective lens, and is provided with the drives necessary for both moving and tilting the specimen so that different areas of 10 μm or less in diameter can be studied at controllable orientations with respect to the microscope axis and the incident electron beam. The *objective* lens forms an image of the specimen at I_1, which is known as the first intermediate image. A sequence of two projector lenses further magnifies the image produced by the objective lens. The first projector lens (otherwise known as the intermediate lens) produces an image of I_1 at I_2, the second intermediate image, and the second projector lens produces an image of I_2 on a fluorescent viewing screen, the fluorescent component of which is usually a zinc phosphor. The final image may be recorded on a photographic film placed immediately beneath the viewing screen. The magnification of the final image can easily be varied in the range from 1000 × to as much as 100,000 × by varying the focal lengths of the two projector lenses, the variation being achieved by adjustment of the energizing current supplied to the lens.

If the image formed by the electron microscope were a perfect image, then any point in the image would correspond to a point in the specimen. That is not the situation in practice because magnetic lenses suffer from defects so that a point in the specimen will give rise to a disc in the image, the so-called disc of confusion. The arrangement of the microscope is such that defects in the objective lens have a greater effect on the final image than do defects in the projector lenses, the principal effects being astigmatism, chromatic aberration, and spherical aberration (the nomenclature of aberrations in electron microscopy corresponds to that used in optical microscopy, but the causes of the aberrations are of course not the same). *Astigmatism* presents no problem in that it can be eliminated in the design and construction of the magnetic lenses. *Chromatic aberration* occurs when there is an energy spread in the electron beam, the focal length of a magnetic lens being dependent on the energy of the electron beam as well as on the magnetic field produced by the lens. Chromatic aberration can be minimized by stabilization of the energizing current of the lenses and of the electron source. Another source of chromatic aberration is inelastic scattering of electrons by the specimen, for which no correction is possible.

The most important defect of magnetic lenses is *spherical aberration*. If a lens is free

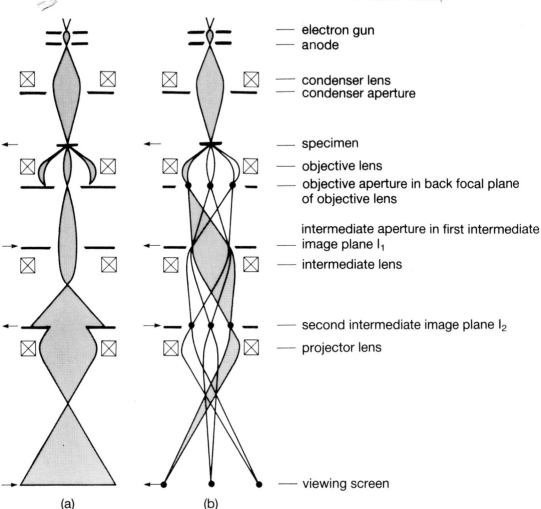

— electron gun
— anode

— condenser lens
— condenser aperture

— specimen
— objective lens
— objective aperture in back focal plane of objective lens

intermediate aperture in first intermediate
— image plane I_1
— intermediate lens

— second intermediate image plane I_2
— projector lens

— viewing screen

(a)   (b)

Fig 10.2 Schematic representation of the parts of an electron microscope. Ray paths for normal incidence are shown for the microscope in the imaging mode in (a) and for selected area diffraction in (b). The arrows indicate the orientation of successive images and of the diffraction pattern relative to the specimen.

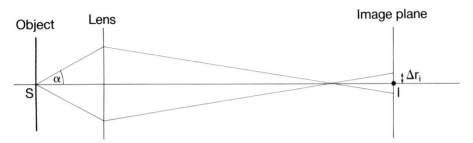

Fig 10.3 Spherical aberration. If the magnetic lens were free of spherical aberration all the electrons leaving the point S on the object would reach the image plane at the point I. Since there is spherical aberration in the magnetic lens, there will be a disc of confusion in the image plane, the radius of which is $\Delta r_i = MC_s\alpha^3$, where M is the magnification, C_s is a constant, and α is the extreme angle of divergence of electrons leaving the point on the object.

from spherical aberration all the electrons leaving the object from the point S will reach the image plane at the point I (Fig 10.3). But where there is spherical aberration the rays passing through the outer parts of the lens will be bent excessively and so brought to a focus at a point closer to the lens than the point I; in consequence there will be a disc of confusion in the image plane of radius Δr_i. It can be shown that $\Delta r_i = MC_s\alpha^3$, where M is the magnification, C_s is the spherical aberration constant of the lens (approximately equal to the focal length, about 2–3 mm), and α is the extreme angle of divergence of electrons leaving the point on the object. Referred back to the object the corresponding disc of confusion is of radius $\Delta r_s = C_s\alpha^3$. There is no way of correcting for the spherical aberration of a magnetic lens except by reducing α by reducing the diameter of the objective aperture. But the theoretical resolving power of the microscope, which is given by the Rayleigh formula as

$$R = \frac{0\cdot61\,\lambda}{\mu\sin\alpha} \simeq \frac{0\cdot61\,\lambda}{\alpha}$$

since the refractive index $\mu \simeq 1$ and α is always small in the electron microscope, will increase as the objective aperture is reduced. It is necessary to strike a compromise between reduction in spherical aberration and improvement in resolution (i.e. smaller resolving power), which will be achieved at an optimum aperture $\alpha_{opt} \simeq (\lambda/C_s)^{\frac{1}{4}}$ and minimum aberration $\Delta R_{min} \simeq (\lambda^3/C_s)^{\frac{1}{4}}$. In most microscopes it is possible to insert an aperture near the back focal plane of the objective lens (Fig 10.2) in order to limit the extreme value of α. When $\alpha = 0\cdot3°$ in a 100 kV microscope a resolution of 5–10 Å should be obtainable. But the specimen may scatter electrons inelastically and so introduce chromatic aberration and reduce the resolving power to no better than 20–30 Å. In this paragraph our objective has been to indicate to the reader in general terms the problems involved in achieving images of good quality with the electron microscope; the relative severity of the various problems and most effective solutions will vary with the model of microscope being used.

The *depth of field D* is the distance between two planes on either side of the plane of the object outside which the blurring due to the object being out of focus becomes comparable with the attained resolution ΔR_a. It can be shown that $D = (2\Delta R_a/\alpha)$. The value of D in electron microscopy is about 7000 Å and so is significantly greater than the thickness, ~ 1000 Å, of the specimen; the image is thus a projection of the three-dimensional detail of the specimen.

The *depth of focus* is the distance over which the image is in focus and is given by $(2\Delta R_a M^2/\alpha)$, where M is the magnification. The depth of focus in electron microscopy is very large; in consequence the position of the photographic film on which the image is to be recorded does not need to be coincident with the fluorescent viewing screen and is usually sited below the screen.

The Abbe theory of image formation (chapter 9) indicates that a periodic specimen gives rise to parallel sets of diffracted beams and that all the beams travelling in a particular direction are focused by the objective lens at a point in the back focal plane (D in Fig 9.4) of the lens to form the diffraction pattern of the specimen. The diffraction pattern can be observed by projecting the back focal plane of the objective lens on to the fluorescent viewing screen; that is simply achieved by reducing the energizing current of the intermediate lens so as to increase its focal length (Fig 10.2). It is evident from Fig 9.4 that all the diffracted beams arising from a particular point in the specimen are focused by the projector lens at a point in the image plane I. For a good image to be formed the objective aperture must be large so that many diffracted beams will be collected and used to form the image. But, as we have seen, in the electron microscope the objective aperture is necessarily small so that only in favourable circumstances, such as are provided by the high resolution imaging techniques which will be discussed later, is it possible to observe atomic detail; usually the observable detail is restricted to fringes which correspond to lattice plane spacings. In the electron microscopy of thin crystals much valuable information can be obtained by use of the techniques known generally as *amplitude contrast*, which will be the subject of the ensuing paragraphs. In such images the crystal is oriented so as to give rise to only one strong Bragg reflexion.

To produce what is known as a *bright field image* an aperture is placed in the back focal plane of the objective lens so as to allow the passage only of the direct beam and

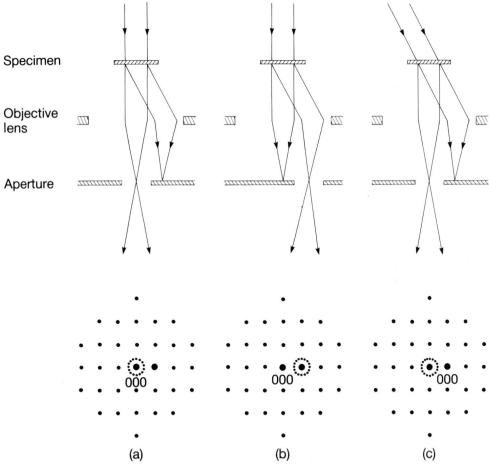

Fig 10.4 Amplitude contrast. (a) illustrates the production of a bright field image by insertion of an aperture in the back focal plane of the objective lens to allow the passage only of the direct beam and the low angle inelastic scattering surrounding it. (b) illustrates the production of a dark field image by insertion of an aperture in the back focal plane of the objective lens to allow the passage of only one diffracted beam and the low angle inelastic scattering surrounding it. (c) illustrates a better way of achieving a dark field image by tilting the incident beam through 2θ so that the diffracted beam selected by the aperture passes down the axis of the microscope. In each case the lower figure indicates how the aperture, represented by a dotted circle, is positioned on the diffraction pattern as observed on the viewing screen.

the low angle inelastic scattering which surrounds it; this is achieved by observation of the diffraction pattern on the fluorescent viewing screen while the aperture is adjusted to centre on the 000 reflexion. No reflected beams can pass through the aperture (Fig 10.4(a)). An image formed in this manner may display intensity variation corresponding to differences in the intensity of Bragg reflexion from different parts of the specimen due to consequent reduction in the intensity of the incident beam passing through the crystal (Fig 10.5).

To produce what is known as a *dark field image* either the aperture is positioned so that only one diffracted beam and the inelastically scattered electrons travelling at similar angles can pass through it (Fig 10.4(b)), or the incident beam is tilted through 2θ so that the selected diffracted beam passes down the axis of the microscope (Fig 10.4(c)). The microscope is then returned to the imaging mode and the dark field image can be observed. Better resolution is achieved by tilting the incident beam than by placing the aperture off the microscope axis, the image then being less affected by spherical aberration. The resultant dark field image will be illuminated only in those areas which correspond to areas of the specimen which are correctly oriented to produce the selected Bragg reflexion (Fig 10.5).

Bright field and dark field imaging are, as we shall see later, highly productive techniques.

Another useful technique is to insert an aperture, known as the *intermediate aperture*, in the image plane of the objective lens. If the aperture has a diameter D and the magnification of the objective lens, assumed to be perfect, is M, then the diameter of the area of the specimen visible in the image will be D/M. The diffraction pattern produced by this selected area of the specimen can also be viewed and recorded (Fig 10.6), by reducing the energizing current to the intermediate lens so that an image of the diffraction pattern is produced on the viewing screen. This technique, known as *selected area diffraction*, enables both the image and the diffraction pattern produced by a small area (diameter $\sim 1\ \mu m$) of the specimen to be observed. Some care is required when the image and the diffraction pattern are compared because electrons follow a helical path as they pass through a magnetic lens: the image and the diffraction pattern will be rotated relative to each other because the focal length of the first projector lens will have been changed in transferring the microscope from the imaging to the diffraction mode. Moreover, the image will be inverted whereas the diffraction pattern is not inverted, as can readily be seen by counting the cross-overs in Fig 10.2; the image and the corresponding diffraction pattern are related by inversion and rotation. The angular relationship is best determined by calibrating the microscope using a crystal of known morphology and unit-cell.

In selected area diffraction there will only be a perfect match between the image and the diffraction pattern when spherical aberration is absent and when the objective lens is correctly focused. When spherical aberration is present and the objective lens is incorrectly focused the image and the diffraction pattern will relate to areas of the specimen which are displaced by a distance which depends on the scattering angle 2θ of the diffracted beam. Let us suppose that the area of the specimen to be examined was selected from a bright field image and is represented by the line PQ in Fig 10.6(a). Electrons scattered by PQ parallel to the axis of the microscope reach the image plane of the objective lens at P_1Q_1, which are the limits of the diffraction aperture situated in this plane. But, since the objective lens has spherical aberration, electrons scattered by PQ at the angle α to the microscope axis will reach the image plane in $P_1'Q_1'$. Tracing this displacement back to the specimen it is seen that the aperture limits P_1Q_1

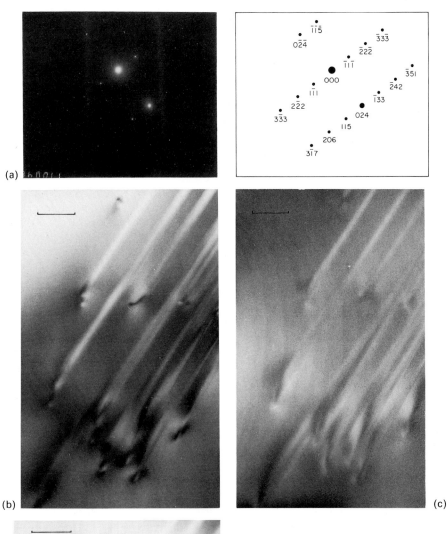

(a)

(b)

(c)

(d)

Fig 10.5 Stacking faults in bustamite, $(Ca, Mn)SiO_3$, which is triclinic with space group $A\bar{1}$ and unit-cell parameters $a = 7.760$, $b = 7.109$, $c = 13.84$ Å, $\alpha = 91.5°$, $\beta = 94.4°$, $\gamma = 103.57°$. The crystal is oriented with $[\bar{3}\bar{2}1]$ parallel to the microscope axis. (a) shows on the left the diffraction pattern and on the right the indices of the observed reflexions. (b) is the bright-field image, which displays faults parallel to (102) and with a displacement vector $\mathbf{R} = \frac{1}{2}[010]$. The faults are terminated by line defects, interpreted as dislocations with $\mathbf{b} = \frac{1}{2}[010]$. (c) is the dark-field image obtained with $2\bar{2}2$ as the operating reflexion; here $\mathbf{g}.\mathbf{R} = 1$ so the contrast is slight. (d) is the dark-field image obtained with $3\bar{3}3$ as the operating reflexion; here $\mathbf{g}.\mathbf{R} \neq n$ so the contrast is enhanced. In (b), (c) and (d) the scale bar corresponds to 500 Å.

correspond not to PQ but to P'Q' in the specimen. The displacement of the image is given by $P_1P_1' = Q_1Q_1' = M_oC_s\alpha^3$, where M_o is the magnification of the objective lens. The diffraction pattern observed by selected area diffraction will thus not arise from precisely the area selected in the bright field image. Since the displacement is proportional to α^3, it will be greatest for higher order reflexions, that is those with larger reciprocal lattice vectors. It is apparent from Fig 10.6(a) that the back focal plane of the objective lens is displaced due to spherical aberration so that the observed pattern will be distorted; but the distortion is small even for high order reflexions.

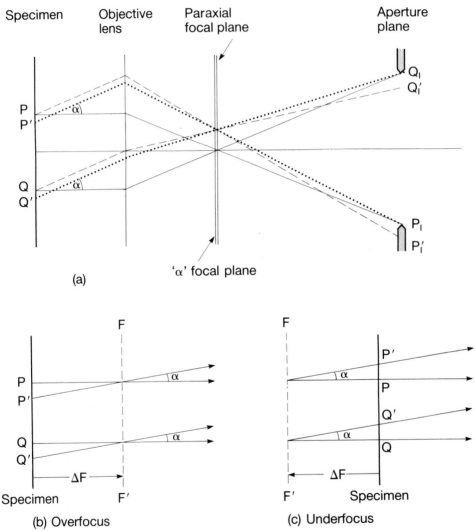

Fig 10.6 Mismatch between image and diffraction pattern. In (a) spherical aberration in the objective lens causes the back focal plane for electrons scattered at an angle α to the axis of the microscope (dashed lines) to be displaced from the paraxial focal plane to the 'α' focal plane. Thus electrons scattered by PQ at an angle α to the microscope axis will reach the image plane at $P_1'Q_1'$. The aperture P_1Q_1 allows the passage of electrons (dotted lines) from P'Q' in the specimen. Incorrect focusing of the objective lens also gives rise to displacement as illustrated in (b) and (c): a diffraction aperture set for observation of PQ in a bright field image will actually correspond to P'Q', the sense of the displacement $|\alpha.\Delta F|$ for PQ being determined by whether there is overfocus (b) or underfocus (c).

Since the plane of the intermediate aperture is fixed in the microscope, the objective lens will only form the image of the specimen in that plane when it is correctly focused. If the objective lens is incorrectly focused, the image in the aperture plane will be of the plane FF' (Fig 10.6(b), (c)), which will be in front of, or behind, the plane of the specimen according to whether the objective lens is underfocused or overfocused. If the diffraction aperture was set for observation of the part of the specimen represented by PQ using a bright field image, then a diffracted beam at a scattering angle $2\theta = \alpha$ would arise, not from PQ, but from P'Q', where $PP' = QQ' = \alpha\,\Delta F$. Thus the observed diffraction pattern will be produced by areas of the specimen displaced by $2\theta\,\Delta F$ anti-parallel to the reciprocal lattice vector for the reflexion concerned.

The combined displacement due to spherical aberration and to focusing error is equal to $C_S\alpha^3 + \Delta F\alpha$. For small values of α a small amount of underfocusing could thus be used to counteract the unavoidable error introduced by spherical aberration. In a 100 kV microscope the displacement at the specimen caused by spherical aberration is of the order of $10^{-2}\ \mu\mathrm{m}$ for the higher values of α passing through the aperture. The use of small apertures may produce a significant lack of correspondence between the observed diffraction pattern and the observed image. It is usual to use diffraction apertures through which an area of the specimen of radius about 1 $\mu\mathrm{m}$ may be viewed.

At this point it is appropriate to mention contamination effects which may affect not only selected area diffraction patterns but all electron diffraction patterns. Even in a vacuum as low as 10^{-5} torr, when subjected to electron bombardment, some metals are prone to develop epitaxial oxide films, which will give rise to diffuse reflexions in the diffraction pattern. In some cases oxide and alloy reflexions will be close together or even overlap so that it becomes important to distinguish diffuse oxide film reflexions from bulk diffraction effects which may be due to ordering or other phenomena. Contamination effects, if any occur, will vary with the nature of the crystal specimen.

Electron diffraction

Before discussing the theory of image formation in the electron microscope it is necessary to give more thorough consideration to the diffraction of electrons by thin crystals. Since electrons are scattered by potential fields their scattering by an atom involves both its nucleus and its electrons. When an electron beam of unit amplitude is incident on an atom of atomic number Z the amplitude of the scattered electrons at a distance R from the atom is

$$\frac{f_e}{R} = \frac{me^2}{2h^2 R}\left(\frac{\lambda}{\sin\theta}\right)^2 (Z-f).$$

where f is the atomic scattering factor of the atom for X-rays and f_e is known as its atomic scattering amplitude for electrons; m is the relativistic mass of the electron. The scattered amplitude thus varies with wavelength λ and with scattering angle θ. It can be shown that for $(\sin\theta)/\lambda$ greater than about $0.4\ \text{Å}^{-1}$ scattering amplitude increases regularly with atomic number, while for $(\sin\theta)/\lambda$ less than about $0.4\ \text{Å}^{-1}$ scattering amplitude does not vary in a regular manner. For elements of the first three periods of the periodic table there is actually a tendency for the scattering amplitude at small angles to decrease with increasing atomic number. On average one can say that at $(\sin\theta)/\lambda = 0$, f_e is proportional to $\sqrt[3]{Z}$ so that light elements are better scatterers relative to heavy elements for electrons than for X-rays for which $f \propto Z$.

Substitution of numerical values in the expressions for the amplitude of the electrons and of the X-rays scattered by a given atom leads to the expression $\{0.85\,\lambda^2(Z-f)10^3\}/(\sin^2\theta)f$ for the ratio of the scattered electron amplitude to the scattered X-ray amplitude for the atom (λ is measured in Å). This ratio is of the order of 10^4 for small values of $(\sin\theta)/\lambda$ so that one can say that atoms scatter electrons very much more strongly than they scatter X-rays.

Since λ for electrons is so much smaller than for X-rays, θ will be very much smaller for a given value of $(\sin\theta)/\lambda$. For X-rays diffracted beams may have appreciable intensity for all values of θ right out to the limiting value of $\theta = 90°$. But for electrons the scattering factor falls off very rapidly and is reduced to one tenth of its value at $(\sin\theta)/\lambda = 0$ at a scattering angle, 2θ, as little as $3°$. Thus in electron diffraction the diffracted beams have appreciable intensity only at rather small Bragg angles.

If a crystal is correctly oriented in an incident electron beam to produce a strong Bragg reflexion, the electron beam may be completely reflected after it has traversed as few as 25 lattice planes. In X-ray diffraction however the possibility of complete reflexion of the incident beam does not arise until at least 10^4 lattice planes have been traversed. It is for this reason that diffracted electron beams may be rescattered during their subsequent path through the crystal and the assumption made in chapter 6 that diffracted X-ray beams are not rescattered is invalid for electron diffraction. As indicated in the introductory remarks at the start of this chapter the full treatment of the diffraction of electrons by crystals requires the application of the dynamical theory of diffraction. We shall here restrict our treatment to thin crystals, of thickness no more than a few thousands of Ångstrom units, for which it can be assumed that the intensity of the incident beam is not significantly reduced during its passage through the crystal and that a diffracted beam is not rescattered; in short we shall apply the kinematical theory of diffraction. In our treatment of electron diffraction we shall make use of the concepts of the reciprocal lattice and the reflecting sphere which were developed in the context of X-ray diffraction in chapter 6.

In applying the kinematical theory of diffraction to the diffraction of electrons it is necessary to assume that the crystal is thin in the direction of the incident beam and in consequence it is not true to say, as it was for X-ray diffraction, that diffraction maxima occur only when the scattering vector $\mathbf{S}$ corresponds to a reciprocal lattice vector, that is to say when the beams scattered by neighbouring lattice points are in phase. Diffracted beams are found to occur in other directions. We investigate this matter by considering a row of N unit-cells along the crystallographic z axis and supposing that the wave $F(\mathbf{S})$ is scattered by the first unit-cell of the row in the direction defined by $\mathbf{S}$. Then the wave scattered by the next unit-cell of the row will have the same amplitude and its phase will be $2\pi\mathbf{c}.\mathbf{S}$ relative to the wave scattered by the first unit-cell (p. 207). So by the principle of superposition the total wave $F_T(\mathbf{S})$ scattered by the row of unit-cells will be

$$F_T(\mathbf{S}) = F(\mathbf{S})\{1 + \exp 2\pi\, i\mathbf{c}.\mathbf{S} + \exp 2\pi\, i2\mathbf{c}.\mathbf{S} + \ldots + \exp 2\pi i(N-1)\mathbf{c}.\mathbf{S}\}.$$

This geometrical series can simply be summed by multiplying by $1 - \exp 2\pi i\mathbf{c}.\mathbf{S}$ to yield the expression

$$F_T(\mathbf{S}) = F(\mathbf{S})\frac{1 - \exp 2\pi i N\mathbf{c}.\mathbf{S}}{1 - \exp 2\pi i\mathbf{c}.\mathbf{S}}.$$

So the total wave scattered by the column of N unit-cells is the wave scattered by one unit-cell modified by the factor $1 - \exp 2\pi i N\mathbf{c}.\mathbf{S}/(1 - \exp 2\pi i\mathbf{c}.\mathbf{S})$. This factor may be rewritten as

$$\frac{\exp \pi i N\mathbf{c}.\mathbf{S}\{\exp(-\pi i N\mathbf{c}.\mathbf{S}) - \exp \pi i N\mathbf{c}.\mathbf{S}\}}{\exp \pi i\mathbf{c}.\mathbf{S}\{\exp(-\pi i\mathbf{c}.\mathbf{S}) - \exp \pi i\mathbf{c}.\mathbf{S}\}}$$

$$= \exp \pi i(N-1)\mathbf{c}.\mathbf{S}\,\frac{\sin \pi N\mathbf{c}.\mathbf{S}}{\sin \pi\mathbf{c}.\mathbf{S}}$$

The intensity of the total wave scattered is therefore

$$I_\mathrm{T}(\mathbf{S}) = F_\mathrm{T}(\mathbf{S})F_\mathrm{T}^*(\mathbf{S})$$

$$= F(\mathbf{S})F^*(\mathbf{S}) \exp \pi i(N-1)\mathbf{c}.\mathbf{S}.\exp(-\pi i(N-1)\mathbf{c}.\mathbf{S})\,\frac{\sin^2 \pi N\mathbf{c}.\mathbf{S}}{\sin^2 \pi\mathbf{c}.\mathbf{S}}$$

$$= I(\mathbf{S})\,\frac{\sin^2 \pi N\mathbf{c}.\mathbf{S}}{\sin^2 \pi\mathbf{c}.\mathbf{S}}$$

The factor relating $I_\mathrm{T}(\mathbf{S})$ to $I(\mathbf{S})$ is the familiar 'grating factor' for one-dimensional diffraction gratings.

The grating factor is zero when $\mathbf{c}.\mathbf{S} = p/N$ where p is an integer and p/N is not an integer. When p/N is an integer, that is when $\sin \pi\mathbf{c}.\mathbf{S} = 0$, recourse must be had to L'Hôpital's Rule[3] to show that the magnitude of the grating factor is then N^2. When N is large all the other maxima between the zero values of the grating factor are negligibly small: this is the situation in X-ray diffraction and in electron diffraction by thick crystals, where the grating factor is effectively equivalent to the third Laue equation (p. 209) and corresponds to the condition that electrons scattered by points $\mathbf{c}$ apart are in phase. But when N is small—the situation in electron diffraction by thin crystals—the maxima which occur between values of $\sin \pi\mathbf{c}.\mathbf{S} = 0$ are not all negligible. The form of the grating factor for different small values of N is illustrated by Longhurst (1973, p. 230). In electron diffraction what is interesting is the intensity distribution close to reciprocal lattice points, that is near those points in reciprocal space where $\mathbf{c}.\mathbf{S} = l$. Writing $\mathbf{g}$ for a reciprocal lattice vector, for which $\mathbf{c}.\mathbf{S} = l$, we can put $\mathbf{S} = \mathbf{g} + \mathbf{s}$, where $\mathbf{s}$ represents the small separation of $\mathbf{g}$ and $\mathbf{S}$. Thus the factor

$$\frac{\sin \pi N\mathbf{c}.\mathbf{S}}{\sin \pi\mathbf{c}.\mathbf{S}} = \frac{\sin \pi N\mathbf{c}.(\mathbf{g}+\mathbf{s})}{\sin \pi\mathbf{c}.(\mathbf{g}+\mathbf{s})},$$

Since $\mathbf{g}$ corresponds to a solution of the Laue equation, $\mathbf{c}.\mathbf{g}$ is an integer and will therefore have no effect on the magnitude of the factor. So we can now write

$$I_\mathrm{T}(\mathbf{S}) = I(\mathbf{S})\,\frac{\sin^2 \pi N\mathbf{c}.\mathbf{s}}{\sin^2 \pi\mathbf{c}.\mathbf{s}},$$

where $|\mathbf{s}|$ is small.

---

[3] When the limiting value of a ratio, $f(x)/g(x)$, assumes the form $0/0$ on substitution of a particular value of the variable x, the limiting value will be given by the ratio of the first derivatives $f'(x)/g'(x)$; and if this ratio yields $0/0$, then by the ratio of the second derivatives $f''(x)/g''(x)$; and so on until a meaningful form appears. This is L'Hôpital's Rule, which is proved (but not under that name) in G. H. Hardy, *A Course of Pure Mathematics*, 10th edition, Cambridge University Press, 1958, and in Thomas and Finney (1984), p. 231.

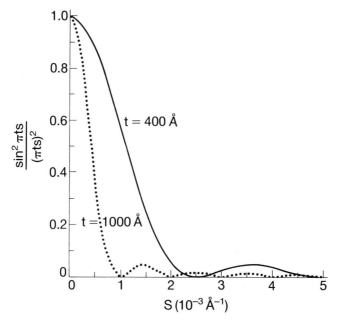

Fig 10.7 Variation of $(\sin^2 \pi ts)/(\pi ts)^2$ with s for two values of t shown as a solid line for $t = 400$ Å and as a dotted line for $t = 1000$ Å. The function is zero for $s = \pm n/t$, where n is a positive integer. Reciprocal lattice points thus become spikes extending to $\pm 1/t$ from the reciprocal lattice point.

For a thin crystal whose smallest dimension is parallel to the z axis and whose dimensions parallel to the x and y axes are relatively large the Laue equations $\mathbf{a} \cdot \mathbf{S} = h$ and $\mathbf{b} \cdot \mathbf{S} = k$ will hold and so $\mathbf{S}$ will be restricted to planes normal to x and to planes normal to y, that is to a set of rods parallel to z^* and passing through points $h\mathbf{a}^* + k\mathbf{b}^*$. The scattered intensity close to the reciprocal lattice point $h\mathbf{a}^* + k\mathbf{b}^* + l\mathbf{c}^*$ will be zero except along the rod parallel to z^*. So the vectors $\mathbf{s}$, representing the difference between $\mathbf{S}$ and the reciprocal lattice vector $\mathbf{g}$, will be parallel to $\mathbf{c}^*$. Therefore

$$\mathbf{c} \cdot \mathbf{s} = \mathbf{c} \cdot \mathbf{c}^* \frac{|s|}{|c^*|} = \frac{|s|}{|c^*|} = sd_{(001)}$$

But $N\mathbf{c} \cdot \mathbf{s} = Nsd_{(001)} = ts$, where t is the thickness of the crystal normal to (001). So the intensity along the rod will vary as

$$I(\mathbf{S}) \frac{\sin^2 \pi ts}{\sin^2 (\pi ts/N)} = I(\mathbf{S})N^2 \frac{\sin^2 \pi ts}{(\pi ts)^2}$$

since s is very small. The function $\sin^2 \pi ts/(\pi ts)^2$ is plotted in Fig 10.7; it is evident from the figure that it is equal to unity when $s = 0$ and first drops to zero at $s = \pm 1/t$. The function then rises to a small subsidiary maximum and drops again to zero at $s = \pm 2/t$ and so on, the maxima decreasing as s increases, the minima being at n/t where n is an integer. So one can say qualitatively that the reciprocal lattice point is drawn out into a 'spike' parallel to z^* extending effectively to $\pm 1/t$ from the reciprocal lattice point.

In electron diffraction we are concerned not only with very thin crystals but also with very short wavelengths: the wavelength of the electron beam in conventional transmission electron microscopy is of the order of 0·05 Å, so that the radius, $1/\lambda$, of

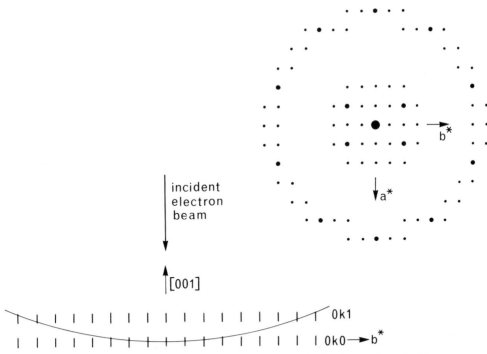

Fig 10.8 The formation of a diffraction pattern by a monochromatic electron beam incident on a stationary crystal. The figure shows the reciprocal lattice rows 0k0 and 0k1 of a very thin orthorhombic crystal whose smallest dimension is parallel to [001]. The electron beam is incident parallel to [001]. The reflecting sphere is seen to intersect three reciprocal lattice spikes on either side of the origin in the zero layer and the seventh and eighth spikes in the first layer. The diffraction pattern obtained is shown on the right: the central circular area containing $hk0$ reflexions is separated from the concentric ring of $hk1$ reflexions by an annular blank area.

the reflecting sphere is of the order of 20 Å^{-1} and the reciprocal lattice constants of the specimen crystal are likely to be of the order of 0·1 Å^{-1}. Thus for a single orientation of the crystal the reflecting sphere is likely to intersect a considerable number of reciprocal lattice spikes of the reciprocal lattice net normal to the incident electron beam. In contrast with X-ray diffraction, a diffraction pattern can be recorded with electrons from a stationary crystal using monochromatic radiation.

The point is illustrated in Fig 10.8, where an electron beam is shown incident along [001] of an orthorhombic crystal plate, the large faces of which are parallel to (001). In the resultant diffraction pattern $hk0$ reflexions lie in a circle centred on the incident electron beam and $hk1$ reflexions lie in a concentric ring. We have illustrated, with simplicity in mind, the case of a thin orthorhombic crystal lying on (001) so that the reciprocal lattice spikes are perpendicular to the $hk0$ reciprocal lattice net. That this is not always so is easily seen by considering a thin monoclinic crystal lying on (001); here the reciprocal lattice spikes are perpendicular to (001), that is parallel to z^*, but the $hk0$ reciprocal lattice net is perpendicular to z so the crystal would have to be rotated through $\beta - 90°$ about its y axis to produce an [001] diffraction pattern centred on the incident beam.

The extent of the observed diffraction pattern may be found in the following manner. Figure 10.9 illustrates a section of the reflecting sphere and of the zero-layer net perpendicular to the incident beam, taken perpendicular to the $hk0$ reciprocal lattice net of an orthorhombic crystal. Since the wave length of electrons is small

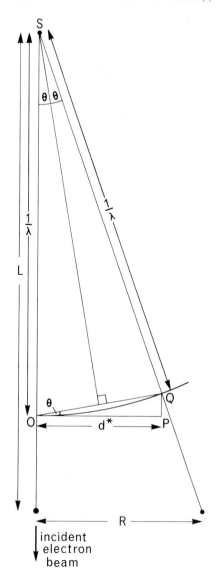

Fig 10.9 The undistorted nature of electron diffraction patterns. In the figure the Bragg angle θ is grossly exaggerated. The diagram is explained in the text.

compared with the spacing of reflecting planes in a crystal the Bragg angles of the lowest angle reflexions are of the order of $1°$, very much less than for X-ray diffraction, and this has significant consequences. Consider a reciprocal lattice point P with reciprocal lattice vectors $\mathbf{d}^* = \mathbf{g}$ and distant $\mathbf{s}$ from the reflecting sphere. Since θ is small $s/g = \text{PQ/OP} = \tan \theta \simeq \theta$ and $g = \text{OP} \simeq \text{OQ} = (2 \sin \theta)/\lambda \simeq 2\theta/\lambda$; therefore $g^2 = 2s/\lambda$. For a thin crystal of thickness t perpendicular to (001) the reciprocal lattice spikes extend effectively to $s = 1/t$ on either side of the reciprocal lattice point. So reciprocal lattice spikes of the zero layer will intersect the reflecting sphere provided $g < \sqrt{2/\lambda t}$. Reciprocal lattice points of the layer $hk1$ (otherwise known as the Laue zone $hk1$) lie in a plane at a perpendicular distance $1/c$, parallel to z, from the $hk0$ layer of reciprocal lattice points. The spikes of the $hk1$ reciprocal lattice points extend from

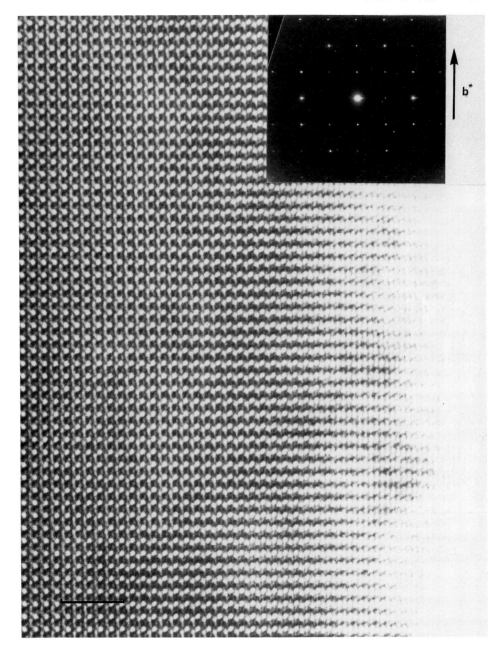

Fig 10.10 High resolution image of parawollastonite, $CaSiO_3$, which is monoclinic with space group $P2_1/a$ and unit-cell parameters $a = 15.43$, $b = 7.32$, $c = 7.07$ Å, $\beta = 95.40°$. The crystal is oriented with [101] parallel to the microscope axis. The diffraction pattern is shown inset top right. The scale bar corresponds to 40 Å. The operating voltage was 100 kV. The image was formed using reflexions within 0.33 Å^{-1} of the origin spot. Each white dot corresponds to one unit-cell in projection. The SiO_3 chains are parallel to the b-axis. The contrast in the image changes as the thickness of the specimen increases towards the left.

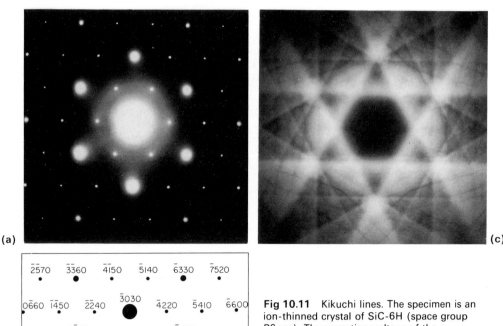

(a)

(c)

(b)

2570	3̄3̄60	4̄150	5̄140	6̄330	7̄520
0̄660	1̄450	2̄2̄40	3̄030	4̄2̄20	5̄410	6̄600
1̄540	0̄330	1̄1̄20	2̄1̄10	3̄300	4̄5̄10
3̄630 2̄4̄20	1̄2̄10	1̄2̄10	2̄4̄20	3̄630
4̄510	3̄300	2̄1̄10	1̄1̄20	0̄330	1̄5̄40
6̄600 5̄4̄10	4̄2̄20	3̄030	2̄2̄40	1̄4̄50	0̄6̄60
7̄5̄20	6̄330	5̄1̄40	4̄150	3̄3̄60	2̄5̄70

Fig 10.11 Kikuchi lines. The specimen is an ion-thinned crystal of SiC-6H (space group $P6_3mc$). The operating voltage of the microscope was 100 kV. The specimen is oriented with [0001] precisely parallel to the incident beam. The spot diffraction pattern (a) was obtained from a very thin part of the specimen where the thickness was <150 Å. The indexing of the diffraction pattern is shown in (b). A thick part of the specimen, thickness >1000 Å, gave rise to the pattern of Kikuchi lines shown in (c), which confirms the precise orientation of the specimen with [0001] parallel to the incident beam. The absence of a central spot in (c) indicates that the incident beam was completely scattered.

$1/c - 1/t$ to $1/c + 1/t$ parallel to z. So reciprocal lattice spikes with

$$\sqrt{\left\{\frac{2}{\lambda}\left(\frac{1}{c}-\frac{1}{t}\right)\right\}} < g < \sqrt{\left\{\frac{2}{\lambda}\left(\frac{1}{c}+\frac{1}{t}\right)\right\}}$$

will intersect the reflecting sphere and give rise to reflexions when the incident beam is parallel to [001]. Here **g** for the $hk1$ reflexion is equal to $h\mathbf{a}^* + k\mathbf{b}^*$. Thus in Fig 10.9 the zero layer reflexions lie within a circle centred on the origin of reciprocal space and the $hk1$ reflexions lie within an annulus with the same centre.

We have already seen that the condition for reflexion is

$$d^* = OQ = \frac{2\theta}{\lambda}$$

Consider a reflexion at a distance R on the film from the intersection of the transmitted electron beam with the film, the film being at a distance L from the specimen S (Fig 10.9). Since θ is very small, $R/L = \tan 2\theta \simeq 2\theta$, so that $d^* = R/L\lambda$. The observed diffraction pattern is thus an undistorted projection of the reciprocal lattice which gives rise to it (Figs 10.10 and 10.11(a)). This property makes the interpretation of electron diffraction patterns rather easy once the effective crystal to film distance L has been determined; L is not a simple length but depends on the magnification of the electron lens system and may be evaluated from measurements of the diffraction pattern of a substance of known unit-cell dimensions.

Since it is not in general possible to control the initial orientation of the crystal in the specimen holder, it is often necessary to index an observed diffraction pattern from previous knowledge of the unit-cell dimensions and lattice type of the crystal.

We start our discussion with the simplest case: the indexing of the diffraction pattern of a cubic crystal. In the cubic system $d^* = a^*\sqrt{(h^2+k^2+l^2)}$ and so, as $d^{*2} \propto h^2+k^2+l^2$, it is always possible to index a cubic reciprocal lattice net without prior knowledge of the unit-cell dimension; but there may be ambiguity if the lattice type is unknown. The first step is to measure the three shortest reciprocal lattice spacings on the observed diffraction pattern, t_1^*, t_2^*, and t_3^* (Fig 10.12); these will be such that $\mathbf{t}_3^* = \mathbf{t}_1^* + \mathbf{t}_2^*$. Since $d^* = a^*\sqrt{(h^2+k^2+l^2)}$, we can immediately evaluate

$$\left(\frac{t_1^*}{t_2^*}\right)^2 = \frac{h_1^2+k_1^2+l_1^2}{h_2^2+k_2^2+l_2^2} = \frac{p}{q}$$

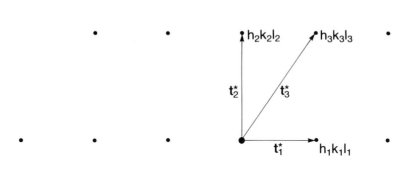

Fig 10.12 Indexing of the electron diffraction pattern of a cubic crystal. The observed reflexions are represented by solid circles. The three shortest reciprocal lattice spacings are denoted t_1^*, t_2^*, and t_3^*.

and

$$\left(\frac{t_1^*}{t_3^*}\right)^2 = \frac{h_1^2 + k_1^2 + l_1^2}{h_3^2 + k_3^2 + l_3^2} = \frac{p}{r},$$

where p, q and r are integers, without prior knowledge of the lattice parameter a or of the scale of the photograph. Once p, q, and r have been determined, the forms $\{hkl\}$ to which the three reciprocal lattice vectors belong will be identifiable. Suppose that $\mathbf{t}_1^*$ is given all positive indices, the other two sets of indices can be found by use of the vector addition $\mathbf{t}_1^* + \mathbf{t}_2^* = \mathbf{t}_3^*$ so that $h_1 + h_2 = h_3$, $k_1 + k_2 = k_3$, and $l_1 + l_2 = l_3$. Once these indices have all been determined, they can be confirmed by comparing the calculated interplanar angles $(h_1 k_1 l_1):(h_2 k_2 l_2)$ and $(h_1 k_1 l_1):(h_3 k_3 l_3)$ with the angles between the corresponding reciprocal lattice vectors measurable on the photographic record of the diffraction pattern. It is then a simple task to determine the zone axis symbol $[UVW]$ of the zone recorded in the observed diffraction pattern. Since cubic crystals have high symmetry there will not be a unique solution: the observed diffraction pattern may be that of the zone $[UVW]$ or that of any symmetry related zone.

By way of example we consider a cubic crystal, Laue Class $m3m$, which gives rise to an electron diffraction photograph on which the following measurements have been made: $t_1^* = 5\cdot 6$ mm, $t_2^* = 12\cdot 5$ mm, $t_3^* = 13\cdot 7$ mm, $\mathbf{t}_1^*:\mathbf{t}_2^* = 90°$, $\mathbf{t}_1^*:\mathbf{t}_3^* = 66°$. Since $(t_2^*/t_1^*)^2 = 5$ and $(t_3^*/t_1^*)^2 = 6$, immediate identification of the forms of $h_1 k_1 l_1$, $h_2 k_2 l_2$, and $h_3 k_3 l_3$ as $\{100\}$, $\{210\}$, and $\{211\}$ respectively is achieved. If we take $h_1 k_1 l_1$ to be 100, then $1 + h_2 = h_3$, $k_2 = k_3$, and $l_2 = l_3$. Therefore $h_2 = 0$, $h_3 = 1$, and the reciprocal lattice points corresponding to $\mathbf{t}_2^*$ and $\mathbf{t}_3^*$ may be either 012 and 112 or 021 and 121, taking both k and l positive. It follows that the reciprocal lattice net is either the zone $[0\bar{2}1]$ or the zone $[0\bar{1}2]$, which are symmetry related. If we choose $[0\bar{2}1]$, then $(100):(012) = 90°$ and $(100):(112) = \cos^{-1}(1/\sqrt{6}) = 65\cdot 9°$ in agreement with observation. All other reflexions on the photograph may now be indexed by vector addition. The decision to take $h_2 k_2 l_2$ as 012 rather than as $0\bar{1}\bar{2}$ implies, for right-handed axes, that the electrons are travelling in the direction $[02\bar{1}]$ rather than $[0\bar{2}1]$. Here and elsewhere we assume that the film is viewed in the direction of the transmitted electron beam.

We have chosen to index the reflexions $h_1 k_1 l_1$ and $h_2 k_2 l_2$ as 100 and 012, but indexing as $\bar{1}00$ and $0\bar{1}\bar{2}$ would likewise yield consistent indices. If there is a diad axis normal to the net, no ambiguity arises because the alternative choice of indices corresponds to rotation through 180° about the normal to the net; but if there is no diad, it is impossible to index the net unambiguously without further information. This point is illustrated by Fig 10.13, which shows the reciprocal lattice net normal to [100]. It is immediately apparent that the distribution of points in the reciprocal lattice layers above and below the zero-layer net for the $[0\bar{2}1]$ zone is not symmetrical about $[0\bar{2}1]$ and therefore, for electrons travelling along $[02\bar{1}]$, it is impossible to decide which zero-layer reflexion is 012 and which is $0\bar{1}\bar{2}$ without additional information. Such information may be obtained from the first layer reflexions, if they are visible, from the observation of Kikuchi lines (pp. 366–375) superimposed on the diffraction pattern, or by rotating the specimen about [100] through $\pm 26\frac{1}{2}°$ or $\pm 18\frac{1}{2}°$. If after $26\frac{1}{2}°$ clockwise rotation the $[0\bar{1}0]$ net is observed and if after $18\frac{1}{2}°$ anti-clockwise rotation the $[0\bar{1}1]$ net is observed, then the reflexions 012 and $0\bar{1}\bar{2}$ have been indexed correctly in the $[0\bar{2}1]$ zone. In the point group $m3m$ there are of course twenty-four directions in the form $\langle 210 \rangle$ giving twelve symmetry related zones, the choice of any one of which would be acceptable for indexing the diffraction pattern.

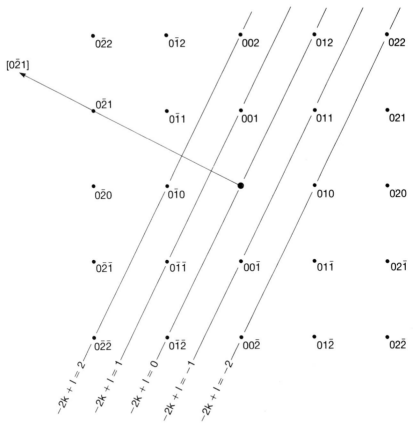

Fig 10.13   The reciprocal lattice net normal to [100] of a cubic crystal in Laue Class *m3m*. The traces of the zero, ±1, and ±2 nets of the [0$\bar{2}$1] zone are represented by solid lines. The distribution of reciprocal lattice points in the +n and −n nets is not symmetrical about the direction [0$\bar{2}$1].

It must of course be borne in mind that systematic absences will arise (p. 194) when the lattice is non-primitive, I or F in the cubic system, so that the choice of indices to correspond to the observed ratios q/p and r/p may not be immediately obvious if the lattice type is unknown. If, of course, the unit-cell dimension a is known and the scale of the diffraction pattern is known, then $h^2 + k^2 + l^2$ is given explicitly by $(d*/a*)^2$ and no ambiguity is possible whatever the lattice type may be.

It is not uncommon for cubic electron diffraction patterns to display some symmetry, which can be *4mm*, or *6mm*, or *2mm* or *m*, with the consequence that two of the three shortest reciprocal lattice vectors will be symmetry related. If the symmetry of the pattern is *4mm* or *6mm*, then the incident beam was along ⟨100⟩ or ⟨111⟩ respectively, the presence of a centre of symmetry in the diffraction pattern causing the zero-layer net normal to ⟨111⟩ to display symmetry *6mm* rather than *3m*. It should be borne in mind that unless the incident beam is accurately parallel to the symmetry axis the intensities of the reflexions will appear to be less symmetrical than the geometry of the net. The observation of symmetry greatly simplifies the task of indexing, which is usually best begun by drawing a sketch stereogram to show all the poles of the forms involved. By way of example we consider the indexing of the diffraction pattern shown diagrammatically in Fig 10.14(a). The symmetry of the

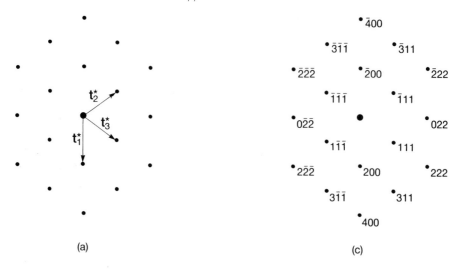

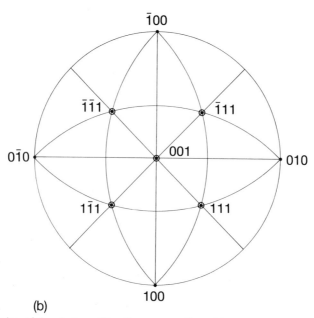

Fig 10.14 Indexing of the electron diffraction pattern of a cubic crystal in Laue Class $m3m$ (a). The three shortest observed spacings are t_1^*, t_2^*, and t_3^*. The symmetry of the diffraction pattern is $2mm$. Measurement yields $t_2^*/t_1^* = t_3^*/t_1^* = \sqrt{3}/2$. Indexing is achieved by reference to a stereogram showing all the planes of the forms $\{100\}$ and $\{111\}$ in point group $m3m$ (b). The indexed electron diffraction pattern is shown in (c): lattice type F, electron beam parallel to [01$\bar{1}$].

diffraction pattern is $2mm$; since a net through the origin of reciprocal space necessarily has a centre of symmetry, all that can be inferred about the symmetry associated with the zone axis normal to this net is that either it lies in a mirror plane, m, or it is a diad at the intersection of two mirror planes, $2mm$. We choose t_1^* along a line of symmetry; and t_2^* and t_3^* such that $t_2^* = t_3^*$ so that $h_2k_2l_2$ and $h_3k_3l_3$ have the same value of $h^2+k^2+l^2$. Measurement of the diffraction pattern yields $t_1^* = 12\cdot4$ mm, $t_2^* = t_3^* = 10\cdot7$ mm. (It is assumed that the scale is unknown, so the reciprocal vector moduli are left in mm as measured on the film). Therefore

$$\left(\frac{t_2^*}{t_1^*}\right)^2 = \left(\frac{t_3^*}{t_1^*}\right)^2 = \frac{3}{4},$$

indicating that $h_1k_1l_1$ is of the form $\{200\}$ while $h_2k_2l_2$ and $h_3k_3l_3$ are of the form $\{111\}$. The easiest way of getting a set of self-consistent indices for these three reciprocal lattice points is by comparing their angular relationship in the diffraction pattern with a sketch stereogram showing all the faces of the forms $\{100\}$ and $\{111\}$ for the Laue Class $m3m$ (Fig 10.14(b)). Suppose that $\mathbf{t}_1^*$ corresponds to 200. Then one of the zones through (100) containing planes of the form $\{111\}$ would be $[(100), (111),$ $(\bar{1}11)]$, in which $(\bar{1}11)$ and (111) are related by the mirror plane parallel to (010). If $\mathbf{t}_2^*$ is $\bar{1}11$, then $\mathbf{t}_3^* = \mathbf{t}_1^* + \mathbf{t}_2^* = 2\mathbf{a}^* + (-\mathbf{a}^* + \mathbf{b}^* + \mathbf{c}^*) = \mathbf{a}^* + \mathbf{b}^* + \mathbf{c}^*$ and $\mathbf{t}_3^*$ is identified as 111. This assignment of indices needs to be confirmed by measurement of the angles $\mathbf{t}_1^* : \mathbf{t}_2^*$ and $\mathbf{t}_1^* : \mathbf{t}_3^*$ and comparison with the calculated angles $\mathbf{t}_1^* : \mathbf{t}_2^* = (100):(\bar{1}11) = \cos^{-1}(-1/\sqrt{3}) = 180° - 54·74° = 125·26°$ and $\mathbf{t}_1^* : \mathbf{t}_3^* = (100):(111) = \cos^{-1} 1/\sqrt{3} = 54·74°$ (by equation (33) on p. 151). Having indexed these three reciprocal lattice points, all others on the diffraction pattern can be indexed immediately by vector addition (Fig 10.14(c)). It is evident that the lattice type is F because h, k, l are all odd or all even. Such a reciprocal lattice net passing through the origin corresponds to a zone of reflexions whose zone axis can be quickly determined by cross multiplication

$$
\begin{array}{c|cccc|c}
1 & 1 & 1 & 1 & 1 & 1 \\
 & & \times & \times & \times & \\
\bar{1} & 1 & 1 & \bar{1} & 1 & 1 \\
\hline
 & 0 & \bar{2} & 2 & & \\
\end{array}
$$

i.e. $[0\bar{1}1]$. Therefore the incident electron beam is travelling parallel or nearly parallel to $[01\bar{1}]$. Since the Laue Class $m3m$ has a diad parallel to $[01\bar{1}]$ the indexing of this pattern is unambiguous. In the cubic system the zone $[01\bar{1}]$ is equivalent to five other zones; there are six zones in the form $\langle 01\bar{1} \rangle$, the choice of any of which could be acceptable for indexing the diffraction pattern.

The unambiguous indexing of the electron diffraction pattern of a cubic crystal is a simple task when the unit-cell dimension a and the scale of the diffraction pattern are known. But if the unit-cell dimension a and the scale are unknown ambiguity may arise unless the lattice type is known because the electron diffraction pattern corresponds only to a single reciprocal lattice net. Consider, for example, the diffraction pattern sketched in Fig 10.15(a), which has a rectangular unit-mesh with $t_2^*/t_1^* = \sqrt{2}$. If the lattice-type is P, then $h_1k_1l_1$ is of the form $\{100\}$ and $h_2k_2l_2$ is of the form $\{110\}$. Taking $h_1k_1l_1$ as 100 and $h_2k_2l_2$ as 011 the diffraction pattern can be indexed as shown in Fig 10.15(b). But if the lattice type is I, there will be systematic absences for $h+k+l$ odd so that the smallest permitted reciprocal lattice vectors will be $t_{(011)}^*$ and $t_{(200)}^*$; $(t_{(200)}^*/t_{(011)}^*) = \sqrt{2}$, whence $h_1k_1l_1$ is 011, $h_2k_2l_2$ is $\bar{2}00$ and the pattern will be indexed as shown in Fig 10.15(c).

In all other crystal systems the task of indexing a diffraction pattern is generally more difficult and is usually achieved by means of a computer program. It will be necessary to input either the moduli of three reciprocal lattice vectors related as $\mathbf{t}_1^* + \mathbf{t}_2^* = \mathbf{t}_3^*$, or the moduli of two reciprocal vectors $\mathbf{t}_1^*$ and $\mathbf{t}_2^*$ and the angle $\mathbf{t}_1^* : \mathbf{t}_2^*$, obtained by measurement of the diffraction pattern, together with unit-cell dimensions and lattice type.

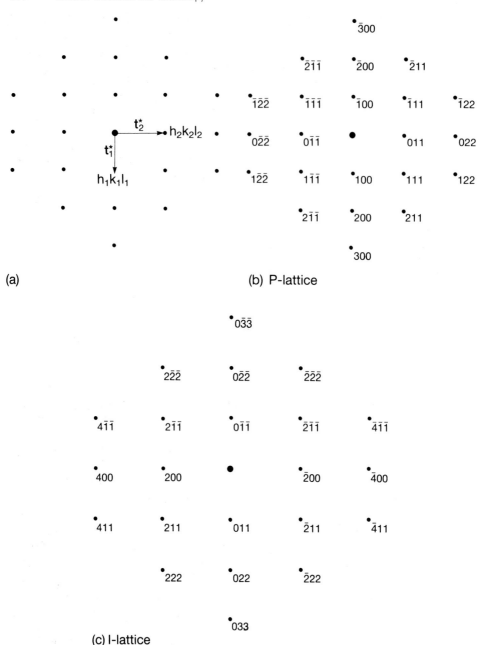

Fig 10.15 Indexing of a cubic electron diffraction pattern when the unit-cell dimension a and the scale are unknown. The observed diffraction pattern is shown in (a): the unit-mesh is rectangular with $t_2^*/t_1^* = \sqrt{2}$. Indexing if the lattice type is P is shown in (b). Indexing if the lattice type is I is shown in (c).

Double diffraction

In indexing electron diffraction patterns it is always necessary to bear in mind the possibility that, even when the specimen is a thin crystal, a strong diffracted beam may be diffracted again. Such *double diffraction* may give rise to a resultant diffracted beam which coincides with a singly diffracted beam and so does not change the geometry of the observed diffraction pattern, or it may give rise to a resultant diffracted beam parallel to a 'forbidden' singly diffracted beam so that a systematically absent reflexion will appear to be present in the observed diffraction pattern.

A strongly diffracted beam from a reciprocal lattice point hkl acts effectively as a second incident beam so that the observed diffraction pattern is the diffraction pattern produced by the strong hkl diffracted beam acting as an incident beam superimposed on the diffraction pattern produced by the incident beam itself. In thin crystals only high intensity reflexions will give rise to double diffraction. That a doubly diffracted beam may give rise to a reflexion coincident with that produced by a singly diffracted beam can be shown very simply. The first Laue equation, which must be satisfied for a reflexion to occur, can be written (Equation (21) of chapter 6) in the form

$$\mathbf{a} \cdot \mathbf{S} = h,$$

where $\mathbf{S} = \mathbf{s} - \mathbf{s}_0$. The vectors $\mathbf{s}_0$ and $\mathbf{s}$ give respectively the directions of the incident beam and the diffracted beam. If the diffracted beam $\mathbf{s}$ is then rescattered in a direction given by $\mathbf{s}'$, the first Laue equation can be applied to this second scattering event $\mathbf{a} \cdot (\mathbf{s}' - \mathbf{s}) = h'$. Adding the two equations we obtain

$$\mathbf{a} \cdot (\mathbf{s} - \mathbf{s}_0 + \mathbf{s}' - \mathbf{s}) = h + h'$$

i.e.

$$\mathbf{a} \cdot (\mathbf{s}' - \mathbf{s}_0) = h + h'$$

Similarly for the other two Laue equations: $\mathbf{b} \cdot (\mathbf{s}' - \mathbf{s}_0) = k + k'$ and $\mathbf{c} \cdot (\mathbf{s}' - \mathbf{s}_0) = l + l'$. So the doubly diffracted beam will appear to be a singly diffracted beam from the reciprocal lattice point $h + h'$, $k + k'$, $l + l'$. If that reciprocal lattice point is already producing a reflexion, then the geometry of the diffraction pattern will remain unchanged, but the intensity of the reflexion will be enhanced. If, however, the reflexion $h + h'$, $k + k'$, $l + l'$ is systematically absent, then the geometry of the diffraction pattern will be changed. By way of example we return to the $[0\bar{1}1]$ diffraction pattern illustrated in Fig 10.14 and now suppose that the 111 reflexion is particularly strong because the net is tilted so that the 111 reciprocal lattice point lies precisely on the reflecting sphere. The possibility arises that the strong 111 diffracted beam will be scattered again by the $1\bar{1}\bar{1}$ reciprocal lattice point to give a doubly diffracted beam $1 + 1$, $1 - 1$, $1 - 1$; the 200 'spot' on the photograph would then represent the superimposition of the 200 diffracted beam and the 111, $1\bar{1}\bar{1}$ doubly diffracted beam. The geometry of the photograph would be unchanged, but the intensity of the 200 'spot' would be enhanced. Alternatively, let us suppose that the crystal under investigation is diamond in which the (001) d-glide (Table 6.4) causes the 200 reflexion to be absent; here the doubly diffracted 111, $1\bar{1}\bar{1}$ reflexion coincides with the absent 200 reflexion and the geometry of the observed diffraction pattern is changed. Such violations of the rules for systematic absences only occur for absences due to the presence of glide planes or screw axes; violations cannot occur for absences due to lattice type because the reciprocal lattice points hkl and $h'k'l'$ must be such as to give rise to permitted reflexions and so $h + h'$, $k + k'$ and $l + l'$ must likewise be a permitted reflexion. When an unexpected reflexion is observed on an electron diffraction photograph, the possibility of double diffraction should be explored, bearing in mind that it is most likely to occur when the primary reflexion hkl is strong.

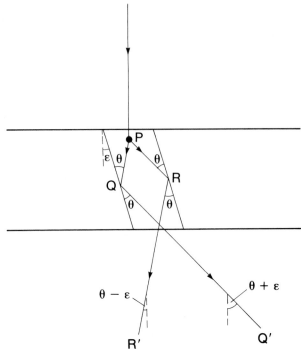

Fig 10.16 Origin of Kikuchi lines. Electrons are scattered inelastically from the point P; electrons travelling along PQ are reflected by the (hkl) planes and those travelling along PR by the $(\bar{h}k\bar{l})$ planes to give Bragg reflexions in the directions QQ' and RR'. Since the angle of divergence of PQ from the incident beam, $\theta-\varepsilon$, is less than that of PR, $\theta+\varepsilon$, the reflected beam QQ' will be of higher intensity than the reflected beam RR'.

In general double diffraction only occurs if two sets of planes (hkl) and $(h+h', k+k', l+l')$ are in the reflecting positions. This situation is likely to arise quite frequently in electron diffraction because the reflecting sphere has a large radius and the reciprocal lattice points are drawn out into spikes so that the three reflexions hkl, $h'k'l'$, and $h+h', k+k', l+l'$ are quite likely to lie in the zone recorded on the observed diffraction pattern. Double diffraction happens only very occasionally in X-ray diffraction where the condition that two reciprocal lattice points lie simultaneously on the reflecting sphere can rarely be satisfied.

When the crystal specimen under investigation consists of two or more phases, the possibility of a diffracted beam produced by one phase being scattered again by another phase arises. Such double diffraction may give rise to observed reflexions which do not coincide with the reciprocal lattice points of either phase. A similar effect may be observed in twinned crystals. A comprehensive account of double diffraction by composite crystals and twins lies outside our scope.

Crystal orientation: Kikuchi lines

The attitude of a crystal with respect to the incident electron beam can only be adjusted to a limited extent in the electron microscope, it being possible to tilt the crystal about two axes perpendicular to the incident beam by no more than $\pm 25°$ on each axis. The techniques of crystal orientation are thus necessarily very different from those employed in X-ray diffraction.

The determination of the orientation of a crystal, and its adjustment, can conveniently be achieved by the use of *Kikuchi lines*, which are the straight bright or dark lines or bands which run through the background scattering on the diffraction pattern of a crystal (Fig 10.11). As the thickness of the crystal increases the Kikuchi lines become stronger and the spot diffraction pattern becomes weaker, so that eventually the spot diffraction pattern has faded into insignificance and only Kikuchi lines are visible; ultimately, when the crystal is very thick, the incident electrons are completely absorbed. In a crystal affected by internal strains the Kikuchi lines are broad and diffuse because the strained crystal scatters the Kikuchi radiation incoherently. For a complete interpretation of Kikuchi lines it is necessary to make use of the dynamical theory of diffraction modified to take account of the elastic scattering of electron beams which have previously been scattered inelastically. We shall use the simpler treatment, due to Kikuchi, which accounts adequately for the main features of the geometry of Kikuchi lines and so enables one to determine the orientation of the crystal to an accuracy of $0.1°$ under optimum conditions; the positions of the Kikuchi lines relative to their associated diffraction maxima can then be utilized to move the crystal to a selected orientation within the rather stringent limits of tilt permitted in the electron microscope.

In the introduction to this chapter it was noted that electrons may be scattered inelastically and incoherently with only a small loss of energy. Such electrons will be scattered in directions close to the incident beam, the intensity of the inelastically scattered electrons decreasing smoothly with increasing angle between the direction of scattering and the incident beam. The smooth decrease in intensity with increasing divergence from the incident beam direction is critical to the arguments which will be developed in the ensuing paragraphs. It is possible that such inelastically scattered electrons, which are mainly responsible for the background intensity observed in diffraction patterns, will be travelling in a direction which will permit them to be scattered elastically so as to give rise to Bragg reflexion from some set of lattice planes (*hkl*). The difference in energy between the inelastically scattered electrons and the incident electrons is of the order of 10 eV, which is negligible compared with the total energy of such electrons. It is consequently reasonable to make the simplifying assumption that the wavelength of the inelastically scattered electrons is equal to that of the incident electron beam.

Now suppose that electrons are scattered inelastically from the point P (Fig 10.16) without significant change of wavelength. Those electrons scattered in the directions PQ and PR will be travelling in directions such that they will be reflected by the (*hkl*) and ($\overline{hkl}$) planes respectively to give Bragg reflexions in the directions QQ′ and RR′ such that QQ′ ∥ PR and RR′ ∥ PQ. For (*hkl*) planes inclined to the incident beam at an angle ε, as shown in Fig 10.16, the electrons travelling in the direction PQ will have a smaller angle of inclination $\theta - \varepsilon$ to the incident beam than will those travelling in the direction PR, whose angle of inclination to the incident beam will be $\theta + \varepsilon$. Therefore the beam parallel to PQ will be of higher intensity than that parallel to PR and it follows that the reflected beam QQ′ will be of higher intensity than the reflected beam RR′. Bragg reflexion of the inelastically scattered beam parallel to PQ occurs because PQ is at the Bragg angle θ to the (*hkl*) planes; if all possible directions of inelastically scattered beams are considered, only those inclined at the angle θ to (*hkl*) will suffer Bragg reflexion from these planes and all possible diffracted beams will lie on the cone of semi-angle $90° - \theta$ about the normal to (*hkl*) as illustrated in Fig 10.17. Similarly all the electron beams such as RR′ diffracted by the planes ($\overline{hkl}$) will lie on the cone of

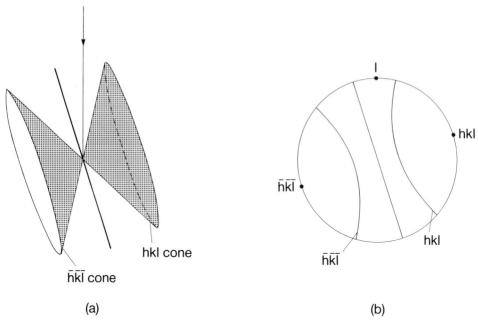

Fig 10.17 Inelastically scattered electrons will only suffer Bragg reflexion from (*hkl*) planes if they are at the Bragg angle θ to the (*hkl*) planes. All possible diffracted beams thus lie on the cone of semi-angle $90° - \theta$ about the normal to (*hkl*); this is illustrated for the (*hkl*) and ($\bar{h}\bar{k}\bar{l}$) planes by a three-dimensional diagram in (a) and by a stereogram in (b), where *l* is the incoming electron beam.

semi-angle $90° - \theta$ about the normal to ($\bar{h}\bar{k}\bar{l}$). Since in electron microscopy the recorded diffraction pattern necessarily lies very close to the incident beam, each of these cones will intersect the image plane of the diffraction pattern in a straight line, so that for this particular set of lattice planes the diffraction pattern will be crossed by two parallel lines. The line which includes the diffracted beam QQ′ (Fig 10.16), will be of higher intensity relative to the background because the intensity of the inelastically scattered electrons is greater along PQ than along PR so that the Bragg reflexions cause more electrons to be transferred to the direction PR than are lost from PR and consequently the background intensity is increased along PR and decreased along PQ. The line which includes the diffracted beam QQ′ is thus known as an *excess* line, while the parallel line which includes the diffracted beam RR′, being of lower intensity than the surrounding background, is known as a *defect* line. The excess line is always further away from the direction of the incident beam than is the defect line, as can be seen in Fig 10.16 where RR′ is inclined at $\theta - \varepsilon$ and QQ′ at $\theta + \varepsilon$ to the incident beam. Since the angle between RR′ and QQ′ is 2θ the distance apart of the two lines on the diffraction pattern corresponds always to 2θ. Such pairs of excess and defect Kikuchi lines will occur for all possible reflecting planes and so the background of the electron diffraction pattern will be crossed by a network of pairs of Kikuchi lines when the crystal is adequately thick.

Since Kikuchi lines are the intersections of the plane of the diffraction pattern with flat cones whose axes are the normals to the reflecting lattice planes, the Kikuchi lines on the diffraction pattern lie parallel to the traces of the relevant lattice planes; therefore each pair of Kikuchi lines is normal to the corresponding *hkl* reciprocal lattice vector of the diffraction pattern and the separation of the pair is $2\theta L$, where L is the effective crystal to film distance. The reciprocal lattice vector $\mathbf{d}^*_{(hkl)}$ has modulus $1/d_{(hkl)} = (2 \sin \theta)/\lambda = 2\theta/\lambda$ for θ small, so that on the film its length is $2\theta L$ (Fig 10.9). Thus the separation of the pair of hkl Kikuchi lines corresponds to the reciprocal lattice vector $d^*_{(hkl)}$.

If the incident electron beam is inclined at an angle ε to the (hkl) plane, the hkl Kikuchi lines on the diffraction pattern will lie at distances from the origin of reciprocal space corresponding to $\varepsilon - \theta$ for the defect line and $\varepsilon + \theta$ for the excess line. The positions of the Kikuchi lines relative to the associated reciprocal lattice points hkl and $\overline{hkl}$ are extremely sensitive to the angle the incident electrons make with the reflecting planes and consequently with the reciprocal lattice vector d^*_{hkl}, the displacement being about 1 cm per degree of ε for $\lambda L \simeq 2$ Åcm. Thus the angle that a particular reciprocal lattice vector $\mathbf{d}^*_{hkl}$ makes with the tangent plane to the reflecting sphere at the origin can be found by observation of the relative positions of the hkl reflexion and the hkl Kikuchi lines. Two relationships are of practical significance: one gives the orientation required to produce a centred diffraction pattern for a zone of reflexions and the other gives the optimum orientation for the production of a dark field image.

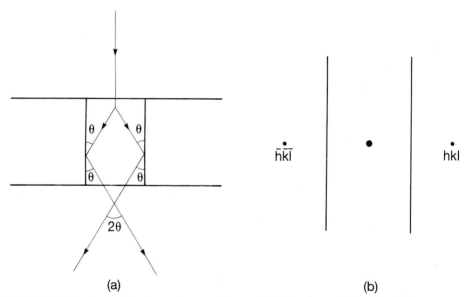

(a)   (b)

Fig 10.18 When the incident beam is accurately parallel to the zone axis the Kikuchi lines are symmetrically disposed at a distance corresponding to θ on either side of the origin of reciprocal space. The diffraction geometry is shown in (a). The Kikuchi lines which bisect lines drawn from the origin to the reflexions hkl and $\overline{hkl}$ are shown in (b).

When the incident beam is accurately parallel to the axis of the zone whose diffraction pattern is to be recorded, the hkl planes of the zone will be parallel to the incident beam so that $\varepsilon = 0$ and the Kikuchi lines will be symmetrically disposed at a distance corresponding to θ on either side of the origin of reciprocal space (Fig 10.18). Moreover the Kikuchi lines will bisect lines drawn from the origin to the reciprocal lattice points hkl and $\overline{hkl}$ respectively; since the geometry is here symmetrical about the incident beam, the two Kikuchi lines will be of equal intensity. When the incident electron beam is parallel to a zone axis, the pairs of Kikuchi lines for all reflexions in the zone will be symmetrical about the incident beam.

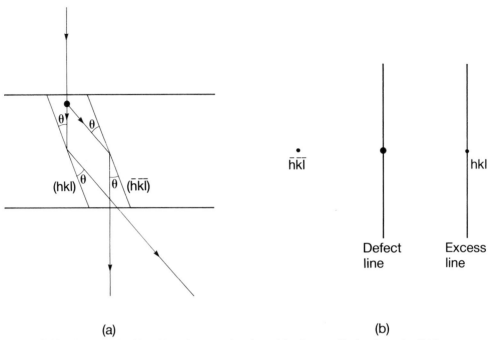

Excess
line

(a)   (b)

Fig 10.19 When the incident beam is correctly oriented for Bragg reflexion from the (*hkl*)
planes, as shown in (a), the Kikuchi defect line passes through the origin of reciprocal space and
the Kikuchi excess line passes through the *hkl* reflexion as shown in (b).

When the incident beam is correctly oriented for Bragg reflexion from the (*hkl*)
planes, that is when it is inclined at the Bragg angle θ to the zone axis and coplanar
with the zone axis and the normal to the (*hkl*) planes (Fig 10.19), the Kikuchi defect
line will pass through the origin of reciprocal space and the excess line will pass
through the reciprocal lattice point *hkl*. In this case $\varepsilon = \theta$; this is the optimum
orientation for dark field imaging.

The geometry of Kikuchi lines can be explained quite simply in terms of the
reciprocal lattice and the reflecting sphere. Since Kikuchi radiation arises by reflexion
in the (*hkl*) and ($\overline{hkl}$) planes, the reciprocal lattice points *hkl* and $\overline{hkl}$ must lie on the
reflecting spheres drawn through the origin of reciprocal space on diameters parallel
to the directions of the inelastically scattered electrons which are reflected into the
Kikuchi lines. In the case where the inelastically scattered electrons are coplanar with
the incident beam and the normals to the reflecting planes these diameters will be
parallel to PQ and PR (Fig 10.16).

Taking the excess line as our example and supposing that the lattice planes (*hkl*) are
inclined at an angle ε to the incident beam, the reciprocal lattice vector $\mathbf{d}^*_{hkl}$ will be
inclined at the angle ε to the plane (OP in Fig 10.20(a)) of observation of the
diffraction pattern, which is normal to the incident beam. The reflecting sphere
corresponding to Kikuchi radiation scattered in the plane of the diagram by the (*hkl*)
lattice planes can be constructed by locating its centre C', which is distant $1/\lambda$ from
both the origin of reciprocal space O and the *hkl* reciprocal lattice point A. The
scattered radiation is in the direction C'A, which makes the angle $\widehat{OCA}$ with the

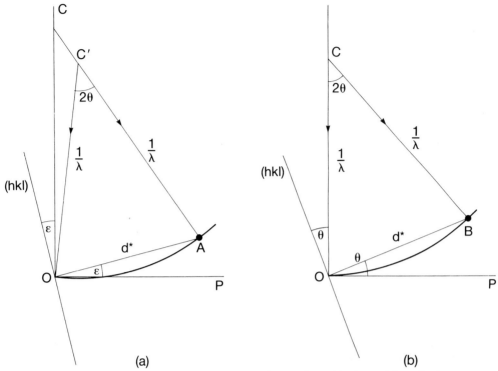

(a)   (b)

Fig 10.20 The geometry of Kikuchi lines in terms of the reciprocal lattice and the reflecting sphere. In (a) the plane of observation of the diffraction pattern, represented by OP, is inclined at the angle ε to d^*_{hkl}, represented by OA; the centre of the reflecting sphere for Kikuchi radiation lies at C' which is distant $1/\lambda$ from O and A. AC' intersects the incident beam in C. The scattered radiation is in the direction CA, which is inclined at $\theta + \varepsilon$ to the incident beam. In the special case, illustrated in (b), where $\varepsilon = \theta$ the reciprocal lattice vector d^*_{hkl}, represented by OB, is inclined at θ to the plane of observation OP and C and C' are coincident; this corresponds to Bragg reflexion of the incident beam as illustrated in Fig 10.19.

incident electron beam. Since

$$\widehat{OC'A} = 2\theta$$

and

$$\widehat{COC'} = 90° - \widehat{C'OA} - \widehat{AOP}$$

$$= 90° - (90° - \theta) - \varepsilon$$

$$= \theta - \varepsilon,$$

then

$$\widehat{OCA} = \widehat{OC'A} - \widehat{COC'}$$

$$= 2\theta - (\theta - \varepsilon)$$

$$= \theta + \varepsilon,$$

a result obtained earlier by a different argument. In the special case when $\varepsilon = \theta$ (Fig 10.20(b)) OC' is parallel to OC and the reciprocal lattice vector, now represented by OB, is inclined at the angle θ to the plane (OP) of observation of the diffraction pattern. This represents Bragg reflexion of the incident beam by the lattice planes (hkl).

It is often necessary to measure the distance between the excess Kikuchi line and the

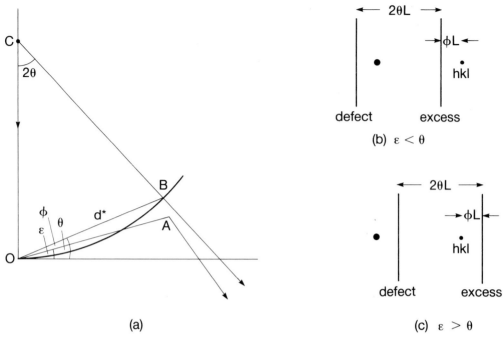

Fig 10.21 Separation of the excess Kikuchi line from the corresponding *hkl* reflexion. In (a) *C* is the centre of the reflecting sphere, *OB* is the orientation of d^*_{hkl} for Bragg reflexion, *OA* is the actual orientation of d^*_{hkl}, $\widehat{COB} = 90° - \theta$, $\widehat{COA} = 90° - \varepsilon$, and $\widehat{AOB} = \phi$. The *hkl* reflexion and the excess Kikuchi line are respectively distant $2\theta L$ and $(\theta + \varepsilon)L$ from the origin of reciprocal space and are separated from one another by ϕL. When *A* lies outside the reflecting sphere the excess Kikuchi line lies between the origin and the *hkl* reflexion as shown in (b): when *A* lies inside the reflecting sphere the excess Kikuchi line lies beyond the *hkl* reflexion as shown in (c).

corresponding *hkl* reflexion on the diffraction pattern. Let ϕ be the angle between the *hkl* reciprocal lattice vector OA for the actual orientation of the crystal and the chord OB of the reflecting sphere drawn on OC (in the direction of the incident beam) as radius (Fig 10.21(a)). The orientation of $\mathbf{d}^*_{hkl}$ for Bragg reflexion of the incident electron beam is thus represented by OB so that $\phi = \theta - \varepsilon$. Since the *hkl* reflexion and the excess Kikuchi line are distant $2\theta L$ and $(\theta + \varepsilon)L$ from the origin of reciprocal space respectively, their separation is $\{2\theta - (\theta + \varepsilon)\}L = (\theta - \varepsilon)L = \phi L$. When the *hkl* reciprocal lattice vector OA for the actual orientation of the crystal lies outside the reflecting sphere $\varepsilon < \theta$ so that the excess Kikuchi line lies between the *hkl* reflexion and the origin of reciprocal space (Fig 10.21(b)); but when OA lies inside the reflecting sphere $\varepsilon > \theta$ and the excess Kikuchi line lies further out than the *hkl* reflexion (Fig 10.21(c)).

Although the positions of Kikuchi lines on a diffraction pattern are sensitive to the precise orientation of the reciprocal lattice vector, the same cannot be said of the spot diffraction pattern itself. Since the radius $(1/\lambda)$ of the reflecting sphere for electrons is very large compared to the dimensions of the reciprocal lattice and since the crystals used for electron diffraction are very thin, the reciprocal lattice points are, as we have shown earlier, replaced by spikes normal to the plane of the flake (Fig 10.8) and the intersection of any part of a spike with the reflecting sphere will give rise to a diffracted beam in effectively the same direction as it would be if the reflecting sphere had passed through the reciprocal lattice point itself (Fig 10.22). While the positions of the

Kikuchi lines depend on the angle ϕ, the positions of the diffracted beams are effectively independent of the precise orientation of the reciprocal lattice net relative to the incident beam; the intensities of the diffracted beams will however be dependent on ϕ (p. 354). The diffracted beams may therefore be assumed to be at the Bragg angles and the angle ϕ may be determined by measuring the perpendicular distance ϕL between the *hkl* reflexion and the excess Kikuchi line.

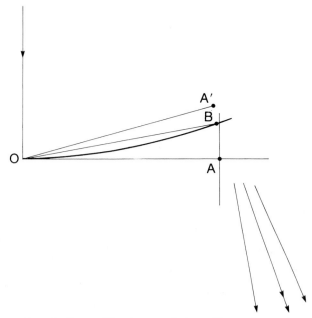

Fig 10.22 The position of a Bragg diffraction spot is insensitive to the precise orientation of the thin crystal because the reciprocal lattice point is drawn out into a spike parallel to the small dimension of the crystal, which is taken to be perpendicular to *OB*. *OB* represents d^* as a chord of the reflecting sphere; this orientation gives rise to the double arrowed diffracted beam. *OA* and *OA'* represent small deviations of d^* from that orientation: reflexion, giving rise to the single arrowed diffracted beams, occurs because the deviation is sufficiently small for the spike still to intersect the reflecting sphere. The reciprocal lattice spike is shown only for the orientation *OA* of d^*_{hkl}.

When imaging a crystal using diffraction contrast techniques it is often convenient to know the distance s of a particular reciprocal lattice point from the reflecting sphere. This distance s is known as the *Bragg deviation parameter* (Fig 10.23). As the angles involved are small $\phi = s/g$, where g is the length of the reciprocal lattice vector. Therefore $s = g\phi = g(p/L)$, where p is the perpendicular distance of the excess Kikuchi line from the *hkl* reflexion. When L is unknown, as will sometimes be the case, it becomes necessary to measure also the perpendicular distance P between the pair of Kikuchi lines. We have already seen that $P = 2\theta L$ (p. 368). Rewriting the Bragg equation $\lambda = 2d \sin \theta$ for small θ as $\lambda = 2d\theta$ and substituting the modulus of the *hkl* reciprocal lattice vector $g = 1/d$ we obtain $g\lambda = 2\theta$. Therefore the separation of the Kikuchi lines on the photograph is $P = g\lambda L$, the scale factor is $L = P/(g\lambda)$, and the Bragg deviation parameter is $s = g^2\lambda(p/P)$. When the reciprocal lattice vector **g** lies inside the reflecting sphere s is taken to be positive and when it lies outside s is negative (Fig 10.23). Thus both the magnitude and sign of s can be determined from the Kikuchi lines. When the excess *hkl* Kikuchi line is further from the origin than the *hkl* reflexion, *s* is positive; when it lies between the origin and the *hkl* reflexion, *s* is negative (Fig 10.21).

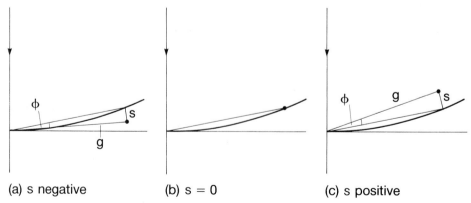

(a) s negative   (b) s = 0   (c) s positive

Fig 10.23   For a reciprocal lattice point whose reciprocal lattice vector is *g* the Bragg deviation parameter s is the distance, measured along the spike through the reciprocal lattice point, of the reciprocal lattice point from the reflecting sphere. (a) shows the reciprocal lattice point outside the reflecting sphere, (b) on the reflecting sphere, and (c) inside the reflecting sphere.

Indexing Kikuchi lines

We have already shown that the *hkl* pair of Kikuchi lines is normal to the reciprocal lattice vector d^*_{hkl} and has a separation $d^*_{hkl}\lambda L$, which is the length of the *hkl* reciprocal lattice vector on the observed diffraction pattern. The Kikuchi lines on a diffraction pattern can thus be indexed by locating the Bragg reflexion corresponding to each pair of Kikuchi lines and indexing the reciprocal lattice points by the methods described earlier (p. 359). Once the spots on the diffraction pattern have been indexed, the corresponding pairs of Kikuchi lines can be identified and indexed. However some pairs of Kikuchi lines may not correspond to reflexions recorded on the photograph because they arise from lattice planes whose normals are not in a reflecting orientation for the incident electron beam. In particular Kikuchi lines associated with upper layer reciprocal lattice points may be recorded on a photograph which shows only the zero layer reciprocal lattice net. For instance in the [001] diffraction pattern of a cubic-close-packed metal the 331 Kikuchi lines, and those related by the [001] tetrad, may be observed. Such lines will be parallel to the traces of {331} on the plane of the diffraction pattern and the separation of each pair will be $d^*_{331}\lambda L$. Since tilting the crystal will simply move pairs of Kikuchi lines across the diffraction pattern, an easy way of indexing Kikuchi lines when there are few Bragg reflexions and when the incident beam is known to be approximately parallel to a zone axis [*UVW*] is to plot the reciprocal lattice net for the relevant Laue zone of reflexions and to superimpose on it the pattern of Kikuchi lines (sometimes called the Kikuchi map). If necessary the reciprocal lattice points of the first layer, for which $hU + kV + lW = 1$, may be added so that the corresponding Kikuchi lines can be drawn. The reader who wishes to explore this technique further is referred to Hirsch *et al.* (1977), pp. 121–124.

When some of the observed Kikuchi lines correspond to reciprocal lattice points which are not on the zero-layer net—as for example the 331 Kikuchi lines on photographs of the diffraction patterns of cubic crystals taken with the incident beam along [001]—such lines can be used to determine the disposition of non-zero layer reciprocal lattice points. They can in consequence be used to provide a means of identifying which spot on the zero layer photograph corresponds to the *hkl* reflexion and which to the $\overline{hkl}$ reflexion, a practical problem to which reference was made earlier (p. 360). Then the Laue symmetry of the crystal about the zone axis which is parallel to the incident beam can be determined. The reader will recall that the symmetry of a zero layer reciprocal lattice net is that associated with the direction normal to the net plus a centre of symmetry.

Kikuchi lines may also be used to determine the precise orientation of the incident beam relative to the crystal (Fig 10.11). Three pairs of Kikuchi lines arising from lattice planes which do not lie in a zone are necessary for the orientation of the crystal to be determined uniquely. We have noted earlier that the cones of Kikuchi radiation associated with the *hkl* and $\overline{hkl}$ reflexions are symmetrically disposed with respect to the (*hkl*) lattice planes (Fig 10.16). On a diffraction pattern therefore the trace of the (*hkl*) plane will lie midway between the defect and the excess Kikuchi line. Suppose that a photograph shows three pairs of Kikuchi lines; then the traces of the three lattice planes $(h_1k_1l_1)$, $(h_2k_2l_2)$, and $(h_3k_3l_3)$ may be drawn and, producing the traces off the photograph if necessary, their points of intersection P, Q, and R located (Fig 10.24). Each point will correspond to the intersection of the axis [UVW] of the zone containing the two planes with the plane of observation. The zone axis symbols can readily be found by cross multiplication, e.g. $[U_R V_R W_R] = [k_1 l_2 - k_2 l_1, l_1 h_2 - l_2 h_1, h_1 k_2 - h_2 k_1]$, care being taken to keep the sense of the lattice vectors consistent. The indices may then be confirmed by calculating the angles between the three zone axes and making comparison with the angles obtained by relationships of the type $RP/L = [U_R V_R W_R]:[U_P V_P W_P]$, where L is the effective crystal to film distance. The poles of the three zone axes may now be plotted on a stereogram. Suppose the incident electron beam intersects the film in the point O, then the angles corresponding to the lengths OP, OQ, OR can be evaluated and small circles of radius OP, OQ, OR drawn about P, Q, R respectively to determine, by their point of mutual intersection, the direction of the incident beam.

Alternatively the position of O relative to the crystallographic axes may be expressed as *p***a**, *q***b**, *r***c**, where *p*, *q*, *r* are not necessarily integral. Evaluation of *p*, *q*, *r* analytically can then be achieved by writing expressions for the angles OP, OQ, OR in terms of *p*, *q*, *r* and $[U_R V_R W_R]$, etc., and solving the three simultaneous equations for *p*, *q*, *r*.

The observation and measurement of Kikuchi lines provides, as has been indicated in the preceding paragraphs, a valuable contribution to the study of crystalline solids with the electron microscope. The use of Kikuchi lines for the calibration of electron microscopes as well as for the characterization and orientation of phases is discussed by Thomas and Goringe (1979).

Imaging

In transmission electron microscopy two types of imaging are in common use. One type of image is formed by phase contrast when two or more of the electron beams leaving the lower (or back) surface of the crystal are recombined to form a high resolution image of the crystal in much the same way as an image is formed in an optical microscope. The other type of image is formed by allowing either the transmitted electron beam alone or just one of the diffracted beams to form an image; such an amplitude contrast image records the variation in intensity of a single beam

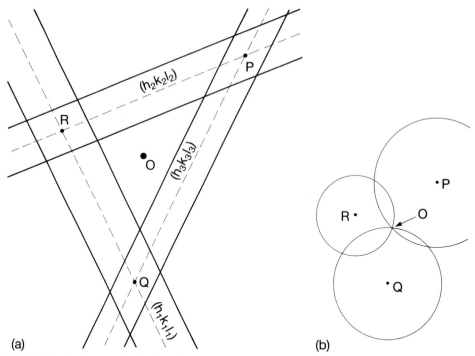

Fig 10.24 A photograph showing three pairs of Kikuchi lines is represented in (a), where *O* represents the intersection of the incident electron beam with the photograph. The traces of the lattice planes corresponding to the pairs of Kikuchi lines are shown as broken lines intersecting in pairs at *P*, *Q*, and *R*. Indexing of the Kikuchi lines enables the zone axes corresponding to *P*, *Q*, *R* to be calculated so that these directions can be plotted on a stereogram. Evaluation of the angles *OP*, *OQ*, *OR* from the diffraction pattern enables small circles of the corresponding radii to be plotted on the stereogram as shown in (b); their point of mutual intersection, *O*, corresponds to the direction of the incident electron beam and so provides a precise determination of the orientation of the crystal in the electron microscope.

across the lower (or back) surface of the crystal. We deal with this latter type of image first.

Amplitude contrast microscopy

The interpretation in detail of images formed by amplitude contrast (sometimes described as diffraction contrast) requires the application of the dynamical theory of electron diffraction. Dynamical theory in its simplest form assumes that only one diffracted beam is produced during the passage of the incident beam through the crystal. This restriction will be satisfied if the crystal is tilted so that only one reciprocal lattice point, other than the origin point, lies on the reflecting sphere (Fig 10.25). The experimental arrangements for amplitude contrast microscopy normally have to be such as to satisfy this *two beam* criterion. In these circumstances the corresponding Kikuchi lines will pass through the two reflexions, the defect line through the 000 reflexion and the excess line through the *hkl* reflexion.

The kinematical theory of electron diffraction is not capable of providing a detailed interpretation of image contrast. It is applicable only if the crystal is very thin and if no strong Bragg reflexion is occurring. In what follows we shall be concerned principally with those aspects of amplitude contrast images which are explicable on the

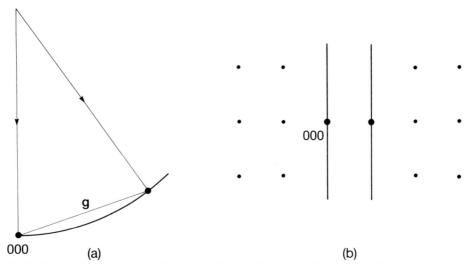

Fig 10.25 Amplitude contrast microscopy. The situation where the crystal is tilted so that only one reciprocal lattice point other than the origin point lies on the reflecting sphere is shown in (a). The corresponding Kikuchi lines pass through the two reflexions as shown in (b).

kinematical theory: the essential geometry of the image, but not necessarily the precise distribution of intensity across the image.

If the objective aperture is so positioned that the image is formed by the incident electron beam (Fig 10.26(a)), the resultant image is described as a *bright-field image*. Areas of the specimen which diffract very little of the incident beam into the selected *hkl* reflexion will appear bright; whereas areas of the specimen which give rise to appreciable diffracted intensity in the selected *hkl* reflexion (the operating reflexion) will appear dark in consequence of the resulting diminution in intensity of the transmitted beam leaving the back surface of those areas of the crystal. The variation in the intensity of the transmitted beam leaving the back surface of the crystal is thus recorded on the image.

If however the image is formed by the diffracted beam from the selected *hkl* reflexion, the parts of the image which appear bright will correspond to those areas of the crystal which give rise to a strong *hkl* diffracted beam passing through the back surface of the crystal. Such an image is described as a *dark field image*. If the incident electron beam were parallel to the axis of the microscope and the objective aperture were positioned to allow the passage only of the *hkl* diffracted beam (Fig 10.26(b)), then the beam passing through the off-centred aperture would be inclined to the microscope axis and so would be significantly affected by spherical aberration. Higher resolution in the image may be achieved by tilting the incident electron beam through $2\theta_{hkl}$ so that the diffracted beam becomes parallel to the microscope axis (Fig 10.26(c)). For the crystal in the orientation already established this will be the $\overline{hkl}$ diffracted beam. In comparing bright-field and dark-field images it is important to appreciate that **g** is reversed in the dark field image with respect to its orientation in the bright field image.

When the incident electron beam is tilted through 2θ in the opposite sense to its tilt for the $\overline{hkl}$ dark-field image, it will be inclined to the relevant reciprocal lattice vector **g** at $90° - 3\theta$ and, since θ is small, the chord of the reflecting sphere in the direction defined by **g** will be of length $3g$ so that the reciprocal lattice point $3h, 3k, 3l$ will lie on the reflecting sphere (Fig 10.26(d)). If the reciprocal lattice spike through the *hkl* reciprocal lattice point is long enough to intersect the reflecting sphere, a weak *hkl* diffracted beam will occur and may be used to form a dark field image. Such *weak-beam dark-field images* can be used to give improved resolution in dark-field images; this topic will not be pursued here and the reader is referred to Thomas and Goringe (1979) for a detailed discussion.

Since contrast in bright-field and in dark-field images is due to variation in the intensity of the electron beam transmitted by the crystal, a first step in interpreting such images can be made by calculating the wave transmitted by the crystal and so determining the variation of intensity across the back face of the crystal for either the transmitted electron beam or a diffracted electron beam. It can be shown by use of the kinematical theory (Hirsch *et al.*, 1977) that when a parallel beam of electrons of unit amplitude is incident on the crystal the intensity of the diffracted beam leaving a point on the back surface of the crystal is $(\sin^2 \pi t s)/(\xi_g s)^2$ where ξ_g is the *extinction distance* for the Bragg reflexion corresponding to the reciprocal lattice vector $\mathbf{g}$, t is the thickness of the crystal, and s is the Bragg deviation parameter. The extinction distance ξ_g is the distance in the crystal at which the intensity of the diffracted beam first falls to zero because it has effectively been rescattered into the transmitted beam; the concept of extinction distance is inherent in dynamical theory. Under two-beam conditions the intensity of the transmitted beam will simply be the intensity of the incident beam less the intensity of the diffracted beam in the absence of absorption and incoherent scattering.

When $s = 0$, $(\sin^2 \pi t s)/(\xi_g s)^2$ becomes $(\pi t/\xi_g)^2$ which is greater than unity when $t > \xi_g/\pi$. However crystals used for electron microscopy may be of thickness $t \geqslant 10\xi_g$; ξ_g varies with $\mathbf{g}$ and for low-angle reflexions in metals may be as small as 150 Å. It is clearly impossible for the diffracted intensity to be greater than the incident intensity. Thus the kinematical theory fails for $s = 0$ and t large; it is however useful here in that it provides a simple explanation of the mechanism of formation of amplitude contrast.

The bright-field and dark-field images of a perfect flat single crystal of uniform thickness should show no contrast as the wave transmitted at every point will be of the same intensity.

We now proceed to discuss amplitude contrast originating in three ways: (1) variation in the orientation of the specimen with respect to the incident beam, (2) variation in the thickness of the specimen, (3) imperfections in the crystal structure, this last being the most important.

Bend contours

Very thin crystals are not always flat; they may be bent or buckled by the heating effect of the electron beam or some other cause. When the crystal is bent the direction of the reciprocal lattice vector $\mathbf{g}$ of the operating reflexion will vary across the crystal. If the orientation of the reciprocal lattice at a particular point in the crystal is such

Fig. 10.26 Bright-field and dark-field images. The formation of a bright-field image is shown in (a), where the operating reflexion *hkl* lies on the reflecting sphere and the objective aperture selects the incident beam. The formation of a dark-field image is shown in (b), where the operating reflexion *hkl* lies on the reflecting sphere and the objective aperture selects the *hkl* diffracted beam. Higher resolution in the dark-field image is obtained by tilting the incident electron beam through $2\theta_{hkl}$ so that the diffracted beam $\overline{hkl}$ is parallel to the microscope axis as shown in (c). The formation of a weak-beam dark-field image is shown in (d), where the tilt is in the opposite sense to that in (c) so that the origin point and the reciprocal lattice point $3h$, $3k$, $3l$ lie on the reflecting sphere.

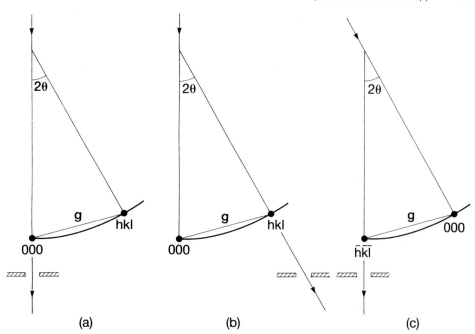

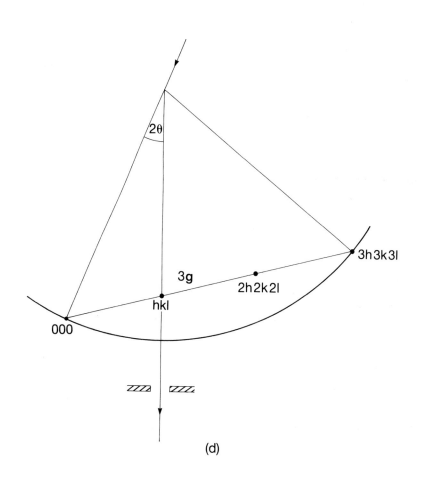

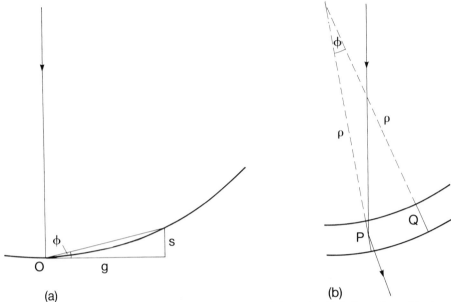

Fig 10.27 Bend contours. (a) serves to define the angle ϕ through which the reciprocal lattice vector **g**, shown normal to the incident beam, has to be rotated to be a chord of the reflecting sphere. If s is the distance of the reciprocal lattice point from the sphere, measured along its spike, then $s = g\phi$. (b) shows a uniformly bent crystal of radius ρ: at P the lattice planes corresponding to **g** are correctly oriented for Bragg reflexion, but at Q, distant $\rho\phi$ from P, the reciprocal lattice point is distant $s = g\phi$ from the reflecting sphere.

that a particular reciprocal lattice vector **g** has to be rotated through the small angle ϕ to become a chord of the reflecting sphere and if s is the distance of the reciprocal lattice point, measured along its spike, from the sphere, then $s = g\phi$ (Fig 10.27(a)) and so the intensity of the diffracted beam, $(\sin^2 \pi ts)/\xi_g^2 s^2$, varies with ϕ.

Suppose the crystal to be uniformly bent so that its surface forms part of a cylinder of radius ρ, whose axis is normal to the incident beam and to the reciprocal lattice vector **g**. Then **g** varies smoothly in direction across the crystal, being always tangential to the surface of the crystal. Suppose that at the point P (Fig 10.27(b)) the lattice planes corresponding to **g** are correctly oriented for Bragg reflexion of the incident beam so that **g** lies on the reflecting sphere. At a point Q distant $\rho\phi$ from P the direction of **g** will have been rotated through the angle ϕ so that the reciprocal lattice point is here distant $s = g\phi$ from the reflecting sphere. As Q becomes more distant from P, ϕ increases smoothly and the intensity in the image varies as $(\sin^2 \pi tg\phi)/\xi_g^2 g^2\phi^2$ relative to P. The image therefore displays a set of fringes. The intensity of the fringe centred on P falls to zero when $s = \pm 1/t$, which corresponds to a point Q in the crystal such that $PQ = \rho\phi = \rho(s/g) = \rho/(gt)$. The central maximum of the fringe system thus has a width $2\rho/gt$. Weaker maxima will, under favourable conditions, be visible on either side of the central fringe with zero intensity between adjacent maxima at points corresponding to integral multiples of ρ/gt from the centre of the central fringe.

In a dark field image **g** will be a chord of the reflecting sphere at the point corresponding to P in the crystal so that bright fringes corresponding to the variation of intensity along the reciprocal lattice spike associated with this reflexion will be seen in the image.

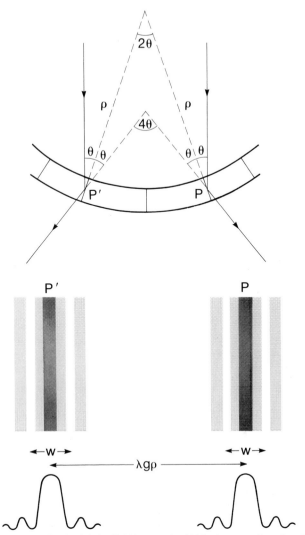

Fig 10.28 Bend contours. In the bright-field image the (hkl) planes at P and at P' give rise respectively to fringe systems associated with the hkl and $\overline{hkl}$ reflexions. The angle between the (hkl) planes at P and at P' is 2θ and, since the radius of curvature of the crystal, ρ, is large $PP' = 2\theta\rho$ in the image plane. For small θ the Bragg equation becomes $\lambda g = 2\theta$, so $PP' = \lambda g\rho$. The separation of the central maxima of the two fringe systems thus corresponds to $\lambda g\rho$. The separation of the zeros of intensity on either side of each central fringe corresponds to $w = 2\rho/gt$. The radius of curvature ρ and the thickness t of the crystal can thus be determined by measurement of the bright-field image.

In the bright field image the variation in intensity is perceptible as a set of dark fringes on a bright background. If the crystal is symmetrical about the incident beam, sets of fringes corresponding to both the hkl and the $\overline{hkl}$ reflexions will be seen (Fig 10.28). If the centre of the fringe system associated with the $\overline{hkl}$ reflexion corresponds to the point P' in the crystal, then the angle between the hkl planes at P and at P' is 2θ. Since the curvature of the crystal is likely to be small, ρ is large and so PP' = $2\theta\rho$; for small θ, always the case in the electron microscope, the Bragg equation becomes $\lambda g = 2\theta$, so PP' = $\lambda g\rho$. It follows that the measurement of PP' on the image yields the radius of curvature ρ of the crystal and measurement of the separation w of the zeros of intensity on either side of each central fringe provides a determination of the thickness t of the crystal since $w = 2\rho/gt$.

This treatment of bend contours is valid only for thin crystals. For thicker crystals the dynamical theory has to be used to interpret the images and it is found that bend contour fringes in dark field images are symmetrical about the position in the image corresponding to Bragg reflexion whereas in bright field images the fringes are asymmetrical. In bright field images the maximum transmitted intensity corresponds to the reciprocal lattice vector $\mathbf{g}$ of the operating reflexion lying just inside the reflecting sphere.

Thickness fringes

The kinematical theory predicts that variations in the thickness of the crystal will not give rise to appreciable variations in the intensity of a diffracted beam when the reciprocal lattice point lies on the reflecting sphere, the intensity being simply related to the volume of the crystal. But when the crystal is so oriented that the reflecting sphere intersects the spike associated with the reciprocal lattice point at a distance s from the reciprocal lattice point itself, then the intensity on the image is diminished by the factor $(\sin^2 \pi t s)/\xi_g^2 s^2$ relative to the intensity when Bragg reflexion occurs. Since s is constant the intensity will depend on the crystal thickness t and will be zero for $t = n/s$. For a crystal wedge of uniformly varying thickness the image will show a fringe system of periodicity corresponding to a change in thickness equal to $1/s$. In general the fringes will correspond to contours of equal t. Such fringes will be quite weak on bright field images, but more clearly visible on dark field images.

The kinematical theory predicts that as the reciprocal lattice vector $\mathbf{g}$ approaches the reflecting sphere, that is as $s \to 0$, the periodicity of the fringe system becomes infinite. That is not so; the dynamical theory indicates that the periodicity at $s = 0$ is finite and equal to the extinction distance ξ_g. In the dynamical theory the extinction distance is an important parameter: when the crystal is oriented so that one diffracted beam is produced, the intensities of the transmitted beam and of the diffracted beam vary with the thickness t of the crystal as $\cos^2 (\pi t/\xi_g)$ and $\sin^2 (\pi t/\xi_g)$ respectively. The extinction distance ξ_g is, as we have indicated earlier, the distance in the crystal at which the intensity of the diffracted beam first falls to zero by its being effectively rescattered into the transmitted beam. The kinematical theory makes satisfactory predictions only when $s \gg 1/\xi_g$; it is unsatisfactory when the crystal is in or near the Bragg reflecting position.

It may be that changes in s and t occur simultaneously, that the crystal varies from point to point in both thickness and orientation as in the case of a bent wedge. Dark field images will generally show contrast, which depends on the variation of the product ts of the thickness and the Bragg deviation parameter from point to point, more clearly than bright field images because the contrast is seen against a dark rather than a bright background.

Contrast due to crystal imperfections

We confine our treatment of this topic to the contrast which arises when some of the atoms in the crystal are displaced from their ideal positions. Atomic displacement will occur when the crystal contains a *stacking fault*: for example an hexagonal close-packed crystal[4] with the stacking sequence ... ABABCACA ... has a stacking fault between the adjacent B and C type planes, the fault being a displacement $\frac{1}{3}, \frac{2}{3}, 0$ of the atoms from the expected A plane to the C plane. Atomic displacement will also occur when the crystal contains *antiphase boundaries*: for example, in the very simple case of an XY structure in which two types of site, α and β, occur throughout the crystal, atoms of X occupy α sites on one side of the antiphase boundary and atoms of Y occupy the α sites on the other side and *vice versa* for the β sites. Both stacking faults and antiphase boundaries have atomic displacements which occur on a plane which may or may not run right across the crystal and such displacements are large, being of the order of magnitude of an interatomic distance. Atomic displacements may also be of the nature of small strains due to the presence of defects in the structure. One sort of defect which gives rise to local strain in a crystal of invariant chemical composition is a *dislocation*. Local strain may also arise from the presence within the host crystal of small volumes of different composition with essentially the same structure as the host but slightly different unit-cell dimensions; in this case, contrast may be due not only to strain but also to the compositional difference between precipitate and matrix phases.

It is not our intention to provide a detailed treatment of contrast in the bright-field and dark-field images of imperfect crystals. We shall confine our treatment to a simple account of the kinematical approach and draw the attention of the reader to certain conclusions of general significance which follow therefrom.

Let an atom with scattering factor f_n be situated ideally at a vector distance $\mathbf{r}_n$ from the origin of a unit-cell and let the origin of that unit-cell be distant $\mathbf{r}_l = U\mathbf{a} + V\mathbf{b} + W\mathbf{c}$ (where U, V, W are integers) from the chosen origin for the whole crystal. Now suppose that the actual position of the atom is displaced by $\mathbf{R}$ from its ideal position at $\mathbf{r}_l + \mathbf{r}_n$ so that it is at $\mathbf{r}_l + \mathbf{r}_n + \mathbf{R}$ relative to the chosen origin for the whole crystal. The wave scattered into the *hkl* reflexion by the atom will be $f_n \exp 2\pi i(\mathbf{r}_l + \mathbf{r}_n + \mathbf{R}) \cdot \mathbf{g}$, where $\mathbf{g}$ is the *hkl* reciprocal lattice vector. The wave ψ scattered by the whole crystal into the *hkl* reflexion when a wave of unit amplitude is incident on the crystal is given by

$$\psi_c = \sum f_n \exp 2\pi i(\mathbf{r}_l \cdot \mathbf{g} + \mathbf{r}_n \cdot \mathbf{g} + \mathbf{R} \cdot \mathbf{g}),$$

where the summation is taken over the whole crystal. Since $\mathbf{r}_l$ is a lattice vector $U\mathbf{a} + V\mathbf{b} + W\mathbf{c}$ and $\mathbf{g}$ is a reciprocal lattice vector $h\mathbf{a}^* + k\mathbf{b}^* + l\mathbf{c}^*$, it follows that $\mathbf{r}_l \cdot \mathbf{g} = hU + kV + lW$, which must be integral so that this scalar product has no effect on the value of the exponential term. If it is assumed that $\mathbf{R}$ has the same value for all the atoms in one unit-cell (this will certainly be so in the extreme case of a primitive unit-cell with one atom per lattice point), then we can put $\psi = \sum f_n \exp 2\pi i \, \mathbf{r}_n \cdot \mathbf{g}$, where the summation is taken over the atoms in one unit-cell, and the expression for the wave scattered by the whole crystal becomes

$$\psi_c = \sum \psi \exp 2\pi i \mathbf{R} \cdot \mathbf{g},$$

where the summation is taken over all unit-cells. In a perfect crystal $\psi_c = \sum \psi$, where the summation is over all unit-cells. In this section we have assumed that the presence of the displacement $\mathbf{R}$ is such that the reciprocal lattice vector $\mathbf{g}$ does not vary in

[4] Close packing will be discussed in volume 2. Here an indication of the geometrical nomenclature suffices. A close packed plane has an hexagonal unit-mesh, which may be one of three types: type A has atoms at 0, 0, type B has atoms at $\frac{2}{3}, \frac{1}{3}$, type C has atoms at $\frac{1}{3}, \frac{2}{3}$. In a three-dimensional close packed structure close packed planes are superimposed with regular interplanar spacing, the origin of each successive plane lying on the normal through the origin of the first plane and the reference axes of each plane being parallel. The two simplest sequences in close packing are ABA ... and ABCA ..., known respectively as hexagonal and cubic close packing.

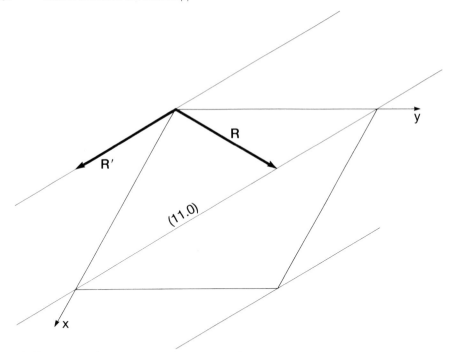

Fig 10.29 Stacking faults: the invisibility criterion. In this hexagonal close packed crystal the fault plane is (0001) and the displacement is $\mathbf{R} = \frac{1}{3}\mathbf{a} + \frac{2}{3}\mathbf{b}$ (which may alternatively be described by $\mathbf{R}' = \frac{1}{3}(\mathbf{a} - \mathbf{b})$). For the invisibility criterion $\mathbf{g}.\mathbf{R} = n$ (where n is an integer) to be satisfied, $h + 2k = 3n$. For the operating reflexion 11.0 the fault is invisible because $\mathbf{g}.\mathbf{R} = 1$ (or $\mathbf{g}.\mathbf{R}' = 0$).

direction within the crystal. This is not necessarily so for small strains; a more rigorous approach allowing for the variation of $\mathbf{g}$ within the crystal leads to substantially the same conclusion, namely that the existence of a strain $\mathbf{R}$ introduces an additional phase $2\pi\mathbf{R}.\mathbf{g}$ into the expression for the wave scattered by the unit-cell affected by the strain both for the position of Bragg reflexion and for a deviation $\mathbf{s}$ from that position.

In reality $\mathbf{R}$ will vary from one atom to another, when the unit-cell contains more than one atom as will normally be the case, and from one unit-cell to another. The calculation of the scattered wave thus presents quite a difficult problem, the more so because for image contrast it is necessary to calculate the scattered wave leaving each point on the back surface of the crystal. However some important conclusions can be drawn without further analysis. If the atomic displacement $\mathbf{R}$ is normal to the reciprocal lattice vector $\mathbf{g}$, then $\mathbf{g}.\mathbf{R} = 0$ and the displacement will have no effect on the wave scattered. In other words an atomic displacement parallel to the (hkl) planes will have no effect on the intensity of the hkl reflexion because the intensity depends only on the positions of atoms relative to the reflecting planes. A stacking fault lying on a plane inclined to the back surface of the crystal can be shown to give rise to a set of fringes parallel to the intersection of this inclined plane with the surface of the crystal (Fig 10.5). No fringes will be observed if the displacement $\mathbf{R}$ is parallel to the lattice planes (hkl) associated with the operating reflexion.

For exemplification we return to the stacking fault in the hexagonal close packed crystal mentioned earlier. The fault plane is (0001) and the displacement across it is $\mathbf{R} = \frac{1}{3}\mathbf{a} + \frac{2}{3}\mathbf{b}$. So $\mathbf{g}.\mathbf{R} = \frac{1}{3}h + \frac{2}{3}k$ and the phase change introduced by the fault is $(2\pi/3)(h + 2k)$. The fault will be invisible when $h = k = 0$, since then $\mathbf{g}.\mathbf{R} = 0$. The fault will also be invisible when $\mathbf{g}.\mathbf{R}$ is equal to an integer n, the phase change then being a multiple of 2π. This will be so when $h+2k = 3n$; for such reflexions the displacement will have no effect on the scattered wave because the vector $\mathbf{R}$ simply moves the atoms by a distance equal to an integral multiple of the $(hk.0)$ interplanar spacing. It is worth noting that the stacking fault could equally well have been described in terms of the displacement vector $\mathbf{R}' = \frac{1}{3}(\mathbf{a}+2\mathbf{b})-\mathbf{b} = \frac{1}{3}(\mathbf{a}-\mathbf{b})$; now $\mathbf{g}.\mathbf{R}' = \frac{1}{3}(h\mathbf{a}^*+k\mathbf{b}^*+l\mathbf{c}^*)$. $(\mathbf{a}-\mathbf{b}) = \frac{1}{3}(h-k)$, which will likewise lead to invisibility when $h+2k = 3n$.

Figure 10.29 illustrates the invisibility of this fault for the operating reflexion 11·0, for which $\mathbf{g}.\mathbf{R} = 1$ and $\mathbf{g}.\mathbf{R}' = 0$. In a different example the magnitude of $\mathbf{g}.\mathbf{R}$ might be so close to an integer that the fault is invisible, not because the phase change is zero but because the phase change is so small and the contrast in the image consequently so small as to be imperceptible. Conversely maximum contrast will be observed when $\mathbf{R}$ is parallel to $\mathbf{g}$ provided that $R \neq n/g$, n being an integer.

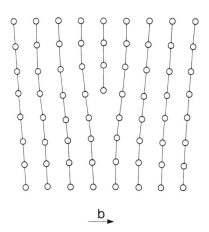

Fig 10.30   Edge dislocation. A plane of lattice points terminates within the crystal in a dislocation line normal to the plane of the diagram. The *Burgers vector* **b** is the shortest lattice vector normal to the incomplete lattice plane.

In the interpretation of images of crystals containing defects which cause atomic displacements in small volumes of the crystal it can be profitable to investigate local variation in the direction of the reciprocal lattice vector $\mathbf{g}$. We take as our example an *edge dislocation*, which is a defect such that a plane of lattice points terminates in a dislocation line within the crystal (Fig 10.30). Such an edge dislocation is character-ized by its *Burger's vector* **b**, which is the shortest lattice vector normal to the incomplete lattice plane. Planes of lattice points parallel to the dislocation line experience small changes in orientation where they pass close to the dislocation line. Now suppose that the incomplete plane of lattice points is normal to the surface of the crystal and that an image is produced when the crystal is oriented so that $\mathbf{g}$ in the perfect crystal lies just inside the reflecting sphere, then at a small distance on one side of the dislocation (to the right in Fig 10.31) $\mathbf{g}$ will make a small angle $-\Delta\theta$, measured anti-clockwise, with the position of $\mathbf{g}$ in the perfect crystal so that s is decreased and the intensity on the image corresponding to such points is locally increased. On the other side of the dislocation $\mathbf{g}$ will make a small angle $+\Delta\theta$ with $\mathbf{g}$ in the perfect crystal so that s is increased and the intensity at corresponding points in the image is locally decreased. So in a bright-field image the dislocation line will appear as a dark line displaced slightly from its true position.

From the symmetry of the edge dislocation illustrated in Fig 10.30 it is apparent that the displacement $\mathbf{R}$ of an atom near to the dislocation line will be in the plane of the diagram and, since $\mathbf{g}$ is effectively normal to the incomplete lattice plane, only atomic displacements parallel to the Burger's vector **b** will contribute to the contrast observable in the image. The dislocation will thus be invisible when $\mathbf{g}.\mathbf{b} = 0$; this *invisibility criterion* provides a powerful tool for the analysis of dislocation arrays in crystals. The detailed interpretation of images of dislocations requires the application of the dynamical theory of diffraction and will not be discussed further here.

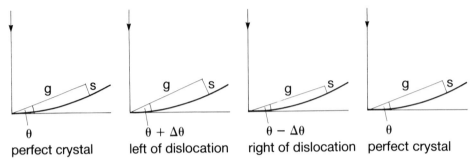

perfect crystal   left of dislocation   right of dislocation   perfect crystal

Fig 10.31   Edge dislocation. The sequence of diagrams from left to right shows the relationship of the reciprocal lattice vector **g** and the Bragg deviation parameter s to the reflecting sphere on passing from the perfect crystal to the left of the dislocation shown in Fig 10.30 to a point close to the dislocation, then to a point on the other side of the dislocation, and finally into perfect crystal again. In a bright-field image the dislocation appears as a dark line displaced slightly from its true position.

Phase contrast microscopy

Hitherto we have discussed image formation in the electron microscope using either the transmitted electron beam or a single diffracted beam. An image analogous to that produced in an optical microscope (Fig 9.4) could be obtained by allowing all the diffracted beams brought to a focus in the back focal plane of the objective lens (Fig 10.2) to contribute to the image. In theory it should be possible in such an image to resolve the atoms in the crystal specimen, but limitations imposed by the design and construction of electron microscopes as well as the difficulty of interpretation of images have generally caused the theoretical limit of resolution to be unattainable. However in certain special cases it has proved possible to image the positions of heavy atoms in a structure. In what follows we indicate the problems encountered in imaging crystal structures and outline the methods used to overcome them.

Lattice fringes

We first consider an extension of bright-field and dark-field imaging by discussing the type of image which would be obtained if both the transmitted beam and one diffracted beam were used to form the image. We assume that the two-beam setting is employed, that the crystal is oriented so that the transmitted beam and the diffracted beam are equally inclined to the axis of the microscope (Fig 10.32), that as they emerge from the back surface of the crystal they have amplitudes A_0 and A_1 respectively, and that the phase of the diffracted beam relative to that of the transmitted beam is $2\pi\phi$. The directions of the transmitted beam and the diffracted beam are represented by the vectors $\mathbf{s}_0$ and $\mathbf{s}$ respectively, where $|\mathbf{s}| = |\mathbf{s}_0| = 1/\lambda$, and the origin is taken on the back surface of the crystal, which is assumed to be parallel to the reciprocal lattice net. The resultant wave at a point distant $\mathbf{r}$ from the origin is then given by

$$\psi(\mathbf{r}) = A_0 \exp\left(-2\pi i \mathbf{r}.\mathbf{s}_0\right) + A_1 \exp 2\pi i \phi \exp\left(-2\pi i \mathbf{r}.\mathbf{s}\right)$$

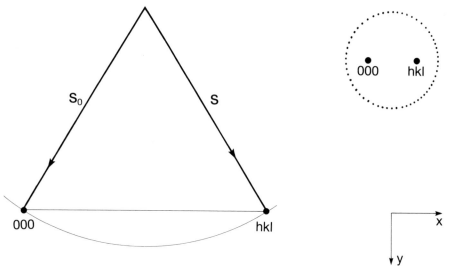

Fig 10.32 Two-beam setting where the transmitted beam, through 000, and the diffracted beam, through the reciprocal lattice point *hkl*, are equally inclined to the axis of the microscope. The objective aperture, upper right diagram, is positioned to allow the passage of only these two beams.

and the intensity of the wave by

$$|\psi(\mathbf{r})|^2 = A_0^2 + A_1^2 + A_0 A_1 \{\exp 2\pi i[\mathbf{r} \cdot (\mathbf{s} - \mathbf{s}_0) - \phi]$$

$$+ \exp 2\pi i[\mathbf{r} \cdot (\mathbf{s}_0 - \mathbf{s}) + \phi]\}$$

$$= A_0^2 + A_1^2 + 2A_0 A_1 \cos 2\pi(\mathbf{r} \cdot \mathbf{g} - \phi),$$

where the reciprocal lattice vector $\mathbf{g} = \mathbf{s} - \mathbf{s}_0$. Writing $\mathbf{r} = \mathbf{x} + \mathbf{y} + \mathbf{z}$, where $\mathbf{x}$ and $\mathbf{z}$ are parallel to and $\mathbf{y}$ is normal to the surface of the crystal and taking $\mathbf{x}$ parallel to $\mathbf{g}$ we obtain $\mathbf{r} \cdot \mathbf{g} = \mathbf{x} \cdot \mathbf{g} + \mathbf{y} \cdot \mathbf{g} + \mathbf{z} \cdot \mathbf{g} = \mathbf{x} \cdot \mathbf{g}$. Therefore

$$|\psi(\mathbf{r})|^2 = A_0^2 + A_1^2 + 2A_0 A_1 \cos 2\pi(xg - \phi).$$

The imaged wave, which is this wave as it leaves the crystal, has an intensity modulated with periodicity $1/g = d$, where d is the spacing of the crystal planes giving rise to the diffracted beam. The image of the crystal will thus show a set of parallel fringes, of spacing corresponding to d, lying parallel to the traces of the corresponding lattice planes. The intensity of the fringes will vary from a maximum of $(A_0 + A_1)^2$ to a minimum of $(A_0 - A_1)^2$. If the intenstiies of the two beams are equal, the fringes will vary in intensity from $4A^2$ to zero. The phase of the diffracted beam relative to that of the transmitted beam, $2\pi\phi$, will determine the position of the fringes relative to the crystal. Such an image enables the spacing of the crystal planes and their orientation in the crystal to be determined.

Another method of imaging crystal planes is to orient a reciprocal lattice net normal to the incident beam and to use the objective aperture to select the transmitted beam and the *hkl* and $\overline{hkl}$ diffracted beams of smallest Bragg angle to form the image (Fig 10.33). Suppose that the direction of the transmitted beam is represented by $\mathbf{s}_0$, the direction of the *hkl* diffracted beam by $\mathbf{s}_+$, and the direction of the $\overline{hkl}$ diffracted beam by $\mathbf{s}_-$. Then if $\mathbf{g}$ is the *hkl* reciprocal lattice vector, the scattering vector for *hkl* is $\mathbf{s}_+ = \mathbf{g} + \mathbf{s}$ and the scattering vector for $\overline{hkl}$ is $\mathbf{s}_- = -\mathbf{g} + \mathbf{s}$, where $\mathbf{s}$ is the Bragg deviation parameter. We make the simplifying assumptions that the crystal is centrosymmetric, that its back surface is normal to the transmitted beam, and that the two diffracted beams have equal amplitude a and equal phase $2\pi\phi$ (relative to the transmitted beam) as they emerge from the back surface of the crystal. We note that the amplitude A_0 of the transmitted beam will be much greater than that of the two diffracted beams. The imaged wave will be the resultant of the transmitted wave and the two diffracted waves. Therefore

$$|\psi(\mathbf{r})|^2 = |A_0 \exp\{-2\pi i\mathbf{r}.\mathbf{s}_0\} + a \exp\{-2\pi i(\mathbf{r}.\mathbf{s}_+ - \phi)\} + a \exp\{-2\pi i(\mathbf{r}.\mathbf{s}_- - \phi)\}|^2$$

$$= A_0^2 + 2a^2 + 2A_0 a\{\cos 2\pi[\mathbf{r}.(\mathbf{s}_+ - \mathbf{s}_0) - \phi] + \cos 2\pi[\mathbf{r}.(\mathbf{s}_- - \mathbf{s}_0) - \phi]\}$$

$$+ 2a^2 \cos 2\pi\mathbf{r}.(\mathbf{s}_+ - \mathbf{s}_-).$$

Substituting $\mathbf{S}_+$ for $\mathbf{s}_+ - \mathbf{s}_0$, $\mathbf{S}_-$ for $\mathbf{s}_- - \mathbf{s}_0$, and $2\mathbf{g}$ for $\mathbf{s}_+ - \mathbf{s}_-$, we obtain

$$|\psi(\mathbf{r})|^2 = A_0^2 + 2a^2 + 2A_0 a\{\cos 2\pi(\mathbf{r}.\mathbf{S}_+ - \phi) + \cos 2\pi(\mathbf{r}.\mathbf{S}_- - \phi)\} + 2a^2 \cos 4\pi\mathbf{r}.\mathbf{g}$$

$$= A_0^2 + 2a^2 + 4A_0 a \cos 2\pi\left(\mathbf{r}.\frac{\mathbf{S}_+ + \mathbf{S}_-}{2} - \phi\right)\cos 2\pi\left(\mathbf{r}.\frac{\mathbf{S}_+ - \mathbf{S}_-}{2}\right) + 2a^2 \cos 4\pi\mathbf{r}.\mathbf{g}$$

and since $\mathbf{S}_+ = \mathbf{g} + \mathbf{s}$ and $\mathbf{S}_- = -\mathbf{g} + \mathbf{s}$

$$|\psi(\mathbf{r})|^2 = A_0^2 + 2a^2 + 4A_0 a \cos 2\pi\{\mathbf{r}.\mathbf{s} - \phi\}\cos 2\pi\mathbf{r}.\mathbf{g} + 2a^2 \cos 4\pi\mathbf{r}.\mathbf{g}.$$

Again putting $\mathbf{r} = \mathbf{x} + \mathbf{y} + \mathbf{z}$, where $\mathbf{x}$ and $\mathbf{z}$ are parallel to and $\mathbf{y}$ is normal to the back surface of the crystal, we obtain for $\mathbf{x}$ parallel to $\mathbf{g}$

$$|\psi(\mathbf{x})|^2 = A_0^2 + 2a^2 + 4A_0 a \cos 2\pi\phi \cos 2\pi xg + 2a^2 \cos 4\pi xg$$

for the wave as it leaves the crystal (i.e. $y = 0$).

The image is thus composed of two sets of fringes parallel to the trace of the (hkl) planes, one set of periodicity $1/g = d_{hkl}$ and the other set of periodicity $1/2g = \frac{1}{2}d_{hkl}$, superimposed on a background of intensity $A_0^2 + 2a^2$. The first set of fringes will have a contrast which depends on the phase $2\pi\phi$ of the diffracted beams relative to the transmitted beam (Fig 10.34). The second set of fringes will have less contrast than the first set because $2a^2 \ll 4A_0 a$; their contrast is not dependent on ϕ. The resultant variation in intensity for integral ϕ is shown in Fig 10.34(b). If however $\cos 2\pi\phi$ is zero, only the second, very weak, set of fringes may be observable (Fig 10.34(c)).

The variation of the resultant wave with $\mathbf{y}$ is given by

$$|\psi(xyz)|^2 = A_0^2 + 2a^2 + 4A_0 a \cos 2\pi(-ys - \phi)\cos 2\pi xg + 2a^2 \cos 4\pi xg$$

and it follows that the contrast of the fringes of spacing d_{hkl} will vary with y; in particular the contrast will be at a maximum when $ys + \phi = n/2$, where n is an integer. We showed earlier (p. 356) that $s = (g^2\lambda)/2 = (2\theta)^2/2\lambda$. Thus by focusing on a plane distant y from the back surface of the crystal, the magnitude of y being chosen appropriately, maximum contrast can be obtained for any value of ϕ. In electron microscopy spherical aberration causes phase changes in all beams except those travelling along the axis of the microscope and this effect has to be taken into consideration when locating the image plane which will give maximum contrast. We

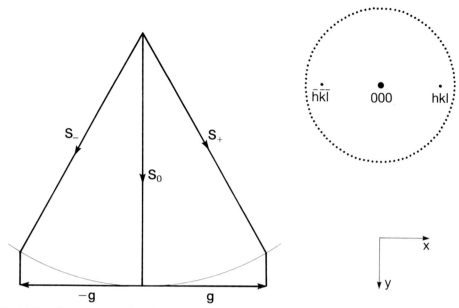

Fig 10.33 Three-beam setting. A reciprocal lattice net is normal to the incident beam and the objective aperture selects the transmitted beam and the *hkl* and *h̄k̄l̄* diffracted beams of smallest Bragg angle to form the image.

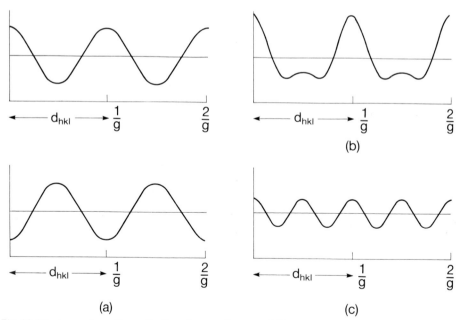

Fig 10.34 Image formation with three beams. One set of fringes parallel to the trace of the (*hkl*) planes has periodicity $1/g = d_{hkl}$ and amplitude $4A_o a \cos 2\pi\phi$ as shown for $\phi = n$ and for $\phi = (2n+1)/2$ in the upper and lower diagrams of (a). The second set of fringes has periodicity $1/2g = \frac{1}{2}d_{hkl}$ and amplitude $2a^2 \ll 4A_o a$; the combination of both fringe systems is shown for integral ϕ in (b). When $\cos 2\pi\phi = 0$ only the second set of very weak fringes will appear, as in (c). These fringe systems are superimposed on a background of intensity $A_o^2 + 2a^2$.

return to this point later when dealing with high resolution images.

With a 100 keV microscope lattice fringes of spacing down to 2 Å can be observed in favourable circumstances. The observation of lattice fringes provides a means of determining the orientation and spacing of the relevant (*hkl*) planes. Such fringes can be used to investigate intergranular boundaries or the boundary between two phases, to study order-disorder transformations and the formation of two phases of different composition from an initially single phase (exsolution or unmixing). The observation of fringes using three beams, the second method considered here, provides the simplest example of phase contrast electron microscopy, where the image plane is deliberately made non-coincident with the back surface of the crystal in order to give increased contrast. A general account of phase contrast microscopy is given in the following section.

High resolution images

So far we have discussed the formation of images in the electron microscope using a single beam (bright-field and dark-field images), two beams, and three beams. In conventional optical microscopy, however, an image is obtained using all the waves passing through the objective aperture (Fig 9.4). We now consider images formed in a similar manner in the electron microscope. The study of such images has in recent years made important contributions to knowledge of the detailed structures of certain metals, minerals and inorganic compounds. Although it is not possible to image individual atoms, it is possible to obtain information about the arrangement of the atoms not only in a perfect crystal (Fig 10.10) but also when the crystal is affected by defects such as stacking faults.

When three-beam lattice fringes are formed using the incident beam normal to a reciprocal lattice plane of the specimen, the amplitude of the two diffracted beams is small compared with that of the transmitted beam because the reciprocal lattice points concerned do not lie on the reflecting sphere. The variation in amplitude and hence the variation in intensity of the wave emerging from the back surface of the crystal is consequently small. We have already shown that the contrast can be improved by defocusing the objective lens to image the wave in a plane parallel to the reciprocal lattice plane so that the variation in phase across the wave is converted into variation in intensity; we shall see that this same technique can be applied to image structures.

Thin crystals, no thicker than 100 Å, are required if a resolution of about 3·5 Å at a magnification in excess of 5×10^5 is to be obtained with 100 keV electrons. The electron wave transmitted by such a thin crystal may be assumed to be of the same intensity as the incident beam, that is to say absorption is negligible. In the course of its passage through the crystal the phase of the electron wave is modified by the periodic potential distribution in the crystal so that the emergent beam differs mainly in phase rather than in amplitude across the back surface of the crystal. Since the intensity of the emergent wave varies little across the surface of the crystal, it is necessary to make use of the techniques of phase contrast microscopy in order to obtain an image of the crystal. This is most easily achieved by defocusing the objective lens. To illustrate the method we suppose that an image of the crystal is formed from a wave $\psi(\mathbf{x})$ of amplitude A and phase $2\pi\phi(\mathbf{x})$, which varies across the crystal, $\mathbf{x}$ being a vector parallel to the back surface of the crystal. The intensity of the wave will be given by

$$|\psi(\mathbf{x})|^2 = |A \exp 2\pi i\phi(\mathbf{x})|^2$$
$$= A^2$$

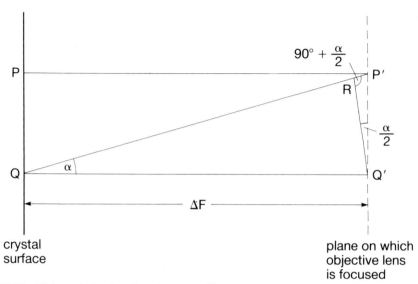

Fig 10.35 High resolution imaging. The path difference for waves travelling from P and from Q towards P' is $QP' - PP' = \Delta F \tan \alpha \tan \frac{1}{2}\alpha = \Delta F \frac{1}{2}\alpha^2$ for small α and their phase difference is $-(2\pi/\lambda)\Delta F \frac{1}{2}\alpha^2$.

and there will be no contrast in the image of the crystal. Now let us suppose that the wave emerging from the back surface of the crystal is the resultant of two waves, ψ_0, of amplitude A_0, travelling normal to the crystal surface, and ψ_1, of amplitude A_1 and phase either $\pi/2$ or $-\pi/2$ relative to ψ_0, travelling at an angle α to the direction of travel of ψ_0. The resultant wave emerging from the crystal is now

$$\psi(\mathbf{x}) = A_0 \pm iA_1$$

If now the phase of ψ_1 can be changed by $\pi/2$ it will become either π or zero, so that $\psi(\mathbf{x}) = A_0 - A_1$ or $A_0 + A_1$ and

$$|\psi(\mathbf{x})|^2 = (A_0 - A_1)^2 \quad \text{or} \quad (A_0 + A_1)^2.$$

The variation of phase across the crystal could then be imaged.

The required phase change of $\pi/2$ may be obtained by focusing not on the back surface of the crystal but on a plane distant ΔF therefrom. Suppose that the objective lens is focused on the plane P'Q' (Fig 10.35) distant ΔF from the back surface PQ of the crystal. In this plane the wave PP', which has travelled from P, will combine with the wave QP', which has travelled from Q and their path difference will be $QP' - PP'$. Since $QQ' = QR = PP'$, the path difference becomes $QP' - QR = RP'$. The angle $\widehat{QQ'R} = 90° - \alpha/2$, so that $\widehat{RQ'P'} = \alpha/2$ and $\widehat{Q'RP'} = 90° + \alpha/2$; therefore $RP'/P'Q' = (\sin \alpha/2)/[\sin (90° + \alpha/2)] = \tan \alpha/2$. But $P'Q' = QQ' \tan \alpha = \Delta F \tan \alpha$. So the path difference is $\Delta F \tan \alpha \tan \alpha/2$, which is equal to $\Delta F(\alpha^2/2)$ for small α. At P' the wave ψ_1 will have travelled further than the wave ψ_0 and in consequence the phase of ψ_1 relative to ψ_0 will be $-(2\pi/\lambda) \Delta F(\alpha^2/2)$. In order to introduce a phase difference of $\pi/2$ it would be necessary to focus on a plane distant $\Delta F = -\lambda/(2\alpha^2)$ from the back surface of the crystal.

The defocus ΔF is positive when the objective lens is focused on a plane between the crystal and the lens, that is when the objective lens is overexcited so that its focal length is shortened; the image is then overfocused. When ΔF is negative, the objective is focused on a plane between the crystal and the electron source; the objective lens is underexcited so that its focal length is lengthened and the image obtained is an underfocus image.

A high resolution electron microscope image (often known as an HREM image) is a representation of the electrostatic potential distribution in the crystal specimen projected down a principal direction of the crystal. Because the image is produced by multiple scattering it cannot be related simply to the object. Image interpretation is usually achieved by comparison of the observed image with a calculated image.

In order to calculate image intensities it is necessary to know the orientation of the crystal specimen relative to the incident electron beam (the production of a recognizable image of the structure requires a principal direction in the crystal to be set parallel to the incident electron beam to an accuracy of 3 milliradians or better), the thickness of the crystal specimen, the amount of defocus of the objective lens, the size of the objective aperture, and the aberrations of the electron optical system of the microscope. When an in-focus image is formed by a perfect lens system the relative phases of the diffracted beams as they emerge from the crystal surface and as they recombine to form the image will be precisely the same. In practice, however, a diffracted beam inclined at an angle α to the microscope axis will suffer a phase change relative to the transmitted beam; the magnitude of the phase change will depend on the amount of defocus of the objective lens and on the aberrations of the lens system, in particular on the spherical aberration of the objective lens. It is found that for small α the phase change introduced into a diffracted beam varies slowly with α, but that for larger values of g, and consequently larger α, the change in phase varies quite rapidly with increasing α. It is usual therefore to limit the diffracted beams used for image formation to those of small α, for which the introduced phase change is not strongly dependent on the magnitude of α. It is found that image quality is best when the underfocus is of the order of 900 Å. The intensity distribution in the image is computed using multi-beam dynamical theory assuming the production of 500–1000 diffracted beams, only 50–200 of which can pass through the objective aperture to form the image. For the formation of a good high resolution image a precisely aligned and very clean microscope with a highly coherent electron beam is necessary; the experimental requirements are very stringent.

High resolution electron microscopy has provided significant information on a nearly atomic scale and in an elegant and convincing form about crystals with two-dimensional faults and defects. Its application to the mixed oxides of tungsten and niobium, the tungsten bronzes, and chain- and band-silicates (biopyriboles) have been particularly notable (Fig 10.10).

The resolution obtainable with the current generation of 100 keV microscopes could be improved if it were possible to design an electron gun which would produce a less divergent incident beam with a smaller energy variation, that is a more nearly parallel and more nearly monochromatic incident electron beam. Slight improvement in resolution could also be achieved with an objective lens designed to have a smaller coefficient of spherical aberration (a good lens currently has $C_s \sim 0.7$ mm). The use of higher accelerating potentials, in 1 MeV and 3 MeV microscopes, enables shorter wavelengths to be used to improve resolution; but in such microscopes the stabilization of the ultra-high voltage presents a severe problem and the transmission of such a high energy electron beam may cause significant radiation damage in the specimen. It is to be expected that improvements in the techniques of high resolution electron imaging will yield a wealth of information on a nearly atomic scale in the immediate future.

Problems

10.1 A single crystal metal foil of uniform thickness 100 Å is mounted in an electron microscope. If the wavelength of the electron beam is 0·037 Å, what is the limiting radius in Å^{-1} of the zero layer diffraction pattern?

10.2 A thin (001) flake of a cubic crystal ($a = 8·21$ Å) is mounted in an electron microscope with z^* parallel to the axis of the microscope. Through what angle must the crystal be tilted to place the 200 reciprocal lattice point on the reflecting sphere for electrons of wavelength 0·037 Å?

10.3 A certain substance, which has a cubic F lattice with $a = 8·2$ Å, gives rise to an electron diffraction pattern which can be indexed on an orthogonal net with A^* $= 0·172$ Å^{-1} and $B^* = 0·122$ Å^{-1}. It is observed that reflexions with H + K odd are absent. Index the diffraction pattern on the conventional unit-cell and determine the direction in the crystal along which the electron beam is incident.

10.4 Since the likelihood of double diffraction increases as the wavelength of the incident radiation decreases, it is much rarer in X-ray diffraction than in electron diffraction. In consequence X-ray examples tend to be easier to analyse than electron examples. This problem is based on one of the earliest observations of X-ray double diffraction.

Diamond is cubic with $a = 3·567$ Å and has an F lattice; two carbon atoms are associated with each lattice point at $\pm(\frac{111}{888})$ from the lattice point. Show that the intensity of the 222 reflexion is zero. A single crystal of diamond is mounted on a four-circle single-crystal diffractometer, which is set to record the 222 reflexion using CuKα radiation ($\lambda = 1·5418$ Å). The intensity of the 222 reflexion is measured as the crystal is rotated about the normal to (111); a diffraction maximum occurs when the incident beam is normal to [$\bar{1}$01]. Draw out the [$\bar{1}$01] reciprocal lattice net for diamond and on it draw the two orientations of the reflecting sphere for which the 222 reciprocal lattice point lies on the reflecting sphere. Show that the observed diffraction maximum can be explained by double diffraction for both orientations of the reflecting sphere. Determine the indices of the reflexions involved. What other directions of the incident beam might be expected to give rise to the 222 reflexion by double diffraction under these experimental conditions? Would the same effect be observable with CoKα ($\lambda = 1·7902$ Å)?

10.5 An electron diffraction pattern was recorded from a body-centred cubic metal (unit-cell dimension a) when the incident electron beam was parallel to [123]. The three pairs of Kikuchi lines of smallest spacing which intersect at the centre of the diffraction pattern have spacings $\sqrt{6}$ a^*, $\sqrt{10}$ a^*, and $\sqrt{12}$ a^*. Assign consistent indices to these three pairs of lines and calculate the angles between them.

10.6 An electron diffraction pattern of a cubic crystal with an F lattice could be indexed on an orthogonal net with axes X^* and Y^* such that $A^*/B^* = \sqrt{8}$. Reflexions occur only when $H = 3n$ and $K = 3n$, and when $H = 3n \pm 1$ and $K = 3n \pm 1$. Draw out the net to scale and show that it is consistent with the [110] net of a matrix crystal twinned by reflexion in (111).

If there were also twinning by reflexion in (111) of the matrix crystal, what reciprocal lattice net of the twinned crystal would be coincident with the [110] net of the matrix crystal? Where would the reflexions of this twinned crystal occur in the [110] net of the matrix crystal?

10.7 An electron diffraction pattern is recorded from an (001) cleavage flake of the mica *muscovite* (monoclinic, space group $C2/c$, $a = 5\cdot19$, $b = 9\cdot04$, $c = 20\cdot08$ Å, $\beta = 95\cdot5°$) with the electron beam normal to the flake. One row of three reflexions $0\cdot22$ Å^{-1} apart is observed. Through what angle and about what direction must the flake be rotated for the centred $hk0$ reciprocal lattice net to be observed? In what direction will the spikes through the reciprocal lattice points be oriented?

10.8 Ordered CuZn is cubic with two atomic sites in the unit-cell; if one is taken at 000, the other is at $\frac{1}{2}\frac{1}{2}\frac{1}{2}$. Faults, known as *antiphase boundaries*, occur in crystals of CuZn such that on one side of the fault the Cu atoms lie at 000 and the Zn atoms at $\frac{1}{2}\frac{1}{2}\frac{1}{2}$ while on the other side of the fault the Cu atoms lie at $\frac{1}{2}\frac{1}{2}\frac{1}{2}$ and the Zn atoms at 000. Deduce the displacement **R** for such an antiphase boundary. What reflexions could be used to observe the antiphase boundary?

10.9 An electron micrograph of an aluminium foil (Al is cubic with $a = 4\cdot04$ Å) shows two parallel bend contours 44.5 mm apart. One contour is associated with the 131 reflexion and the other with the $\bar{1}3\bar{1}$ reflexion. Both bend contours have weak subsidiary fringes on either side of them at 2·8 mm from the principal contour. Assuming that the foil is uniformly bent and that the electron wavelength is 0·037 Å, what is the thickness of the foil?

10.10 What reflexion would you select if you were attempting to observe the lattice fringes of largest possible spacing under two-beam conditions from (i) the [110], (ii) the [111] reciprocal lattice net of a foil of (a) a body-centred cubic, (b) a face-centred cubic metal?

Appendices

Appendix A: Constructions in the Stereographic Projection

I: To project a pole P at $\rho°$ from N along a given diameter QOR (Fig A.1)

The pole P lies in the plane PNS, which intersects the plane of projection in the diameter QOR of the primitive circle. Imagine the sphere of projection to be rotated through 90° about QOR (Fig A.1(a)): S will move to S′, N to N′, and P to P′. S′ lies on the primitive circle 90° away from Q and R.

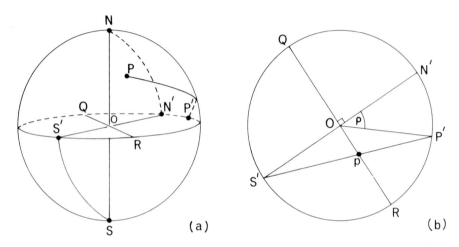

Fig A.1   To project a pole P at $\rho°$ from N along a given diameter QOR. In the spherical projection (a) P is located in the plane NRSQ; on rotation through 90° about QOR this plane becomes N′RS′Q and P moves to P′. The plane diagram (b) shows the location of p at the intersection of S′P′ with QOR. Rotation back to the original orientation does not move p; p is the stereographic projection of P.

The projection of P is then made (Fig A.1(b)) in the following manner:
1. Draw the diameter N′OS′ perpendicular to QOR,
2. Draw OP′ such that $\widehat{N'OP'} = \rho°$,
3. Join S′P′ to intersect QOR in p.

   Imagine now that the sphere of projection is rotated back through 90° about the diameter QOR so that S′ moves to S, N′ to N, and P′ to P; p remains unchanged in position and is therefore the projection of the pole P.
   Of course we could more simply have placed p at a distance $r \tan \frac{1}{2}\rho$ from O along OR, but we have described this construction to emphasize its principle, which is made use of in subsequent more complicated constructions. There too the reader will find apparently unnecessary elaborations, which are introduced primarily to establish principles.

II: To find the projection of the opposite P_0 of the pole P (Fig A.2)

The opposite of a pole P is the pole situated at the opposite extremity of the diameter of the sphere of projection through P. P_0 thus represents the direction at 180° to the direction represented by P. If P represents the normal to a face (hkl), then P_0 represents the normal to the face $(\bar{h}\bar{k}\bar{l})$.
   It is evident from the section of the sphere of projection containing P_0, S, and P drawn in Fig A.2(a) that if P is projected using the south pole and P_0 using the north pole as projection point, then $Op = Op'_0$. It is therefore a simple matter to plot p'_0 by drawing the diameter through p and marking on it a point at an equal distance from O on the other side of O from p.
   If the south pole is used as the projection point for both P and P_0 then P_0 projects outside the primitive circle and its position can be found by making use of the fact that PP_0 is a diameter and so the angle $PSP_0 = 90°$. If the primitive circle is imagined to be the section of the sphere of

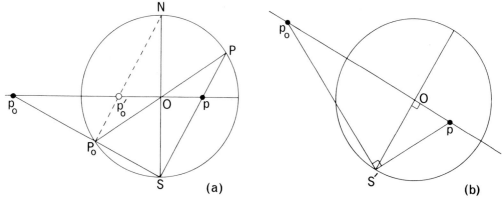

Fig A.2   To find the projection of the opposite P_0 of the pole P. (a) is the section of the sphere of projection containing POP_0 and perpendicular to the plane of projection; p and p_0 are respectively the stereographic projections of P and P_0 from the south pole; p_0' is the stereographic projection of P_0 from the north pole. (b) shows the sphere of projection rotated about the diameter through p to bring the south pole to S'; the intersection of Op with the perpendicular to S'p through S' locates p_0.

projection containing P, S, and N, the projection of the opposite of the pole P can be constructed in the following manner (Fig A.2(b)):

1. Draw OS' perpendicular to Op,
2. Draw $S'p_0$ perpendicular to pS',
3. p_0 lies at the intersection of $S'p_0$ with pO produced.

III: To construct a great circle through two poles (Fig A.3)

We shall first discuss the general case and then pass on to a special case.

a To construct a great circle through two general poles

A great circle is defined by specifying two non-opposite poles, P and Q (Fig A.3(a)), which lie on the great circle; the great circle also passes by definition through the centre of the sphere of projection. However, to draw a circle—and in general the projection of a great circle is an arc of a circle—it is necessary to locate three points lying on the circle. Now since the plane of a great circle passes through the centre, the opposite of any pole lying on the great circle also lies on the great circle. The construction is made in the following manner (Fig A.3(b)):

1. Construct the opposite p_0 of the pole p by construction II,
2. Construct the perpendicular bisectors gt and gt_0 of qp and qp_0,
3. The geometrical centre of the projected great circle is then g.

b To construct a great circle through two poles, one of which lies on the primitive

If the pole P through which the great circle is to pass lies on the primitive, its opposite P_0 lies on the primitive at the opposite end of the diameter of the primitive through P. The geometrical centre of the projected great circle therefore lies on the perpendicular bisector of this diameter POP_0, that is on the diameter perpendicular to the diameter POP_0. The great circle is constructed in the following manner (Fig A.3(c)):

1. Draw the diameter through P,
2. Draw the diameter ROR_0 perpendicular to POP_0,
3. Construct the perpendicular bisector gt of Pq,
4. The geometrical centre of the required great circle lies at the intersection g of gt and ROR_0.

Some part of the great circle will project outside the primitive unless the point of projection is changed from the south to the north pole for this part. That part of the great circle which lies within the primitive when projected from the north pole will contain the opposite of every pole on the other part of the great circle which lies within the primitive circle when projected from the south pole; the opposite of every pole lying on the arc PmP_0 lies on the arc P_0m_0P, POP_0 being

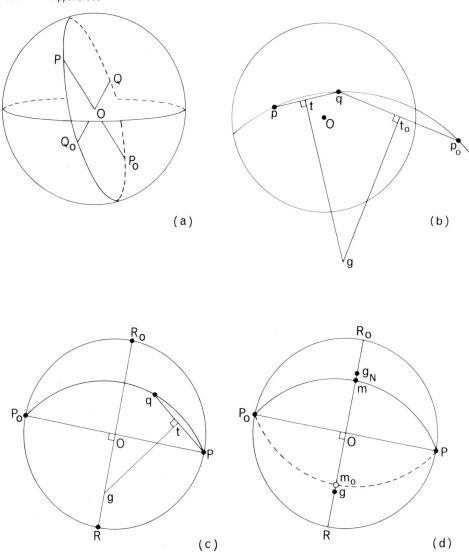

Fig A.3 To construct a great circle through two poles. (a) shows the sphere of projection with the plane containing P, Q, and their opposites P_0, Q_0 outlined. (b) illustrates the construction for the general case where p and q lie within the primitive; p_0 is the opposite of p; gt and gt_0 are the perpendicular bisectors of pq and qp_0 respectively; g is the geometrical centre of the required great circle. (c) illustrates the construction for the special case where one pole, P, lies on the primitive; ROR_0 is the diameter of the primitive perpendicular to OP; gt is the perpendicular bisector of qP. (d) illustrates the construction of that part of the great circle which has to be projected from the north pole in order to be brought within the primitive; the geometrical centre of this part of the great circle lies on the diameter ROR_0 of the primitive perpendicular to OP at g_N such that $Og = Og_N$. A great circle projected from the north pole is shown conventionally by a broken arc.

the diameter of the primitive in which the great circle intersects the equatorial plane in Fig A.3(d). In particular if m lies on the diameter perpendicular to POP_0 then m_0 also lies on that diameter on the opposite side of O so that $Om = Om_0$. It is apparent from the figure that the geometrical centres g and g_N of the great circle projected from the south and north poles respectively lie on the diameter ROR_0 on either side of O so that $Og = Og_N$.

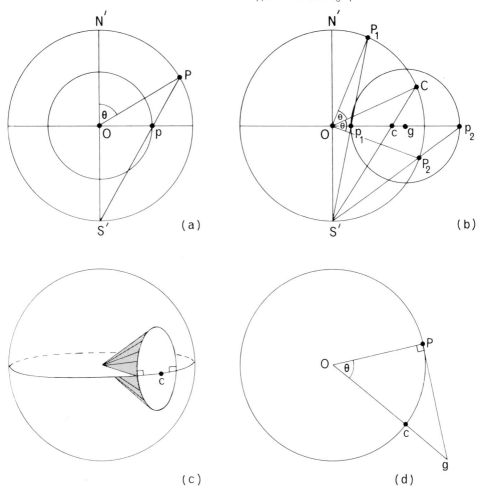

Fig A.4 To construct a small circle of radius $\theta°$ about a given pole. (a) illustrates the special case where the given pole is the north pole; the required small circle has its geometrical centre at O and passes through p where $\widehat{OS'p} = \frac{1}{2}\theta$. (b) illustrates the general case where the stereographic centre of the small circle is c; rotation through 90° about Oc brings the south pole to S'; S'c produced meets the primitive in C; $\widehat{COP_1} = \widehat{COP_2} = \theta°$; p_1 and p_2 are the stereographic projections from S' of P_1 and P_2; the geometrical centre of the small circle lies on OC at g such that $p_1g = gp_2$. When the stereographic centre of the small circle lies on the primitive (c) a simple construction, shown in (d), becomes available: $\widehat{COP} = \theta$ and g lies at the intersection of the normal to OP through P with OC produced.

IV: To construct a small circle of radius $\theta°$ about a pole (Fig A.4)

a When the pole is the north pole

In this case the plane of the small circle is parallel to the equatorial plane and both the stereographic and geometrical centre of the projected small circle will coincide at N. All that is required to determine the radius of the projected circle is a pole plotted at $\theta°$ from N. The construction is performed in the following manner (Fig A.4(a)):

1. Plot the projection p of a pole P at $\theta°$ from N along any diameter by construction I,
2. Draw a circle with centre O and radius Op.

b When the stereographic centre lies within the primitive circle

In this case the geometrical centre g and the projection c of the stereographic centre of the small circle will not coincide (as we showed in chapter 2), but will both lie on a diameter of the primitive circle. The projected small circle is therefore defined by two points lying at $\theta°$ on either side of the stereographic centre of the small circle and on the diameter of the primitive circle through the stereographic centre. To construct such a small circle it is necessary to imagine that the sphere of projection is rotated about the diameter Oc (Fig A.4(b)) through $90°$ so that the projection point moves into the plane of the diagram. The construction can then be performed as follows:

1. Draw the diameter N'OS' perpendicular to the diameter Oc.
2. Draw S'c to meet the primitive circle at C; OC is then the direction of the stereographic centre of the small circle.
3. Construct the radii of the primitive circle OP_1 and OP_2 on either side of OC such that $\widehat{COP_1} = \widehat{COP_2} = \theta$.
4. Draw $S'P_1$ and $S'P_2$ to intersect Oc in p_1 and p_2 respectively. (Notice that in the diagram P_2 lies in the southern hemisphere and so its projection, p_2, lies outside the primitive circle.)
5. Bisect p_1p_2 to give the geometrical centre g of the projected small circle.
6. Imagine the sphere of projection to be rotated back to its original attitude. This will leave the geometrical centre of the projected small circle unchanged in position at g and p_1 and p_2 also unaffected. Draw the circle with centre g and radius gp_1.

c When the stereographic centre lies on the primitive circle

In this case construction IVb is applicable, but a simpler construction is available. This depends on the property of the stereographic projection that the angle between two arcs, each of which is the projection of a plane, is equal to the angle between the two planes (chapter 2). This property will not be proved here; the reader is referred to Terpstra and Codd (1961, p. 12) for a proof. A small circle whose stereographic centre is on the primitive circle is the intersection of a plane perpendicular to the equatorial plane with the sphere of projection (Fig A.4(c)). Therefore at their points of intersection the projection of the small circle and the primitive circle will be mutually perpendicular. Therefore the tangents to the primitive circle at the points of intersection are radii of the projected small circle. The geometrical centre and the stereographic centre of the projected small circle must lie on the same diameter of the primitive circle.

The construction can be made as follows (Fig A.4(d)):

1. Draw the diameter through the stereographic centre C of the small circle.
2. Draw OP such that $\widehat{COP} = \theta$. P lies on the primitive circle.
3. Draw the perpendicular to OP through P to meet OC produced in g.
4. Draw a circle with centre g and radius gP.

V: To construct a small circle passing through the pole p when the stereographic centre C of the small circle lies on the primitive circle (Fig A.5)

Since the plane of the small circle is perpendicular to the equatorial plane the small circle will be symmetrical across the equatorial plane. It follows that if the small circle is projected using both north and south poles so that its projection lies wholly within the primitive circle, then that part projected from the north pole will be coincident with that projected from the south pole. The pole whose projection is coincident with p when the north pole is used as the projection point will, like p, lie on the small circle; to draw the projection of the small circle this pole has to be re-projected using the south pole as the projection point.

Since the geometrical centre of the projected small circle lies on OC the construction can be effected by imagining the sphere of projection to be rotated through $90°$ about its diameter Op and proceeding as follows (Fig A.5):

1. Draw the diameter through Op.
2. Draw the diameter N'OS' perpendicular to Op.
3. Draw S'p to meet the primitive circle in P.
4. Draw N'P to meet Op produced in p_1 and imagine the sphere to be rotated back to its original attitude.
5. Draw the perpendicular bisector of pp_1 to meet OC produced in g.
6. Draw the circle with centre g and radius gp.

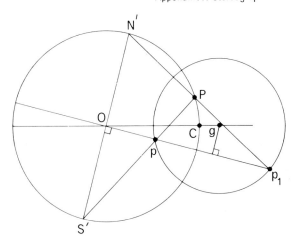

Fig A.5   To construct a small circle passing through the pole p when its stereographic centre C lies on the primitive circle. The sphere of projection is imagined to be rotated about Op to bring the north and south poles to N′ and S′. S′p meets the primitive in P and N′P meets Op produced in p_1. The perpendicular bisector of pp_1 meets OC produced in g, which is the geometrical centre of the required small circle.

VI: To find the pole of a great circle (Fig A.6)

By definition the pole of a great circle lies 90° from every point on the great circle and so represents the normal to the plane of the great circle. In Fig A.6(a) a great circle meets the primitive circle at the opposite ends of the diameter AC. Any direction perpendicular to the directions represented by the poles A and C must lie in a plane perpendicular to the line AC. This plane, which is perpendicular to the equatorial plane, will pass through the south pole and will therefore project as the diameter bOp of the primitive circle perpendicular to the diameter AOC. The pole of the great circle must also be perpendicular to the pole B, which lies at the intersection of the great circles ABC and BP.

   The projection of the pole of a great circle can therefore be found by drawing the diameter of the primitive circle which is perpendicular to the diameter through the points of intersection of the great circle with the primitive circle and then measuring a distance equivalent to 90° along

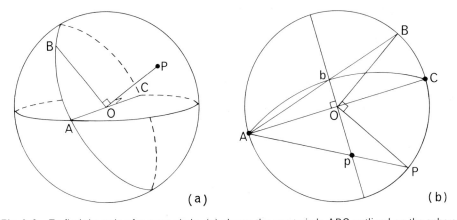

(a)   (b)

Fig A.6   To find the pole of a great circle. (a) shows the great circle ABC outlined on the sphere of projection; OP is the pole of this great circle. The stereographic projection (b) shows the great circle AC which intersects the diameter of the primitive perpendicular to AC in b; rotation of the sphere of projection through 90° about bO brings the south pole to A so that b is the stereographic projection of B; P lies on the primitive and $\widehat{BOP} = 90°$; AP intersects bO produced in p, which is the pole of the great circle. This figure serves also to illustrate the construction of a great circle whose pole is p.

the diameter from b, its point of intersection with the great circle. The construction is performed by imagining the sphere of projection to be rotated through 90° about its diameter bOp so as to bring the point of projection into the plane of the diagram and coincident with the point A; and proceeding as follows (Fig A.6(b)):

1. Draw the diameter through the points of intersection, A and C, of the great circle with the primitive circle.
2. Draw the diameter perpendicular to AOC, to intersect the great circle in b.
3. Draw Ab to meet the primitive in B.
4. Join OB.
5. Draw OP such that $\widehat{BOP} = 90°$.
6. Draw AP.
7. The intersection p of bO produced with AP is the required pole of the great circle. Rotation of the sphere of projection back to its original attitude does not affect the position of p.

VII: To draw a great circle given the position of its pole (Fig A.6)

This construction follows directly from VI. The three projected poles lying on the projected great circle that can be found most easily are one at each of the two points of intersection of the great circle with the primitive circle (these are at the ends of the diameter perpendicular to the diameter through the projection of the pole) and one on the diameter through the projection of the pole P. If the sphere of projection is imagined to be rotated through 90° about Op the construction can be performed as follows:

1. Draw the diameter Op.
2. Draw the diameter AC perpendicular to Op.
3. Draw the line Ap to meet the trace of the sphere of projection in P.
4. Draw OB such that $\widehat{BOP} = 90°$.
5. Draw AB.
6. The point of intersection b of AB with pO produced is the projection of a pole lying on the great circle.
7. Imagine the sphere of projection to be rotated back to its original attitude and draw the great circle through A, b, and C by construction IIIb.

VIII: To measure the angle between two poles (Fig A.7)

The angle between the directions represented by the poles P_1 and P_2 is $P_1\widehat{O}P_2$ (Fig A.7(a)). On the stereogram this angle is represented by the arc p_1p_2 of the great circle through p_1 and p_2; this

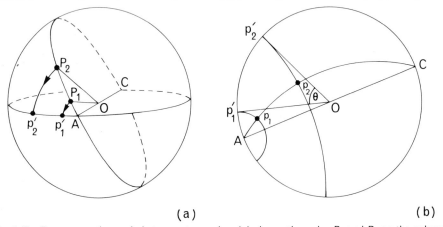

(a)   (b)

Fig A.7   To measure the angle between two poles. (a) shows the poles P_1 and P_2 on the sphere of projection; the great circle P_1P_2 intersects the plane of projection in the diameter AOC of the primitive; rotation about AOC brings P_1 and P_2 on to the primitive at p_1' and p_2'; $p_1'\widehat{O}p_2' = P_1\widehat{O}P_2$. (b) illustrates the construction: p_1 and p_2 are the stereographic projections of the given poles P_1 and P_2, and the great circle on which they lie intersects the primitive in AOC; a small circle with stereographic centre A is drawn through p_1 to intersect the primitive in p_1' and another small circle with the same stereographic centre through p_2 intersects the primitive in p_2'; the required angle is $p_1'\widehat{O}p_2'$.

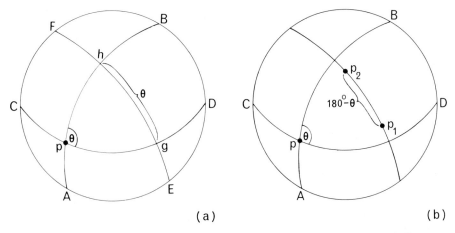

(a)   (b)

Fig A.8   To find the angle between two great circles, ApB and CpD as shown in (a). The great circle whose pole is p intersects the two great circles in h and g respectively. Measurement of the angle between h and g by the method illustrated in Fig A.7 gives the required angle. (b) illustrates an alternative construction: the poles p_1 and p_2 of the great circles ApB and CpD are located by the method illustrated in Fig A.6 (b) ; the angle between p_1 and p_2, found by the method illustrated in Fig A.7, is then the supplement of the required angle.

arc cannot be measured to give the angle directly. One method of measuring the angle is to rotate the poles on the sphere of projection about the diameter AC, keeping the equatorial plane and the projection points fixed, until the great circle containing p_1 and p_2 becomes the primitive circle; the poles P_1 and P_2 will then coincide with their projections and direct measurement of the angle P_1OP_2 becomes possible. During rotation the path of each pole will be a circle centred on the diameter that is the axis of rotation. Or in other words each pole will describe an arc of a small circle whose plane is perpendicular to the axis of rotation and whose stereographic centre is the axis of rotation. Thus to measure the angle between the projected poles p_1 and p_2 the steps in the construction are (Fig A.7(b)):

1. Construct the great circle Ap_1p_2C by construction IIIa.
2. Draw small circles, each with stereographic centre A to pass through p_1 and p_2 respectively by construction V.
3. Measure the angle $p_1'Op_2'$ between the points of intersection of the two small circles with the primitive. This is equal to the required angle P_1OP_2.

IX: To find the angle between two great circles (Fig A.8)

The angle between two great circles can be measured by drawing the tangents to the projections of the great circles at their point of mutual intersection (chapter 2), but this is neither an accurate nor a convenient way of measuring the angle. An alternative and better method is to draw in the great circle which is perpendicular to both great circles. The angle between the planes of the great circles is then given by the angle between their lines of intersection with this plane.

The steps in the construction needed to measure the angle between the two great circles ApB and CpD (Fig A.8(a)) are:

1. Construct the great circle EghF whose pole p is the intersection of the two great circles by construction VII.
2. Measure the angle between g and h, which are the intersections of the great circle EghF with the great circles AphB and CpgD respectively by construction VIII.

Yet another method is to find the poles of the two great circles and then to measure the angle between them. The angle between two planes is equal to the supplement of the angle between their normals or poles. The steps of the construction are (Fig A.8(b)):

1. Find the poles p_1 and p_2 of the two great circles, ApB and CpD, by construction VI.
2. Measure the angle between p_1 and p_2 by construction VIII.

Appendix B: Two simple devices for measuring interfacial angles

The two devices we describe here are intended as aids in the teaching of the elements of morphological crystallography and symmetry, enabling the student to obtain very quickly the angular data necessary for the plotting of a stereogram. In careful hands the measured angles are quite accurate enough for plotting on a stereogram of $2\frac{1}{2}$ inch radius. For greater accuracy of angular measurement it is necessary to use more refined optical goniometers, with a correspondingly greater expenditure of time; such instruments are well described by Terpstra and Codd (1961). The two-circle optical goniometer still has occasional uses in research.

The *contact goniometer* was invented by Carangeot in 1780. It is illustrated in Fig B.1 and fully described in the caption of that figure. It is a very simple instrument suitable for measuring large crystals and the wooden crystal models used in elementary teaching. Its accuracy (about $\pm 2°$) is limited by the ability of the operator to hold the crystal edge perpendicular to the plane of the goniometer while keeping the crystal firmly against the edge of the protractor and the adjustable arm.

The simple *optical goniometer* shown in Fig B.2 was designed by Dr. J. V. P. Long of the Department of Mineralogy and Petrology, Cambridge. It is quite accurate enough (about $\pm 1°$) for elementary teaching and for this purpose has two advantages over more elaborate instruments: its simplicity enables the student to set the crystal and obtain angular measurements very quickly, and its construction is such that it can be made very cheaply. The instrument, which is fully described in the caption to Fig B.2, is so designed that an image of the lamp filament is visible through the viewing tube when a reflecting surface lies normal to the axis of the viewing tube. Thus if a crystal is set with a zone axis parallel to the horizontal axis of the instrument, successive faces in the zone will be brought into the reflecting position as the shaft is rotated. By noting the angular readings on the dial r at which reflexions coincide with the cross-wires interfacial angles in this zone may be measured.

The procedure for setting a crystal on the simple optical goniometer may be described with reference to Fig B.2(c). (1) Set the instrument so that the axis A of the universal joint t is vertical. (2) Mount the crystal with plasticine so that face 1 in the selected zone is approximately horizontal and so that an edge between two adjacent faces in this zone is parallel to the axis of the horizontal shaft p. (3) Locate the image of the filament on the cross-wires by turning the horizontal shaft and adjusting the tilt of the crystal with the axis B of the universal joint. (4) Turn the horizontal shaft so as to bring the next face 2 into the reflecting position and locate the image of the filament on the cross-wires by adjusting the axis A of the universal joint. (5) Slide the magnet v on the steel block q so that the crystal remains in the centre of the field of view throughout $360°$ rotation of the horizontal shaft. When the crystal is set, measurements of interfacial angles

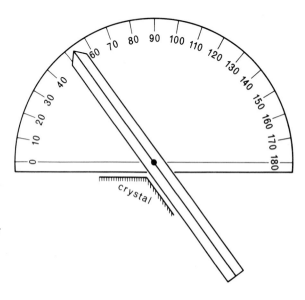

Figure B.1   The contact goniometer is constructed from a perspex protractor to which a perspex arm is attached so as to pivot about the centre of the protractor. The crystal is placed between the edge of the protractor on the 0° side and the arm so that the crystal edge is perpendicular to the plane of the protractor and the angle between face normals is read off. The goniometer is calibrated in degrees or half degrees (only 10° divisions are shown on the figure for simplicity). The interfacial angle of the crystal shown in exploded relationship to the goniometer is thus measured as $52\frac{1}{2}°$.

in the selected zone can be made. The crystal may then be remounted for measurement of another zone. The crystal should have dimensions in the range 5–10 mm and bright faces for convenient use of this instrument; suitable examples are octahedra of magnetite and cleavage rhombohedra of calcite.

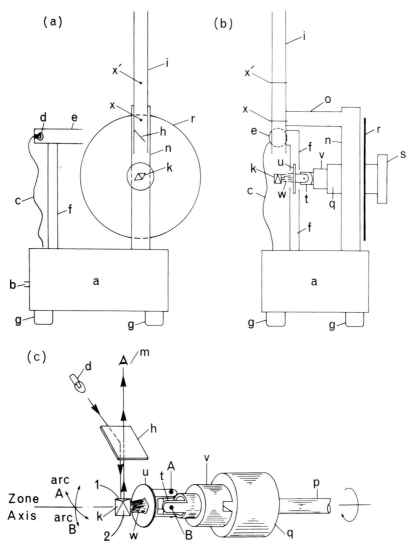

Fig B.2   The simple optical goniometer. (a) and (b) are respectively front and side elevations and (c) shows the detail of the crystal adjustment and optical paths. The transformer, contained in an alloy box *a*, is fed from the mains at *b* and its output passes through the flex *c* to the torch bulb *d*, which is set in the cylindrical tube *e* mounted adjustably on the rod *f*; the transformer box sits on rubber feet *g*. The incident light beam is reflected by the half-silvered cover-slip *h*, mounted in the square-section tube *i*, to impinge on the crystal *k*. When a face of the crystal is set horizontal the light beam is reflected up the viewing tube *i*, passing undeviated through the half-silvered plate to be observed at *m*. The viewing tube *i* is rigidly attached by the arm *o* to the pillar of square-section tube *n*. A horizontal shaft *p*, carrying at one of its ends the steel cylinder *q* and at the other the graduated dial *r* (a 360° protractor) and attached knob *s*, rotates on bearings set in the pillar *n*. The orientation of the crystal *k* can be adjusted by manual manipulation of the universal joint *t* which is cemented with epoxy resin to the steel disc *u* and to the magnet *v*. The crystal is centred by manual adjustment of the position of the magnet *v* on the steel cylinder *q*. The crystal is attached to the steel disc *u* by the plasticine *w*. The viewing tube carries cross-wires *x*, *x'*.

Appendix C: Rules for selecting standard settings of space groups in *International Tables for X-ray Crystallography* and in *Crystal Data*

International Tables for X-ray Crystallography, vol. 1 (1969) gives two conventional diagrams (symmetry elements and general equivalent positions) and a listing of coordinates of general and special equivalent positions for the standard setting of each of the 230 space groups, except for those of the cubic system where coordinates only are given. *Crystal Data* (1973) provides a listing, exhaustive at the time of its publication, of the unit-cell dimensions of crystalline substances, system by system in terms of the a/b ratio for the triclinic, monoclinic, and orthorhombic systems, the c/a ratio for the trigonal, tetragonal, and hexagonal systems, and the magnitude of the cell edge a for the cubic system. In *Crystal Data* the alternative name anorthic is used in preference to triclinic; the trigonal system is treated as a subdivision of the hexagonal system, both being referred to hexagonal axes ($a = b \neq c, \alpha = \beta = 90°, \gamma = 120°$). These modifications lead to a simple means of indexing the data for each substance by the initial letter of its crystal system (A, M, O, T, H, C) followed by the axial ratio or, in the case of a cubic substance, by the unit-cell edge. Thus M–1·020 represents monoclinic, $a/b = 1·020$; T–0·690 represents tetragonal, $c/a = 0·690$; and C–8·29 represents cubic, $a = 8·29$ Å.

Generally accepted conventions for the choice of reference axes lead to unique settings in the tetragonal, hexagonal, and cubic systems as well as in the trigonal system whether of hexagonal or rhombohedral lattice type. *International Tables* and *Crystal Data* therefore make the same choice of unit-cell in these systems.[1] An ambiguity arises in space groups of these systems where two sets of symmetry elements are interleaved and when this happens there is no unique space group symbol even where the lattice is primitive; for example the space group $I\bar{4}c2$ could just as well be described as $I\bar{4}b2$ or as $I\bar{4}a2_1$. In such cases the choice of standard space group symbol is arbitrary and need concern us no further here.

In the remaining three systems, triclinic (anorthic), monoclinic, and orthorhombic, the orientation of the x, y, and z axes is not completely determined by symmetry: in the orthorhombic system the reference axes are taken parallel to the orthogonal diads of the lattice, in the monoclinic system the y-axis is conventionally taken parallel to the diad of the lattice,[2] and in the triclinic system there can be no symmetry control at all. The conventions which *International Tables* and *Crystal Data* use for their choice of reference axes in these systems differ in principle and sometimes in practice. In *Crystal Data* reference axes which are not fixed by symmetry are selected so as to correspond to the shortest possible lattice translations and labelled as a right-handed axial system with $c < a < b$, α obtuse, and β obtuse. Restrictions on the choice of setting imposed by symmetry considerations (i.e. that y is parallel to the diad in the monoclinic system) take precedence over these dimensional conditions. The conventions used in *International Tables* for determining the choice of setting and the standard space group symbol cannot be usefully generalized and are now discussed system by system.

Orthorhombic System. The primary convention adopted by *International Tables* is that for space groups in the point groups 222 and $mm2$ which have one axis distinct from the others; then that axis is designated the z-axis. Thus $P222_1$ and $P2_12_12$ and $Pmm2$ are the standard symbols for these three space groups. A secondary convention applies to all space groups of the system which have one-face centred lattices: that the C-lattice is preferred to either the A- or the B-lattice. In the four space groups of point group $mm2$ where this convention conflicts with the primary convention, the A-lattice is preferred to the B-lattice. A third convention is required for space groups which have non-primitive lattices and two sets of symmetry elements interleaved: the standard symbol is chosen so as to show that symmetry element which appears first in the following sequence, $m, a, b, c, n, d, 2, 2_1$. Thus the symbol $Cmcm$ is preferred to $Cbnn$. Apart from the application of these three conventions *International Tables* are not consistent in their choice of standard setting, the arbitrary standard settings of the old International Tables (1935) being preferred to the formulation of elaborate new rules.

The dimensional convention $c < a < b$ is rigidly adopted by *Crystal Data* for the orthorhombic system. In consequence A-, B-, and C-lattices are all permissible and the distinct axis in space

---

[1] For rhombohedral space groups in the trigonal system *International Tables* give space group symbols, conventional diagrams, and coordinates of general and special equivalent positions referred to hexagonal axes and *in addition* coordinates of general and special equivalent positions referred to rhombohedral axes.

[2] *International Tables* provide a description of each monoclinic space group also in the alternative setting with the z-axis parallel to the diad of the lattice.

groups of the point groups 222 and *mm* may be *x*, or *y*, or *z*. For all three point groups of the system the space group symbols produced by application of this convention may be in non-standard form.

Monoclinic System. The convention that the non-primitive lattice is a C-lattice is consistently adopted in *International Tables*. Since space groups with a C-lattice have two sets of symmetry elements interleaved the same convention is adopted as for the analogous situation in the orthorhombic system. A further convention is applied to space groups with a primitive lattice and a glide plane: the glide plane is taken to be a *c*-glide.

The convention $c < a$ and β obtuse is rigidly adopted by *Crystal Data*. In consequence the non-primitive lattice type may be C, A, or I and so some space group symbols will be non-standard.

Triclinic (Anorthic) System. No conventions are required for the choice of reference axes for the two space groups P1 and P$\bar{1}$ in *International Tables*. However in *Crystal Data*, where a/b is tabulated for triclinic substances, conventions are required to determine which lattice repeats are to be labelled *a* and *b*. The conventions adopted are that the three shortest non-coplanar lattice repeats are taken as the directions of the reference axes so as to form a right-handed axial system with $c < a < b$, α obtuse, β obtuse. For a discussion of conventions for choosing a unit-cell in the triclinic system the reader is referred to Kelsey and McKie (1964).

Appendix D: Spherical trigonometry: the equations for a general triangle

Let a sphere, centre O, intersect three of its radii in A, B, and C, the angles between the radii being $\widehat{BOC} = a$, $\widehat{COA} = b$, $\widehat{AOB} = c$ (Fig D.1). The sphere intersects the planes BOC, COA, AOB in the great circle arcs BC, CA, AB which form the sides of the *spherical triangle* ABC.

Select a point D on OB produced such that AD $\perp$ OA and a point E on OC produced such that AE $\perp$ OA. From the plane triangle ODE,

$$DE^2 = OE^2 + OD^2 - 2OE \cdot OD \cos \widehat{DOE}$$

and from the plane triangle ADE,

$$DE^2 = AE^2 + AD^2 - 2AE \cdot AD \cos \widehat{DAE}.$$

Thus $OE^2 + OD^2 - 2OE \cdot OD \cos \widehat{DOE} = AE^2 + AD^2 - 2AE \cdot AD \cos \widehat{DAE}.$

But AD $\perp$ OA and AE $\perp$ OA, therefore

$$OD^2 - AD^2 = OA^2 = OE^2 - AE^2,$$

whence $OA^2 = OE \cdot OD \cdot \cos \widehat{DOE} - AE \cdot AD \cdot \cos \widehat{DAE}$

i.e. $\cos \widehat{DOE} = \dfrac{OA}{OE} \cdot \dfrac{OA}{OD} + \dfrac{AE}{OE} \cdot \dfrac{AD}{OD} \cdot \cos \widehat{DAE}$

i.e. $\cos \widehat{DOE} = \cos \widehat{AOE} \cdot \cos \widehat{AOD} + \sin \widehat{AOE} \cdot \sin \widehat{AOD} \cdot \cos \widehat{DAE}.$

Thus $\cos a = \cos b \cdot \cos c + \sin b \cdot \sin c \cdot \cos A.$

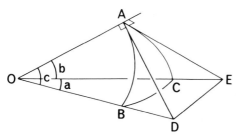

Fig D.1   A, B, C are points on the surface of a sphere of centre O, and define the spherical triangle ABC. The plane ADE is perpendicular to OA. The inter-radial angles a, b, c correspond to the angular lengths of the sides BC, CA, AB of the spherical triangle.

C.S.—U

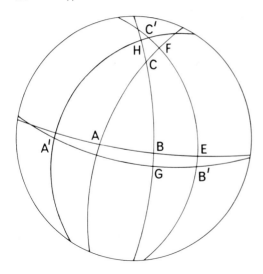

Fig D.2 Stereographic projection showing the bounding great circles of the spherical triangle ABC. The great circles whose poles are A, B, C are respectively B'C', C'A', A'B'; their intersections define the polar triangle A'B'C' of the triangle ABC. ABC is itself the polar triangle of A'B'C'. The two great circles through A intersect the great circle whose pole is A in E and F. The two great circles through A' intersect the great circle whose pole is A' in G and H.

By exactly similar argument we can derive the expressions

$$\cos b = \cos c \,.\, \cos a + \sin c \,.\, \sin a \,.\, \cos B$$

and $\qquad \cos c = \cos a \,.\, \cos b + \sin a \,.\, \sin b \,.\, \cos C.$

In order to derive the remaining equations for the general spherical triangle it is necessary to construct the *polar triangle* A'B'C' of the triangle ABC, which is defined as the triangle formed by the intersection of the great circles whose poles are A, B, and C (Fig D.2). Since A is the pole of the great circle B'C' and C the pole of the great circle A'B',

$$A{:}B' = C{:}B' = \tfrac{1}{2}\pi.$$

Therefore B' is the pole of the great circle AC. The triangle ABC is therefore the polar triangle of its own polar triangle A'B'C'.

Let the great circles AB and AC intersect the great circle B'C' in E and F respectively. Then E:F = A and

$$C'{:}E = F{:}B' = \tfrac{1}{2}\pi,$$

whence $\qquad a' = B'{:}C' = B'{:}F + F{:}C' = B'{:}F + E{:}C' - E{:}F.$

Thus $\qquad a' = \pi - A.$

Similarly if the great circles A'B' and A'C' intersect the great circle BC in G and H respectively.

$$G{:}C = B{:}H = \tfrac{1}{2}\pi$$

and $\qquad A' = G{:}H = G{:}C + C{:}H = G{:}C + B{:}H - B{:}C.$

Thus $\qquad A' = \pi - a.$

By analogous arguments it can be shown that

$$b' = \pi - B \qquad \text{and} \qquad c' = \pi - C,$$
$$B' = \pi - b \qquad \text{and} \qquad C' = \pi - c.$$

Now for the general spherical triangle A'B'C',

$$\cos a' = \cos b' \,.\, \cos c' + \sin b' \,.\, \sin c' \,.\, \cos A'.$$

Therefore $\quad \cos(\pi - A) = \cos(\pi - B) \,.\, \cos(\pi - C) + \sin(\pi - B) \,.\, \sin(\pi - C) \,.\, \cos(\pi - a)$

i.e. $\qquad \cos A = -\cos B \,.\, \cos C + \sin B \,.\, \sin C \,.\, \cos a$

and similarly

$$\cos B = -\cos C \,.\, \cos A + \sin C \,.\, \sin A \,.\, \cos b$$

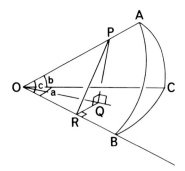

Fig D.3   A, B, C are points on the surface of a sphere of centre O. PQ is perpendicular to the plane BOC and QR is perpendicular to OB. The angle $\widehat{PRQ}$ is thus equal to the angle at the corner B of the spherical triangle ABC.

and $\cos C = -\cos A . \cos B + \sin A . \sin B . \cos c.$

To prove the remaining equations for the general triangle it is convenient to use a different construction. Fig D.3 shows three radii from the centre of the sphere O intersecting the surface of the sphere in A, B, and C. As before the angles $\widehat{BOC}, \widehat{COA}, \widehat{AOB}$ are a, b, c respectively. From a point P on OA a perpendicular is drawn to the plane BOC to meet the plane in Q and QR is drawn perpendicular to OB to intersect OB in R. Since $\widehat{OQP}, \widehat{ORQ},$ and $\widehat{PQR}$ are right angles

$$OP^2 = OQ^2 + PQ^2,$$

$$OQ^2 = OR^2 + RQ^2,$$

$$PR^2 = RQ^2 + PQ^2.$$

Thus $OP^2 = OR^2 + PR^2.$

Therefore $\widehat{ORP} = \tfrac{1}{2}\pi.$

Now $\widehat{PRQ} = B$

therefore $PQ = PR \sin B$

$$= OP \sin c . \sin B.$$

In a precisely similar manner it can be shown that

$$PQ = OP \sin b . \sin C.$$

Thus $\dfrac{\sin B}{\sin b} = \dfrac{\sin C}{\sin c}.$

By an analogous argument it can be shown that

$$\dfrac{\sin A}{\sin a} = \dfrac{\sin B}{\sin b}.$$

Appendix E: Three-dimensional analytical geometry and matrix methods

I: Derivation of the expression $l^2 + m^2 + n^2 = 1$, where l, m, n, are the direction cosines of a line referred to orthogonal axes

The *direction cosines* of a line are defined as the cosines of the angles that the line makes with the positive direction of each of the three reference axes, x, y, and z.

We consider a line defined by its direction cosines l, m, n and draw a line OP parallel to it and passing through the origin O (Fig E.1). From the point P a perpendicular is dropped to each axis to intersect the x-axis in A, the y-axis in B, and the z-axis in C. The direction cosines of OP are then by definition,

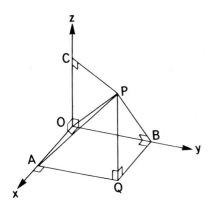

Fig E.1 The reference axes x, y, z are mutually perpendicular. From the point P the lines PA, PB, PC are drawn perpendicular to the x, y, z axes respectively, which they meet in A, B, C. The direction cosines of OP are then $l = \cos \widehat{POA}$, $m = \cos \widehat{POB}$, $n = \cos \widehat{POC}$.

$$l = \cos \widehat{POA} = \frac{OA}{OP}, \qquad m = \cos \widehat{POB} = \frac{OB}{OP}, \qquad n = \cos \widehat{POC} = \frac{OC}{OP}.$$

But for orthogonal axes by Pythagoras' Theorem (Fig E.1),

$$OA^2 + OB^2 + OC^2 = OP^2.$$

Therefore

$$\left(\frac{OA}{OP}\right)^2 + \left(\frac{OB}{OP}\right)^2 + \left(\frac{OC}{OP}\right)^2 = 1$$

and so

$$l^2 + m^2 + n^2 = 1.$$

II: Derivation of the expression for the angle between two lines given their direction cosines referred to orthogonal axes

We consider two lines OP_1 and OP_2 defined by their direction cosines $l_1 m_1 n_1$ and $l_2 m_2 n_2$ respectively, P_1 and P_2 being each at a distance r from the origin O (Fig E.2). The coordinates of P_1 are then $x_1 = rl_1$, $y_1 = rm_1$, $z_1 = rn_1$ and of P_2 are $x_2 = rl_2$, $y_2 = rm_2$, $z_2 = rn_2$. If the origin is moved from O to P_1, the coordinates of P_2 referred to P_1 as origin will be $x_2' = x_2 - x_1$, $y_2' = y_2 - y_1$, $z_2' = z_2 - z_1$. Therefore

$$\begin{aligned}(P_1 P_2)^2 &= (x_2 - x_1)^2 + (y_2 - y_1)^2 + (z_2 - z_1)^2 \\ &= r^2\{(l_2 - l_1)^2 + (m_2 - m_1)^2 + (n_2 - n_1)^2\} \\ &= r^2\{2 - 2(l_1 l_2 + m_1 m_2 + n_1 n_2)\}.\end{aligned}$$

Now

$$\cos \theta = 1 - 2 \sin^2 \tfrac{1}{2}\theta$$

i.e.

$$\cos \theta = 1 - 2 \left(\frac{P_1 P_2}{2r}\right)^2.$$

Thus

$$\cos \theta = l_1 l_2 + m_1 m_2 + n_1 n_2.$$

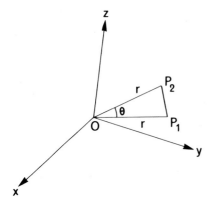

Fig E.2 The reference axes x, y, z are mutually perpendicular. P_1 and P_2 are placed at the same distance r from the origin O so as to define two lines OP_1 and OP_2, which make an angle θ with one another.

III: Matrix methods

Matrix methods provide a convenient means of representing equations and of manipulating arrays of numbers. The treatment given here is restricted to one-dimensional and two-dimensional matrices; extension to three or more dimensions is not difficult. A two-dimensional $M \times N$ matrix is a rectangular ordered array of numbers arranged in M rows and N columns. A matrix is conventionally enclosed in brackets—we shall use square brackets—to distinguish it from a determinant, which is conventionally bounded by vertical lines. If $M = N$, then the matrix is a *square* matrix. If $M = 1$ the matrix is a one-dimensional *row* matrix, but if $N = 1$ the matrix is a one-dimensional *column* matrix.

The *transpose* of an $M \times N$ matrix is an $N \times M$ matrix of N rows and M columns obtained by interchanging the rows and columns of the original matrix. Thus the transpose of matrix I, a 4×3 matrix, is II, a 3×4 matrix:

$$
\text{I} \qquad
\begin{bmatrix}
a_1 & b_1 & c_1 \\
a_2 & b_2 & c_2 \\
a_3 & b_3 & c_3 \\
a_4 & b_4 & c_4
\end{bmatrix}
\qquad
\text{II} \qquad
\begin{bmatrix}
a_1 & a_2 & a_3 & a_4 \\
b_1 & b_2 & b_3 & b_4 \\
c_1 & c_2 & c_3 & c_4
\end{bmatrix}
$$

The transformation of data from one set of reference axes to another set of reference axes is most conveniently achieved by matrix multiplication. The multiplication of two matrices P and Q is represented as $R = PQ$ where R is the product matrix. If P is an $m \times n$ matrix and Q is an $M \times N$ matrix, then the product matrix R will be a matrix of m rows and N columns, an $m \times N$ matrix. The number in the ith row and the jth column of R is the sum of the product of the ith row of matrix P and the jth column of matrix Q taken term by term. By way of example we take the multiplication of matrix IV with matrix III to yield the product matrix V:

$$
\text{III} \qquad \text{IV} \qquad\qquad \text{V}
$$

$$
\begin{bmatrix}
a_1 & b_1 & c_1 \\
a_2 & b_2 & c_2 \\
a_3 & b_3 & c_3
\end{bmatrix}
\begin{bmatrix}
A_1 & B_1 \\
A_2 & B_2 \\
A_3 & B_3
\end{bmatrix}
=
\begin{bmatrix}
a_1A_1+b_1A_2+c_1A_3 & a_1B_1+b_1B_2+c_1B_3 \\
a_2A_1+b_2A_2+c_2A_3 & a_2B_1+b_2B_2+c_2B_3 \\
a_3A_1+b_3A_2+c_3A_3 & a_3B_1+b_3B_2+c_3B_3
\end{bmatrix}
$$

So the number in the 3rd row and 2nd column of V is the sum of the product of the 3rd row of III, a_3 b_3 c_3, with the 2nd column of IV, B_1 B_2 B_3, taken term by term, $a_3B_1+b_3B_2 +c_3B_3$. Necessarily the number of columns in matrix P ($=$III in the example) must match the number of rows in matrix Q ($=$IV) for the multiplication $R = PQ$ to be possible: in the example III is a 3×3 matrix and IV is a 3×2 matrix. In general the product PQ can be formed if P is an $m \times n$ matrix and Q is an $M \times N$ matrix only if $n = M$. Formation of the product QP is only possible when the number of columns N in Q is equal to the number of rows m in P, i.e. $N = m$. So the commutative law does not hold for matrix multiplication and in general $PQ \neq QP$.

Returning to the example it is clear that multiplication of matrix III with matrix IV is impossible as matrix IV has 2 columns and matrix III has 3 rows. It is however possible to multiply the transpose of matrix III, denoted $\widetilde{\text{III}}$, which has 3 rows with the transpose of matrix IV, denoted $\widetilde{\text{IV}}$, which has 3 columns to give the product matrix VI:

$$
\widetilde{\text{IV}} \qquad\qquad \widetilde{\text{III}}
$$

$$
\begin{bmatrix}
A_1 & A_2 & A_3 \\
B_1 & B_2 & B_3
\end{bmatrix}
\begin{bmatrix}
a_1 & a_2 & a_3 \\
b_1 & b_2 & b_3 \\
c_1 & c_2 & c_3
\end{bmatrix}
$$

$$
\text{VI}
$$

$$
=
\begin{bmatrix}
A_1a_1+A_2b_1+A_3c_1 & A_1a_2+A_2b_2+A_3c_2 & A_1a_3+A_2b_3+A_3c_3 \\
B_1a_1+B_2b_1+B_3c_1 & B_1a_2+B_2b_2+B_3c_2 & B_1a_3+B_2b_3+B_3c_3
\end{bmatrix}
$$

It is immediately apparent that the product matrix VI is the transpose of matrix V. In general if $R = PQ$, then $\tilde{R} = \tilde{Q}\tilde{P}$.

We leave it to readers to satisfy themselves that the associative law holds for matrix multiplication, that is to say that the product of three matrices is independent of the sequence of multiplication: if $S = PQR$, then $S = S_1 R$ where $S_1 = PQ$ and $S = PS_2$ where $S_2 = QR$. The number of columns in P must be equal to the number of rows in Q and the number of columns in Q must be equal to the number of rows in R. The matrix S will have the same number of rows as P and the same number of columns as R. So the product of the double matrix multiplication

$$[a_1 \quad b_1 \quad c_1] \begin{bmatrix} p_1 & q_1 & r_1 \\ p_2 & q_2 & r_2 \\ p_3 & q_3 & r_3 \end{bmatrix} \begin{bmatrix} U_1 \\ V_1 \\ W_1 \end{bmatrix}$$

will be a 1×1 matrix equal to $a_1(p_1 U_1 + q_1 V_1 + r_1 W_1) + b_1(p_2 U_1 + q_2 V_1 + r_2 W_1) + c_1(p_3 U_1 + q_3 V_1 + r_3 W_1)$.

Appendix F: Crystal setting

Any account of X-ray diffraction by single crystals, such as that provided in chapter 8, would be incomplete without some description of the techniques used for mounting and setting a crystal in a particular orientation. No attempt is made here to give a comprehensive survey of all the techniques available for mounting and setting single crystals; we restrict ourselves to those techniques which we ordinarily use and which we find both easy to teach to others and suitable for setting a crystal very quickly in a chosen orientation with respect of the camera geometry. In this appendix we assume an understanding of the reciprocal lattice and the reflecting sphere, without which it would be pointless to engage in practical single crystal X-ray diffraction studies.

The first task is always to select a single crystal and to identify in it some prominent zone axis. For this purpose optical goniometry, in the case of well shaped (euhedral) crystals, may be useful and polarized light microscopy, in the case of transparent crystals whether euhedral or not, is particularly useful. By the use of either of these techniques it is usually possible, for instance, to identify the principal axis of a uniaxial crystal; we shall assume initially that it is desired to mount and set such a crystal about such an axis. Transparent crystals of lower symmetry and opaque shapeless crystals of any system present a more difficult problem, the solution of which is achieved by the use of just the same methods but takes more time.

Crystal mounting

We start with a crystal lying in a pool of refractive index oil on a microscope slide on the stage of a binocular microscope. The magnification should be in the range $\times 10$ to $\times 40$. The crystal is eased to the edge of the pool of oil and beyond with the aid of a fine needle point mounted in a wooden handle (a fine artists' brush from which the camel hairs have been removed makes a very convenient handle). A drop of amyl acetate—or some other volatile solvent in which the crystal is insoluble—is placed close to the crystal and the crystal is pushed with the needle point through the pool of amyl acetate and out towards the edge of the slide. The crystal is moved with the needle point until the selected zone axis is approximately perpendicular to the edge of the microscope slide.[1] When the last trace of amyl acetate has evaporated and the crystal is quite dry, a small drop of a slow setting adhesive (Durofix thinned with amyl acetate is eminently suitable) is placed on the slide close to the crystal. A glass fibre, about 15 mm long, < 0.05 mm diameter, and with one of its ends pushed into a pea-sized blob of plasticine, is then dipped in the drop of adhesive and quickly brought into contact with the crystal so that the glass fibre is approximately parallel to the selected zone axis of the crystal. If this rather delicate operation

---

[1] Instead of a hand-held needle point a 'micro-manipulator' may be used. Micro-manipulators, which are available commercially, are in essence three-dimensional pantographs; they are useful for dealing with exceptionally small crystals but are an unnecessary luxury for persons with steady hands dealing with crystals of the sort of size usually used for X-ray diffraction.

has been correctly performed the crystal will be stuck firmly on the tip of the glass fibre within rather less than one minute of making contact (if the operation is performed too slowly, the crystal will merely be held by surface tension to the surface skin of the adhesive droplet and will fall off if tapped against the slide; it is then necessary to start again). The crystal and its fibre are then mounted on a set of crystallographic arcs (Fig 8.1) by pressing the blob of plasticine against the column at the top of the arc assembly.

Crystal setting with a zone axis parallel to the spindle axis of the arcs

For both oscillation and Weissenberg photography it is necessary to align a zone axis in the crystal parallel to the camera axis. Initially we shall consider situations in which the selected zone axis diverges by no more than a few degrees of arc from the camera axis; later on we shall discuss gross mis-setting.

The crystal is first centred at the intersection of the camera axis with the incident X-ray beam. The spindle carrying the arcs is then rotated manually until the plane of one arc is approximately parallel to the incident X-ray beam, the plane of the other arc being then perpendicular to the incident X-ray beam; the spindle is locked in this position and a Laue photograph is taken. If the selected zone axis has been set precisely parallel to the camera axis, the reciprocal lattice net corresponding to the normals to planes lying in this zone will be horizontal and so the reflexions corresponding to the intersection of this equatorial net with the reflecting sphere will lie in a horizontal plane; thus when the cylindrical Laue photograph is laid flat after development this prominent 'zone' of reflexions will be seen to lie on a straight (horizontal) line perpendicular to the (vertical) camera axis (Fig F.1(a)). If, however, the selected zone axis is inclined to the camera axis, the reciprocal lattice nets normal to it will not be horizontal. The reflections in the equatorial layer will then lie on the surface of a shallow cone, whose axis is the zone axis and whose apex is at the centre of the reflecting sphere. The incident X-ray beam will lie on the surface of this cone. The cone will thus intersect the cylindrical film in a curve which passes through the point of intersection of the forward direction of the incident beam with the film. The shape of such a curve is, in general, complex and it is convenient to discuss it with respect to two simple situations: where the zone axis is in the plane perpendicular to the incident X-ray beam containing the camera axis (Fig F.1(b)) and where the zone axis is in the plane containing the incident X-ray beam and the camera axis (Fig F.1(c)).

In Figs F.1(a)–(d) the film is shown in each case with the observer imagined to be looking *towards* the X-ray tube and the edge of the film which was uppermost in the camera uppermost in the figure. When removing the film from the cassette for development it is necessary to indicate its orientation in the camera; this can conveniently be done by holding the cassette as it was in the camera with the exit-hole towards oneself and scratching a distinctive mark (such as X or √) on the top right-hand corner of the film. Some means of indicating the intersection of the film with the plane perpendicular to the camera axis containing the incident X-ray beam (the 'horizontal' plane) is required and this is achieved by slipping a brass 'mushroom' (Fig F.1(e)) over the collimator when setting photographs are being taken; the 'mushroom' casts an arcuate shadow on each side of the film, each shadow being symmetrical about the equatorial line (Figs F.1(a)–(d)).

If the arc whose plane is perpendicular to the incident X-ray beam (i.e., its axis is parallel to the incident beam) is observed looking towards the X-ray tube, then the mis-setting illustrated in Fig F.1(b) will require for its correction an anticlockwise adjustment of that arc only. The mis-setting illustrated in Fig F.1(c) is such that the equatorial net of the reciprocal lattice (which of course passes through the origin O) slopes upwards towards the X-ray tube; in this case the equatorial reflexions lie on a curve which is symmetrical about the vertical through the exit-hole of the film. Correction of this mis-setting is achieved by adjustment of the arc whose plane is parallel to that of the incident X-ray beam and the camera axis (i.e., its axis is perpendicular to the incident beam) in a clockwise sense when the observer moves to the right through 90° from his previous position.

We now consider the general case where adjustments on both arcs are required (Fig F.1(d)). We shall show how the amount of adjustment on each arc can be deduced from measurements on the film in this general case and in the special cases considered qualitatively in the preceding paragraph. The problem is complicated by the dependence of the inclination of the upper arc (Fig 8.1) to the camera axis on the inclination of the lower arc to the camera axis; only when the scale reading of the lower arc is zero is the axis of the upper arc perpendicular to the camera axis.

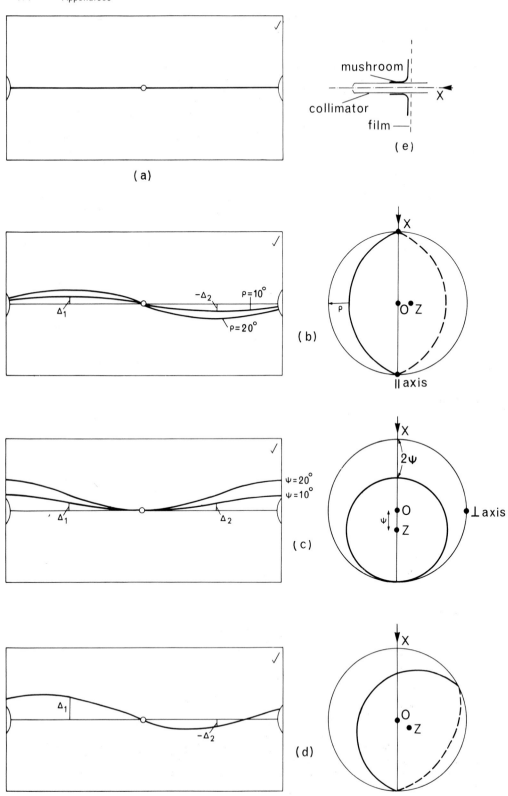

Fig F.1   Setting a crystal on an oscillation camera so that a selected zone axis Z is parallel to the spindle axis of the camera. (b)–(d) show diagrammatically the appearance of the film for various mis-settings; each film carries a tick in the top right-hand corner to indicate its orientation in the camera during exposure; the shadows cast by the 'mushroom', illustrated in (e), are indicated by arcs at the sides of each film, and serve to define the equatorial line; no reflexions are shown on the diagrams, only a continuous curve representing the zero layer reflexions. In (a) the crystal is perfectly set. (b) shows the curves of the zero layer when the zone axis Z is inclined at an angle $\rho = 10°$ (or 20°) to the spindle axis O in the plane normal to the incident X-ray beam; the stereogram on the right shows the great circle corresponding to the zero layer reciprocal lattice net. (c) shows the curves of the zero layer when the zone axis Z is inclined at an angle $\psi = 10°$ (and 20°) to the spindle axis in the plane of the incident beam and the spindle axis; the stereogram on the right shows the zero-layer reciprocal lattice net projected as a small circle of radius $90° - \psi$ passing through the opposite of X. (d) shows a general case of mis-setting and the orientation of the corresponding zero-layer reciprocal lattice net. In (b)–(d) the displacement of the zero-layer curve at points corresponding to $\theta = \pm 45°$ are indicated as Δ_1 on the left and Δ_2 on the right-hand side of the film: in (b) $\Delta_2 = -\Delta_1$ and $\rho = 2\Delta_1$, in (c) $\Delta_2 = \Delta_1$ and $\psi = 2\Delta_1$, and in (d) $\rho = \Delta_1 - \Delta_2$ and $\psi = \Delta_1 + \Delta_2$ where Δ_1, Δ_2 are measured in mm (positive above the equatorial line) and ρ, ψ are in degrees for a camera of radius 28·65 mm.

In Fig F.2(a), the zone axis is represented by the pole Z, the camera or oscillation axis by the pole O at the centre of the stereogram, and the incident X-ray beam by the poles XX'; the lower arc is taken to be set with its axis parallel to the incident beam and the axis of the upper arc is represented by the pole A_1. To bring the zone axis into coincidence with the camera axis it is necessary to move the crystal through the angle ψ in a clockwise sense about the axis of the upper arc, where ψ is the angle between the great circles $A_1 Z$ and $A_1 O$. The effect of this adjustment is to transfer the pole of the zone axis to Z', which lies in the plane normal to the axis XX' of the lower arc; Z' thus lies at the intersection of the small circle through Z whose stereographic centre is A_1 with the great circle whose pole is X. The crystal is then rotated through the angle ρ in an anticlockwise sense about the axis of the lower arc to transfer the zone axis from Z' to O; this adjustment of

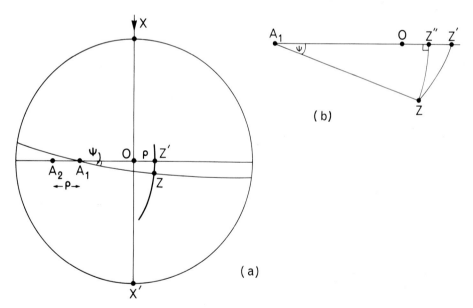

(a)

Fig F.2   In the stereogram (a) the forward direction of the incident beam is XX', the spindle axis of the camera is O, and the selected zone axis is Z, and the initial attitude of the axis of the upper arc is A_1. Movement of the crystal through the angle ψ on the upper arc causes Z to move on the small circle (shown bold) through Z, whose stereographic centre is A_1, until it reaches Z' on the great circle normal to XX'. After bringing the zone axis to Z' it is moved into coincidence with O by a movement ρ on the lower arc, which will move the axis of the upper arc from A_1 to A_2. (b) exaggerates the distinction between the small circle ZZ' and the great circle XZ'', which intersects the great circle normal to XX' in Z''. In calculating the corrections to be made to the arcs it is assumed that Z' and Z'' are coincident.

the lower arc moves the axis of the upper arc from A_1 to A_2 through the angle ρ in the plane normal to X. The great circle XZ is normal to the plane A_1O which it intersects in Z" (Fig F.2(b)). We shall assume for simplicity that Z' and Z" are coincident, which is tantamount to assuming that the zone axis Z is normal to the axis A_1 of the upper arc; if this were precisely so, then ZZ" = ψ and OZ" = ρ. This simplifying assumption leads to underestimation of the adjustments, ρ and ψ, of the arcs but greatly simplifies the calculation of the magnitudes, necessarily approximate, of ρ and ψ from measurements of the setting photograph.

We now consider the general case of a reflexion R produced by a lattice plane in the zone whose axis is Z (Fig F.3(a)). Since the direction R of the reflected beam and the forward direction X' of the incident X-ray beam lie on the surface of a cone whose axis is the zone axis Z, ZX' = ZR = η, where η is the semi-angle of the cone, and RX' = 2θ, where θ is the Bragg angle of the reflexion. Therefore for the general spherical triangle RX'Z,

$$\cos ZR = \cos RX' . \cos ZX' + \sin RX' . \sin ZX' . \cos \widehat{RX'Z}$$

i.e. $\cos \eta = \cos 2\theta . \cos \eta + \sin 2\theta . \sin \eta . \cos \varepsilon$

where $\widehat{RX'Z} = \varepsilon;$

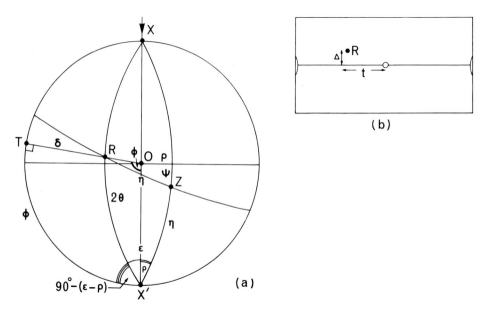

(b)

(a)

Fig F.3 The stereogram (a) illustrates the general case of a reflexion R produced by a lattice plane in the zone whose zone axis is Z; the spindle axis of the camera is O; and the forward direction of the incident beam is XX'. The angular distance of the intersection of the great circle X'Z with the great circle normal to XX' from Z is ψ and from O is ρ. The definition of the coordinates of the reflexion R on the film are shown in (b): $\Delta = r\tan\delta$ and $t = \phi r$, where r is the camera radius. Angular relations in (a) are grossly exaggerated for clarity.

whence $2 \sin^2 \theta \cos \eta = 2 \sin \theta \cos \theta . \sin \eta . \cos \varepsilon$

so that $\cos \varepsilon = \cot \eta . \tan \theta.$

Adoption of the convention that ψ is positive when $\eta < 90°$ (i.e. the correction of magnitude ψ is made in the clockwise sense when the observer moves to the right through 90° from the position facing the X-ray tube) leads to the equation $\eta = 90° - \psi$, so that

$$\cos \varepsilon = \tan \psi . \tan \theta.$$

The reflected beam R intersects the cylindrical film at a point with coordinates t, Δ (Fig F.3(b)), where t is the horizontal distance of the reflexion from the centre of the exit hole of the main beam and Δ is the vertical distance of the reflexion above the equatorial line. Now the coordinate

$t = \phi r$, where r is the camera radius and $\phi = \widehat{X'OR} = X'T$ (Fig F.3(a)); and $\Delta = r \tan \delta$, where $\delta = TR$. From the triangle $TX'R$, which has a right angle at T, equation (7) of chapter 5 gives

$$\sin \delta = \sin 2\theta \cos (\varepsilon - \rho),$$

equation (1) of chapter 5 gives

$$\cos \delta = \cos 2\theta \cos \phi + \sin 2\theta \sin \phi \sin (\varepsilon - \rho),$$

and equation (2) of chapter 5 gives

$$\cos 2\theta = \cos \phi \cos \delta + \sin \phi \sin \delta \cos 90°$$

$$= \cos \phi \cos \delta.$$

Therefore $\dfrac{\cos 2\theta}{\cos \phi} = \cos 2\theta \cos \phi + \sin 2\theta \sin \phi \sin (\varepsilon - \rho).$

Multiplying by $\cos \phi$ and putting $\sin^2 \phi = 1 - \cos^2 \phi$,

$$\cos 2\theta \sin^2 \phi = \sin 2\theta \sin \phi \cos \phi \sin (\varepsilon - \rho)$$

and so $\tan \phi = \tan 2\theta \sin (\varepsilon - \rho).$

Since we already have the relationship $\cos \varepsilon = \tan \psi . \tan \theta$, we can calculate δ and ϕ for any scattering angle 2θ if ψ and ρ are known for the zone and so the shape of the zero layer curve of a mis-set crystal can be calculated. But the converse procedure of calculating ρ and ψ from the observed shape of the zero layer curve, which is the problem with which we are concerned here, is laborious; it may be simplified by considering merely the coordinates of the zero layer curve for $\theta = \pm 45°$. When $2\theta = 90°$, $\cos \varepsilon = \tan \psi$ and

$$\sin \delta_1 = \cos (\varepsilon - \rho)$$

$$= \cos \varepsilon . \cos \rho + \sin \varepsilon . \sin \rho$$

$$= \tan \psi . \cos \rho + \sin (\cos^{-1} \tan \psi) . \sin \rho.$$

Then if ρ and ψ are small, $\tan \psi \to \psi$, $\cos \rho \to 1$, $\sin \rho \to \rho$, $\sin (\cos^{-1} \tan \psi) \to 1$, and, since δ is necessarily also small, $\sin \delta_1 \to \delta_1$ so that

$$\delta_1 = \psi + \rho$$

When $2\theta = -90°$, $\sin 2\theta = -1$ so that

$$\sin \delta_2 = -\cos \varepsilon . \cos \rho - \sin \varepsilon . \sin \rho$$

and $\cos \varepsilon = -\tan \psi.$

Then if ρ and ψ are small,

$$\delta_2 = \psi - \rho.$$

Thus $2\psi = \delta_1 + \delta_2$

and $2\rho = \delta_1 - \delta_2.$

Moreover, if δ is small the film coordinate Δ (Fig F.3(b)) becomes $\Delta = r\delta$ and, when $r = 28.65$ mm, Δ measured in mm is equal to 2δ where δ is measured in degrees, so that

$$\psi = \Delta_1 + \Delta_2$$

and $\rho = \Delta_1 - \Delta_2,$

where Δ_1 and Δ_2 are the heights of the zero layer in millimetres above the equatorial line at $\theta = 45°$ and $-45°$ respectively; ψ and ρ are then given in degrees. These relationships remain good approximations for cylindrical cameras of 30 mm radius, always assuming that ρ and ψ are small angles.

It is, in practice, often convenient to be able to derive the corrections to be made to the two arcs directly from the observed shape of the zero layer curve. The displacement of the curve from the equatorial line at $2\theta = 180°$ is dependent only on the arc whose plane is parallel to the incident X-ray beam. Thus if the zero layer curve lies above the equatorial line at the edges of the film (the exit-hole or the shadow of the back-stop being in the centre of the film), then the zone axis is tilted away from the collimator (as illustrated in Fig F.1(c)). The curve close to the intersection of the forward direction of the incident beam with the film (i.e. the exit-hole or the shadow of the back-stop) is approximately a straight line; the arc whose plane is normal to the incident beam has to be rotated to achieve the required correction in the sense which will cause this line to become parallel to the equatorial line. Thus in the cases illustrated in Fig F.1(b) an anticlockwise rotation of $\rho°$ is required to correct the mis-setting of the crystal.

The zero layer curve intersects the equatorial line in general in two points, at $\theta = 0$ and at a point whose θ value depends on the relative magnitudes of the mis-settings on the two arcs. If $|\psi| = |\rho|$ the second point of intersection is at $2\theta = +90°$ or at $2\theta = -90°$. If the second point of intersection lies at $|2\theta| > 90°$, the greater error is in the arc whose plane is perpendicular to the incident beam (as illustrated in Fig F.1(d)); but if the second intersection is at $|2\theta| < 90°$ the greater error is in the arc whose plane is parallel to the incident beam. It is thus only necessary to measure the displacements, Δ_1 and Δ_2, of the zero layer curve from the equatorial line at $2\theta = 90°$ and at $2\theta = -90°$; to take their sum and difference $\Delta_1 + \Delta_2$ and $\Delta_1 - \Delta_2$; and so to deduce the necessary corrections, ρ and ψ, for the mis-setting of the crystal.

The equations derived earlier, $\psi = \Delta_1 + \Delta_2$ and $\rho = \Delta_1 - \Delta_2$, apply only for small mis-settings and are only good approximations when the angle between the axis of the upper arc and the camera axis approaches 90°. They do however always lead to corrections in the right sense so that by use of these simple equations successive correction will achieve the desired result of bringing the zone-axis into coincidence with the camera axis. Moreover the range of possible angles between the axis of the upper arc and the camera axis, $90° \pm 30°$, has little effect on the accuracy of the applied corrections, provided the corrections are small. In practice the initial correction of a grossly mis-set crystal can only be very approximate, but thereafter the simple equations become quite accurate so that the crystal can be accurately set in a small number of operations.

For precision setting of a crystal, as is necessary before moving film photographs are taken, the precise positioning of the equatorial line on the film is essential. This can readily be achieved when the crystal is very nearly set by use of the *double Laue photograph*. A Laue photograph is taken with the arcs as nearly as possible parallel and perpendicular to the incident beam, then the crystal is rotated anti-clockwise about the camera axis through 179° and a second Laue photograph of about one half of the exposure time of the first is taken on the same film. The reflexions of the first exposure will lie outside the weaker reflexions of the second exposure on the right-hand side of the film and relatively inside on the left-hand side (Fig F.4). The displacement of the reflexions coupled with their intensity difference enables the zero layer curves corresponding to the two exposures of the film to be readily recognized. Had the rotation between the first and second exposures of the film been precisely 180° the two curves would have been mirrored in the equatorial line; but making the rotation just less than 180° serves to separate the reflexions and so make the interpretation of the double Laue photograph easier. The relative displacement is however, small, so that the equatorial line can be precisely located and, in particular,

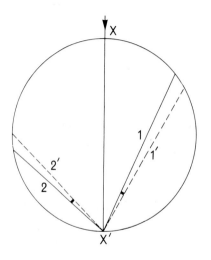

Fig F.4 The double Laue method. The figure shows the reflecting circle for two positions of the crystal. The directions labelled 1 and 2 represent reciprocal lattice rows in the zero layer net for the initial orientation of the crystal; those labelled 1' and 2' are the corresponding directions after the crystal has been rotated about the spindle axis (normal to the plane of the figure) in the anticlockwise sense through 179° (the angles between 1 and 1', 2 and 2' are grossly exaggerated for clarity). The directions of the reflected beams are given by the corresponding radii of the reflecting circle.

measurement of the separation of the two curves at $2\theta = \pm 90°$ yields precise values of $2\Delta_1$ and $2\Delta_2$, which are both necessarily small, so that the final corrections ψ and ρ can be made accurately. Successive use of double Laue photographs until the two curves become linear and coincident with the equatorial line enables the selected zone axis to be set precisely parallel to the camera axis. We emphasize the point that the double Laue technique is only profitably applicable when the rough setting methods described earlier have brought the selected zone axis almost into coincidence with the camera axis. Rough setting is usually adequate for the taking of oscillation photographs; precision setting by the double Laue method is a necessary preliminary to the taking of Weissenberg photographs.

In the initial stages of the setting of a grossly mis-set crystal it is often advantageous to use oscillation rather than Laue photographs even though fewer reflexions are recorded. Layer lines are discernible on an oscillation photograph of even a grossly mis-set crystal; and from even the very rough estimate that can be made of the layer line spacing in such a case it is usually possible to see whether these layer lines refer to the zone axis one is trying to set parallel to the camera axis. Against this advantage must be balanced the information—or rather at this stage of setting, the suggestions—about the symmetry of the crystal that can be gleaned from a Laue photograph. In practice it is usually a question of deciding whether it will be easier to set the crystal by recognition of a characteristic, usually large, layer line spacing on an oscillation photograph, or by recognition of some characteristic symmetry, usually a mirror plane, on a Laue photograph. In making the decision one bears in mind what one already knows from morphology and optics about the probable orientation of the crystal. If one approach is not rapidly fruitful, one tries the other.

When the selected zone axis has been set parallel to the camera axis, it may then be necessary to locate a particular symmetry direction within the zero layer. This is most simply done by inspecting a setting Laue photograph for symmetry, adjusting the orientation of the arcs by rotation of the camera spindle until the selected symmetry axis is parallel to the incident beam, and taking another Laue photograph to confirm. If the selected symmetry axis lies at a distance s millimetres from the centre of the exit-hole on the zero layer line of the setting Laue, it will be necessary to adjust the camera spindle in the appropriate sense through $s(90/\pi r)$ degrees, where r is the camera radius in millimetres, i.e., $0.955s$ for a camera of 30 mm radius and s for a camera of 28·65 mm radius. The Laue photograph taken after the adjustment has been made will display the Laue symmetry of the crystal along this direction. At this stage a further fine adjustment may be necessary.

For *Weissenberg* photography the crystal is set on an oscillation camera so that the selected zone axis is parallel—within about 1°—to the camera axis. The arcs are then transferred to the Weissenberg camera. Transference of arcs from one camera spindle to another may affect the alignment of the crystal by as much as 1° and anyway very precise setting is needed before a Weissenberg photograph is taken so the final stage of setting the crystal is performed on the Weissenberg camera. For this purpose double Laue photographs are used, the screens of the Weissenberg camera being moved back to a separation of at least 20 mm so that the two zero layer curves of the double Laue photograph are clearly displayed. It is usually necessary to take at least two double Laue photographs before coincidence of the two zero layer lines is achieved. It is then necessary to return the screens to their normal positions, with a separation of 2–4 mm, before taking a $\pm 20°$ oscillation photograph to check that the screens are allowing free passage of the diffracted beams of the zero layer line and that the back-stop is correctly placed to trap the undeviated X-ray beam. This oscillation photograph should show the zero layer reflexions symmetrically placed between the shadows cast by the screens and there should be no excessive blackening of the film, which would indicate incorrect positioning of the back-stop. Before taking upper layer Weissenberg photographs it is likewise necessary to check with a $\pm 20°$ oscillation photograph that the screens have been correctly positioned so that the reflexions of the upper layer line are symmetrically disposed between the shadows cast by the screens and that the back-stop has been correctly positioned.

For *precession* photography it is necessary to set a zone axis parallel to the incident beam so that the spindle axis has to be parallel to some direction in the corresponding zone. It is moreover convenient to have a reciprocal lattice row of the selected zone set parallel to the spindle axis. Where the selected reciprocal lattice row is parallel to a zone axis (e.g. the reciprocal lattice row $h00$ is parallel to [100] in an orthorhombic crystal) it is often convenient to set the crystal with that zone axis parallel to the spindle axis on an oscillation camera. The arcs are then transferred to a precession camera so that the selected reciprocal lattice row is approximately

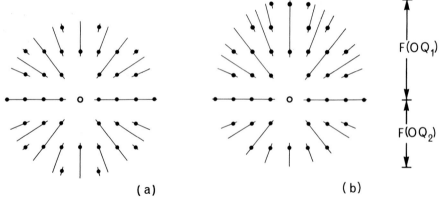

Fig F.5 Precession setting. The diagrams show schematically the appearance of 10° precession photographs taken with unfiltered radiation and without a screen. In (a) the crystal is precisely aligned with a reciprocal lattice row parallel to the spindle axis and a zone axis parallel to the incident beam; the distant ends of the white radiation streaks lie on a circle. In (b) the crystal is mis-set so that the distant ends of the white radiation streaks lie on a non-circular figure; the long and short axes of this figure measured on the film perpendicular to the line parallel to the spindle axis are $F(OQ_1)$ and $F(OQ_2)$; the quantity required for correcting the mis-setting is $F\Delta = F(OQ_1) - F(OQ_2)$. To avoid confusion in these diagrams non-zero layer streaks have been omitted.

parallel to the spindle axis of the precession camera. In the course of setting the crystal on the oscillation camera it may become clear which direction in the zero layer will have to be set parallel to the incident beam on the precession camera; this will usually be so when the direction is a symmetry axis.

If the crystal is not accurately aligned with a reciprocal lattice row parallel to the spindle axis and a zone axis parallel to the incident X-ray beam, it can be brought into alignment by taking a precession photograph with $\bar{\mu} = 10°$, white radiation, and no screen. If the crystal is precisely aligned the zero layer reflexions will lie within a circle of radius $2F \sin \bar{\mu}$, which will be clearly defined by the disposition of the associated white radiation streaks (Fig F.5(a)). If the crystal is slightly mis-set, the circle is distorted into a loop (Fig F.5(b)), the asymmetry of which can be utilized to correct the mis-setting. Consider the section (Fig F.6(a)) through the reflecting sphere containing the incident beam XIO and the direction IN_1, which is inclined at $\bar{\mu}$ to IO. Then OP_1 represents the line of intersection of the zero layer reciprocal lattice net with the plane of the figure when the crystal is perfectly set; OP_1' represents the line of intersection of the zero layer reciprocal lattice net with the plane of the figure for the mis-set crystal. Let IN_1' be the normal from I to OP_1' and ε the magnitude of the angle N_1IN_1'. Then since $\widehat{OIN_1} = \widehat{N_1IP_1} = \bar{\mu}$ and $\widehat{OIN_1'} = \widehat{N_1'IP_1'} = \bar{\mu} + \varepsilon$, $\widehat{P_1IP_1'} = 2\varepsilon$ and $\widehat{IP_1N_1} = 90° - \bar{\mu}$. Let Q_1 be the intersection of IP_1' produced with OP_1 produced. Then in the triangle IP_1Q_1 (Fig F.6(b)) $\widehat{IQ_1P_1} = 90° - \bar{\mu} - 2\varepsilon$ and by the sine rule

$$\frac{P_1Q_1}{IP_1} = \frac{\sin 2\varepsilon}{\cos (\bar{\mu} + 2\varepsilon)}.$$

Similarly it can be shown that when the normal to the tangent plane at O for $\bar{\mu} = 0$ lies below XIO (Fig F.6(c), (d)), then

$$\frac{P_2Q_2}{IP_2} = \frac{\sin 2\varepsilon}{\cos (\bar{\mu} - 2\varepsilon)}.$$

The difference $F\Delta$ in the distances of the opposite edges of the limiting loop of the zero layer reflexions from the point of intersection of the undeviated beam with the film can be measured on the film. This difference is given by

$$F\Delta = F(OQ_1 - OQ_2)$$
$$= F(OP_1 + P_1Q_1 - OP_2 + P_2Q_2)$$
$$= F(P_1Q_1 + P_2Q_2),$$

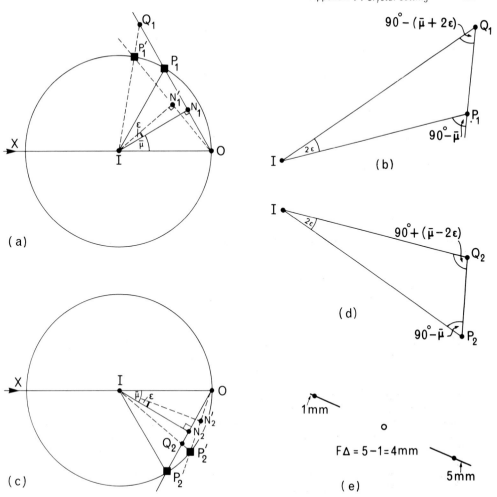

Fig F.6 Precession setting. (a) is a section through the reflecting sphere containing the incident beam XIO and the normal IN_1 to the reciprocal lattice row OP_1 for the perfectly set crystal; when the crystal is mis-set (exaggerated on the figure), P_1 becomes P_1' and N_1 becomes N_1'; Q_1 is the intersection of IP_1' produced with OP_1 produced. The plane triangle IP_1Q_1 shown in (b) is an enlargement of the same triangle in (a). (c) and (d) are the corresponding diagrams when P_2 lies below XIO. (e) illustrates a convenient means of evaluating $F\Delta$ by measurement of the white radiation streaks beyond a pair of centrosymmetrically related Kα reflexions lying in an appropriate direction; the centre of the film is indicated by the open circle.

since $\quad OP_1 = OP_2 = 2 \sin \bar\mu.$

And since $IP_1 = IP_2 = 1$,

$$F\Delta = F \sin 2\varepsilon \left(\frac{1}{\cos(\bar\mu+2\varepsilon)} + \frac{1}{\cos(\bar\mu-2\varepsilon)} \right)$$
$$= F \sin 2\varepsilon \left(\frac{\cos(\bar\mu-2\varepsilon)+\cos(\bar\mu+2\varepsilon)}{\cos(\bar\mu+2\varepsilon).\cos(\bar\mu-2\varepsilon)} \right)$$
$$= \frac{2F \sin 2\varepsilon.\cos\bar\mu.\cos 2\varepsilon}{\cos^2\bar\mu.\cos^2 2\varepsilon - \sin^2\bar\mu.\sin^2 2\varepsilon}$$
$$= \frac{F \sin 4\varepsilon.\cos\bar\mu}{\cos^2 2\varepsilon - \sin^2\bar\mu}.$$

From this expression ε can be calculated from a measured value of $F\Delta$, F and $\bar{\mu}$ being known. Table F.1 gives a tabulation of $F\Delta$ and ε for $F = 60$ mm, $\bar{\mu} = 10°$. As the variation of ε with $\bar{\mu}$ for a particular value of $F\Delta$ is quite small, it is convenient in practice to use a precession angle $\bar{\mu}$ of $10°$. A graph of $F\Delta$ against ε provides a quick means of determining ε from a measured value of $F\Delta$.

Measurements of $F\Delta$ can be utilized to correct the settings of the arcs by adopting the following procedure. We make the assumption that a reciprocal lattice row is known to be nearly parallel to the spindle axis. The spindle axis is then rotated until the plane of the lower arc is parallel to the incident beam and a $10°$ precession photograph is taken with white radiation. This precession photograph should show at least this reciprocal lattice row. The magnitude of $F\Delta$ measured along this row will give the angular correction ε to be made to the lower arc.[2] If the tilt on the lower arc is fairly small, the plane of the upper arc will be approximately perpendicular to the incident beam so measurement of the angle which the reciprocal lattice row makes with the horizontal on the same precession photograph will give the correction to be applied to the upper arc directly. This correction can be obtained more accurately by rotating the spindle axis through $90°$, taking another $10°$ precession photograph, measuring $F\Delta$ and so determining the correction, ε, to be applied to the upper arc. As for setting on an oscillation camera it is usually necessary to take several $10°$ precession photographs in each of these orientations and so by making successive corrections bring the selected reciprocal lattice row into precise alignment with the spindle axis of the camera; in particular when the reading on the lower arc departs far from zero, the calculated correction ε to the upper arc is not accurate and so the upper arc can only be corrected by successive approximation.

When a reciprocal lattice row has been set accurately parallel to the spindle axis, any desired reciprocal lattice net containing this row can be recorded by setting the spindle dial at the appropriate angle. Errors in this setting may be determined by measuring $F\Delta$ in the 'vertical' direction of the film to determine the angle ε through which the setting of the dial must be rotated in order to bring a prominent reciprocal lattice net parallel to the plane of the film.

If the orientation of the crystal with respect to the reciprocal lattice is unknown, then a series of $10°$ precession photographs has to be taken with white radiation at $5°$ intervals of the dial setting until a major reciprocal lattice net is located. The same procedure may be used to locate reciprocal lattice directions in a crystal fragment of unknown orientation. The crystal is mounted on arcs, both arcs being set at zero, and $10°$ precession photographs are taken at $5°$ intervals of dial reading until a reciprocal lattice net with a high density of points is located. The crystal is then remounted so that this net is parallel to the plane of one of the arcs and the normal setting procedure is begun.

Table F.1. Angular setting error in precession photographs. The table gives values of ε corresponding to measured values of $F\Delta$ for $\bar{\mu} = 10°$ and $F = 60$ mm. An extended table is given in *International Tables*, vol 2 (1959), p. 200.

F (mm)		F (mm)	
0	0	10	2°20′
1	0°14′	11	2°34′
2	0°28′	12	2°47′
3	0°42′	13	3°01′
4	0°56′	14	3°15′
5	1°10′	15	3°29′
6	1°24′	16	3°43′
7	1°38′	17	3°56′
8	1°52′	18	4°10′
9	2°06′	19	4°24′

[2] It is often convenient to evaluate $F\Delta$ by selecting two centrosymmetrically related $K\alpha$ reflexions and then measuring the difference in the lengths of the white radiation streaks extending outwards from these $K\alpha$ spots (Fig F.6(e)).

Appendix G: Units and constants

In this book we have not attempted to make use of SI units systematically for two reasons. Firstly, the Ångstrom is in crystallography, and especially in crystal chemistry, a more useful unit than the nanometre (1 nm = 10 Å). Secondly, almost all the literature of thermochemistry is written in terms of the calorie rather than the joule (1 J = 0·239 cal). For these reasons of convenience and readability of the literature we have preferred a system of mixed units in the text. In this appendix we relate these units to SI units and give the values of some useful constants.

Summary of SI units

length	metre	m (nanometre = 10^{-9} m)
mass	kilogramme	kg
time	second	s
electric current	ampere	A
thermodynamic temperature	kelvin	K
energy	joule	$J = kg\,m^2\,s^{-2}$
force	newton	$N = kg\,m\,s^{-2} = J\,m^{-1}$
electric charge	coulomb	$C = A\,s$
electric potential difference	volt	$V = kg\,m^2\,s^{-3}\,A^{-1} = J\,A^{-1}\,s^{-1}$
customary temperature, t	degree Celsius	$°C \quad t\,°C = T\,K - 273·15$
pressure	newton per square metre	$N\,m^{-2}$

Conversion factors

length:	Ångstrom unit	$1\,Å = 10^{-8}\,cm = 10^{-1}\,nm$
	micron	$1\,\mu = 10^{-4}\,cm = 1\,\mu m$
density:	grammes per cc	$1\,g\,cm^{-3} = 10^3\,kg\,m^{-3}$
pressure:	atmosphere	$1\,atm = 1·0133\,bar = 101·33\,kN\,m^{-2}$
	bar	$1\,bar = 10^5\,N\,m^{-2}$
energy:	erg	$1\,erg = 10^{-7}\,J$
	calorie	$1\,cal = 4·184\,J = 41·84\,bar\,cm^3$
	electron volt	$1\,eV = 1·6021.10^{-19}\,J = 23·061\,kcal\,mole^{-1}$
free energy:	$1\,kcal\,deg^{-1}\,mole^{-1} = 4·184.10^3\,J\,K^{-1}\,mol^{-1}$	
entropy:	entropy unit	$1\,e.u. = 1\,cal\,deg^{-1}\,mole^{-1} = 4·184\,J\,K^{-1}\,mol^{-1}$

Constants

$\pi = 3·14159$	$\pi^{-1} = 0·31831$
$1° = 0·01745$ radian	1 radian $= 57·296°$
$e = 2·71828$	
$\ln 10 = 2·30259$	$\log_{10} e = 0·43429$
Planck's constant	$h = 6·6256.10^{-27}\,erg\,s = 6·6256.10^{-34}\,J\,s$
Avogadro's number	$N = 6·02252.10^{23}\,mole^{-1}$
Boltzmann's constant	$k = 1·3805.10^{-16}\,erg\,deg^{-1} = 1·3805.10^{-23}\,J\,K^{-1}$
gas constant	$R = kN = 1·9872\,cal\,deg^{-1}\,mole^{-1} = 8·314\,J\,K^{-1}\,mol^{-1}$
electron rest mass	$m = 0·911.10^{-27}\,g$
unit of atomic mass	$1\,a.m.u. = 1·66042.10^{-24}\,g$
charge on the proton	$e = 4·8030.10^{-10}\,e.s.u. = 1·602.10^{-19}\,C$
velocity of light *in vacuo*	$c = 2·9979.10^{10}\,cm\,s^{-1} = 2·9979.10^8\,m\,s^{-1}$

Answers to problems

Chapter 1

1.1 All z-coordinates are changed by $\frac{1}{4}$. The structures are identical.

1.2 Sr $\frac{111}{222}$; Ti 000; O $\frac{1}{2}$00, 0$\frac{1}{2}$0, 00$\frac{1}{2}$.

1.3 (i) [001], (ii) [021], (iii) [0$\bar{1}$1], (iv) [100], (v) [2$\bar{1}$0], (vi) [120].

1.4 [12$\bar{4}$].

1.5 (i) [031], (ii) [302], (iii) [$\bar{1}$20].

1.6 [1$\bar{1}$1]; $p = 2$, $q = -1$.

1.7 (i) [1$\bar{1}$0], (ii) [$\bar{1}\bar{1}$1]. (112) and ($\bar{1}\bar{1}\bar{2}$).

1.8 65·17°; 90°.

1.9 ($\bar{1}$1), (13), (10), (2$\bar{1}$), (1$\bar{2}$), ($\bar{5}$1).

Chapter 2

2.1 304° E of N.

2.2 Toronto is distant 5704 km and New York 5558 km.

2.3 The bearing and elevation of P are 225°E of N and 25°. L_3: horizontal = 59°.

2.4 [0$\bar{2}$1], (i) 18$\frac{1}{2}$°, (ii) 24°.

2.5 (i) (111), (011), ($\bar{1}$11), ($\bar{1}$00), ($\bar{1}\bar{1}\bar{1}$), (0$\bar{1}\bar{1}$), (1$\bar{1}\bar{1}$). (ii) (1$\bar{1}$1), ($\bar{1}\bar{1}$0) ($\bar{1}$1$\bar{1}$). ($\bar{1}$23) lies at the intersection of the zones [(011), ($\bar{1}$01)] and [($\bar{1}$11), (101)].

2.6 (412).

2.7 49°, 51°, 65°.

2.8 Faces a to g are (210), (311), (101), (1$\bar{1}$3), (0$\bar{1}$2), ($\bar{1}\bar{1}$1), ($\bar{2}\bar{1}$0) respectively. [$\bar{1}$21].

2.9 On 35° small circle (011), (101), (110). On 55° small circle (100), (010), (001).

Chapter 3

3.1 $S_1 = \bar{2}\,(=m)$, $S_2 = \bar{1}$, $S_3 = \bar{6}$, $S_4 = \bar{4}$, $S_6 = \bar{3}$.

3.2 (i) 23, $\bar{4}3m$, (ii) $m3$, 432, $m3m$.

3.3 The form $\{hkil\}$ encloses space only in $\bar{6}$, $6/m$, 622, $\bar{6}m2$, $6/mmm$.

3.4 (100) and (001) mirror planes, [010] diad. The point group is $m2m$, conventionally $mm2$.

3.7 $522(A = 2, B = 2, C = 5$ with $\widehat{BC} = \widehat{CA} = 90°$, $\widehat{AB} = 36°)$ and $532\,(A = 2, B = 3, C = 5$ with $\widehat{BC} = 37·38°$, $\widehat{CA} = 31·72°$, $\widehat{AB} = 20·91°)$.

3.8 Six.

Chapter 4

4.1 Tetragonal I-lattice. $a = 3·87$ Å, $c = 6·37$ Å. Ca at 000; C at $\frac{1}{2}$, $\frac{1}{2}$, 0·10 and $\frac{1}{2}$, $\frac{1}{2}$, $\overline{0·10}$.

4.2 $\frac{111}{444}$, $\frac{311}{444}$, $\frac{131}{444}$, $\frac{113}{444}$, $\frac{133}{444}$, $\frac{313}{444}$, $\frac{331}{444}$, $\frac{333}{444}$. (i) F, $A(000)$, $X(\frac{111}{444}, \frac{113}{444})$, (ii) F, $A(000)$, $X(\frac{111}{444})$, (iii) P, $A(000, 0\frac{1}{2}\frac{1}{2}, \frac{1}{2}0\frac{1}{2}, \frac{1}{2}\frac{1}{2}0)$, $X(\frac{111}{444}, \frac{333}{444})$.

4.3 C. P, $a = 3·46$ Å, $c = 6·10$ Å, Pt$(000, \frac{111}{222})$, S$(\frac{1}{2}0\frac{1}{4}, \frac{1}{2}0\frac{3}{4})$.

4.4 $Pmcb$.

4.5 O^{2-} at 0·399, 0·467, $\frac{1}{4}$; 0·601, 0·533, $\frac{3}{4}$; 0·696, 0·802, 0·449; 0·304, 0·198, 0·551; 0·196, 0·698, 0·551; 0·804, 0·302, 0·449. There are three independent Fe-O distances.

4.7 $Abm2$, $B2cm$, $C2mb$, $Bma2$, $Cm2a$.

4.8 Au and Cu: F, 1 atom per lattice point. Cu_3Au: P, 3 atoms of Cu and 1 of Au per lattice point.

4.9 $n \parallel (100)$ *and* $m \parallel (010)$ through the origin and $2_1 \parallel [001]$ through 0, $\frac{1}{4}$ etc.

GEP's: $4(1)$ xyz; $x\bar{y}z$; $\bar{x}, \frac{1}{2}-y, \frac{1}{2}+z$; $\bar{x}, \frac{1}{2}+y, \frac{1}{2}+z$.

SEP's: $2(m)$ $x0z$; $\bar{x}, \frac{1}{2}, \frac{1}{2}+z$.

Cu: $2(m)$ with $x = \frac{1}{6}$, $z = \frac{1}{2}$.
    $4(1)$ with $x = \frac{1}{3}$, $y = \frac{1}{4}$, $z = 0$
As: $2(m)$ with $x = \frac{5}{6}$, $z = 0$
S: $2(m)$ with $x = \frac{5}{6}$, $z = \frac{3}{8}$
    $2(m)$ with $x = \frac{1}{6}$, $z = \frac{7}{8}$
    $4(1)$ with $x = \frac{1}{3}$, $y = \frac{1}{4}$, $z = \frac{3}{8}$.

4.10 The only two possible structures are exemplified by NaCl and ZnS. The NaCl structure has atoms of one kind at 000 and of the other at $\frac{111}{222}$; its space group is $Fm3m$. These SEP's also occur in $F432$ and $F\bar{4}3m$, but when no other positions are occupied additional symmetry elements are present and the space group becomes $Fm3m$. The ZnS structure has atoms of one kind at 000 and of the other at $\frac{111}{444}$; the space group is $F\bar{4}3m$.

Chapter 5

5.1 7·76 Å.

5.2 1·368 Å.

5.3 $a^* = 0·1292$, $b^* = 0·1723$, $c^* = 0·1414$ Å^{-1}, $\beta^* = 74·37°$. $d_{111} = 4·206$ Å.

5.4 $a = 9·78$, $c = 9·36$ Å.

5.5 (i) 54·74°, (ii) 64·44°, (iii) 33·27°.

5.6 $\begin{bmatrix} 94·673 & 0 & -14·040 \\ 0 & 79·388 & 0 \\ -14·040 & 0 & 27·984 \end{bmatrix}$; 61·38°

5.7 70·37°; 37·94°.

5.8 (213).

5.9 $\begin{bmatrix} 0 & 0 & \bar{1} \\ 0 & 1 & 0 \\ 1 & 0 & 1 \end{bmatrix}$; $\begin{bmatrix} 1 & 0 & 1 \\ 0 & 1 & 0 \\ \bar{1} & 0 & 0 \end{bmatrix}$; $\begin{bmatrix} 1 & 0 & \bar{1} \\ 0 & 1 & 0 \\ 1 & 0 & 0 \end{bmatrix}$.

5.10 $\begin{bmatrix} \frac{1}{2} & 0 & \frac{\bar{1}}{2} \\ 0 & \frac{1}{2} & 0 \\ \frac{1}{2} & 0 & \frac{1}{2} \end{bmatrix}$; 000, $00\frac{1}{2}$, $\frac{1}{2}0\frac{1}{2}$, $\frac{111}{222}$; Pr ($\frac{1}{2}\frac{1}{4}0$), Al (000), O ($0\frac{1}{4}0$; $\frac{1}{4}0\frac{1}{4}$; $\frac{1}{4}0\frac{3}{4}$).

Real		Ideal
	$(000; \frac{111}{222})+$	
Pr: 0·002, $\frac{1}{4}$, $\frac{1}{2}$		$0\frac{11}{42}$
0·002, $\frac{3}{4}$, $\frac{1}{2}$		$0\frac{31}{42}$
Al: 000		000
$0\frac{1}{2}0$		$0\frac{1}{2}0$
O(1): $\frac{1}{4}$, 0·02, $\frac{1}{4}$		$\frac{101}{404}$
$\frac{3}{4}$, 0·02, $\frac{3}{4}$		$\frac{303}{404}$
$\frac{1}{4}$, 0·02, $\frac{3}{4}$		$\frac{103}{404}$
$\frac{3}{4}$, 0·02, $\frac{1}{4}$		$\frac{301}{404}$
O(2): 0·04, $\frac{1}{4}$, 0		$0\frac{1}{4}0$
0·04, $\frac{3}{4}$, 0		$0\frac{3}{4}0$

Chapter 6

6.1 $h+k+l = 2n$.

6.2 $F(hkl) = 4\{f_{Zn} + f_S \cos \dfrac{\pi}{2}(h+k+l) + i f_S \sin \dfrac{\pi}{2}(h+k+l)\}$

$I(111) = 18496; I(200) = 3136; I(220) = 33856.$

$I(111) = 6272; I(200) = 0; I(220) = 12544.$

$F(hkl) = 8f_{Si} \cos \dfrac{\pi}{4}(h+k+l).$

6.3 $mmmIbca$; $Ibca$.

6.4 $mmmPn$--; $Pnm2_1$ (or $Pn2_1m$) or $Pnmm$.

6.5 Crystal $4.53°$ or $57.66°$. Detector $117.81°$.

6.6 $I(331) = 1486$. $I(420) = 79$.

6.7 CuKα: $\theta = 13.324°$. MoKα: $\theta = 6.067°$.

6.8 $\theta(460) = 83.07°$. $I: x = 49.38°$ and $D: x = 63.24°$,

$a\{\cos D: x - \cos(180° - I: x)\} = 4\lambda$. $I: y = 40.62°$ and $D: y = 26.76°$,

$b\{\cos D: y - \cos(180 - I: y)\} = 2\lambda$. $I: z = D: z = 90°$.

6.9 (i) $\Delta\theta = 0.249°$, (ii) $\Delta\theta = 2.047°$.

Chapter 7

7.1 100, 110, 111, 200, 210, 211, 220, 221 and 300, 310, 311, 222, 320, 321, 400, 322 and 410, 330 and 411, 331; P; $a = 3.835$ Å.

7.2 110, 200, 211, 220, 310, 222, 321, 400, 330 and 411, 420, 332, 422, 431 and 510, 521, 440; I; $a = 3.150$ Å.

7.3 111, 200, 220, 311, 222, 400, 331, 420; F; $a = 5.537$ Å.

7.4 111, 220, 311, 400, 331, 422, 511 and 333; F; $a = 5.432$ Å.

7.5 111, 220, 311, 222, 400, 422, 511 and 333, 440; F; $a = 8.35$ Å; doublet has $N = 72$ (69 and 70 are impossible for F, 71 is impossible for all lattice types, 73 and 74 are impossible for F); 822 and 660; $a = 8.397 \pm 0.002$ Å.

7.6 Since the multiplicities are $m(200) = 4$ and $m(002) = 2$, $a = 3.99$ Å and $c = 4.03$ Å.

7.7 2θ for Si(111) = $28.466°$; $2\theta_{130} = 32.191°$; d_{130} (specimen) = 2.7806 Å, $d_{130}(Mg_2SiO_4)$ = 2.7661 Å, $d_{130}(Fe_2SiO_4)$ = 2.8295 Å; mol % Mg_2SiO_4 in specimen = 77%.

7.8 111, 200, 220, 311, 222, 400, 331, 420, 422, 333 and 511; F; 4.21 Å; α_1 line gives $a = 4.209$ Å; 4MgO per unit-cell; for the NaCl structure $F(hkl) = 4\{f_{Mg} + f_O(-1)^{h+k+l}\}$ and for the ZnS

structure $F(hkl) = 4\left\{f_{Mg} + f_O \cos \dfrac{\pi}{2}(h+k+l) + if_O \sin \dfrac{\pi}{2}(h+k+l)\right\}$; $m(420) = m(331) = 24$;

$I(420)/I(331)$ for the NaCl structure = 6.8 and for the ZnS structure = 0.2; periclase has the NaCl structure.

7.9 $F(hkl) = f_{Cs+} + f_{Cl-}(-1)^{h+k+l}$; 100, 110, 111, 200, 210, 211, 220, 300, 310; $a = 4.14$ Å; 110, 200, 211, 220, 310; $a = 4.59$ Å; f_{Cs+} is significantly greater than f_{Cl-}, but almost equal to f_{I-}.

7.10 $F(hkl) = f_{Au} + f_{Cu}\{(-1)^{k+l} + (-1)^{h+l} + (-1)^{h+k}\}$; (i) $F(hkl) = f_{Au} + 3f_{Cu}$; (ii) $F(hkl) = f_{Au} - f_{Cu}$; 100(w), 110(w), 111(s), 200(s), 210(w), 211(w), 220(s), 300(w), 310(w); in the powder pattern of disordered Cu_3Au lines of type (i) remain strong, those of type (ii) go to zero intensity so only 111, 200, 220, ... will be present.

Chapter 8

8.1 15.18 Å.

8.2 $\bar{3}m$; $a = 4.77$, $c = 15.65$ Å.

8.3 $a = 8.20$ Å; I.

8.4 $a = 5.24$, $c = 10.30$ Å; I.

8.5 $a = 4.97$, $b = 9.64$, $c = 7.71$ Å, $\beta = 93.0°$.

8.6 $a = 12.1$, $c = 10.8$ Å, $\beta = 106°$; $\mu = 8.73°$; screen shift = 3.53 mm, cassette shift = 4.40 mm.

8.7 00l from $l = \bar{5}$ to $l = 5$; 10.16 mm; 24.2 mm; 20l from $l = \bar{6}$ to $\bar{2}$ and 2 to 6.

8.8 [110]; $\theta = 17.97°$ for inner pair, $\theta = 26.13°$ for outer pair; inner $2\bar{4}0$ and $4\bar{2}0$, outer $2\bar{6}0$, $6\bar{2}0$; inner pair 0.747 Å, outer pair 0.754 Å.

8.9 (i) $6/mmm$, [0001]; (ii) $4/mmm$, [001]; $m3m$, $\langle 100 \rangle$; (iii) $\bar{3}m$, [0001]; $m3m$, $\langle 111 \rangle$; (iv) mmm, [100] or [010] or [001]; $4/mmm$, $\langle 100 \rangle$ or $\langle 110 \rangle$; $6/mmm$, $\langle 2\bar{1}\bar{1}0 \rangle$ or $\langle 10\bar{1}0 \rangle$; $m3$, $\langle 100 \rangle$; $m3m$, $\langle 110 \rangle$; (v) $6/m$, [0001]; (vi) $4/m$, [001]; (vii) $\bar{3}$, [0001]; $m3$, $\langle 111 \rangle$; (viii) $2/m$, [010]; $\bar{3}m$, $\langle 2\bar{1}\bar{1}0 \rangle$ if $\bar{3}m1$ or $\langle 10\bar{1}0 \rangle$ if $\bar{3}1m$.

Chapter 9

9.1 Zr: 2 sets of x, y, z. Pb: 2 sets of x, y, z. O: 2 sets of x, y, z and 2 sets of z. In all 26 independent parameters.

9.2 $I(511)/I(\bar{5}1\bar{1}) = 57/108 = 0.53$; dextro-rotatory.

9.3 Pbnm; 4; Mn must be on 4(m) rather than 4($\bar{1}$) sites because c is very small; O is on two sets of 4(m) sites; 3 sets of x, y, six parameters in all are required.

9.4 The sign of $f_{Cs} - f_{Cl}$ for each reflexion for each possible origin is:

	(1)	(2)	(3)	(4)	(5)	(6)	(7)	(8)
$F(100)$	+	−	+	+	−	+	−	−
$F(010)$	+	+	−	+	−	−	+	−
$F(001)$	+	+	+	−	−	−	−	+
$F(111)$	+	−	−	−	−	+	+	+
$F(200)$	+	+	+	+	+	+	+	+

$F(100) = F(010) = F(001)$ only for (1) and (5). $F(200)$ is always positive and independent of choice of origin; reflexions with all indices even are known as *structure invariants*.

9.5 $s(\bar{2}11) = +1$; $s(413) = s(241)$.

9.6
$$\begin{bmatrix} \tfrac{1}{2} & \bar{\tfrac{1}{2}} & 0 \\ \tfrac{1}{2} & 0 & \bar{\tfrac{1}{2}} \\ 0 & \tfrac{1}{2} & \tfrac{1}{2} \end{bmatrix}$$

9.7
$$\begin{bmatrix} \tfrac{1}{2} & \bar{\tfrac{1}{2}} & 0 \\ \tfrac{1}{2} & \tfrac{1}{2} & 0 \\ 0 & 0 & 1 \end{bmatrix} ; \quad \begin{bmatrix} 1 & \bar{1} & 0 \\ 1 & 1 & 0 \\ 0 & 0 & 1 \end{bmatrix}$$

9.8 $\theta = 11.10°$; ± 0.03 Å; 0.55 ± 0.02 Å; 0.80 ± 0.02 Å.

9.9 Ratios of $I(111)/I(200)$:

for X-rays

	LiH	NaH
NaCl type	25/67	1014/1242
ZnS type	52/25	1202/971

for neutrons

	LiH	LiD	NaH	NaD
NaCl type	0.5/5.0	11.8/3.7	8.3/0.01	1.6/16.7
ZnS type	2.8/0.5	7.8/11.8	4.2/8.3	9.1/1.6

X-rays could be used for LiH. Neutrons could be used for LiD or NaD.

Chapter 10

10.1 0.735 Å^{-1}.

10.2 $0.26°$.

10.3 $A^* = d^*_{110}$, $B^* = d^*_{001}$; $[\bar{1}10]$.

10.4 $313 + \bar{1}1\bar{1}$ and $\bar{1}1\bar{1} + 313$; for incident beam $\perp [0\bar{1}1]$, $133 + 1\bar{1}\bar{1}$ and $1\bar{1}\bar{1} + 133$ and for incident beam $\perp [1\bar{1}0]$, $331 + \bar{1}\bar{1}1$ and $\bar{1}\bar{1}1 + 331$; CoKα does not give rise to double diffraction.

10.5 $1\bar{2}1$, $\bar{2}22$, and $\bar{3}01$; $(1\bar{2}1):(\bar{2}22) = 61.87°$, $(\bar{2}22):(\bar{3}01) = 43.09°$, $(1\bar{2}1):(\bar{3}01) = 104.96°$.

10.6 [114]; reflexions in the (111) twin lie on an orthogonal net with $(A')^* = t^*(1\bar{1}0)$, $(B')^* = t^*(\bar{2}\bar{2}1) = t^*(003)$, and $H' + K'$ even, so that every reflexion from the twin coincides with a reflexion from the matrix crystal.

10.7 $5.5°$ about the y-axis; spikes $\| z^*$.

10.8 $R = \tfrac{1}{2}a + \tfrac{1}{2}b + \tfrac{1}{2}c$; reflexions with $h + k + l$ odd.

10.9 $t = 956$ Å.

10.10 (i) (a) $1\bar{1}0$, (b) $1\bar{1}1$; (ii) (a) $1\bar{1}0$, (b) $2\bar{2}0$.

428

Bibliography

Chapter 1
Gillispie, C. C. (1972). *Dictionary of Scientific Biography*, Scribners, New York.
Phillips, F. C. (1971). *An Introduction to Crystallography*, 4th edn, Oliver and Boyd, Edinburgh.

Chapter 3
Bloss, F. D. (1971). *Crystallography and Crystal Chemistry*. Holt, Rinehart and Winston, New York.

Chapter 4
Buerger, M. J. (1956). *Elementary Crystallography*, Wiley, New York.
Crystal Data 3rd edn, vols 1–4 (1973–8).
Hilton, H. (1963). *Mathematical Crystallography*, Dover, New York.
International Tables for X-ray Crystallography, vol. 1, 3rd edn, Kynoch Press, Birmingham (1969). Recently republished with much additional material as the next reference cited.
International Tables for Crystallography, vol. A, Reidel, Dordrecht (1983).

Chapter 5
Buerger, M. J. (1942). *X-ray Crystallography*, Wiley, New York.
Phillips, F. C. (1971). *An Introduction to Crystallography*, 4th edn, Oliver and Boyd, Edinburgh.
Riley, K. (1974). *Mathematical Methods for the Physical Sciences*, Cambridge University Press.
Stephenson, G. (1973). *Mathematical Methods for Science Students*, 2nd edn, Longman, London.
Thomas, G. B. and R. L. Finney (1984). *Calculus and Analytic Geometry*, 6th edn, Addison-Wesley, Reading, Mass.

Chapter 6
International Tables for X-ray Crystallography, vol. 1, 3rd edn (1969).
International Tables for X-ray Crystallography, vol. 3 (1962); 2nd edn (1968).
James, R. W. (1967). *The Optical Principles of the Diffraction of X-rays*, Bell, London.
Jenkins, F. A., and H. E. White (1976). *Fundamentals of Optics*, McGraw-Hill, New York.
Lipson, H., and C. A. Taylor (1958). *Fourier Transforms and X-ray Diffraction*, Bell, London.
Longhurst, R. S. (1973). *Geometrical and Physical Optics*, Longmans, London.
Smith, F. G. and J. H. Thomson (1971). *Optics*, Wiley, London.

Chapter 7
Christophe-Michel-Levy, M., and A. Sandrea (1953). La hogbomite de Frain (Tchécoslovaquie). *Bull. Soc. franc. Minér. Crist.*, **76**, 430–433.
Fang, J. H., and F. D. Bloss (1966). *X-ray Diffraction Tables*, Southern Illinois University Press, Carbondale and Edwardsville, Ill.

Gandolfi, G. (1967). Discussion upon methods to obtain X-ray 'powder patterns' from a single crystal. *Miner. Petrogr. Acta*, **13**, 67–74.

Klug, H. P., and L. E. Alexander (1974). *X-ray Diffraction Procedures*, Wiley, New York.

Lipson, H., and H. Steeple (1970). *Interpretation of X-ray Powder Diffraction Patterns*, Macmillan, London.

McKie, D. (1963). The högbomite polytypes. *Min. Mag.*, **33**, 563–580.

Yoder, H. S., and Th. G. Sahama (1957). Olivine X-ray determinative curve. *Amer. Min.*, **42**, 475–491.

Chapter 8

Alcock, N. W., and G. M. Sheldrick (1967). The determination of accurate unit-cell dimensions from inclined Weissenberg photographs. *Acta Cryst.*, **23**, 35–38.

Buerger, M. J. (1942). *X-ray Crystallography*, Wiley, New York.

Buerger, M. J. (1964). *The Precession Method in X-ray Crystallography*, Wiley, New York.

Henry, N. F. M., H. Lipson, and W. A. Wooster (1960). *The Interpretation of X-ray Diffraction Photographs*, 2nd edn, Macmillan, London.

Jeffery, J. W. (1971). *Methods in X-ray Crystallography*, Academic Press, London and New York.

Main, P., and M. M. Woolfson (1963). Accurate lattice parameters from Weissenberg photographs. *Acta Cryst.*, **16**, 731–733.

Nuffield, E. W. (1966). *X-ray Diffraction Methods*, Wiley, New York.

Woolfson, M. M. (1970). *An Introduction to X-ray Crystallography*, Cambridge University Press, London.

Chapter 9

Bacon, G. E. (1975). *Neutron Diffraction*, 3rd edn, Oxford University Press.

Cowley, J. M. (1981). *Diffraction Physics*, 2nd edn, North-Holland.

Dougherty, J. P., and S. K. Kurtz (1976). A second harmonic analyzer for the detection of non-centrosymmetry. *J. Appl. Cryst.*, **9**, 145–158.

Dunitz, J. D. (1979). *X-ray Analysis and the Structure of Organic Molecules*, Cornell University Press.

International Tables for X-ray Crystallography, vol. 3, 2nd edn (1968).

International Tables for X-ray Crystallography, vol. 4 (1974).

Rietveld, H. M. (1969). A profile refinement method for nuclear and magnetic structures. *J. Appl. Cryst.*, **2**, 65–71.

Woolfson, M. M. (1970). *An Introduction to X-ray Crystallography*, Cambridge University Press.

Chapter 10

Grundy, P. J., and G. A. Jones (1976). *Electron Microscopy in the Study of Materials*, Arnold, London.

Hirsch, P. B., A. Howie, R. B. Nicholson, D. W. Pashley and M. J. Whelan (1977). *Electron Microscopy of Thin Crystals*, Butterworths, London.

Longhurst, R. S. (1973). *Geometrical and Physical Optics*, Longman, London.

Thomas, G., and M. J. Goringe (1979). *Transmission Electron Microscopy of Materials*, Wiley, New York.

Thomas, G. B., and R. L. Finney (1984). *Calculus and Analytic Geometry*, 6th edn, Addison-Wesley, Reading, Mass.

Appendices

Crystal Data. 3rd edn, vols 1–4 (1973–8).

International Tables for X-ray Crystallography, vol. 1, 3rd edn (1969), republished with much additional material as the next reference cited.

International Tables for Crystallography, vol. A, Reidel, Dordrecht (1983).

International Tables for X-ray Crystallography, vol. 2, 3rd edn (1972).

Kelsey, C. H., and D. McKie (1964). The unit-cell of aenigmatite. *Min. Mag.*, **33**, 986–1001.

Terpstra, P., and L. W. Codd (1961). *Crystallometry*. Academic Press, New York.

Index